Excited States of Biological Molecules

WILEY MONOGRAPHS IN CHEMICAL PHYSICS

Editors

John B. Birks, *Reader in Physics, University of Manchester*

Sean P. McGlynn, *Professor of Chemistry, Louisiana State University*

Photophysics of Aromatic Molecules:
J. B. Birks, Department of Physics,
University of Manchester

Atomic & Molecular Radiation
Physics: *L. G. Christophorou,*
Oak Ridge National Laboratory,
Tennessee

The Jahn–Teller Effect in
Molecules and Crystals:
R. Englman, Soreq Nuclear
Research Centre, Yavne

Organic Molecular Photophysics:
Volumes 1 and 2: *edited by J. B. Birks,*
Department of Physics, University
of Manchester

Internal Rotation in Molecules:
edited by W. J. Orville-Thomas,
Department of Chemistry and Applied Chemistry,
University of Salford

Excited States of Biological Molecules:
edited by J. B. Birks,
Department of Physics, University
of Manchester

Excited States of Biological Molecules

Based on the Proceedings of the International Conference at the Calouste Gulbenkian Foundation Centre, Lisbon, Portugal, on April 18–24, 1974

Editor

J. B. Birks

The Schuster Laboratory, University of Manchester, Manchester, U.K.

A Wiley–Interscience Publication

JOHN WILEY & SONS

London · New York · Toronto · Sydney

Library of Congress Cataloging in Publication Data:

Main entry under title:

Excited states of biological molecules.

(Wiley monographs in chemical physics)
"A Wiley-Interscience publication."
1. Excited state chemistry—Congresses. 2. Molecules—
Congresses. 3. Biological chemistry—Congresses. I. Birks, John Betteley. II. International
Conference on the Excited States of Biological Molecules,
Lisbon, 1974. [DNLM: 1. Biochemistry. 2. Molecular biology.
QU4 159e 1974]

QD461.5.E915   547 7   75–6985

ISBN 0 471 074136 6

Filmset in Great Britain by
Technical Filmsetters Europe Limited, 76 Great Bridgewater Street, Manchester M1 5JY,
and printed by The Pitman Press Ltd., Bath

To our Portuguese friends

Preface

The International Conference on the Excited States of Biological Molecules was held at the Calouste Gulbenkian Foundation Centre, Lisbon, Portugal on April 18–24 1974. It was held under the auspices of the Calouste Gulbenkian Foundation who provided generous financial support and excellent facilities, and under the scientific sponsorship of the European Physical Society and the European Photochemical Association. The organizing committee consisted of R. B. Cundall (Nottingham) chairman/treasurer, M. D. Lumb (Manchester) secretary, J. B. Birks (Manchester) papers secretary/editor, F. W. J. Teale (Birmingham), H. Menano (Oeiras) and S. J. Formosinho (Coimbra).

The Conference was divided into six sessions, each introduced by a plenary lecturer, as follows:

A. Excited states of organic molecules (Sir George Porter, London), which was subdivided into
    A.1. Primary processes, and
    A.2. Photochemistry and photobiology.
B. Excited states of DNA and nucleotides (C. Hélène, Orleans).
C. Excited states of photosynthetic pigments (H. T. Witt, Berlin).
D. Excited states of proteins and amino-acids (G. Weber, Illinois).
E. Excited states of visual pigments (B. Rosenberg, Michigan).
G. Energy transfer in biological molecules (J. Eisinger, New Jersey).

The Proceedings include the full texts of the 6 plenary lectures and of most of the 73 contributed papers presented at the Conference. The latter were, however, subject to editorial limitations on length imposed to reduce the overall size of the volume. The shorter length of the contributed papers in sessions A and A.1 compared with those in the other sessions is a direct result of this editorial axe; it does not reflect either the relative importance of the subject matter or the literary capacity of the authors. In a few cases the authors have preferred to publish an abstract only. A simple alphabetical and numerical code is used to designate the papers: the occasional gap in the sequence corresponds to a cancelled session or to a paper that was not presented.

This is the fourth book in the series to be devoted to the photophysics of organic molecules. In the present volume the experimental and theoretical methods developed in the study of simple aromatic molecules are extended to more complex biological molecules such as DNA, the proteins, the porphyrins and the retinols. Its contents provide a broad topical survey of this branch of biochemical physics.

This volume is dedicated to our Portuguese friends. On April 25 1974, the day after the Conference, some of us witnessed the quiet popular revolution by which they regained their freedom. Long may they continue to enjoy it.

J. B. BIRKS

List of Delegates

Delegate	Institution
ALEN, Mr N.	*Ortec Ltd., Luton, U.K.*
ALPERT, Dr B.	*Institut de Biologie Physico-chimique, Paris, France*
AMOUYAL, Dr E.	*Laboratoire de Chimie-Physique, Orsay, France*
AVIS, Mrs E.	*The Royal Institution, London, U.K.*
BADLEY, Dr A.	*Unilever Ltd., Welwyn, U.K.*
BEAVEN, Dr G. H.	*National Institute for Medical Research, London, U.K.*
BEDNAR, Dr T. W.	*University of Milwaukee, U.S.A.*
BERENDS, Prof. W.	*Delft University of Technology, The Netherlands*
BERNS, Dr D. S.	*New York State Department of Health, Albany, U.S.A.*
BEYER, Dr C. F.	*The Rockefeller University, New York, U.S.A.*
BIRKS, Dr J. B.	*University of Manchester, U.K.*
BOTTIROLI, Dr G.	*Institute of Comparative Anatomy, Pavia, Italy.*
BRAGA, Dr C. L.	*University of Minho, Portugal*
BRAND, Prof. L.	*Johns Hopkins University, Baltimore, U.S.A.*
BRETON, Dr J.	*CEN-Saclay, Gif-sur-Yvette, France*
BURROWS, Dr H. D.	*University of Coimbra, Portugal*
CAMPBELL, Dr J.	*Max-Planck-Institut, Mülheim a.d. Ruhr, Germany*
CARBO, Prof. R.	*Instituto Quimico de Sarvia, Barcelona, Spain*
CHAMMA, Dr A.	*Centre Universitaire de Perpignon, France*
CHECCUCCI, Dr A.	*CNR Laboratory, Pisa, Italy*
CHRISTOPHOROU, Prof. L. G.	*Oak Ridge National Laboratory, Tennessee, U.S.A.*

Delegate	Institution
CLARKE, Prof. R. H.	*Boston University, U.S.A.*
CONTE, Dr J. C.	*Instituto Superior Técnico, Lisbon, Portugal*
COVA, Prof. S.	*Milan Polytechnic, Italy*
CUNDALL, Dr R. B.	*University of Nottingham, U.K.*
DAS-GUPTA, Dr D. K.	*University College of North Wales, Bangor, U.K.*
DA SILVA, Miss M. F.	*University of Manchester, U.K.*
DA SILVA, Dr I. L.	*Instituto Superior Técnico, Lisbon, Portugal*
DE B. COSTA, Dr S. M.	*Instituto Superior Técnico, Lisbon, Portugal*
DEUME, Dr M. J.	*Centre Universitaire de Perpignon, France*
DUCLA-SOARES, Prof. E.	*Lisbon University, Portugal*
EDELHOCH, Dr H.	*National Institutes of Health, Bethesda, U.S.A.*
EISINGER, Dr J.	*Bell Laboratories, Murray Hill, New Jersey, U.S.A.*
EWALD, Dr M.	*University of Bordeaux, France*
FEITELSON, Dr J.	*Hebrew University, Jerusalem, Israel*
FORMOSINHO, Dr S. J.	*University of Coimbra, Portugal*
FRACKOVIAK, Prof. D.	*Poznan Technical University, Poland.*
GALIAZZO, Dr G.	*Istituto Chimica Organica, Padua, Italy*
GEACINTOV, Prof. N. E.	*New York University, U.S.A.*
GHIRON, Prof. C.	*University of Missouri, Columbia, U.S.A.*
GIACOMETTI, Dr G.	*Istituto di Chimica Biologica, Rome, Italy*
HÄGELE, Dr W.	*University of Stuttgart, Germany*
HAYON, Dr E.	*U.S. Army Natick Laboratory, U.S.A.*
HÉLÈNE, Prof. C.	*CNRS, Orleans, France*
HEMENGER, Dr R. P.	*Oak Ridge National Laboratory, Tennessee, U.S.A.*
HEVESI, Dr J.	*Szeged Institute of Biophysics, Hungary*
HODGKINSON, Dr K. A.	*University of Manchester, U.K.*
ILANI, Dr A.	*Hebrew University Medical School, Jerusalem, Israel*
INGHAM, Dr K. C.	*National Institutes of Health, Bethesda, U.S.A.*
JACQUIGNON, Dr P. C.	*Laboratoire CNRS, Gif-sur-Yvette, France*
KAZZAZ, Prof. A. A.	*University of Baghdad, Iraq*
KLASINC, Prof. L.	*Institute Ruder Boskovic, Zagreb, Yugoslavia*
KLEIBEUKER, Dr J. F.	*Agricultural University of Wageningen, The Netherlands*
KLEINKAUF, Prof. H.	*Technical University of Berlin, Germany*
KNOPP, Prof. J. A.	*North Carolina State University, Raleigh, U.S.A.*
KNOWLES, Dr A.	*University of Sussex, U.K.*
KRALJIĆ, Dr I.	*Physical Chemistry Laboratories, Faculté des Sciences, Orsay, France*
LABHART, Prof. H.	*University of Zurich, Switzerland*

Delegate	*Institution*
Lami, Prof. H.	*Faculté de Pharmacie, Strasbourg, France*
Land, Dr E. J.	*Paterson Laboratories, Christie Hospital and Holt Radium Institute, Manchester, U.K.*
Latt, Dr S. A.	*Clinical Genetics Division of the Children's Hospital, Boston, U.S.A.*
Laustriat, Prof. G.	*Faculté de Pharmacie, Strasbourg, France*
Leite, Dr M. S.	*University of Coimbra, Portugal*
Lenci, Dr F.	*CNR Laboratory, Pisa, Italy*
Lewis, Dr C.	*The Royal Institution, London, U.K.*
Lewis, Prof. T. J.	*University College of N. Wales, Bangor, U.K.*
Lindenberg, Prof. K.	*University of California at St. Diego, La Jolla, U.S.A.*
Lindqvist, Dr L.	*University of Paris-Sud, France*
Longworth, Dr J. W.	*Oak Ridge National Laboratory, Tennessee, U.S.A.*
Low, Prof. W.	*Hebrew University, Jerusalem, Israel*
Lumb, Dr M. D.	*U.M.I.S.T., Manchester, U.K.*
McCaffery, Dr A. J.	*University of Sussex, U.K.*
McFarlane, Dr R. M.	*I.B.M. Research Laboratory, San Jose, U.S.A.*
McKellar, Dr J. F.	*University of Salford, U.K.*
Menano, Dr H.	*Gulbenkian Foundation Biological Research Institute, Oeiras, Portugal*
Montenay-Garestier, Dr T.	*Museum National d'Histoire Naturelle de Paris, France*
Muel, Dr B.	*Fondation Curie Institut de Radium, Orsay, France*
Ottolenghi, Prof. M.	*Hebrew University, Jerusalem, Israel*
Pantos, Dr E.	*University of Manchester, U.K.*
Pearlstein, Dr R. M.	*Oak Ridge National Laboratory, Tennessee, U.S.A.*
Pereira, Dr L. C.	*University of Lourenco Marques, Mozambique*
Pethig, Dr R.	*University College of North Wales, Bangor, U.K.*
Phillips, Dr D.	*University of Southampton, U.K.*
Porter, Prof. Sir George	*The Royal Institution, London, U.K.*
Prenna, Prof. G.	*Institute of Comparative Anatomy, Pavia, Italy*
Rosenberg, Prof. B.	*Michigan State University, East Lansing, U.S.A.*
Rosenfeld, Miss T.	*Hebrew University, Jerusalem, Israel*
Saperas, Dr B.	*Centre Universitaire de Perpignon, France*
Schaafsma, Prof. T. J.	*Agricultural University of Wageningen, The Netherlands*
Schaap, Dr A. P.	*Wayne State University, Detroit, U.S.A.*
Schaffner, Prof. K.	*University of Geneva, Switzerland*
Schram, Prof. E.	*University of Brussels, Belgium*

Delegate	*Institution*
SCHULTE-FROHLINDE, Prof. D.	*Max-Planck-Institut, Mülheim a.d. Ruhr, Germany*
SMAGOWICZ, Dr W. J.	*Institute of Biochemistry and Biophysics, Warsaw, Poland*
SONG, Prof. P.-S.	*Texas Technical University, Lubbock, U.S.A.*
SOUSA, Miss T. R.	*Instituto Superior Técnico, Lisbon, Portugal*
STEEN, Dr H. B.	*Norsk Hydro's Institute for Cancer Research, Oslo, Norway*
STEVENS, Prof. B.	*University of South Florida, Tampa, U.S.A.*
SUTHERLAND, Prof. J. C.	*University of California, Irvine, U.S.A.*
SWENBERG, Dr C. E.	*University of Manchester, U.K.*
SZABO, Dr A. G.	*National Research Council, Ottawa, Canada*
TATISCHEFF, Mrs I.	*Institut du Radium, Laboratoire Curie, Paris, France*
TAVARES, Miss M. A. F.	*University of Coimbra, Portugal*
TEALE, Dr F. W. J.	*University of Birmingham, U.K.*
THOMAZ, Dr M. F.	*University of Lourenco Marques, Mozambique*
TRISSL, Dr H. W.	*University of Constance, Germany*
TURNBULL, Dr J. H.	*Royal Military College of Science, Shrivenham, U.K.*
VIALLET, Prof. P.	*Centre Universitaire de Perpignon, France*
VIGNY, Dr P.	*Institut du Radium, Paris, France*
VON SCHÜTZ, Dr J.	*University of Stuttgart, Germany*
VON SONNTAG, Dr C.	*Max-Planck-Institut, Mülheim a.d. Ruhr, Germany*
WAHL, Dr Ph.	*Centre de Biophysique Moleculaire, Orleans, France*
WARE, Prof. W. R.	*University of Western Ontario, London, Canada*
WEBER, Prof. G.	*University of Illinois at Urbana-Champaign, U.S.A.*
WEIBEL, Dr O.	*Swiss Federal Institute of Technology, Zurich, Switzerland*
WELLER, Prof. A.	*Max-Planck-Institut, Göttingen, Germany*
WERNER, Dr A.	*Bar-Ilan University, Israel*
WILLIAMS, Dr W. P.	*Chelsea College, London, U.K.*
WITT, Prof. H. T.	*Max-Volmer-Institut, Technical University of Berlin, Germany*
WRIGHT, Dr T. R.	*Kodak Ltd., Harrow, U.K.*
ZUCLICH, Dr J.	*Technology Inc., San Antonio, Texas, U.S.A.*

Contents

xiii

Session A.2. Photochemistry and Photobiology

Session C. Excited States of Photosynthetic Pigments

Session D. Excited States of Proteins and Amino-acids

Session E. Excited States of Visual Pigments

Session G. Energy Transfer in Biological Molecules

Session A
Excited states of organic molecules

A. Excited states of organic molecules

Sir George Porter, F.R.S.

The Royal Institution of Great Britain
21 Albemarle Street, London W1X 4BS, U.K.

SUMMARY

Recent developments in the methods available for the direct observation of excited states are briefly reviewed; with special reference to nanosecond and picosecond laser flash photolysis. Some problems of current interest in excited state processes of organic molecules are discussed. In particular, recent work in the photochemistry of quinones of biological importance is described.

1 INTRODUCTION

A molecule in an excited electronic state must be regarded as a new species with physical and chemical properties quite distinct from those of the same molecule in its ground state. In spite of its higher energy, the reactivity in the excited state may be greater or less than that of the ground state.

The physical and chemical properties of excited states are becoming more and more accessible to direct observation as new, rapid recording techniques are developed. From absorption spectroscopy of the ground state only, it is possible to determine ionization potentials and electron affinities of the excited state and, with less reliability, dipole moments and acidity constants. For non-equilibrium properties, such as the rate coefficients of physical and chemical processes, some form of time-resolved physical measurement is necessary. Progress in this field has been exceedingly rapid over the last few years.

2 EXPERIMENTAL TECHNIQUES

Until about six years ago, when lasers began to be used successfully as sources for flash photolysis, the lower limit of time resolution was a few micro-

3

seconds. Since then this limit has been reduced by a factor of a million. Much effort had been put into the problem of reducing the duration of the conventional electrical discharge through a gas, but with only moderate success. Owing to the rather low efficiency of energy conversion to light in this type of tube and the difficulty of collecting the light efficiently from the large area of the lamp, stored energies of at least a joule, and often many hundreds of joules, are necessary to give enough excitation for the direct detection of intermediates in a single experiment. Such energies cannot be discharged through a gas in times much less than a microsecond; one of the best achievements recently being 120 J in 1 microsecond (half pulse-height) and 6 J in 0·2 microseconds, which made it possible to detect and study the triplet state of acetone in solution.[1]

Much shorter pulses are obtainable from an electric spark discharge, though the energy of a single spark may then be of the order of microjoules with a duration of one or two nanoseconds. Nevertheless, this source is extremely useful in the study of fluorescence lifetimes. The pioneering work was done by the Brodys on chlorophyll fluorescence and, recently, sensitivity and precision of the technique have been revolutionized by the use of single-photon counting methods (see, for example, the review by Knight and Selinger.[2])

Although averaging methods can also be applied to absorption work and have been so applied, particularly by Witt, they are not as useful as in fluorescence work, since the signal which has to be stored is very large. Laser single-pulse methods are therefore generally used.

2.1 Laser flash photolysis

Solid-state lasers were the first operational lasers and the first to be used for flash photolysis. They are also still the most useful for this purpose. The principal ones are the ruby laser, which emits at 694 nm and may be frequency doubled to 347 nm, and the neodymium laser which emits at 1060 nm and may be doubled to 530 nm and quadrupled to 265 nm. The pulse consists of a series of spikes of total duration equal to that of the pumping flash, i.e. about a millisecond, but this is Q-switched, e.g. by a saturable dye, to produce a pulse which, for ruby, is slightly less than 20 ns in duration and of energy 1·5–2 J at 694 nm and approximately $\frac{1}{10}$ of this when doubled to 347 nm. These lasers may be mode-locked, and neodymium is usually found to be the most reproducible for this purpose.

Given a laser pulse, monochromatic and of sufficient energy to produce a measurable photochemical effect, the problem which remains is to record the time-resolved spectrum immediately after the pulse and for some period afterwards. As in normal flash photolysis, two methods are available in principle, (a) photoelectric recording over a narrow wavelength band at all times, and (b) photographic recording over a wide range of wavelengths in a narrow time interval after a delay. The former is sensitive and more convenient for kinetic work but is limited to the response time of the photoelectric detection system, which is usually greater than 10 ns. The latter is more convenient for a preliminary survey or accurate spectroscopic work and is essential for subnanosecond work. In principle, the two methods can be combined, by using an image converter tube with nanosecond response time, but, apart from its expense, this method is not very suitable for quantitative work in kinetics or in spectroscopy.

2.2 Photoelectric detection

The signal-to-noise ratio is proportional to the time available for recording the signal and therefore, for rapid changes, it is necessary to compensate for the short time available by increasing the number of observations, n, the monitoring light intensity, I, or the flash energy. The signal-to-noise ratio increases as the half-power of n and of I but increases directly with signal, which is, in many cases, approximately proportional to the energy of the flash. Increasing flash intensity is therefore usually desirable up to the point where large perturbations change the system being studied, as may happen in some *in vivo* studies for example.[9] As already mentioned, averaging techniques are less suitable for absorption work than for emission studies owing to the large signals which have to be stored and the difficulties of the repetitive use of flashes of high energy. It is therefore more usual to resort to increasing the monitoring light intensity I, which may be achieved by using a second, longer flash or by pulsing an arc source which normally runs continuously at much reduced energy.

2.3 Photographic recording

This was the method first used in conventional flash photolysis and is still the preferred method for the recording of spectra at high resolution. Furthermore, in the subnanosecond time region now becoming available, photographic methods are usually essential.

Laser flash photolysis using photographic recording incorporates three new principles additional to those used in conventional flash photolysis:

(1) Synchronization of the photolysis flash and monitoring flash is achieved, almost perfectly, by splitting a single light pulse into two parts.

(2) Delays between two parts of the flash are provided optically by increasing the light path of the monitoring beam by a distance $l = c\Delta t$ where Δt is the required delay and c is the velocity of light.

(3) The monitoring beam is converted to a wavelength suitable for recording the absorption spectrum either by frequency multiplication, by Raman shifting or, more generally and more usefully, by frequency broadening into a continuum.

The first laser flash apparatus using these principles and photographic recording was described by Porter and Topp.[3] Full details are available from the original publication and only the main features will be described here.

The ruby laser flash is passed through a passive Q-switch of vanadyl phthalocyanine giving a pulse of 1·5 J at 69 nm, 17 ns in duration. For most purposes this is frequency doubled, by passing through an index-matched crystal of ammonium dihydrogen phosphate, to give a pulse of energy 0·1 J at 347 nm. This pulse is now split into two parts by a partially reflecting double prism. The first part passes to the reaction vessel and causes the photochemical or other changes; the second part, which is the monitoring pulse, is first delayed by traversing a variable optical path and then focused onto a cell containing a suitable fluorescent dye. The dye should have a fluorescent lifetime considerably shorter than the pulse duration and a fluorescence spectrum which has a broad and flat intensity distribution over the region to be investigated. One of the most useful, 1,1′,4,4′-tetraphenylbutadiene, has a continuum extending from 390 to 590 nm.

The monitoring flash now passes through the reaction vessel and into the spectrograph; the monochromatic photolysis pulse can be removed if troublesome, by interposing a filter between reaction vessel and spectrograph.

The time resolution of this technique is limited only by the duration of the flash. It was first used to record the absorption spectra of higher singlet states.

2.4 Picosecond flash photolysis

This technique has been developed by Rentzepis and his colleagues at the Bell Laboratories and by Euan Reid in our laboratory at the Royal Institution. Whilst nanosecond laser flash photolysis is now in routine use in many laboratories, and is not much more difficult than conventional methods, picosecond flash methods introduce many difficulties of reproducibility and interpretation and are still at the development stage.

The picosecond pulse (usually about 5 picoseconds in duration) is first produced as one of a train of pulses from a mode-locked laser. The only suitable laser for this purpose at present is neodymium. A single pulse is switched out of the train by a Pockels cell which is pulsed *via* a spark gap, onto which the pulse train is focused. The single pulse must now be amplified by passage through at least one further laser and then frequency doubled (to 530 nm) or quadrupled (to 265 nm). The reproducibility of extracting a single pulse of reasonably constant energy is rather poor at present.

The flash photolysis apparatus now uses the principles already described for nanosecond flash photolysis, with two important modifications:

(1) As a source of continuum a fluorescent material is now unsuitable since lifetimes of strongly fluorescent substances are one nanosecond or longer. Fortunately, the very high fields associated with picosecond pulses produce new effects which can be used to yield continuous spectra of duration comparable with that of the monochromatic exciting pulse. The effect, discovered by Alfano and Shapiro[4] and still not fully understood, is one of self-focusing and self-phase-modulation and all that is necessary is to focus a beam of very high

intensity into a material such as glass or water. The high field intensity produces non-linear effects, the refractive index becomes very dependent on the field and self-focusing in filaments occurs which further increases the field. The result is a broadening of the monochromatic line into a continuum which extends over as much as 100 nm in both directions and which has a duration no greater than that of the exciting pulse. There may be some discontinuities in the continuum caused by stimulated Raman and inverse Raman scattering, the latter being of some potential interest in itself for the study of transient species.

(2) A second modification is necessary, not because of the shorter time resolution but because of the irreproducibility of the intensity from pulse to pulse. This can best be overcome by using one pulse to record the whole sequence of time-resolved spectra. The pulse is split into some ten parts, each of which is delayed by a different amount and this is very simply achieved by using the method introduced by Rentzepis and Topp.[5] This employs a reflecting echelon (a step system of mirrors) or a transmission echelon, which in practice is a pile of microscope slides, arranged in steps, so that different positions of the monitoring beam pass through different numbers of slides, producing a time-resolved sequence of spectra. The use of an optical multichannel analyser has many advantages over a photographic method for recording these.

Laser flash photolysis is therefore now possible with time resolution down to a few picoseconds. Its principal limitation at present is in the wavelengths of excitation which are available, particularly in the ultraviolet region, and the irreproducibility of single-pulse-picosecond lasers. There can be little doubt that these difficulties will be overcome by the rapid developments of laser technology which are taking place at the present time.

3 EXCITED STATE CHEMISTRY OF CARBONYL COMPOUNDS

The photochemistry of compounds having one or more carbonyl groups has received more attention than that of any other class of compounds. The early organic photochemists, notably Ciamician, studied mainly the aromatic ketones in solution whilst somewhat later the physical chemists devoted their attention to simple aliphatic aldehydes and ketones in the vapour phase. In both cases the interest was in the unexcited products, including transient free radicals, rather than in the excited state.

The influence of the electronic structure of the excited state on the chemistry of the state has been studied only relatively recently, though the principles on which it is based, particularly the distinction between singlet, triplet, $n-\pi^*$ and $\pi-\pi^*$ levels in carbonyl derivatives, were clearly stated over twenty years ago.[6]

The influence of solvent and substituents on the relative energies of these states was first clearly established with substituted benzophenones. The striking decrease in chemical reactivity with respect to hydrogen abstraction when the

π–π^* replaces the n–π^* as lowest state, as well as concomitant changes in luminescence properties, led to a convincing clarification of reactivities of these excited states.[7]

Here we shall review briefly more recent work of a similar nature on the benzoquinones with particular attention to those of biological interest.

3.1 Flash photolysis of the benzoquinones

Four of the quinones of biological importance, particularly in photosynthesis, are shown in Figure 1. Three of them are alkyl or alkoxy substituted benzoquinones with one of the alkyl substituents a long phytyl or isoprenyl chain. The fourth, vitamin K, is a similarly substituted naphthaquinone.

Plastoquinone A
(I)

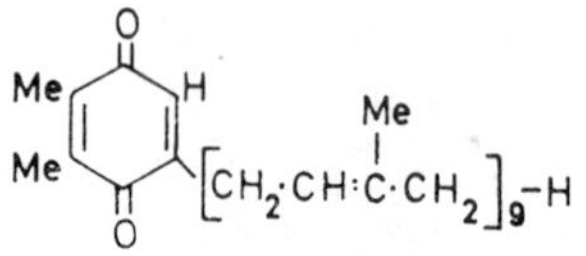

α-Tocopherylquinone
(II)

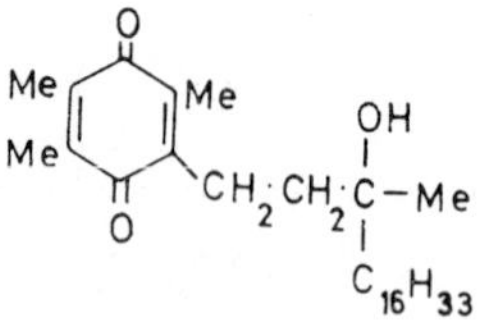

Vitamin K_1
(III)

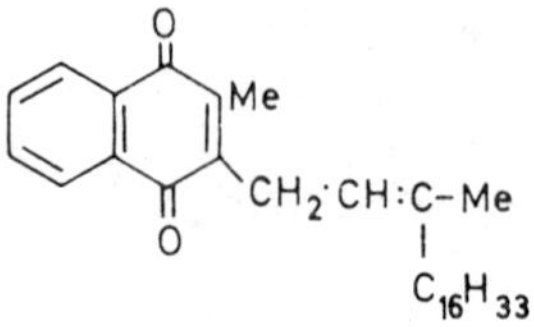

Ubiquinone
(IV)

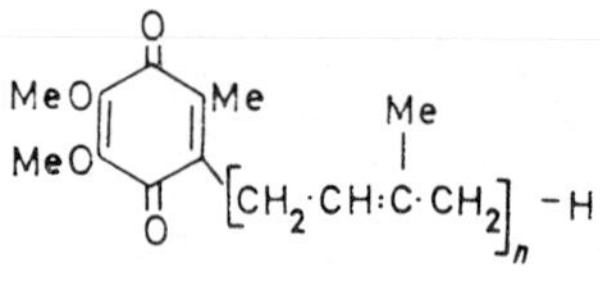

Figure 1 Some quinones of biological importance

As a first model for the understanding of the photochemistry of these quinones one turns naturally to the methyl substituted derivatives of benzoquinone and it happens that these compounds, and duroquinone in particular, were studied quite early by conventional flash photolysis. Indeed, duroquinone was the subject of the first detailed study by flash photolysis of an organic photochemical reaction in solution.[10] Three transients were observed and assigned: (1) the triplet state at 490 nm, (2) the semiquinone radical at 410 nm,

(3) the semiquinone radical-ion at 430 nm. Kinetic studies of the triplet were limited by its short lifetime and it was only possible later to show that the triplet state was the species responsible for semiquinone formation by hydrogen abstraction.

With other benzoquinones, and in less viscous solvents, it was no longer possible to observe the triplet state, although the semiquinone radicals were still formed with comparable yield. It was suspected that this was a consequence of the shorter triplet lifetime and, when nanosecond laser photolysis became available, it became possible to reinvestigate the problem.[8] Striking differences were found in the triplet lifetime as a function of the number of methyl substituents and the results are summarized in Table 1.

Table 1 Lifetimes of the triplet states of alkyl benzoquinones in solution

Quinone	Solvent	Lifetime (ns)		
		ethanol	ethanol + water (50:50 by volume)	water
Benzoquinone		<10[a]	<10	<10
Toluquinone		<10	<10	300
p-Xyloquinone		<10	30	320
o-Xyloquinone		45	600	1200
Cumoquinone		450	2400	1900
Duroquinone		9000	7700	2900

[a] <10 denotes no transient absorption observed.

First, consider the triplet lifetimes in ethanol, in which the principal reaction is hydrogen abstraction to give semiquinone radicals. The reactivity is least in duroquinone and increases by more than an order of magnitude on successive removal of methyl substituents, until, with benzoquinone and toluquinone, the lifetime of the triplet is too short for detection by nanosecond techniques.

The substitution effect on reactivity is interpreted as follows. The lowest triplet state of benzoquinone itself is undoubtedly of n–π type. This has an electron distribution with excess (over the ground state) negative charge on the ring and a deficit on the oxygens, making the oxygens electrophilic and therefore reactive with respect to hydrogen abstraction reactions. On substituting methyl groups, which are electron repelling, into the ring, this n–π type of electron distribution is less favoured energetically and the energy of the n–π triplet increases until eventually, in duroquinone, the π–π triplet becomes the lower. In the π–π triplet the electron distribution change is reversed and the oxygens become more negative, with a consequent reduction in reactivity with respect to hydrogen abstraction and a longer triplet lifetime, as observed. It should be noted that the quantum yield of abstraction is still high since, if there are no efficient competing reactions, the hydrogen abstraction takes place eventually even though it may be of a lower rate.

We must now explain the effect of adding water to the alcoholic solvent. With the benzoquinones having few methyl substituents and consequently triplet states which are principally $n-\pi$ in character, the effect of water is to increase the energy of the $n-\pi$ state with respect to the $\pi-\pi$ state and so to lengthen the triplet lifetime as observed. With duroquinone the lower state is essentially pure $\pi-\pi$, even in pure ethanol, and addition of water *shortens* the triplet lifetime. This must be due to the appearance of a competing reaction of the $\pi-\pi$ triplet in aqueous solvents. Remembering that the electronic distribution of this $\pi-\pi$ triplet is one having a positive charge excess in the ring we have proposed that the new reaction is loss of a proton from one of the methyl groups which have now greatly increased acidity over those in the ground state. This type of reaction is known to occur, even in the ground state, in strongly alkaline solutions. The exact mechanism of the change is not yet known; the initial reaction may be addition of a proton to the oxygen or loss of a proton from the methyl group but the final result is probably formation of an *ortho*-quinone methide.

$$(1)$$

With the biological quinones, *ortho*-quinone methides seem to be stabilized by the long hydrocarbon side chain and are readily observed as relatively long-lived transients. However, the mechanism of formation seems to be different in each case. With tocopheryl quinone the first quinone methide to be observed is the anion, suggesting that the first step is proton loss as shown above. On the other hand, plastoquinone always forms first the unionized *ortho*-quinone methide and there is evidence that in this case the first step is hydrogen abstraction. The case of plastoquinone is particularly important in view of its postulated involvement in an early electron transfer step of the photosynthetic process and it may therefore be interesting to consider the recent work on the photochemistry of this compound by Creed, Hales and Porter.[11]

3.2 Plastoquinone

The principal products of the steady state photolysis of plastoquinones (PQ) are dihydrobenzofurans and chromenols. Spectroscopic evidence indicates that the triplet state has significant $n-\pi$ character, similar to that of the xyloquinones.

Nanosecond flash photolysis of plastoquinone-9 in ethanolic solution led to the identification of at least six distinct transient substances apart from permanent products. Two, absorbing at 418 nm and 435 nm were assigned to the semiquinone radical and its anion respectively. The other four transients were as follows:

P_{420} λ_{max} = 420 nm and 440 nm

First-order decay with lifetime of 100 ns. Lifetime unaffected by oxygen, naphthalene or piperylene.

P_{510} λ_{max} = 510 nm

Decay approximately independent of solvent and pH with lifetime 0·3 s. Rate of formation identical with decay of P_{420}.

P_{490} λ_{max} = 490 nm

Decay solvent and pH dependent with lifetime of 2×10^3 s in ethanol and much faster in acid solution. Rate of formation identical with decay of P_{420}.

P_{590} λ_{max} = 590 nm

Formed from P_{490} in basic solution and reconverted to P_{490} in acidic solution. Clearly the base in equilibrium with its acidic form P_{490}.

On the basis of this and other evidence P_{490} and P_{510} are assigned to the transoid (QM_T) and cisoid (QM_C) o-quinone methides:

$$QM_T \ (P_{490}) \qquad QM_C \ (P_{510}) \tag{2}$$

whilst P_{580} is clearly the anion of the o-quinone methide P_{490}.

The assignment of P_{420} presents a more difficult problem. It cannot be assigned to the normal triplet in view of its resistance to known triplet quenchers. It is presumably formed from the triplet in less than 10 ns. The most likely assignment is to a diradical formed by intramolecular hydrogen abstraction and this diradical then undergoes an H-atom shift *via* hydrogen abstraction within a solvent cage:

$$PQ \ \text{(TRIPLET)} \longrightarrow \quad P_{420} \quad \longrightarrow \ QM_C + QM_T \tag{3}$$

The quinone methides react to give the observed final products. QM_C isomerizes to give the chromenol whilst QM_T reacts with solvent to give the substituted dihydrobenzofurans. That this is indeed the case was shown by irradiating a solution containing the quinone methides with light absorbed only by these intermediates. Analysis showed that the irradiated solution contained less dihydrobenzofuran and more chromenol than another portion of the same solution which had not been irradiated.

The overall mechanism proposed is summarized below:

$$PQ \xrightarrow{h\nu} PQ_T^{\cdot} \xrightarrow{<10ns} P_{420} \xrightarrow{100ns} P_{490}(QM_T) \; + \; P_{510}(QM_C) \underset{H^+}{\overset{-H^+}{\rightleftharpoons}} P_{590} \tag{4}$$

DIHYDROBENZOFURAN   CHROMENOL

One further observation is of particular interest in connection with the biological implications of this work. It was found that quinone methides are also formed by disproportionation of the semiquinone radicals. Clearly, direct photolysis of plastoquinone is not relevant to the photosynthetic mechanism but this observation suggests a means whereby the *ortho*-quinone may be formed, nevertheless. In other work we have shown that the benzo-quinones, including plastoquinone, readily quench the excited singlet state of chlorophyll to form semiquinone radical-anions by electron transfer. Dis-proportionation of two such radicals, formed by two excited chlorophyll molecules would now result in the formation of an *ortho*-quinone methide. This is of some interest in view of Witt's hypothesis of the involvement of a 'plastoquinone pair' in the photosynthetic process.

REFERENCES

1. G. Porter, R. W. Yip, J. M. Dunstan, A. J. Cessna and S. E. Sugamori, *Trans. Faraday Soc.*, **67**, 3149 (1971).
2. A. E. W. Knight and B. K. Selinger, *Aust. J. Chem.*, **26**, 1 (1973).

3. G. Porter and M. R. Topp, *Proc. Roy. Soc. Lond.*, **A315**, 163 (1970).
4. R. R. Alfano and S. L. Shapiro, *Phys. Rev. Letters*, **24**, 592, 1217 (1970).
5. M. R. Topp, R. M. Rentzepis and R. P. Jones, *J. Appl. Phys.*, **42**, 3415 (1971).
6. M. Kasha, *Disc. Faraday Soc.*, **9**, 14 (1950).
7. G. Porter, *Reactivity of the Photoexcited Molecule*, p. 79, Interscience, New York, 1967.
8. D. R. Kemp and G. Porter, *Proc. Roy. Soc. Lond.*, **A326**, 117 (1971).
9. H. T. Witt, *Z. Naturforsch*, **246**, 1031 (1969).
10. N. K. Bridge and G. Porter, *Proc. Roy. Soc. Lond.*, **A244**, 259, 276 (1958).
11. D. Creed, B. J. Hales and G. Porter, *Proc. Roy. Soc. Lond.*, **A334**, 505 (1973).

A.1. Fluorescence probes and nanosecond fluorometry

Ludwig Brand, William R. Laws, J. Hamilton Easter and Ari Gafni

Biology Department, The Johns Hopkins University
Baltimore, Maryland 21218, U.S.A.

ABSTRACT

Fluorescence probes are small molecules which undergo changes in one or more of their fluorescent properties as a result of interaction with a protein or other macromolecule. Our aim has been to utilize probes whose fluorescence changes can be ascribed to known excited-state reactions. Nanosecond fluorescence methods have been used to investigate molecules capable of undergoing excited-state solvent relaxation and excited-state proton transfer. Solvent relaxation has been observed with TNS in viscous solvents such as glycerol using nanosecond time-resolved emission spectroscopy. Time-resolved emission shifts have also been observed with this type of dye bound to model membrane systems. The single-photon counting method has been used to measure rate constants for excited-state proton transfer with molecules such as acridine and β-naphthol. In the case of β-naphthol, reversible excited-state proton transfer is expected in the pH region 2 to 5. Theory predicts that in this pH region the fluorescence decay of the ionized species should be described by a double exponential with equal amplitudes opposite in sign. This has been experimentally verified. Changes in rates of excited-state proton transfer have been observed when chromophores of this type are bound to proteins. Studies of this type can be used to measure conformational changes in macromolecules. Problems associated with the single-photon counting method for obtaining fluorescence decay times and difficulties associated with analysis of the data will be discussed.

A.2. Photophysics of fluorescent and non-fluorescent exciplexes

W. R. Ware, D. Watt and J. D. Holmes

University of Western Ontario, London, Canada

ABSTRACT

The problem of extracting rate constants from fluorescence decay and steady-state measurements for exciplex systems will be discussed in general because of the wide range of applicability of these techniques to fluorescence quenching in photochemistry and photobiology. The specific case of cyanonaphthalene quenched by olefins will then be discussed to illustrate these techniques and emphasize the importance of obtaining individual rate constants as a critical step in understanding the interaction of quencher and fluorophor.

A.3. *Excited state electron transfer processes*

M. Schulz, H. Staerk and A. Weller

Max-Planck-Institut für biophysikalische Chemie, Göttingen, Germany

ABSTRACT

Kinetic studies of electron transfer fluorescence quenching reactions in aceto-nitrile show that the actual electron transfer processes can occur (adiabatically) with a rate of about $10^{11}\,\text{s}^{-1}$ when the electron acceptor (A) and donor (D) molecules (one of them in the excited singlet state) form a loose encounter complex (e.g. $^1A^* \ldots D$) with a centre-to-centre distance of 7 ± 1 Å. Diffusional separation of the solvated radical ion pair thus formed ($^2A^{\bar{\cdot}} \ldots {}^2D^{\dot{+}}$) produces the free radical ions which can be observed by nanosecond flash spectroscopy and which recombine to give (by reverse electron transfer) triplet and/or ground state molecules.

Nanosecond-flash spectroscopic studies carried out with 1-(9-anthryl)-3-(p-dimethylaminophenyl)-propane: $A-(CH_2)_3-D$ (serving as a model for the encounter complex between 9-methylanthracene and dimethyl-p-toluidine) gave no detectable transients in acetonitrile although the fluorescence is almost completely quenched. Evaluation of these data gave for the rate constants of the intramolecular electron transfer processes:

$$^1A^*-(CH_2)_3-{}^1D \rightarrow {}^2A^{\bar{\cdot}}-(CH_2)_3-{}^2D^{\dot{+}} \qquad k_{si} \approx 1 \times 10^{11}\,\text{s}^{-1}$$

$$^2A^{\bar{\cdot}}-(CH_2)_3-{}^2D^{\dot{+}} \rightarrow {}^3A^*-(CH_2)_3-{}^1D \qquad k_{it} < 5 \times 10^6\,\text{s}^{-1}$$

$$^2A^{\bar{\cdot}}-(CH_2)_3-{}^2D^{\dot{+}} \rightarrow {}^1A-(CH_2)_3-{}^1D \qquad k_{ig} > 5 \times 10^8\,\text{s}^{-1}$$

These values are interpreted on the basis of the multiplicity changes and Franck–Condon factors involved.

A.4. Dissociative oxciplex relaxation in photoperoxidation reactions

B. Stevens

Department of Chemistry, University of South Florida, Tampa, Florida 33620, U.S.A.

ABSTRACT

The experimental evidence indicates that the self-sensitized addition of molecular oxygen to a dissolved aromatic hydrocarbon M involves a reaction sequence

$$M(S_1) + O_2{}^3\Sigma \overset{1}{\rightarrow} M(T_1) + O_2{}^3\Sigma \overset{2}{\rightarrow} M(S_0) + O_2{}^1\Delta \overset{3}{\rightarrow} MO_2$$

in which the products of processes 1 and 2 are the reactants in processes 2 and 3, rather than the direct addition of $O_2{}^3\Sigma$ to either the excited singlet $M(S_1)$ or triplet $M(T_1)$ states of the acceptor.

An orbital correlation scheme presented for the concerted addition process 3 and based on anthracene as a model acceptor with a 'planar peroxide' complex configuration of highest symmetry shows that the complex or oxciplex states $^3(S_1{}^3\Sigma)$ and $^{1,3}(T_1{}^3\Sigma)$ must correlate endothermically with electronically excited states of the peroxide MO_2. Arguments are given in support of the exchange mechanism for process 2 in non-polar solvents in which, for a similar oxciplex configuration, the final state correlates with that component of the $O_2{}^1\Delta$ state which does not lead to formation of the MO_2 ground state $^1A_1$. Direct formation of $MO_2{}^1A_1$ in processes 1 or 2 therefore requires electronic relaxation of the initially formed oxciplex states $^3(S_1{}^3\Sigma)$ and $^1(T_1{}^3\Sigma)$ which is either spin or orbitally forbidden but is achieved by dissociation and recombination of the molecular components.

A.5. Compound-negative-ion-resonant states and threshold-electron-excitation spectra of monosubstituted benzene derivatives†

L. G. Christophorou,‡ D. L. McCorkle and J. G. Carter

Health Physics Division, Oak Ridge National Laboratory,
Oak Ridge, Tennessee 37830

Threshold-electron-excitation (TEE) spectra for fluorobenzene, benzaldehyde and benzoic acid have been obtained and compared with photoabsorption

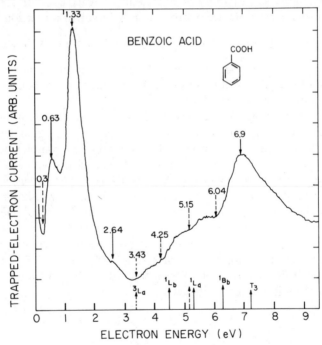

Figure 1   Threshold-electron excitation spectrum of benzoic acid

† Research sponsored by the U.S. Atomic Energy Commission under contract with Union Carbide Corporation.
‡ Also Department of Physics, University of Tennessee, Knoxville, Tennessee.

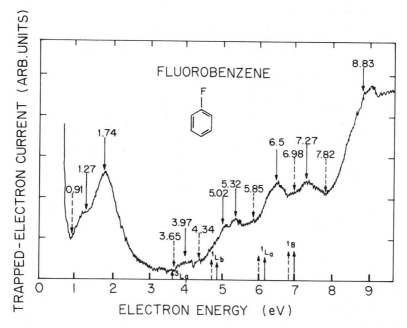

Figure 2 Threshold-electron excitation spectrum of fluorobenzene

spectra. Figures 1 and 2 show representative TEE spectra for benzoic acid and fluorobenzene. The solid arrows above the spectrum indicate the positions of the maxima in the TEE spectrum and the broken arrows indicate the positions of the onsets. The numbers which accompany each arrow give the position (in eV) of a peak or an onset, and they have been obtained by an averaging of the data for a large number of spectra similar to the ones presented. The broken and solid arrows perpendicular to the energy axis give, respectively, the positions of the band onsets ($0 \rightarrow 0$ transitions) and band maxima obtained with photophysical methods. The excitation of optically forbidden states is clearly evident in the TEE spectra.

The two distinct peaks at 1·27 and 1·74 eV for fluorobenzene and at 0·63 and 1·33 eV for benzoic acid are due to excitation of two compound-negative-ion-resonant (CNIR) states. Actually, two CNIR states were observed below the first-excited electronic state, as opposed to only one for benzene, for all ten monosubstituted benzene derivatives studied. The positions of the maxima of these double resonances are shown in Figure 3. The bracketed numbers accompanying the formula for each compound are the photoionization or photoabsorption values (in eV) of the first ionization potentials of the parent molecules. The numbers below each level give (in eV) the position of the maxima of the observed CNIRs. The maxima of the $C_6H_5NO_2$ resonances were determined on the basis of an earlier study[1] on dissociative attachment to $C_6H_5NO_2$

20

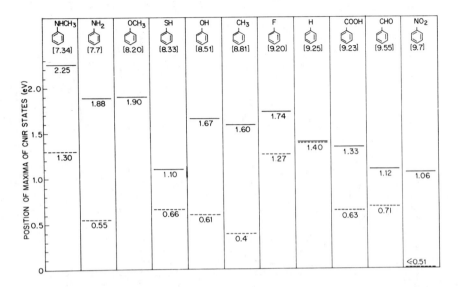

Figure 3 Position of maxima of the double compound-negative-ion resonant states for benzene and ten monosubstituted benzene derivatives in descending order of the first ionization potential of the neutral molecule

where the lowest lying resonance above thermal energies peaked at $1.06\,eV$ and led to the production of NO_2^-.

The number and the positions of these double CNIRs have been understood[2] in terms of the net π-electron charge transfer between the substituent and the benzene ring. For the benzene molecule the two lowest unfilled π-orbitals, $e_{2u,1}$ and $e_{2u,2}$, are degenerate; this accounts for the observation of only one CNIR for benzene since electron capture in either orbital will yield a resonance at exactly the same energy. For the benzene-derivative molecules, however, the two lowest unoccupied π-orbitals are no longer degenerate due to the perturbation introduced by the substituent; thus, two CNIRs arise from the quasi-trapping of slow electrons in the $e_{2u,1}$ and $e_{2u,2}$ orbitals. The present results distinctly show the effects of intramolecular charge transfer between the substituent and the benzene ring on the energies of the lowest unoccupied π-orbitals involved in the quasi-trapping of slow electrons responsible for the observed CNIRs. A full discussion of this and a detailed exposition of the results can be found in Reference 2.

REFERENCES

1. L. G. Christophorou, R. N. Compton, G. S. Hurst and P. W. Reinhardt, *J. Chem. Phys.*, **45**, 536 (1966).
2. L. G. Christophorou, D. L. McCorkle and J. G. Carter, *J. Chem. Phys.*, **60**, 3779 (1974).

Session A.1
Primary processes

A.1.1. Strange absorption and emission shifts due to pH and medium effects

A. Chamma, M. Deumie, B. Saperas and P. Viallet

Laboratoire de Chimie Physique, Centre Universitaire 66000, Perpignan, France

P. Jacquignon and F. Perrin

Laboratoire CNRS, 91190 Gif-Sur-Yvette, France

ABSTRACT

In a series of aromatic molecules containing both a pyrrolic and a pyridinic nitrogen atom, there is a difference between the pK values of the S_1 state calculated from the shifts due to changes in pH of the absorption spectra and the fluorescence spectra.

To account for this phenomenon, we propose a mechanism which involves a proton transfer in the excited state. This transfer from pyrrolic to pyridinic nitrogen, in a neutral medium, can be described as follows:

$$ABH + hv \rightarrow ABH^*$$

$$ABH \quad \rightarrow AHB^*$$

$$AHB \quad \rightarrow hv' + AHB$$

$$AHB \quad \rightarrow ABH$$

According to this hypothesis, the apparent anomaly is because the two types of measurements involve different species. The same hypothesis also accounts for differences observed in the solvent effect on the absorption and fluorescence spectra of the same molecules. However, in the case of aromatic nitrogen molecules, the possibility of band-crossing among the energy levels when protonation occurs must be considered. This band-crossing occurs in the indole molecule and it would explain the experimental results observed in one of our molecules. It is hoped that the explanation of these energy-level shifts will be confirmed by the further experiments in progress.

A.1.2. Unique properties of the lowest excited singlet state of 4-dialkylaminopyrimidines

J. Smagowicz, K. Berens and K. L. Wierzchowski

Institute of Biochemistry and Biophysics, Polish Academy of Sciences, Rakowiecka 36, 02-532 Warszawa, Poland

SUMMARY

On the basis of an analysis of the ground and the lowest excited singlet state dipole moments of 4-dimethylaminopyrimidine and its sterically hindered 5-methyl derivative, evidence is presented that the appearance of two fluorescence bands in the unhindered molecule is due to reorientation of the $N(CH_3)_2$ group in the excited state.

1 INTRODUCTION

This paper is a part of a study aimed at elucidation of the excited state properties and mechanisms of radiative and radiationless transitions in amino-pyrimidines and aminopurines, being fluorescent models of natural nucleic acid bases. Previously it has been demonstrated that in symmetrically sub-stituted 2- and 5-aminopyrimidines[1,2] as well as in 2-aminopurine[3] the lowest excited singlet state is of (l, a_π^*) character, this influencing the mechanism of $S_1 \rightsquigarrow T_1$ intersystem crossing. Much less information, in this respect, is available for unsymmetrically substituted 4-aminopyrimidine because of its low quantum yield of fluorescence. N-alkylation of this compound increases, however, the emission efficiency and makes it possible to undertake more systematic studies, particularly on the effect of conformation of the $N(CH_3)_2$ group on the mech-anism of electronic transitions. So far, similar studies have been performed only on aromatic amines[4,5].

24

2 RESULTS AND DISCUSSION

The absorption and fluorescence spectra of 4-dimethylaminopyrimidine (4DMApy) and of its *ortho*-methyl derivative (4DMA-*o*Mepy) were measured at room temperature in a series of solvents of different polarity. Changes in the molecular dipole moment on excitation to the lowest singlet state were evaluated by the solvent perturbation method[6] under the assumption of a spherical cavity model with 5 Å radius (Figure 1). The increase in dipole moment of the *ortho*-methylated, sterically hindered compound ($\Delta\mu = 11\cdot2$ D) is almost twice as large as that of the unhindered one ($\Delta\mu = 5\cdot6$ D). Furthermore, 4DMApy exhibits two-component fluorescence in most polar solvents (Figure 2). The

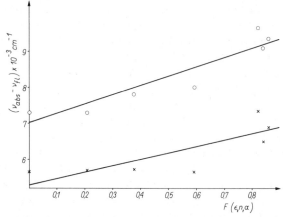

Figure 1 Plot of fluorescence Stokes shifts of 4-dimethylaminopyrimidine (crosses) and 4-dimethylamino–5-methylpyrimidine (circles) *vs* the Bilot–Kawski function F: cyclohexane ($F = 0\cdot001$), trichloroethylene ($F = 0\cdot210$), ethyl ether ($F = 0\cdot428$), dicholomethane ($F = 0\cdot595$), ethanol ($F = 0\cdot820$), dimethylformamide ($F = 0\cdot840$), acetonitrile ($F = 0\cdot862$)

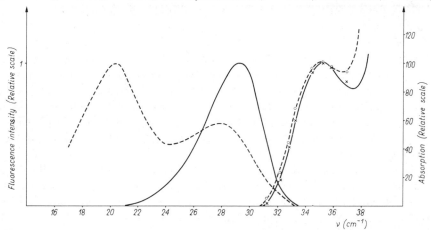

Figure 2 Absorption and fluorescence spectra of 4-dimethylaminopyrimidine in ethyl ether ——— and in ethanol – – –. Fluorescence spectra are normalized with respect to the maximum of the corresponding absorption bands. Experimental points on the absorption curve correspond to the excitation spectra

two experimental points in Figure 1, corresponding to large Stokes shifts of the long-wavelength component of the dual emission in ethanol and acetonitrile, fit very well the Kawski plot for the single fluorescence of the sterically hindered derivative.

At low temperature (90 K), in ethanol glass, only one fluorescence band appears in the spectrum of 4DMApy. It is shifted to the region where room-temperature emission of a hydrocarbon solution of this compound is observed. The excitation spectra of the two components of 4DMApy fluorescence are identical (Figure 2), but the lifetimes are not, since with increasing oxygen pressure over the sample quenching of the long-wavelength part of fluorescence spectrum is stronger than that of the short-wavelength one. Examination of purified samples of 4DMApy by thin-layer and gas chromatography did not reveal any heterogeneity of the material.[7] All this indicates that 4DMApy has the unique property of exhibiting in the lowest excited singlet state two equilibrium configurations widely differing in polarity. The more polar configuration is apparently the same as that into which the sterically hindered o-methyl derivative relaxes during the lifetime of its excited state.

The CNDO/2 calculations of the total energy of N,N-dimethyl cytosine (4-dimethylamino-2-oxo-pyrimidine) as a function of angle α of twist of the $N(CH_3)_2$ group relative to the ring indicate that the quasi-coplanar ($\alpha = \pm 30°$) as well as the perpendicular ($\alpha = 270°$) structures of this molecule are energetically favourable in the ground state, with the former one more stable by about 5 kcal/mole. In the ortho-methylated derivative of this compound only perpendicular ($\alpha = 90°$, $270°$) structures are possible. These conformational predictions are most probably valid also for 4DMApy and 4DMA-oMepy. If so, then the blue and red components of the 4DMApy emission may reasonably be attributed to the quasi-coplanar (q.c) and perpendicular (p) conformers, respectively.

No evidence for the occurrence of an equilibrium of rotamers in the ground state is provided by either absorption studies or chromatographic findings. The identity of the excitation spectra of both emissions of 4DMApy, taken together with the small difference in fluorescence quantum yield of compounds with perpendicular and quasi-coplanar orientations of the $N(CH_3)_2$ group, also indicates that no ground state equilibrium is involved in this case. On the other hand, the shorter lifetime of the blue emission compared with the red one can be taken as a manifestation of excited state reorientation of the $N(CH_3)_2$ group from the quasi-coplanar conformation, giving rise to the blue emission, to the perpendicular one, which is characteristic of the red emitting species.

This solvent-induced rearrangement of the dimethylamino group in 4DMApy can be rationalized in terms of different stabilities of both solvated rotamers in the ground and excited states. Dipole moments of 4DMApy and 4DMA-oMepy are equal in the ground state,[8] and probably so are those of both rotamers of 4DMApy: $\mu°(q.c) \cong \mu°(p)$. Consequently, the energies of solvent stabilization of both forms of the molecule are expected to be similar, and the stability sequence is: $E(q.c) \cong E(p)$, as indicated by CNDO/2 calculations.

Excitation of the molecule is accompanied by an increase in dipole moment of the perpendicular rotamer, twice that of the quasi-coplanar one: $\mu^*(q\,c) < \mu^*(p)$. As a result, in the presence of a polar solvent the stability sequence can be drastically changed in the excited state, favouring the perpendicular form of 4DMApy. Further investigations on the nature of the highly polar excited state of 4DMA-oMepy are in progress.

REFERENCES

1. K. Berens, J. Smagowicz and K. L. Wierzchowski, *I Europ. Biophys. Congress*, Vienna, Abstracts, 63 (1971).
2. K. Berens, Ph.D. Thesis, Institute of Biochemistry and Biophysics, Warsaw, 1971.
3. J. Smagowicz and K. L. Wierzchowski, *J. Luminescence*, **8**, 210 (1974).
4. M. Kasha and R. Rawls, *Photochem. Photobiol.*, **7**, 561 (1968).
5. E. C. Lim and S. C. Chakrabarti, *J. Chem. Phys.*, **47**, 4726 (1967).
6. A. Kawski, *Acta Phys. Pol.*, **29**, 507 (1966).
7. Z. Proba and K. L. Wierzchowski, *Acta Biochem. Polon.*, **22**, 131 (1975).
8. I. Kulakowska, M. Geller, B. Lesyng, K. Bolewska and K. L. Wierzchowski, *Biochim. Biophys. Acta,* in press.

A.1.3. Magnetic circularly polarized fluorescence of tetraphenylporphine

John Clark Sutherland, George D. Cimino and John T. Lowe

*Department of Physiology, California College of Medicine,
University of California, Irvine, California, U.S.A.*

SUMMARY

We have measured the magnetic-field-induced difference in the intensities of the left and right circularly polarized components of the fluorescence emission spectra of free base and diacid *meso*-tetraphenylporphine (TPP) in benzene solution at room temperature. For the diacid TTP (D_{4h} symmetry) we observe a single band with roughly the same shape as the emission intensity spectrum. We classify this spectrum as the sum of 'A' and 'C' type terms with the 'C' type dominant. We observe weaker MCPF signals from the $0 \leftarrow 0$ and $1 \leftarrow 0$ emission bands of the TPP free base which has only twofold symmetry. These signals represent the weaker 'B' type MCPF spectra. This report of MCPF at room temperature from a compound with less non-degenerate excited states suggests that it will be possible to observe MCPF from a wide variety of fluorescent molecules.

1 INTRODUCTION

Magnetic circularly polarized fluorescence (MCPF) is the magnetic-field-induced difference in the intensities of left and right circularly polarized components of fluorescent light. MCPF is thus the emission analogue of magnetic circular dichroism (MCD). Therefore the extensive body of theory relating MCD spectra to molecular structure can also be used to interpret MCPF. Yet information provided by MCPF is not merely a duplication of that obtained from MCD, since the former may reflect photophysical processes occurring during the life of the excited state. Further, MCPF will permit the powerful theory of magnetic optical activity to be applied to exciplexes and other systems which cannot be probed by absorption techniques such as MCD.

28

We chose a porphyrin for our initial investigation, since it was expected to give large effects and because the optical properties of porphyrins are well understood. Our experiments were performed at room temperature in non-viscous solvents so that we can assume a random orientation of the emitting molecules. Random orientation eliminates artifacts due to linear polarization of the fluorescent light which has been reported by Steinberg and Gafni.[1]

Random orientation also facilitates the determination of the degree of splitting of degenerate states and thus comparison of data obtained from MCD. However, theory[2] suggests that for degenerate excited states, larger signals will be observed at low temperatures. Recent low temperature experiments on zinc octaethyl porphyrin seem to confirm this hypothesis.[3]

2 EXPERIMENTAL PROCEDURE

Materials. *meso*-Tetraphenylporphine (TPP) was dissolved in reagent grade benzene. A small quantity of concentrated HCl was added to form diacid TPP.

Apparatus. Absorption spectra were measured with a Perkin–Elmer model 124 Spectrophotometer. MCPF and MCD spectra were measured with an instrument constructed in our laboratory which will be described elsewhere. Calibration of the MCD instrument was achieved by measuring the natural CD of a sample of d-10-camphorsulphonic acid. The strength of the magnetic field was measured with an RFL Laboratories Model 750 gaussmeter and by measuring the MCD of cytochrome c.

Both MCPF and fluorescence intensity are measured in arbitrary units. However, both quantities are expressed in the *same* arbitrary units since the processing of both signals are the same up to the input to the lock-in amplifier which imparts a known gain to the a.c. component. The magnetic anisotropy, the ratio of MCPF to fluorescence intensity, is thus an absolute quantity.

The fluorescence of TTP and diacid TTP was excited by 633 nm light from a Hughes model 3078HR helium–neon laser. The laser beam was polarized with its electric vector parallel to the sample–detector axis to minimize effects of linearly polarized emission or scattering. All measurements were performed with the sample at room temperature.

3 THEORY

The lowest energy-allowed electronic transition of a D_{4h} porphyrin is from a non-degenerate ground state to a doubly degenerate excited state.[4] The degenerate state is split by a magnetic field into two states one of which absorbs only left circularly polarized light (LCPL) while the other absorbs only RCPL. If the plane of each molecule was perpendicular to the optical axis, the energy splitting of the excited state would be $2\beta HM_z$, where β is the Bohr magneton,

H is the strength of the magnetic field and M_z is the 'orbital angular momentum' of the excited state.

For most organic molecules in solution at room temperature and for easily achievable magnetic fields, the magnetic splitting of the degenerate state is much less than the width of the absorption band. Thus the MCD will resemble the derivative of the absorption spectrum and is referred to as an 'A' type MCD spectrum.[2]

MCPF is the fluorescence analogue of MCD. Since $\beta H M_z \ll kT$ at room temperature, after excitation and vibrational relaxation, both components of the split degenerate pair are populated according to Boltzmann's law. The emission from the two levels will be LCPL and RCPL respectively and will thus tend to produce a derivative or 'A' type MCPF spectrum similar to that observed in MCD.

A consequence of the Boltzmann distribution of the populations of the two states, however, is that there will be more intensity of the polarization associated with the lower energy state. This contribution to the MCPF resembles the shape of the emission band and is called a 'C' type spectrum. The 'C' term is proportional to $(kT)^{-1}$. The observed MCPL will be the sum of the contributions of the A and C terms and the shape of spectrum will depend on their relative magnitudes.

There may be other differences between MCD and MCPL due to physical of chemical processes occurring during the life of the excited state. Thus we label the 'angular momentum' of the occupied excited state as M_z^*.

4 RESULTS AND DISCUSSION

Figure 1 shows the absorption spectrum and MCD of diacid TTP. The MCD typical of a D_{4h} porphyrin molecule with a non-degenerate ground state resembles the derivative of the absorption spectrum and is thus an 'A' type. The $1 \leftarrow 0$ vibrational bands are weak and overlapping in absorption. Thus they appear as a very weak MCD band showing only a small residual effect.

Figure 2 shows the fluorescence emission spectrum and MCPF of diacid TTP. The emission spectrum is approximately the mirror image of the absorption spectrum. The MCPF resembles the shape of the emission spectrum indicating that the 'C' term is dominant even at room temperature. No signal was recorded with the magnetic field turned off as expected for a planar molecule. No signal was detected at twice the modulation frequency (where linear polarization effects would appear) with or without the magnetic field. No MCPF signal was detected at the wavelength of the $1 \leftarrow 0$ vibrational emission bands. This is not surprising since these bands were also much weaker than the $0 \leftarrow 0$ band in MCD and would be lost in the noise. The approximate magnitude of the noise is indicated by the error bar in the figure.

While both fluorescence and MCPF are reported in arbitrary units, they are measured in the *same* arbitrary units and their ratio is an absolute quantity.

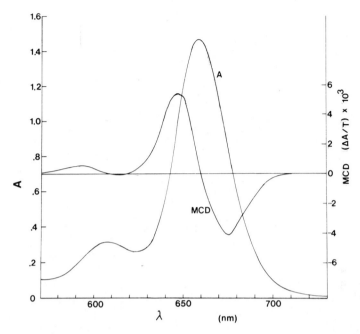

Figure 1   The absorption spectrum and MCD of diacid TPP. The units of MCD are absorbance per unit magnetic field with field strength expressed in teslas

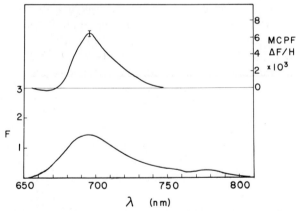

Figure 2   The fluorescence emission spectrum and MCPF of acid TPP. Fluorescence and MCPF were measured in arbitrary units. The MCPF was normalized to a field of one tesla

The data in Figure 4 give $\Delta fm/f = 4 \cdot 55 \times 10^{-3}$ (T^{-1}). Assuming that the effect of random orientation of the emitting molecules is the same as in MCD and ignoring the contribution of the A term to MCPF we have $\Delta fm/f = \beta M_z^*/2kT$. This formula yields a value of M_z^* of $3 \cdot 7$ Bohr magnetons and is (within experimental error) equal to the value of M_z of $3 \cdot 5\beta$ Bohr magnetons estimated from the MCD.

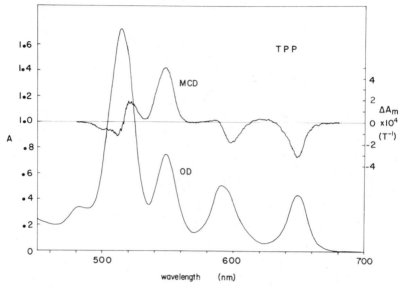

Figure 3 The absorption spectrum and MCD of TPP free base which has D_{zh} symmetry. The four bands are, in order of decreasing wavelength, Q_{x-0}, Q_{x-1}, Q_{y-0}, and Q_{y-1}

Free base TPP has only twofold symmetry and thus cannot have exactly degenerate states. The Q_{0-1} bands both split to form four bands labelled Q_{x-0}, Q_{x-1}, Q_{y-0}, Q_{y-1} in order of increasing energy and absorption intensity (see Figure 3). The MCD of these bands are 'B' type and thus resemble the shape of the corresponding absorption spectrum as shown in Figure 3.

Figure 4 shows the emission spectrum and MCPF of TPP free base. Both the $0 \leftarrow 0$ and $1 \leftarrow 0$ bands are observed. These transitions are also observed

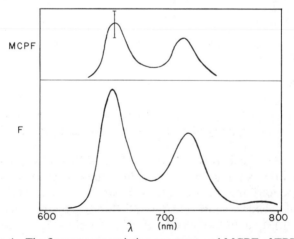

Figure 4 The fluorescence emission spectrum and MCPF of TPP free base

in the MCPF spectrum. As expected from MCD they are 'B' type, resemble the shape of the emission spectrum and are weaker in amplitude than the MCPF observed for diacid TPP (compare the error bars).

5 CONCLUSION

The MCPF of room temperature solutions can be measured and the results explained within the theoretical framework developed to treat MCD with the obvious changes resulting from the interchange of ground and excited states. Within the accuracy of our measurements the magnetic splitting of the excited state of TTP in benzene is the same when measured in emission as in absorption.

ACKNOWLEDGEMENTS

We thank Dr Stephen H. White for the use of his laser, Dr Everly Fleischer for the samples of TPP and Dr P. J. Stephens, University of Southern California, for helpful discussions. This research was supported by a grant from the Research Corporation.

REFERENCES

1. I. Z. Steinberg and A. Gafni, *Rev. Sci. Inst.*, **43**, 409 (1972); A. Gafni and I. Z. Steinberg, *Photochem. Photobiol.*, **15**, 93 (1972).
2. P. J. Stephens, *J. Chem. Phys.*, **52**, 3489 (1970); A. D. Buckingham and P. J. Stephens, *Ann. Rev. Phys. Chem.*, 399 (1966).
3. R. A. Shatwell and A. J. McCaffery, *J.C.S. Chem. Comm.*, 546 (1973).
4. M. Gouterman, *J. Molec. Spect.*, **6**, 138 (1961).

A.1.4 On the primary reactions in sensitized photooxidations

S. El Mohsni and I. Kraljić‡

*Laboratoire de Physico-Chimie des Rayonnements,† Université de Paris-Sud,
91405 Orsay, France*

M. Arvis and F. Barat

*Département Recherche et Analyse, Section des Recherches et de l'Intéraction
du Rayonnement avec la Matière, CEN-Saclay B.P. n° 2,
91110 Gif sur Yvette, France*

SUMMARY

In order to establish the nature of primary reactions and to identify the reacting intermediates in sensitized photooxidations in general and in photodynamic action in particular, a number of systems have been investigated in neutral aqueous solutions. In photostationary experiments phenosafranine, rhodamine B and some other dyes served as sensitizers (S) while biological and other compounds were used as substrates (A). The flash photolysis technique has been used to determine rate constants of quenching of the triplet state of phenosafranine by l-tryptophan, indole, 3-indole acetic acid, l-histidine, imidazole, allylthiourea, thiourea, allylurea, N_3^- and O_2. On the basis of these results the mechanism of photodynamic action is discussed (free radical vs singlet oxygen ($^1O_2$) mechanism).

1 INTRODUCTION

Two reaction mechanisms, playing the major role in sensitized photooxidation, are the free radical mechanism and the singlet oxygen ($^1O_2$) mechanism. This subject was reviewed recently.[1,2] In the first case the triplet sensitizer (^3S) reacts with A producing free radicals:

$$S_0 + hv \rightarrow {}^1S^* \rightarrow {}^3S \tag{1}$$

$$^3S + A \rightarrow S^- + A^+ \tag{2}$$

† Laboratoire associé au C.N.R.S.

‡ To whom correspondence should be addressed.

$$A^+ + {}^3O_2 \rightarrow AO_2^+ \rightarrow AO_2 \tag{3}$$

S_0 is the ground state of sensitizer, ${}^1S^*$ is its first excited singlet state; S^- is the semireduced form of S; AO_2 is a final oxygenation product of A.

In the singlet oxygen or Kautsky mechanism,[3] the ${}^3S$ transfers its energy to ${}^3O_2$ producing thus the ${}^1O_2$ which can then react with a substrate:

$${}^3S + {}^3O_2 \rightarrow S_0 + {}^1O_2 \tag{4}$$

$${}^1O_2 + A \rightarrow (AO_2) \rightarrow AO_2 \tag{5}$$

(AO_2) is an intermediate oxygen adduct which may rearrange or decompose into the final oxygenation product AO_2.

In order to establish the nature of primary reactions in photosensitized oxidations and oxygenations of various substrates with phenosafranine (PS) and rhodamine B (Rh B) as sensitizers, photostationary and flash photolysis experiments were performed in neutral aqueous solutions.

2 RESULTS AND DISCUSSION

In photostationary experiments conducted on aerated aqueous solutions we applied the technique of interception (scavenging) of reactive intermediates by p-nitrosodimethyl-aniline[4] (RNO), which can be bleached selectively by oxidizing intermediates in the presence of oxygen. In the systems with PS (or Rh B) and allylthiourea (ATU) as substrate RNO was not measurably bleached, while ATU was oxygenated. However, under these conditions (without and with 5×10^{-5} M RNO) both sensitizers were partially consumed (bleached). With some other substrates we observe a net bleaching of RNO as measured at 440 nm, its absorption maximum. This effect is obtained in the presence of l-histidine and imidazole with both PS and Rh B as well as with some other sensitizers. Such behaviour indicates that the bleaching of RNO in the presence of l-histidine may be due to a mechanism which is different from that leading to bleaching of sensitizer in the presence of ATU. Indeed, the addition of l-histidine to the system PS–ATU $(10^{-2}$ M$)$–RNO $(5 \times 10^{-5}$ M$)$ is without influence on the bleaching of sensitizer and at the same time it induces the bleaching of RNO (Figure 1). Such results seem to be best explained by assuming that the bleaching of S is caused via the primary reaction ${}^3S + ATU$ (free radical mechanism) whereas the destruction of RNO occurs via the ${}^1O_2$ mechanism (reaction 5 and 6):

$$(AO_2) + RNO \rightarrow -RNO + products \tag{6}$$

In this work we determined the quenching rate constants of ${}^3S$ of PS by a number of substrates. The triplet–triplet absorption spectrum of PS has been determined previously.[5] We followed the decay of ${}^3S$ absorption at 822 nm. From experimental decay curves the quenching rate constants were calculated by means of equation (7)

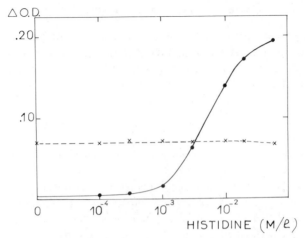

HISTIDINE (M/ℓ)

Figure 1   Effect of *l*-histidine on the bleaching of PS and RNO in the system PS
$(2 \times 10^{-5}M)$–ATU $(1 \times 10^{-2}M)$–RNO $(5 \times 10^{-5}M)$ at pH 7·1 in air saturated solution.
Irradiation time: 10 min (at 550 ± 10 nm); bleaching of PS followed at 550 nm (broken
line); bleaching of RNO followed at 440 nm (full line); ΔO.D. represents the difference of
O.D. between the non-irradiated and irradiated sample

$$k_q = (k_A - k_0)/[A] \tag{7}$$

k_q = quenching rate constant of ^3S comprising the physical quenching and
oxido-reduction process, such as a reaction (2); k_A = decay rate constant of ^3S
in the presence of A; k_0 = decay rate constant in the absence of A.

The results on quenching rate constants are presented in Table 1.

Table 1   Rate constants of triplet phenosafranine and rhodamine
B by substrates at pH 7·2; r = redox reaction (reaction 2),
q = physical quenching

Compound	Phenosafranine $k_q\,M^{-1}\,s^{-1}$	Rhodamine B $k_r\,M^{-1}\,s^{-1}$
Imidazole	$<10^5$	$<10^5$ r
l-Histidine	$<10^5$	$<10^5$ r
Indole	$4\cdot2 \times 10^7$ r + q	$<10^8$ r
l-Tryptophan	$6\cdot5 \times 10^7$ r + q	$\sim10^8$ r
3-Indole acetic acid	$7\cdot8 \times 10^8$ r + q	$>10^8$ r
Allylthiourea	$3\cdot3 \times 10^5$ r + q	10^5–10^6 r
Thiourea	$<2\cdot7 \times 10^5$	
Allylurea	$<10^5$	$<10^5$ r
N_3^-	$1\cdot5 \times 10^6$	
O_2	$7\cdot0 \times 10^8$	

The yield of ^3S of Rh B was too low[6] to enable us to observe its absorption
spectrum. For this reason, we could not measure directly its rate constants.

However, we were able to make an estimation of some of them by observing the absorption of the semireduced form of this sensitizer produced via the reaction (2). From such experiments we conclude that l-tryptophan, 3-indole acetic acid and indole react rapidly, ATU slowly and l-histidine, imidazole and allylurea do not react measurably with ^{3}S of Rh B by an oxido-reduction mechanism. These rate constants relate only to the oxido-reduction process and do not comprise the physical quenching of ^{3}S by these substrates.

Now we can draw several conclusions concerning the primary reactions in photosensitized oxidation of substrates used in this work. First, l-histidine and imidazole react so slowly with ^{3}S of PS and Rh B that they cannot participate in a free radical mechanism in aerated solutions, where the interaction between ^{3}S and ^{3}O$_2$ is a rapid process. This means that these substrates are oxygenated via the ^{1}O$_2$ mechanism. This supports our view that the bleaching of RNO in the presence of these substrates in photostationary experiments is due to a mechanism involving the ^{1}O$_2$.

Indole derivatives react rapidly with ^{3}S of both sensitizers partially by an oxido-reduction process. Thus, they can participate in a free radical mechanism on the level of primary reactions. Tryptophan reacts also with singlet oxygen[7] and it is likely that indole and 3-indole acetic acid have about the same reactivity towards this excited species. This would mean that these substrates can take part in both mechanisms on a primary level, without being necessarily oxygenated by both reaction pathways. At tryptophan and 3-indole acetic acid concentrations above 5×10^{-3} and 10^{-3} M respectively in air-saturated solutions, these substrates will react with or quench more ^{3}S than does oxygen. Accordingly, under such conditions relatively small amounts of ^{1}O$_2$ would be formed (via reaction 4), thus diminishing its participation in the oxygenation process.

Concerning other compounds used in this work (see Table 1), we can conclude that ATU can participate in a free radical mechanism when present in higher concentrations, above 10^{-2} M, in air-saturated solutions. The fact that allylurea does not react measurably with ^{3}S of PS indicates that it is the thiourea moiety in ATU which takes part in the oxido-reduction process represented by the reaction (2). N$_3^-$ ion can also quench (or react partially with) ^{3}S of PS if present in concentration above 10^{-2} M. Since this ion is a good quencher of ^{1}O$_2$[8,9] it can also interfere with the normal course of reactions in a singlet oxygen mechanism.

REFERENCES

1. K. Gollnick, *Adv. Photochem.*, **6**, 1 (1968).
2. L. I. Grossweiner and A. G. Kepka, *Photochem. Photobiol.*, **16**, 305 (1972).
3. H. Kautsky, *Trans. Faraday Soc.*, **35**, 216 (1939).
4. (a) I. Kraljić and C. N. Trumbore, *J. Amer. Chem. Soc.*, **87**, 2547 (1965).
   (b) I. Kraljić, *Int. J. Radiat. Phys. Chem.*, **2**, 59 (1970).
5. A. K. Chibisov, B. V. Skvortsov, A. N. Karyakin and L. N. Rygalov, *Khim. Vys. Energ.*, **3**, 210 (1969).

6. A. K. Chibisov, H. A. Kezle, L. V. Levshin and T. D. Slavnova, *J.C.S. Chem. Comm.*, 1292 (1972).
7. R. Nilsson, P. B. Merkel and D. R. Kearns, *Photochem. Photobiol.*, **16**, 117 (1972).
8. N. Hasty, P. B. Merkel, P. Radlick and D. R. Kearns, *Tetrahedron Letters*, 49 (1972).
9. K. Gollnick, D. Haisch and G. Schade, *J. Amer. Chem. Soc.*, **94**, 1747 (1972).

A.1.6. Photoselection studies of the luminescence of ketones and triplet–triplet energy transfer

Andrew Adamczyk,† Stuart W. Beavan and David Phillips‡

Department of Chemistry, The University, Southampton SU9 5NH, U.K.

SUMMARY

Some preliminary results on the emission properties, including polarization measurements using the method of photoselection, of testosterone in poly-(methyl methacrylate), PMMA, at 298 K and 77 K are reported. Similar data are given for benzophenone and phenanthrene and evidence presented that in triplet–triplet electronic energy transfer between the last two molecules randomly distributed in PMMA, any effect of orientation upon the efficiency of the transfer process must be a slower moving function than the square of the sine of the angle between the $C{=}O$ axis of benzophenone and the normal to the molecular plane of the phenanthrene.

1 TESTOSTERONE

The emission and absorption spectroscopy of $\alpha\beta$ unsaturated ketones is of importance in that it permits the elucidation of the ordering of $n\pi^*$ and $\pi\pi^*$ states, and their energy levels, which determine the reactivity and lifetimes of the species. Several studies have been carried out on the spectroscopy of steroidal enones and related compounds,[1–4] including derivatives of testosterone[1] (I), but the parent compound has hitherto not been investigated. Phosphorescence from such species at low temperatures is easily observed, but no comparable information is available at room temperature. Recently fluorescence has been observed from the related compound,[2] bicyclo-3,3,0 oct-1(5)en-2-one, and this is the first observation of such luminescence from

---

† *Present address*: Du Pont de Nemours (Deutschland) GmbH, Geschaftsbereich Fotoprodukte, Werk Neu Isenberg, Neu Isenberg, W. Germany.

‡ To whom correspondence should be addressed.

this type of compound. It seemed of interest therefore to study the fluorescence and phosphorescence of testosterone, the polarization of these emissions, and the decay times of the excited states for comparison with related molecules.

Experimental details of the use of the spectrofluorometer have been given elsewhere.[5] Figure 1(a) shows the total luminescence spectrum of testosterone in PMMA at 77 K, and this can be compared with that obtained at 298 K (Figure 1b), which is identifiable as fluorescence only. The phosphorescence spectrum is shown in Figure 1(a). It is evident that the fluorescence is strongly polarized parallel to the absorption, the degree of polarization being $+0.45$, close to the theoretical maximum of $+0.50$ for a parallel transition. The degree of polarization of the phosphorescence is also positive, but much smaller (generally $< +0.1$). There is considerable structure in the polarization curves for the phosphorescence which may reflect vibronic activity in the phosphorescence transition. Excitation spectra for total emission are also given in Figure 1. Information relating to testosterone is summarized in Figure 2.

2 PHENANTHRENE

Phosphorescence spectra for phenanthrene were obtained by excitation of the $0 - 0$ band of the ^1L_b transition whch is polarized along the short axis of

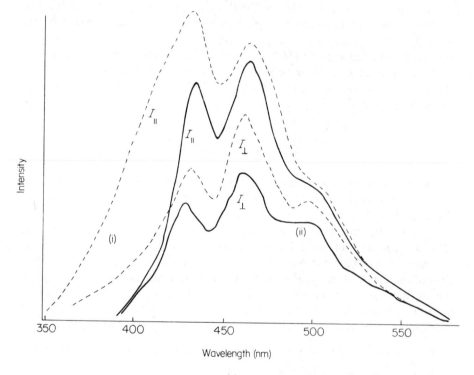

Wavelength (nm)

(a)

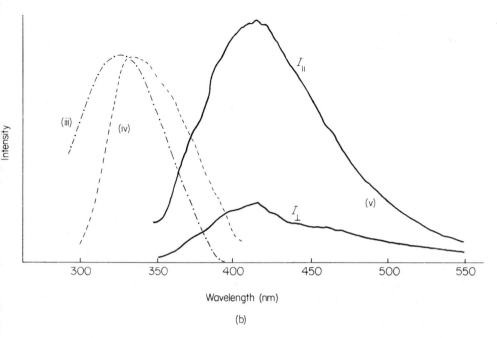

(b)

Figure 1   Luminescence of testosterone.

(a) (i) Total emission spectra of testosterone in PMMA at 77 K, excitation at 330 nm. Intensities of emission parallel $I_{\parallel}$ and perpendicular $I_{\perp}$ to plane of excitation radiation shown. (ii) Phosphorescence component of total emission.

(b) (iii) UV absorption spectrum of testosterone in MMA monomer at 298 K. (iv) Uncorrected fluorescence excitation spectrum of testosterone in PMMA at 298 K. Emission monitored at 420 nm. Excitation spectrum normalized to UV absorption spectrum at 330 nm. (v) Uncorrected fluorescence spectrum of testosterone in PMMA at 298 K· 330 nm. Parallel and perpendicular components are shown. No correction has been made for the inherent polarization of the grating instrument

the molecule[9-11] and the emission was found to be polarized perpendicular to the molecular plane (Table 1). Considerable concentration depolarization

Table 1[5]   Average degree of polarization, $\bar{P}$, of benzophenone, B, and phenanthrene, P, phosphorescence in PMMA

Substrate (with concentration in M)	$\bar{P}(B)$	$\bar{P}(P)$	Temp. K	Excitation wavelength, nm
$B(10^{-2})$	$+0.11$	—	298	370
$B(10^{-1}-10^{-3})$	$+0.10$	—	77	370
$P(10^{-3})$	—	-0.17	298	345
$P(10^{-3})$	—	-0.17	77	345
$P(10^{-1})$	—	-0.14	298	345
$P(10^{-1})$	—	-0.13	77	345
$B^a(10^{-2}) + P(10^{-1})$	$+0.10$	$+0.01$	298	370
$B^a(10^{-2}) + P(10^{-1})$	$\mid 0.09$	0.00	77	370
$B^a(10^{-2}) + P^a(10^{-1})$	$+0.10$	-0.13	77	345

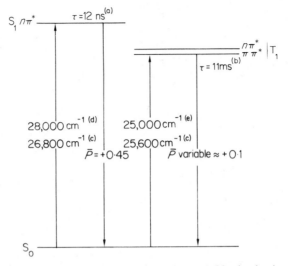

Figure 2   Summary of testosterone luminescence data. (a) Obtained using single-photon-counting method with testosterone in PMMA at 298 K. (b) Obtained by monitoring phosphorescence decay of testosterone in methylcyclohexane at 77 K. (c) From G. Marsh, D. R. Kearns and K. Schaffner, *Helv. Chim. Acta*, **51**, 1890 (1968). (d) From the onset of fluorescence in these experiments. (e) From the onset of phosphorescence in these experiments

of the phenanthrene phosphorescence was observed, which may be due to triplet–triplet energy transfer between phenanthrene molecules. This would be expected to be of lesser importance in benzophenone because of the much shorter lifetime of the triplet state of the ketone compared with the aromatic molecule.

3 TRIPLET–TRIPLET ENERGY TRANSFER

Triplet–triplet electronic energy transfer occurs generally through the exchange mechanism[12] and since this involves overlap of the donor and acceptor molecular orbitals, there is good reason to expect the efficiency of such transfer processes to be sensitive to the relative mutual orientation of the chromophores in the donor–acceptor pair. Recent studies using photoselection methods[13] on random distributions of benzophenone donor and phenanthrene-d_{10} acceptor molecules have yielded conflicting results. Thus a magneto-photoselection study showed that if any such orientation dependence exists, then the functional dependence must be a slower moving function than the square of the cosine of the angles between the molecular planes of the donor and acceptor molecules.[14] By contrast, in a simple photoselection study results appeared to show that energy transfer between the same pair is optimized when the C=O axis of the benzophenone lies parallel to the molecular plane of the phenanthrene. The results of a careful repeat of the optical photoselection study using phenanthrene rather than phenanthrene-d_{10} are given in Table 1.[5] It can be seen that, whereas direct excitation of the benzophenone in benzophenone–phenanthrene mixtures leads to positively polarized benzophenone phosphorescence, and direct excitation of phenanthrene in the same mixtures leads to negatively polarized phenanthrene phosphorescence, the benzophenone-sensitized phenanthrene phosphorescence is completely depolarized. This indicates that under the conditions of these experiments the triplet–triplet energy transfer process does not have a measurable dependence upon orientation of chromophores. A more complete analysis shows that if the overall probability of energy transfer p_t is expressable as a separable function of mutual orientation of the two molecules $f(\delta)$ and the intermolecular separation $f(R)$ such that $p_t = f(\delta)f(R)$, in the case where $f(\delta)$ is a delta function such that transfer occurs only when the $C{=}O$ axis of the benzophenone is parallel to the plane of the phenanthrene the expected value of the degree of polarization of sensitized phenanthrene phosphorescence is -0.33. The degree of polarization is significantly reduced for $f(\delta) = \sin^2 \delta$, where δ is the angle between the $C{=}O$ axis and the normal to the plane of the phenanthrene molecule, being -0.125, and is of course zero if energy transfer is independent of orientation. From the present results, taking into account experimental errors, it can be said that the dependence is a less sensitive function than $\sin^2 \delta$ and evidently in a randomly distributed system of donor–acceptor pairs in the solid it is the distance dependence which is the dominant parameter in determining energy transfer efficiency. Attempts to observe effects of orientation will be most successful in systems in which the donor–acceptor chromophores are maintained at optimal fixed distances apart.[16–21]

ACKNOWLEDGEMENTS

We thank the Science Research Council and Monsanto Chemicals Ltd. for financial support.

REFERENCES

1. G. Marsh, D. R. Kearns and K. Schaffner, *J. Amer. Chem. Soc.*, **93**, 3129 (1971).
2. R. L. Cargill, W. A. Bundy, D. M. Pond, A. B. Sears, J. Saltiel and J. Winterle, *Molec. Photochem.*, **3**, 123 (1971).
3. C. R. Jones, D. R. Kearns and R. M. Wing, *J. Chem. Phys.*, **58**, 1370 (1973).
4. R. O. Loufty and J. M. Morris, *Chem. Phys. Letters*, **22**, 584 (1973).
5. A. Adamczyk and D. Phillips, *J. Chem. Soc. (Faraday II)*, **70**, 537 (1974).
6. W. D. Chandler and L. Goodman, *J. Chem. Phys.*, **45**, 4088 (1966).
7. V. G. Krishna, *J. Molec. Spectry.*, **13**, 296 (1964).
8. R. Shimida and L. Goodman, *J. Chem. Phys.*, **43**, 2027 (1965).
9. D. S. McClure, *J. Chem. Phys.*, **25**, 481 (1956); *J. Chem. Phys.*, **22**, 1256 (1954).
10. J. B. Gallivan and J. S. Brinen, *Molecular Luminescence*, p. 93 (ed. E. C. Lim), W. A. Benjamin Inc., New York, 1969.
11. T. Azumi and S. P. McGlynn, *J. Chem. Phys.*, **37**, 2413 (1962).

12. D. L. Dexter, *J. Chem. Phys.*, **21**, 836 (1953).
13. A. C. Albrecht, *J. Molec. Spectry.*, **6**, 84 (1961).
14. H. S. Judeikis and S. Siegel, *J. Chem. Phys.*, **53**, 3500 (1970).
15. K. B. Eisenthal, *J. Chem. Phys.*, **50**, 3120 (1969).
16. S. Chandrasekhar, *Rev. Mod. Phys.*, **15**, 1, Section VIII (1943).
17. A. A. Lamola, P. A. Leermakers, G. W. Byers and G. S. Hammond, *J. Amer. Chem. Soc.*, **87**, 2322 (1965).
18. D. E. Breen and R. A. Kellar, *J. Amer. Chem. Soc.*, **90**, 1935 (1968).
19. R. A. Kellar, *J. Amer. Chem. Soc.*, **90**, 1940 (1968).
20. J. A. Hudson and R. M. Hedges, *Molecular Luminescence*, p. 667 (ed. E. C. Lim), W. A. Benjamin Inc., New York, 1969.
21. N. Filipescu, *Molecular Luminescence*, p. 697 (ed. E. C. Lim), W. A. Benjamin Inc., New York, 1969.

A.1.7. Application of photoelectron spectrometry to biologically active molecules and their constituent parts. I. Indoles

H. Güsten†, L. Klasinc‡, J. V. Knop§, and N. Trinajstić‡

†*Institut für Radiochemie, Kernforschungszentrum Karlsruhe, B.R. Deutschland*

‡*Institute 'Ruder Bošković' and Faculty of Science, University of Zagreb, Croatia, Yugoslavia*

§*Rechenzentrum der Universität Düsseldorf, B.R. Deutschland*

SUMMARY

Photoelectron (PE) spectra of indene, indole (I), benzofuran, 1-methyl indole (1-Me-I), 2-methyl indole (2-Me-I), 3-methyl indole (3-Me-I), 2-methyl-5-methoxy indole (2-Me-5-OMe-I), indole-3-acetic acid (I-3-acetic acid) and carbazole have been measured using He I excitation. The results for indene, I, and benzofuran agree with ones previously reported, but because of the higher resolution of our instrument they enable closer insight into the nature of the excited states produced. Spectra of Me-Is show that such substitution at position 1, 2 and 3 causes a negligible change in the shape of the PE spectrum, but it results in a nearly constant shift of about 0·3 eV of I band systems to lower ionization potentials. The PE spectrum of I-3-acetic acid is very similar to that of 3-Me-I, whereas considerable changes are observed in the PE spectrum of 2-Me-5-OMe-I in comparison with 2-Me-I. Theoretical calculations in the framework of PPP, CNDO/2 and MINDO/2 methods did not give satisfactory results.

1 RESULTS AND DISCUSSION

Indole (I) and benzofuran appear as fundamental constituent parts of many biologically active molecules and knowledge of their electronic structure is of interest. The aim of this work was to study the electronic structure and the effect of substitution in I by means of PE spectrometry. Two groups of molecules were investigated. In the first group we wished to establish the role of the NH group in I by comparison with indene, benzofuran and carbazole. The second group contained some Is with substituents (mainly the methyl group) at the biologically interesting positions 1, 2, 3 and 5: 1-Me-I, 2-Me-I, 3-Me-I, 3-Me-5-OMe-I and I-3-acetic acid. Although the PE spectra of indene, I and benzofuran have been reported previously[1,2] their reinvestigation seemed of interest, because of the higher resolution of our instrument, thus allowing better comparison with other compounds.

The adiabatic and vertical ionization potentials IP (in eV), together with some observed prominent vibrational progressions (in cm^{-1}) in the band systems found in the PE spectra of indene, I, benzofuran, carbazole, I-3-acetic acid and 2-Me-5-OMe-I are given in Table 1. In cases of diffuse or overlapping bands adiabatic potentials cannot be measured accurately and in such cases only vertical potentials are quoted. Doubtful values are shown in parentheses. As observed earlier,[2] a striking feature is the similarity of the PE spectra of indene, I and benzofuran, whereas the spectrum of carbazole also resembles that of anthracene. In agreement with Eland[2] we assign the four lowest energy band systems in the first three compounds, and the five in carbazole, to excitation from π-levels.

The effect of methyl substitution is easy to see in the PE spectra reproduced in Figure 1. The observed band systems are designated by numbers. The effect consists of a nearly constant shift by about 0·3 eV of the practically unchanged spectrum, to lower ionization energies. The changes in the shape of the spectrum are observed only in the region IP > 12 eV.

The PE spectrum of the I-3-acetic acid could be compared with the very similar spectrum of 3-Me-I. The greatest effect is observed in the IP4 which is shifted by about 0·5 eV to lower energy, whereas IP5 and IP6 are not resolved giving together a broad peak. The highest peak arising from IP8 and IP9 is somewhat shifted to lower energies, IP10 and IP11 are much broader but their maxima are at the same energies as in 3-Me-I.

The introduction of a methoxy group in the 5-position of 2-Me-I significantly changes the shape of the PE spectrum (Figure 2.). The IP1 and IP2(?) band systems are unresolved and shifted to lower energies; at the position of IP3 there are two overlapping band systems. There are some difficulties in assignment of the band systems (one of the overlapping systems could be the strongly shifted IP2) although the first four seem to arise from π-excitations. Standard theoretical calculations have been performed within the framework of PPP, CNDO/2 and MINDO/2 methods.[3] All gave unsatisfactory results regarding the energy and sequence of π- and σ-orbitals, as well. The energies tend to be too high by at least 1 eV and two σ-levels are predicted after the first two π-levels, and several prior to the fourth π-band system by the CNDO/2 and MINDO/2 methods. Especially the former gave unreliably high values for the molecular orbital energies—only five with values lower than 16 eV.

Table 1

	IP (eV)	1	2	3	4	5	6	7	8	9	10	11
Indene	Adiabatic	8·16	8·79	10·20	10·80	11·59	12·88	(13·38)	14·57	16·28	17·53	
	Vertical	8·16	8·94	10·29	11·34	12·21	13·04	13·78	14·83	16·28	18 sh[a] 19 sh[a]	
	Vibrations[b] (cm^{-1})	540; 1370 1530	560	650 1410	1210	(900)		490 1130				
Indole	Adiabatic	7·76	(7·99)	9·78	10·88	(11·33)	(11·99)	12·72	13·30	14·04	(15·05)	(16·7)
	Vertical	7·91	8·37	9·78	11·03	11·52	12·26	13·16	13·77	14·25	15·33	17·03
	Vibrations[b] (cm^{-1})	(640) 1370		480 1370	970							
Benzofuran	Adiabatic	8·37	(8·81)	10·40	11·33	(12·23)	(12·92)	(13·47)	14·26		15·44	16·56
	Vertical	8·37	8·99	10·40	11·73	12·60	13·02 13·25	14·0	14·26	14·51sh[a]	15·53	16·76
	Vibrations[b] (cm^{-1})	400; 600 1130	(400)	565; 770 1530								
Carbazole	Adiabatic	7·60	7·99	9·06	9·75	10·79	11·18	11·96	13·56	14·16	14·75	(15·82)
	Vertical	7·60		9·06	9·75	10·79	11·58	12·40	13·77			16·10
	Vibrations[b] (cm^{-1})			1200 1370	650							
Indole-3-acetic acid	Adiabatic	7·64	(8·04)	9·59	10·42	(10·87)	(11·47)	12·61	(13·14)		15·02	16·45
	Vertical	7·76	8·26	9·59		11·02	11·87	12·97	13·75	14·42	15·20	16·17
	Vibrations (cm^{-1})	960		550	1200							
2-Methyl-5-methoxy indole	Adiabatic	7·23	7·55?	9·46	9·96	10·51	11·56	12·69	(13·25)	14·61	16·39	17·80
	Vertical	7·41		9·61		10·90	11·84		13·60	14·88	16·71	

[a] sh = shoulder
[b] ± 30 cm^{-1}

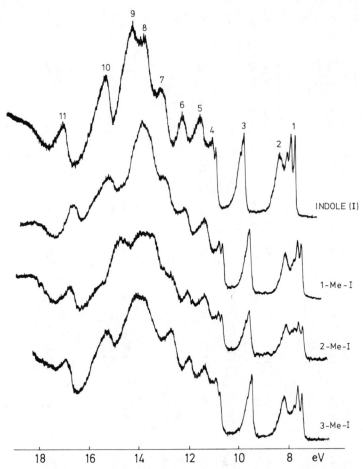

Figure 1 He I photoelectron spectra of indole (I), l-methyl indole (l-Me-I), 2-methyl indole (2-Me-I) and 3-methyl indole (3-Me-I). Slit 2 mm, scan 300 s

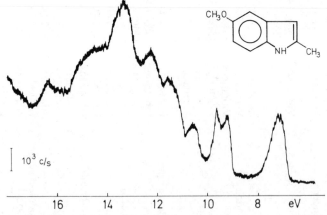

Figure 2 He I photoelectron spectrum of 2-methyl-5-methoxy indole. Slit 2 mm, scan 300 s

2 EXPERIMENTAL

All PE spectra have been recorded on a Vacuum Generators UVG3 instrument specifically designed for gas phase studies of organic compounds,[4] using the HeI line (21·22 eV) for ionization. Enhanced temperatures of the sample inlet system (I 80 °C, carbazole 180 °C, 1-Me-I, 2-Me-I and 3-Me-I 50 °C, I-3-acetic acid 160 °C, 2-Me-5-OMe-I 100 °C) and rare gases (Ar, Xe) for the energy scale control have been employed. All substances were of high purity, redistilled, recrystallized or sublimated before use.

ACKNOWLEDGEMENT

This work was performed on the basis of a German–Yugoslav scientific cooperation program. The financial support of the International Bureau Jülich and the Republic Fund of Scientific Work of Croatia is acknowledged.

REFERENCES

1. J. H. D. Eland and C. J. Danby, *Z. Naturforsch.*, **23A**, 355 (1968).
2. J. H. D. Eland, *Int. J. Mass Spectry. Ion Phys.*, **2**, 471 (1969).
3. QCPE programs adapted for the UNIVAC 1106 at University of Zagreb.
4. L. Klasinc, B. Kovač and B. Ruščić, *Kem. Ind.* (*Zagreb*), **23**, 569 (1974).

A.1.8 Primary photophysical and photochemical processes of indole in polar and non-polar solvents, studied by nanosecond laser photolysis at 265 nm

Christiane J. Pernot and Lars Lindqvist

*Laboratoire de Photophysique Moléculaire du C.N.R.S.,
Université de Paris-Sud, 91405 Orsay, France*

ABSTRACT

A laser photolysis study of indole in cyclohexane, methylcyclohexane, aceto-nitrile, *l*-propanol, methanol and water was undertaken with the purpose of establishing the photoinduced primary processes of this compound in solvents of different polarities. The fourth harmonic (264·5 nm) of a Nd^{3+}-doped glass laser was used as pulse excitation source (FWHM 30 ns). Transient absorption changes were measured at selected wavelengths as a function of time.

The triplet state of indole was found to be populated on photoexcitation in all the solvents enumerated. Its identity was checked using triplet quenchers (*cis*-piperylene, oxygen). The triplet absorption has a maximum at 430 nm with $\varepsilon_{max} \approx 3000$ M^{-1} cm^{-1}. The triplet lifetime was 16 μs in cyclohexane at 25 °C. The activation energy of the triplet decay, measured in methylcyclohexane and *l*-propanol between -100 °C and $+25$ °C was found to be equal to that of diffusion in these solvents.

In *l*-propanol, methanol and water the solvated electron was detected by its absorption spectrum. The solvated electron, formed by ionization from the excited singlet state of indole, disappeared by reaction with indole. A very weak absorption with maximum at 550 nm was attributed to the indole cation.

Studies of the triplet and solvated electron yields in methylcyclohexane, *l*-propanol and methanol showed that these were independent of the tempera-ture between -100 °C and $+50$ °C. This result indicates that internal con-version was negligible.

The results extend previous flash photolysis studies made in particular by Grossweiner and coworkers. Thanks to the higher time resolution of the present study, more detailed data were obtained.

A.1.9. Migration of electron excitation energy in dye–detergent systems

J. Hevesi and Zs. Rózsa

Institute of Biophysics, József Attila University, Szeged, Hungary

ABSTRACT

In dye–detergent solutions of methylene blue and thionine the migration of electron excitation energy from thionine (donor) to methylene blue (acceptor) can be observed. This process was studied by the examination of the fluorescence spectra of 2×10^{-6} to 1×10^{-4} equimolar solutions with sodium lauryl sulphate detergent. For each dye concentration there exists an 'optimal' detergent concentration at which the energy transfer is most efficient. This transfer is observed at the critical micelle concentration ($3 \cdot 5 \times 10^{-3}$ M detergent concentration) in 2×10^{-6} M solution and it is shifted to higher detergent concentration. This result proves that the effectivity of the energy transfer in dye–detergent systems depends on the dye concentration.

A.1.10. The effect of excimer formation on the energy migration in binary liquid systems

J. C. Conte and T. R. Sousa

Laboratório de Química Física Molecular—Complexo Interdisciplinar— I.A.C.—Instituto Superior Técnico—Lisboa—Portugal

SUMMARY

Measurements have been made of the fluorescence emission from pyrene (P) + 1,6 diphenylhexatriene (DPH) solutions in benzene. It was found that, although pyrene is an excimer-forming molecule, only the monomer can transfer its energy to DPH. However, the rate constants for transfer were found to depend on the P concentration which can be explained by assuming that energy migrates among P molecules. Different theoretical models can be invoked to explain the results qualitatively. It is concluded that accurate determinations of the monomer–excimer parameters for different pyrene concentrations are needed for a quantitative discussion.

1 INTRODUCTION

In a binary liquid system where non-radiative energy transfer between a donor and an acceptor is likely to occur[1,2] the excitation energy can migrate among donor molecules.[3] In systems where the energy donors form excimers,[4,5] successive excimer formation and dissociation offers a means of explaining how energy migration may take place.[4,6,7] According to this model migration occurs among excited monomers and excimers alike. A different view has been proposed by Voltz and coworkers, who propose that energy migrates among monomer molecules only while material diffusion is operative among monomer and excimer.[8,9,10,11]

2 EXPERIMENTAL RESULTS AND ANALYSIS

The fact that the Stern–Volmer constants for the transfer from the monomer sidered of special interest.[12,13] In the present paper the results obtained when 1,6-diphenylhexatriene (DPH) is used as an acceptor are presented.

The experimental arrangement has been described in previous publications.[13,14] Some of the measurements were repeated on a different experimental arrangement to be described elsewhere.

Results were obtained for the fluorescence emission of P + DPH solutions in benzene for various concentrations of P (c_X) and DPH (c_Y). Deoxygenation was achieved by nitrogen bubbling.

Under steady state conditions of excitation and for deoxygenated samples when X (the donor) is assumed to form excimers and is able to transfer its energy to Y (the acceptor), it is possible to obtain expressions for the intensities of monomer (I_{MX}) and excimer (I_{DX}) donor emission. It can easily be shown[13] that, only if there is no transfer from the excimer, will linear variations $1/I_{MX}$ and $1/I_{DX}$ versus c_Y be observed. This was the case with the present measurements. From the straight lines it was then possible to evaluate the Stern–Volmer transfer constants for the monomer $\sigma_{YX}^m = k_{YX}^m/k_X$ where k_{YX}^m is the corresponding transfer rate as a function of the donor concentration c_X.

The values so obtained are represented in Figure 1. They may be fitted by a curve of the form

$$\sigma_{YX}^m = Y_0 + \frac{\alpha c_X}{1 + \beta c_X} \tag{1}$$

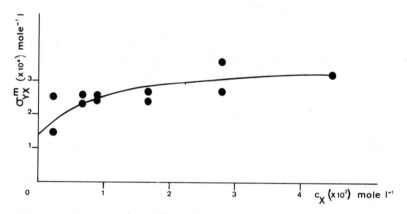

Figure 1. Concentration (c_X) dependence of Stern–Volmer constant (σ_{YX}^m). ● expt; solid curve, from (1) with $Y_0 = 1.375 \times 10^4$ mole^{-1} l; $\alpha = 2.281 \times 10^6$ mole^{-2} l^2; $\beta = 10^2$ mole^{-1} l.

3 CONCLUSIONS

The fact that the Stern–Volmer constants for the transfer from the monomer depend on the donor concentration c_X seems to indicate that an energy migration effect takes place among donor molecules.

It is interesting to note that both theoretical descriptions of the migration process indicated above lead to an expression for the energy transfer rate constant of the form

$$k_{YX}^m = \frac{A + Bc_X}{1 + Cc_X}$$

where obviously the constants A, B and C have different meanings for each of the models. However this shows that both can be invoked to explain the experimental results qualitatively. A final decision on the validity of any of the models can only be made if by independent means one can accurately evaluate some of the quantities appearing in the equation.

Incidentally it can be shown that in any case the value that can be extracted from these measurements for the rate constant of excimer formation exceeds that obtained for dilute solutions. The same had already been found for benzene and derivatives[10] and can be explained qualitatively by assuming that energy migration also affects the process of excimer formation.

A detailed discussion of the present results including additional data on oxygenated samples will be published elsewhere.

ACKNOWLEDGEMENTS

Financial support from NATO (Research Grant no. 242) and I.A.C. (Research project 11625, 4) is gratefully acknowledged. The authors would like to thank Mrs E. C. Mendes for help with the experimental measurements.

REFERENCES

1. J. B. Birks, *Photophysics of Aromatic Molecules*, p. 518, Wiley–Interscience, London, 1970.
2. F. Wilkinson, *Adv. Photochem.*, **3**, 241 (1964).
3. H. Kallmann and M. Furst, *Liquid Scintillation Counting*, p. 3 (ed. C. G. Bell and F. N. Hayes, Pergamon Press, Oxford, 1958.
4. J. B. Birks, J. C. Conte and G. Walker, *I.E.E.E. Trans. Nucl. Sci.*, **NS 13**, 148 (1966).
5. C. L. Braga, M. D. Lumb and J. B. Birks, *Trans. Faraday Soc.*, **62**, 1830 (1966).
6. J. B. Birks, and J. C. Conte, *Proc. Roy. Soc.*, **A303**, 85 (1968).
7. J. B. Birks, *Energetics and Mechanisms in Radiation Biology*, p. 203 (ed. G. O. Phillips), Academic Press, 1968.
8. R. Voltz, *Rad. Res. Rev.*, **1**, 301 (1968).
9. J. Klein, Thesis, University of Strasbourg, 1968.
10. R. Voltz and F. Heisel, *J. Chim. Phys.*, **67**, 169 (1970).
11. J. Klein, R. Voltz and G. Laustriat, *J. Chim. Phys.*, **67**, 704 (1970).
12. J. C. Conte, *Mem. Acad. Ciências (Lisbon)*, **13**, 199 (1969).
13. J. C. Conte, *Rev. Port. Quim*, **11** 169 (1969).
14. J. C. Conte, *Rev. Port. Quim.*, **9**, 13 (1967).

A.1.11. Temperature and deuteration effects on fluorescence of benzenoid solutions

I. C. Ferreira, L. C. Pereira and Marília F. Thomaz

Secção de Física Molecular, Departamento de Física Universidade de Lourenço Marques, Moçambique

ABSTRACT

The knowledge of the environmental and structural effects on the radiative and non-radiative transitions in simple aromatic molecules is a matter of increasing importance. This can be used to select convenient conditions for studying the luminescence arising from complex systems with biological importance, such as proteins and nucleic acids. In this work we report the deuterium isotope substitution and temperature effects on fluorescence of toluene and p-xylene solutions. Values of fluorescence quantum yields were measured between $-80°C$ to $70°C$ for p-xylene, p-xylene $\cdot d_{10}$, toluene and toluene $\cdot d_8$ and also fluorescence decay-times at room temperature. The influence of several experimental factors on the fluorescence intensity at various temperatures is described (e.g. transparency of the solutions, reflectivity of the optical arrangement and optical density and refractive index variations) and corrections are applied to obtain the true quantum yields.

The results seem compatible with an absence of normal deuteration effect, independent of temperature, for both molecules, and show the existence of a non-radiative transition from the first excited singlet state to the ground state.

A.1.12. The phosphorescence excitation spectra of solid and matrix-isolated benzene in the region of 2650–1900 Å

E. Pantos, T. D. S. Hamilton and I. H. Munro

Atomic, Molecular and Polymer Physics Group, The Schuster Laboratory, The University, Manchester M13 9PL, U.K.

SUMMARY

The phosphorescence excitation spectra of thin benzene films and benzene in Ar, Kr and Xe matrices have been recorded at ~ 10 K. The effect of the environment on the quantum efficiency of internal conversion to the first singlet excited state is examined. The previously reported phosphorescence excitation spectrum attributed to a $^1E_{2g} \leftarrow {}^1A_{1g}$ transition is reexamined.

1 INTRODUCTION

The quantum efficiency β_λ of internal conversion to the lowest excited singlet state ($^1B_{2u}$) of benzene depends on the exciting wavelength and environment.[1] In the vapour phase at low pressures the quantum yield of resonance fluorescence drops sharply to zero with excitation of vibrational levels of energy > 3000 cm^{-1} above the zero-point level showing the sudden onset (at $\lambda < 2435$ Å) of an efficient competing intramolecular radiationless process.[2] A similar threshold for non-radiative relaxation has been found from the onset of diffuseness of the vibronic levels in absorption.[3]

The exclusion of intersystem crossing[2] or predissociation as the cause of this non-radiative relaxation threshold has led to the postulation of a mechanism for non-radiative relaxation (channel 3) which results from the intersection of the potential surface of the $^1B_{2u}$ state with that of either (a) the ground state of a valence isomer of benzene[4] or (b) an unidentified excited state lying between the $^1B_{2u}$ and $^1B_{1u}$ states.[3]

It has recently been reported[5] that the phosphorescence excitation spectra of benzene in Xe and Kr matrices exhibit three bands near 2180 Å attributed to a $^1E_{2g} \leftarrow {}^1A_{1g}$ transition not observed in absorption and lying between the $^1B_{2u} \leftarrow {}^1A_{1g}$ and $^1B_{1u} \leftarrow {}^1A_{1g}$ transitions.

In this context we have examined the excitation spectra of solid benzene and matrix-isolated benzene in order to establish (a) whether a drop in β_λ similar to that in the vapour phase takes place and (b) whether another singlet $\pi-\pi^*$ electronic state exists between the $^1B_{2u}$ and $^1B_{1u}$ states.

2 EXPERIMENTAL RESULTS

Eastman–Kodak spectrograde benzene was used with no further purification. It was degassed by several freeze–pump–thaw cycles. Benzene vapours or benzene–rare gas (Air Products research grade) mixtures were deposited on a liquid helium cooled LiF disc at an estimated deposition rate of 100–200 Å/min. Light from a hydrogen discharge lamp was dispersed by a McPherson 225 excitation monochromator. The emission was observed at 90° to the exciting beam through a Bausch and Lomb 0·25 m analysing monochromator and an EMI 6256QB photomultiplier.

Both fluorescence and phosphorescence were detected from the solid benzene films, whilst the matrices exhibited only phosphorescence. Crystal benzene is not expected to phosphoresce because of the efficient triplet–triplet annihilation and the large triplet lifetime.[6] However, phosphorescence may be observed when structural defects or chemical impurities are present.[7,8] In the emission spectra of our solid benzene films toluene emission was detected superimposed on benzene fluorescence and benzene defect phosphorescence. The toluene bands were predominant in the spectra of films deposited at ~ 130 K and measured at ~ 10 K, while the spectra of films deposited and measured at ~ 10 K were characteristic of benzene emission. Comparisons were made with the emission spectra of mixed crystals.[9] The ratio of fluorescence to phosphorexcence excitation spectra deposited and measured at ~ 10 K was constant over 2650–1900 Å.

From Figure 1 we see that excitation at wavelengths $\lambda < 2435$ Å does result in light emission, i.e. β_λ has a nonzero value. To calculate β_λ from the spectra of Figure 1 we need to know the incident light losses due to scattering and reflection on the film surface. Scattered light losses are difficult to measure. An estimate can be made from the transmission losses in the non-absorbing region (60–80%) but a strong wavelength dependence must be expected (Rayleigh scattering $\alpha\ 1/\lambda^4$). Reflection losses are also wavelength-dependent, especially for the solid benzene films.[10] Light penetration depth in regions of strong absorption and surface quenching will affect the calculated value of β_λ. We found that $\beta_\lambda = 1$ for the solid benzene films at 2650–1900 Å. In the matrices $\beta_\lambda = 1$ at 2650–2300 Å, whilst below 2300 Å β_λ appears to slowly decrease to $\beta_\lambda \simeq 0.5$ at 2100 Å.

The vibronic structure corresponding to the structure found in absorption (Figure 2) is lost in the thicker films at $\lambda < 2150$ Å where absorption is nearly 100%, while the intensity maximum is shifted to longer wavelengths. The

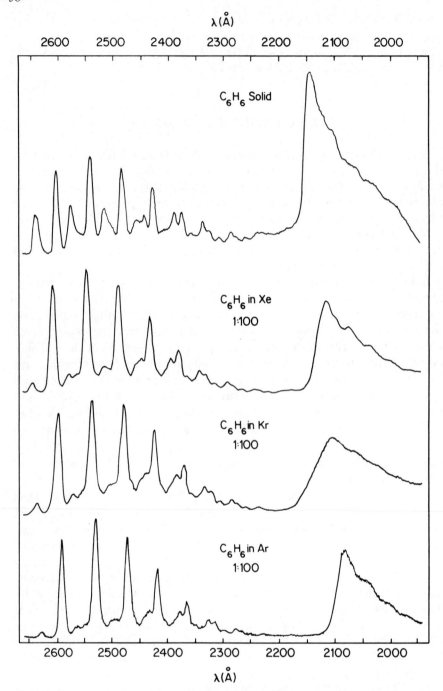

Figure 1   The phosphorescence excitation spectra of C_6H_6 in solid Xe, Kr and Ar films deposited and measured at ~ 10 K

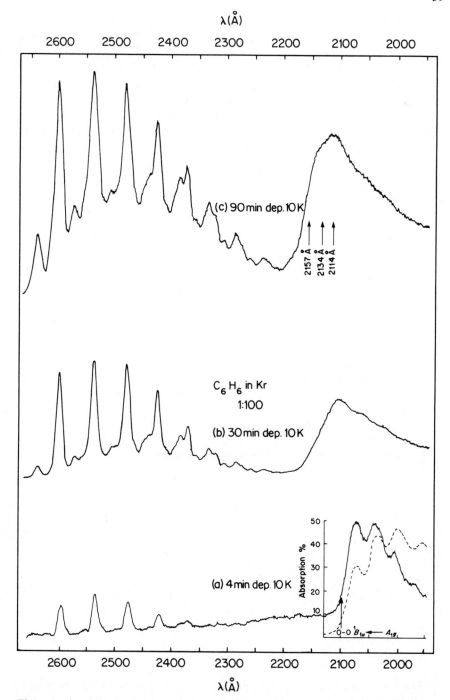

Figure 2   The phosphorescence excitation spectra of C_6H_6 in Kr matrices of different thickness. Deposition and measurement at ~ 10 K. The arrows indicate the position of the three bands reported in Reference 5

position of the three bands reported in the phosphorescence excitation spectrum of a thick C_6H_6/Kr matrix film[5] are indicated by arrows in the spectrum of the 90 min deposit film. In Figure 3 we see that the first of the bands reported by Morris and Angus[5] at 2155 Å in a C_6H_6/Xe matrix is in the region where vibronic bands belonging to progressions of the $^1B_{2u} \leftarrow {}^1A_{1g}$ transition are expected to be. Exactly the same situation is encountered in a C_6H_6/Kr matrix of the same concentration. At shorter wavelengths a sharp rise in intensity takes place which as can be seen from Figures 1 and 2 precedes the onset of the $^1B_{1u} \leftarrow {}^1A_{1g}$ spectrum.

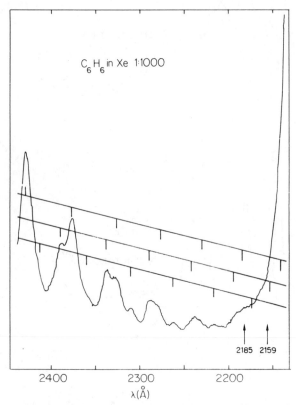

Figure 3 The phosphorescence excitation spectra of C_6H_6 in Xe deposited and measured at ~ 10 K. The two arrows indicate the position of the first two of the bands reported in reference 5

3 DISCUSSION

The mechanism of channel 3 appears to be at least partially conserved in condensed phases where some quenching sets in below 2300 Å.[11,12] When benzene is embedded in an inert solid solution the excited vibronic states are subjected to vibrational relaxation originating from an interaction with the dense medium.[13] The constant ratio of fluorescence to phosphorescence with excitation wavelength indicates that no intersystem crossing takes place from the highly excited singlet states into the triplet manifold of benzene in the solid. In the matrices we assume that the triplet state is populated solely via intersystem crossing from the lowest singlet. The non-zero $(1 \geq \beta_\lambda > 0)$ values of β_λ for $\lambda < 2300\text{Å}$ indicate that medium-induced vibrational relaxation competes very effectively with channel 3. The effectiveness of the medium in inducing vibrational relaxation depends on the coupling between the molecular vibrations of the guest and the host vibron–phonon states. Theory predicts[13] that rare-gas matrices will enhance vibrational relaxation to a lesser degree than a polyatomic matrix because they do not provide high frequency vibrons (of the order of the guest vibration frequencies). The smaller value of β_λ we find in the matrix rather than in the pure solid is consistent with this view. However, the uncertainties involved in our β_λ measurements make it imperative that precise, simultaneous measurements are made of the incident, transmitted, emitted, scattered and reflected light for a given film.

With regard to the reported observation of a $^1E_{2g} \leftarrow \, ^1A_{1g}$ transition,[5] we consider that although our observations do not exclude the existence of another singlet electronic state just below $^1B_{1u}$, unscrambling of the vibronic bands of such a transition from the neighbouring bands of the $^1B_{2u} \leftarrow \, ^1A_{1g}$ and $^1B_{1u} \leftarrow \, ^1A_{1g}$ transition is not possible from the single photon excitation spectra. The gas, matrix and solid film spectra of benzene derivatives with C_{2v} or D_{3h} symmetry, for which $^1E_{2g} \leftarrow \, ^1A_{1g}$ is symmetry allowed, did not show any absorption attributable to such a transition.[14]

REFERENCES

1. J. B. Birks, *Photophysics of Aromatic Molecules*, Wiley–Interscience, London and New York, 1970.
2. C. S. Parmenter, *Adv. Chem. Phys.*, **22**, 365 (1972).
3. J. H. Callomon, J. E. Parkin and R. Lopez-Delgado, *Chem. Phys. Lett.*, **13**, 125 (1972).
4. D. Phillips, J. Lemaire, C. S. Burton and W. A. Noyes, *Adv. Photochem.*, **5**, 329 (1968).
5. G. C. Morris and J. G. Angus, *J. Mol. Spectr.*, **45**, 271 (1973).
6. H. Sternlicht, G. C. Nieman and G. W. Robinson, *J. Chem. Phys.*, **38**, 1326 (1962).
7. T. Azumi and Y. Nakano, *J. Chem. Phys.*, **51**, 2515 (1969).
8. D. S. Tinti, W. R. Moomaw and M. A. El-Sayed, *J. Chem. Phys.*, **50**, 1035 (1969).
9. D. M. Harland and G. C. Nieman, *J. Chem. Phys.*, **59**, 4435 (1973).
10. M. Brith, R. Lubart and I. T. Steinberger, *J. Chem. Phys.*, **54**, 5104 (1970).
11. M. D. Lumb, C. Lloyd-Braga and L. C. Pereira, *Trans. Faraday Soc.*, **65**, 1992 (1969).
12. C. W. Lawson, F. Hirayama and S. Lipsky, *J. Chem. Phys.*, **51**, 1590 (1969).
13. A. Nitzan and J. Jortner, *J. Chem. Phys.*, **58**, 2412 (1973).
14. E. Pantos, Ph.D. Thesis, Manchester University, 1973.

A.1.13. Time-resolved spectroscopy using a conventional pulse fluorometer

M. F. Thomaz, Ma Isabel Barradas and J. A. Ferreira

Secção de Física Molecular, Departamento de Física,
Universidade de Lourenço Marques, Moçambique

ABSTRACT

Time-resolved emission and excitation spectroscopy has been used recently in the study of molecular interactions involving biological molecules. In most applications the single-photon technique has been employed. This is necessary when dealing with: (i) low-intensity light levels, and/or (ii) excited states with short lifetimes (1 ns). However, the method presents some difficulties such as the long times that are required to accumulate enough counting pulses to build a spectrum.

In the present paper we show that in certain conditions of the free-running spark the flashlight intensity, spectrum and time duration are suitable to obtain time-resolved spectra in good conditions of time and spectral resolution, with a conventional sampling fluorometer. The following examples are considered to illustrate the applications of the fluorometer in time-resolved spectroscopy:

  (i) monomer-excimer kinetics of pyrene and benzene solutions,
 (ii) analysis of aromatic hydrocarbon mixtures,
(iii) vapour phase fluorescence of pyrene.

A.1.14. Exciplexes of aromatic hydrocarbons with amines

M. Amalia F. Tavares

Laboratório de Física da Universidade, Coimbra, Portugal

SUMMARY

Measurements of fluorescence spectra of solutions in non-polar solvents have been used to study the nature of intermolecular interaction in exciplexes of aromatic hydrocarbons with amines. The results obtained with diethylaniline confirm those previously obtained with dimethylaniline: (a) the main interaction has a charge-transfer character; (b) when there is an overlap between the emission spectrum of the excited molecule and the absorption spectrum of the other, the energy of the exciplex is much lower than would correspond to a charge-transfer state.

Exciplexes of aromatic hydrocarbons with aromatic amines have first been studied by Weller and his coworkers,[1-5] and by Mataga and his group.[6-8] In these excited complexes there is an interaction between the π-electron system of the aromatic hydrocarbon and the donor group of the aromatic amine. The main forces responsible for the stabilization of exciplexes are charge-transfer forces. From the exciplexes with aromatic amines, Weller concluded that the conditions for the formation of charge-transfer complexes in the excited state are:

$$I(A) \geq I(D) \quad \text{and} \quad E_A(A) \geq E_A(D)$$

where $I(A)$ and $I(D)$ are the ionization potentials of the acceptor and donor, respectively, and $E_A(A)$ and $E_A(D)$ are their electron affinities. Mataga also suggested that exciton-type interaction is necessary for the stabilization of the exciplex.

To study the nature of the interaction in these exciplexes, fluorescence spectra of triphenylene, chrysene, pyrene, anthracene, 1:2-benzanthracene and perylene with diethylaniline (DEA) were recorded. Dilute solutions in non-polar solvents were used to compare monomer and exciplex fluorescence.

The excitation wavelength was chosen so as to excite only the aromatic hydro-carbon. The spectra with and without DEA have been normalized and sub-tracted, and the resulting band plotted against a wavenumber scale to obtain the maximum of the exciplex band. The results are compared with those previously obtained with dimethylaniline (DMA).[9]

The energy of a charge-transfer state corresponding to a transfer of a whole electronic charge from one molecule to the other is given by $I(D) - E_A(A) - C$, where $I(D)$ is the DEA (donor) ionization potential, $E_A(A)$ the electron affinity of the aromatic hydrocarbon (acceptor) and C is the coulombic interaction between the positive and negative ions. Therefore a plot of the energy of the charge-transfer state against E_A should yield a straight line with slope -1, if C does not change much for the different molecules.

In Figure 1 the observed energies of the exciplex maxima are plotted against the electron affinities of the aromatic hydrocarbons taking account of the uncertainty in the values of the latter. The data for all the molecules, except triphenylene and chrysene, lie on a straight line. The slope of the line is not -1 because the degree of charge-transfer is less than unity.

In DMA exciplexes[9] only the triphenylene data do not fit on the same straight line as the data for the other molecules. Triphenylene is the only molecule whose

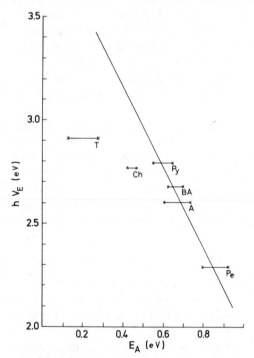

Figure 1   Energy of DEA exciplex maximum plotted against electron affinity. T = tripheny-lene; Ch = chrysene; Py = pyrene; A = anthra-cene; BA = 1:2-benzanthracene; Pe = perylene

fluorescence spectrum has a small overlap with the DMA absorption spectrum, so that a resonance interaction with DMA is possible in the excited state.

The absorption spectrum of DEA is shifted to the red relative to that of DMA, so that a resonance interaction with DEA is possible for both triphenylene and chrysene: the data for these two molecules do not fit on the straight line and the energy of their exciplexes is lower than for a charge-transfer state.

In Table 1 the energies of excimers, DMA exciplexes and DEA exciplexes

Table 1 Energy of excimer (hv_D), DMA exciplex (hv_M) and DEA exciplex (hv_E) maxima (in cm^{-1}). S_1 (in cm^{-1}) is the $0 - 0$ molecular fluorescence transition energy

	S_1	hv_D	$S_1 - hv_D$	hv_M	$S_1 - hv_M$	hv_E	$S_1 - hv_E$
Triphenylene	29,200	—	—	23,600	5600	23,500	5700
Chrysene	27,500	—	—	23,500	4000	22,400	5100
Pyrene	27,000	20,800	6200	22,300	4700	22,500	4500
Anthracene	26,400	—	—	20,800	5600	21,000	5400
1:2-Benzanthracene	25,800	19,600	6200	21,500	4300	21,600	4200
Perylene	22,900	—	—	19,200	3700	18,500	4400

are compared. Of the molecules studied only pyrene and 1:2-benzanthracene show excimer fluorescence in solution. The values of $S_1 - hv$ suggest that the interaction potential has a smaller value for the DMA and DEA exciplexes than for the excimers.

As the ionization potential of DEA has a smaller value than for DMA, according to the formula $I - E_A - C$ the energy of the DEA exciplex should be lower than that of the DMA exciplex, if C remains unchanged. This is so for perylene. The decrease in chrysene exciplex energy is too large to be explained only by the decrease in the ionization potential. For pyrene, anthracene and 1:2-benzanthracene the situation is reversed, it is the DMA exciplex that has the higher binding energy. It is possible that this fact is due to a change in the geometry of the complex because the ethyl group is longer than the methyl group and could cause steric hindrance. Theoretical calculations of the change of energy with distance may help to explain the behaviour of these molecules.

REFERENCES

1. H. Leonhardt and A. Weller, *Ber. Bungsenges. Phys. Chem.*, **67**, 791 (1963).
2. H. Knibbe, D. Rehm and A. Weller, *Z. Phys. Chem. N.F.*, **56**, 95 (1967).
3. H. Beens, H. Knibbe and A. Weller, *J. Chem. Phys.*, **47**, 1183 (1967).
4. H. Leonhardt and A. Weller, *Luminescence of Organic and Inorganic Materials* (eds. H. Kallman and G. Spruch), John Wiley and Sons, 1962.
5. H. Knibbe and A. Weller, *Z. Phys. Chem. N.F.*, **56**, 99 (1967).
6. N. Mataga, T. Okada and K. Ezumi, *Mol. Phys.*, **10**, 203 (1966).
7. N. Mataga, T. Okada and N. Yamamoto, *Chem. Phys. Letters*, **1**, 119 (1967).
8. N. Mataga, T. Okada and N. Yamamoto, *Bull. Chem. Soc. Japan*, **39**, 2562 (1966).
9. M. A. F. Tavares, *Trans. Faraday Soc.*, **66**, 2431 (1970).

A.1.15. Excited state electrostatic molecular potentials of formaldehyde

Rosa Caballol, Ramón Gallifa, Miguel Martín and Ramón Carbó

Sección de Química Cuántica, Departamento de Química Orgánica,
Instituto Químico de Sarriá, Barcelona-17 (Spain)

SUMMARY

The electrostatic molecular potential is calculated for the ground and first excited states of formaldehyde. A fast computation technique is discussed in order to evaluate efficiently the molecular potential. All valence electron INDO wavefunctions are used. The excited state results are obtained through a compact SCF open shell framework, which is also briefly explained.

Formaldehyde shows an inversion of reactivity, when excited. In the ground state a potential well is found in the oxygen neighbourhood, the methylene region being repulsive. The excited states, contrarily, present a potential well in the vicinity of the carbon atom, the oxygen region being repulsive, Experimental facts are consistent with the theoretical results.

1 ELECTROSTATIC MOLECULAR POTENTIALS

Electrostatic molecular potentials have appeared recently as a powerful tool to deal with protonation studies of organic molecules.[1-9]

The interaction energy of a positive point charge, located at the position r, with a molecule, can be obtained by means of

$$V(r) = R(r) - A(r) \tag{1}$$

where

$$R(r) = \sum_I Z_I |r - r_I|^{-1} \tag{2}$$

is a repulsion term due to the nuclear charges $\{Z_I\}$, centred at the points $\{r_I\}$; and within the LCAO formalism, the attractive term $A(r)$, can be obtained as

$$A(r) = \sum_{\mu} \sum_{v} P_{\mu v} \, (\mu|1/r|v) \tag{3}$$

The double sum in equation (3) runs over all the AOs; $P_{\mu v}$ is the element, belonging to AOs μ and v, of the first-order density matrix, or

$$P_{\mu v} = \sum_{i} \omega_i c_{i\mu} c_{iv} \tag{4}$$

the sum runs over all the occupied MOs; ω_i is the ith MO occupation number; c_i are the LCAO coefficients of this MO, associated to AOs μ and v. Finally, $(\mu|1/r|v)$ is an attraction integral between the former AOs and the positive point charge placed at r. The Mulliken approximation[10] has been tested on 'ab initio' SCF wavefunctions with success. [11] However, when using all valence electron methods, the zero differential overlap approach (ZDO), is more suitable and consistent with the nature of this kind of method.[11b] The main purpose of this work is to use an INDO framework;[12] then the ZDO approximation will be supposed in the evaluation of the attractive term (3). Within the ZDO approximation, expression (3) becomes

$$A(r) \simeq \sum_{\mu} P_{\mu\mu}(\mu|1/r|\mu) \tag{5}$$

because under ZDO, the attraction integrals can be simplified through

$$(\mu|1/r|v) \simeq \delta_{\mu v}(\mu|1/r|\mu) \tag{6}$$

where $\delta_{\mu v}$ is the Kronecker delta.

Equation (5), when used in the potential evaluation, needs only the calculation of n integrals for each point charge position; thus, the gain in computation speed is significant.

2 EXCITED STATES IN THE SCF FRAMEWORK

The computational scheme used here for the ground and excited state wavefunctions is summarized as follows. Starting from an electronic energy expression as

$$E = \sum_{i} \omega_i H_{ii} + \sum_{i,j} (\alpha_{ij} J_{ij} - \beta_{ij} K_{ij}) \tag{7}$$

where H_{ii}, J_{ij} and K_{ij} are the core, coulomb and exchange integrals over MOs, and α_{ij}, β_{ij} are parameters characteristic of each molecular state; then, a set of Fock operators can be defined

$$F_i = \omega_i H + \sum_{j} (\alpha_{ij} J_j - \beta_{ij} K_j) \tag{8}$$

for each MO, and one has to solve the Hartree–Fock equations

$$F_i|i\rangle = \sum_j \varepsilon_{ij}|j\rangle \tag{9}$$

where $\{|i\rangle\}$ are the occupied MOs, and $\{|j\rangle\}$ the Lagrange multipliers, associated to the orthonormality constraints of the MOs. To reduce the equations (9) to a unique pseudosecular equation, a coupling operator can be defined[13]

$$R = \sum_i T_i^+ F_i T_i \tag{10}$$

which fulfils

$$R|i\rangle = \varepsilon_{ii}|i\rangle \tag{11}$$

if the projection operator T_i is defined through

$$T_i = I - \sum_{j \neq i} |j\rangle\langle j| \tag{12}$$

where I is the identity operator.

INDO simplifications[12] are easy to use in (8); in this manner, a very compact algorithm can be structed.

3 RESULTS: FORMALDEHYDE ELECTROSTATIC POTENTIALS

(a) General considerations

Experimental geometries for each formaldehyde state have been used[14,15]. In any case the methylene group lies in the XZ plane, and the Z-axis bisects the CH_2 angle. Atomic units (a.u.) of length are used in each axis and contour lines in the figures are given in kcal/mol. The ground state is a $^1A_1$ state; the excited states studied correspond to the $n \rightarrow \pi^*$ transitions, their symmetry being $^1A''$ and $^3A''$. The ground state has a planar structure, but excited configurations have the oxygen atom displaced symmetrically out of the methylene plane.

(b) Ground $^1A_1$ state

Figure 1 shows the XZ and YZ potential planes of formaldehyde, calculated with the INDO framework.

A full 'ab initio' calculation, which has been published recently by Giessner-Prettre and Pullman,[5] shows that the present approach is sufficient to give a good description of the molecular potentials. The electrophilic attack of the molecule will be located in the oxygen neighbourhood, according to the well-known experimental carbonyl reactivity. A positively charged species will be directed towards the mushroom-shaped negative region, which surrounds the oxygen atom.

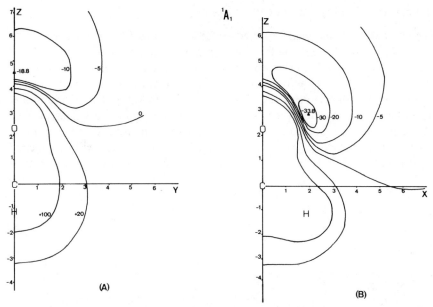

Figure 1   INDO electrostatic potential map of the formaldehyde ground state. (a) Molecular plane (XZ), (b) Perpendicular plane (YZ)

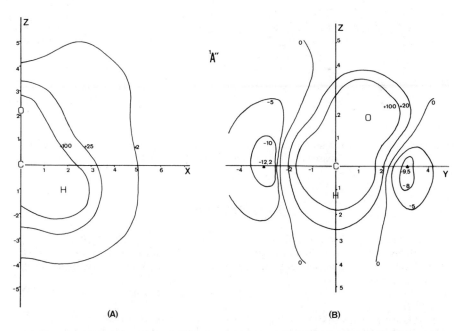

Figure 2   INDO electrostatic potential map of formaldehyde first excited singlet state. (a) CH_2 plane (XZ), (b) Perpendicular plane (YZ)

(c) Excited $^1A''$ and $^3A''$ states

INDO calculations on the excited states are shown in Figures 2 and 3. In both cases the methylene plane becomes repulsive, the oxygen neighbourhood loses the ground state characteristics and the potential well appears in the carbon vicinity. Two attractive regions appear above and below the CH_2 plane; at 3 a.u. in the direction of the negative Y-axis lies the potential minimum. According to this situation, the reactivity of the excited molecule will be reversed with respect to the ground state. That is, electrophilic attack will take place into carbon atom instead of oxygen, or, in other words, carbon will be attached to positive reaction sites. Experimental evidence confirms this theoretical result.[16]

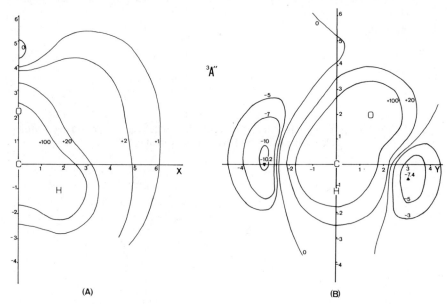

Figure 3   INDO electrostatic potential map of formaldehyde first triplet state. (a) CH_2 plane (XZ), (b) Perpendicular plane (YZ)

REFERENCES

1. R. Bonaccorsi, E. Scrocco and J. Tomasi, *J. Chem. Phys.*, **52**, 5270 (1970).
2. R. Bonaccorsi, E. Scrocco and J. Tomasi, *Theoret. Chim. Acta (Berl.)*, **21**, 17 (1971).
3. R. Bonaccorsi, C. Petrongolo, E. Scrocco and J. Tomasi, *Theoret. Chim. Acta (Berl.)*, **20**, 331 (1971).
4. R. Bonaccorsi, A. Pullman, E. Scrocco and J. Tomasi, *Chem. Phys. Letters*, **12**, 622 (1972).
5. C. Giessner-Prettre and A. Pullman, *Theoret. Chim. Acta (Berl.)*, **25**, 83 (1972).
6. G. Berthier, R. Bonaccorsi, E. Scrocco and J. Tomasi, *Theoret. Chim. Acta (Berl.)*, **26**, 101 (1972).

7. R. Bonaccorsi, A. Pullman, E. Scrocco and J. Tomasi, *Theoret. Chim. Acta (Berl.)*, **24**, 51 (1972).
8. C. Petrongolo and J. Tomasi, *Chem. Phys. Letters*, **20**, 201 (1973).
9. C. Ghio and J. Tomasi, *Theoet. Chim. Acta (Berl.)*, **30**, 151 (1973).
10. R. S. Mulliken, *J. Chim. Phys.*, **46**, 500 (1949).
11. (a) R. Carbó and M. Martin, *Intl. J. of Q. Chem.*, **9**, 193 (1975);
    (b) R. Caballol, R. Gallifa, M. Martín and R. Carbó, *Chem. Phys. Letters*, **25**, 89 (1974).
12. J. A. Pople and D. L. Beveridge, *Approximate MO Theorey*, McGraw-Hill, New York, 1970.
13. R. Caballol, R. Gallifa, J. M. Riera and R. Carbó, *Intl. J. of Q. Chem.*, **8**, 373 (1974).
14. G. Herzberg, *Electronic Spectra of Polyatomic Molecules*, Van Nostrand, Princeton N.J., 1966.
15. G. H. Kirby and K. Miller, *J. Mol. Structure*, **8**, 373 (1971).
16. N. J. Turro, J. C. Dalton, K. Daws, G. Farrington, R. Hautala, D. Morton, M. Niemczyk and N. Schore, *Accounts Chem. Res.*, **5**, 92 (1972).

A.1.17. Demonstration of anisotropic molecular rotations by differential polarized phase fluorometry

Gregorio Weber and George W. Mitchell

Roger Adams Laboratory, University of Illinois, Urbana, Illinois 61801, U.S.A.

1 INTRODUCTION

Although Perrin's theory of fluorescence depolarization by anisotropic rotations[1] is nearly forty years old, the direct demonstration by real time methods of the effects that he predicted has yet to be accomplished. Clearly the demonstration of the existence and character of anisotropic rotations must be carried out in a system with no more than three degrees of rotational freedom. Therefore, only certain simple aromatic molecules would seem plausible objects for this study. When the fluorescence polarization of such molecules is studied over a sufficiently long range of viscosities, the plots of reciprocal of polarization against absolute temperature/viscosity are concave towards the latter axis indicating the presence of more than one rotational rate.[2] If the fluorescence polarization is observed on excitation at different wavelengths, the principal rates of rotation of the molecules are weighted differently,[3] and the change of such average with wavelength of excitation is a further indication of rotational anisotropy. For solutions of perylene and of 1-naphthylamine in propylene glycol, in the interval of temperatures of -20 to $+25°C$ observations at different wavelengths of excitation show that the average rotational relaxation time can change by a factor of 2 to 3.

2 THEORY AND INSTRUMENTATION

We present here some measurements of average rate of rotation in real time by differential polarized phase fluorometry, which demonstrate directly the existence of anisotropic rotations in aromatic molecules. The apparatus is essentially the cross-correlation phase fluorometer of Spencer and Weber,[4,5] with a minor addition: light modulated at 14 or 28 MHz by an ultrasonic

72

standing wave is monochromatized, polarized and focused on the sample. Two gated photomultipliers—as opposed to one in the original instrument—receive the fluorescence through polarizers ($P_\parallel$ and $P_\perp$). The plane of polarization of both of these polarizers is first made parallel to the excitation polarizer ($P_\parallel$) and the phase difference between the cross-correlation photocurrents is made null. Then one of the polarizers is rotated 90° and the phase difference between photocurrents is measured. This phase difference contains information only as regards the change in orientation of the emitting elements in the solution between excitation and emission.

If the modulated photocurrents from the photomultipliers illuminated through the polarizers $P_\parallel$ and $P_\perp$, respectively, have phase differences $\delta_\parallel$ and $\delta_\perp$ with the exciting light, the phase difference Δ between them is given by the relation,

$$\tan \Delta = \frac{\tan \delta_\parallel - \tan \delta_\perp}{1 + \tan \delta_\parallel \tan \delta_\perp} \tag{1}$$

Spencer and Weber[6] have derived values for $\tan \delta_\parallel$ and $\tan \delta_\perp$ *for spherical molecules*. From equation (16) of their paper, and the above relation, one obtains after simplification the relation,

$$\alpha^2 + \frac{2\alpha}{1 - P_0^2}\left(1 - \frac{P_0}{\tan \Delta}\omega\tau_0\right) + \frac{1 + \omega^2\tau_0^2}{1 - P_0^2} = 0 \tag{2}$$

with

$$\alpha = \frac{\tau_0}{\rho}m$$

$$m = (3 - P_0)/(1 - P_0^2) \tag{3}$$

In the above equations $\omega\,(= 2\pi \times$ modulation frequency) is the circular frequency of the exciting light, ρ the rotational relaxation time of the fluorophore, τ_0 the fluorescence lifetime and P_0 the limiting polarization, which depends upon the exciting wavelength.

Figure 1, as well as equation (2), shows that the observed relaxation time is a function of the frequency of light modulation. When more than one relaxation time is present they will be differently weighted at the different light modulation frequencies and the changes observed in the average will demonstrate the heterogeneity of the rotations.

3 EXPERIMENTAL RESULTS:
ROTATIONAL ANISOTROPY IN AROMATIC MOLECULES

We have investigated the dependence of the average rotational relaxation time with excitation wavelength at a fixed temperature, and with temperature at a fixed wavelength of excitation in some aromatic molecules in which there

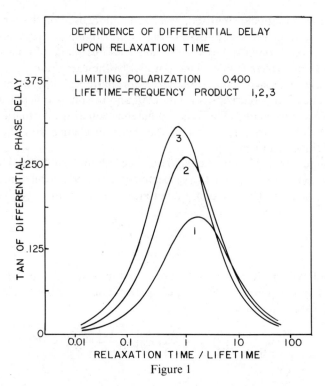

Figure 1

is certainty that no more than three degrees of rotational freedom are active during the fluorescence lifetime.

Figure 2 shows the values of tan Δ as a function of P_0 for methyl acridinium chloride in propylene glycol at $-5.6°C$, a solvent and temperature that were chosen to obtain appropriately large values of tan Δ at the modulation frequency employed (14·2 MHz). For $P_0 = 0$, tan Δ is seen to diverge significantly from zero, and numerical analysis of the results shows that over an appreciable range of values of P_0, $P_0\omega\tau_0/\tan \Delta$ is less than unity, with values of tan Δ sufficiently large to make this relation significant. In conclusion, the rotations of methyl acridinium chloride are clearly anisotropic. A similar experiment was carried out with 9-amino acridine. Although the two acridine derivatives are similar in molecular shape and in spectral properties, the existence of anisotropic rotations is not evident in 9-amino acridine as it is in methyl-acridinium chloride. Figure 3 shows the variation of tan Δ with temperature for 9-amino acridine, excited at a unique wavelength (400 nm). The rotations follow quite closely those expected for a sphere of molar volume of 130 $\pm$ 10 ml.

4 CRITIQUE AND SOME CONCLUSIONS

The method of differential polarized phase fluorometry that we have presented provides a rapid and accurate method of determination of the average rate of

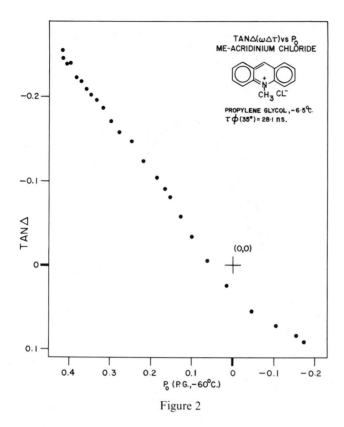

Figure 2

molecular rotation in real time, at one fixed temperature and environment. It has therefore the features necessary to commend it for observations of biological materials. By its use it has been possible to demonstrate conclusively the anisotropic character of the rotations of some small molecules, but a demonstration of the physical origin of the anisotropy has not yet proved possible, and does not seem an easily attainable goal for the next few years. This conclusion should not surprise those familiar with the subject. The requirements for a complete solution in the case of small molecules are indeed formidable. They include a theory able to account for the molecular motions in a medium where the Stokes–Einstein approximation is not rigorously valid, and an accurate knowledge of the directions of the transition moments in the molecule and of their possible changes in the course of the emission on account of solvent relaxation processes, a cause that may also produce an appreciable change of the effective molecular volume and shape during the emission. The rigorous application to macromolecules, where some of these difficulties are minimized, seems more promising, but even here much remains to be done before we can have the certainty of detection—let alone a complete experimental description—of the anisotropic rotations.

76

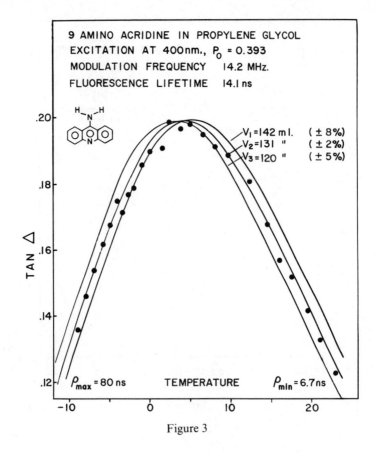

Figure 3

REFERENCES

1. F. Perrin, *J. Physique*, **7**, 1 (1936).
2. G. Weber, 'Polarized fluorescence', in *Fluorescence Techniques in Cell Biology* (eds. A. A. Thaer and M. Sternetz), Springer, Berlin, 1973.
3. B. Witholt and L. Brand, *Biochem.*, **9**, 1948 (1970).
4. R. D. Spencer and G. Weber, *Ann. N. Y. Acad. Sci.*, **158**, 361 (1969).
5. R. D. Spencer, Doctoral Thesis, University of Illinois at Urbana, 1970.
6. R. D. Spencer and G. Weber, *J. Chem. Phys.*, **52**, 1654 (1970).

Session A.2
Photochemistry and photobiology

A.2.1. Production of singlet molecular oxygen by the adrenodoxin reductase–adrenodoxin enzyme system

A. Paul Schaap, Kiyoshi Goda and Tokuji Kimura

*Department of Chemistry, Wayne State University,
Detroit, Michigan 48202, U.S.A.*

SUMMARY

The production of singlet molecular oxygen by the bovine adrenodoxin reductase–adrenodoxin system is reported. Singlet oxygen formation is monitored by the attendant chemiluminescence which is attributed to emission from the singlet oxygen dimol species. The chemiluminescence is markedly inhibited by the addition of 1,4-diazabicyclo[2.2.2]octane, a singlet oxygen quencher, or superoxide dismutase. These results are compared to those obtained in similar experiments with the xanthine oxidase–xanthine system.

1 INTRODUCTION

Enzyme-mediated peroxidation reactions *in vivo* have been actively investigated in recent years. The intermediacy of singlet molecular oxygen ($^1O_2$), a metastable, excited form of oxygen, in biological oxidations has been considered.

The photosensitized formation of singlet oxygen *via* energy transfer from electronically excited dye molecules to oxygen and the chemical generation of this species have both been extensively studied.[1] Singlet oxygen reacts with many types of organic molecules including such biologically important substrates as unsaturated fatty acids, amino-acids, steroids and deoxyribonucleic acid.[2]

Recent reports from our laboratory and others have indicated that singlet molecular oxygen can be produced by certain enzyme systems. Stauff[3] and Arneson[4] have considered the role of singlet oxygen in the chemiluminescence produced by the xanthine oxidase–xanthine system. Howes and Steele have described the singlet oxygen-mediated chemiluminescence obtained from rat liver microsomes, NADPH and O_2^5. Allen, Stjernholm and Steele have discussed the chemiluminescence evoked from human polymorphonuclear leukocytes upon phagocytosis in terms of the production of $^1O_2$.[6]

We have reported that singlet molecular oxygen is produced by the bovine adrenodoxin reductase–adrenodoxin system.[7] The formation of $^1O_2$ was monitored by the attendant chemiluminescence in the presence of the fluorescer, luminol. The peroxidation of lipids prepared from adrenal mitochondria was examined. The chemiluminescence reaction was markedly inhibited by the addition of the lipid fraction. The formation of malondialdehyde from the unsaturated fatty acid portion of phospholipids was shown to be dependent on the NADPH-oxidation reaction. Superoxide dismutase and 1,4-diazabicyclo[2.2.2]octane both exhibited a strong inhibitory effect on this oxidation. Cytochrome c stimulates both the chemiluminescence intensity and the oxidation of the fatty acids.

2 MATERIALS AND METHODS

Pure adrenodoxin and adrenodoxin reductase were prepared from bovine adrenal cortexes as described previously.[8] Superoxide dismutase, xanthine oxidase, cytochrome c and catalase were purchased from Miles and Sigma. Xanthine oxidase activity was determined spectrophotometrically at 22° by the method of Kalcker.[9] Units of the xanthine oxidase activity were defined as described by Arneson.[4]

The formation of singlet oxygen was followed by monitoring the attendant chemiluminescence in a method similar to that of Vorhaben and Steele.[10] All chemiluminescence measurements were made on the Beckman liquid scintillation counter LS-100 at ambient temperature. For the experiments reported here, the coincidence circuit was turned off (out of coincidence mode).

3 RESULTS AND DISCUSSION

We have observed that chemiluminescence is obtained directly from the adrenodoxin reductase–adrenodoxin (AR–AD) enzyme system without the addition of luminol (Figure 1). This luminescence is attributed to the 'dimol' emission from excited oxygen molecular pairs.

Khan and Kasha have shown that singlet oxygen generated from the decomposition of superoxide yields this dimol emission.[11]

$$KO_2 \rightarrow K^+O_2^- \tag{1}$$

$$2O_2^- \rightarrow {}^1O_2 + O_2^{2-} \tag{2}$$

$$2{}^1O_2 \rightarrow (O_2 \cdots O_2)^* \rightarrow 2{}^3O_2 + h\nu \tag{3}$$

dimol complex

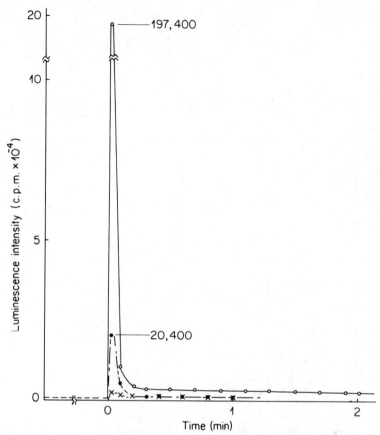

Figure 1 Effect of 1,4-diazabicyclo[2.2.2]octane (DABCO) and superoxide dismutase (SOD) on the chemiluminescence produced by the adrenodoxin reductase–adrenodoxin system. Direct chemiluminescence from $^1O_2$ was monitored with the coincidence circuit of the scintillation counter off. (○) The AR–AD reaction solution was contained in a total volume of 15 ml: AR ($1\cdot1 \times 10^{-7}$ M), AD ($1\cdot1 \times 10^{-6}$ M), tris-glycine buffer (pH 8·8), NADPH ($3\cdot2 \times 10^{-5}$ M), cytochrome c ($1\cdot8 \times 10^{-6}$ M) and catalase (13·3 mg/l). (●) With SOD (16·7 mg/l) added to reaction solution. (×) With DABCO (0·07 M) added to the reaction solution

Singlet oxygen is produced by the dismutation reaction of superoxide anion radicals (equation (2)).

1,4-Diazabicyclo[2.2.2]octane (DABCO) is an efficient singlet oxygen quencher.[12] The addition of DABCO markedly reduced the chemiluminescence intensity from the AR–AD system (Figure 1).

Superoxide dismutase (SOD) catalyses the dismutation reaction of O_2^- to give O_2^{2-} and ground state oxygen, $^3O_2$ (equation (4)).[13] We find that SOD also effectively inhibits the direct chemiluminescence from the AR–AD system (Figure 1).

$$2O_2^- \xrightarrow[\text{dismutase}]{\text{superoxide}} O_2^{2-} + {}^3O_2 \qquad (4)$$

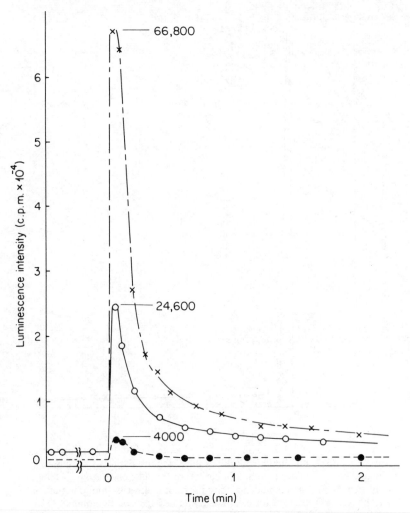

Figure 2 Effect of 1,4-diazabicyclo[2.2.2]octane (DABCO) and superoxide dismutase (SOD) on the chemiluminescence produced by the xanthine oxidase–xanthine system. Direct chemiluminescence from $^1O_2$ was monitored with the scintillation counter off. ($\times$) The XO–X reaction solution was contained in a total volume of 15 ml: XO (0·073 units), X (4·0 μ mol, 2·67 $\times$ 10^{-5} M), tris-glycine buffer (pH 8·3) and catalase (13·3 mg/l). ($\bullet$) With SOD (16·7 mg/l) added to reaction solution. ($\bigcirc$) With DABCO (0·14 M) added to the reaction solution

These quenching results strongly support the mechanism proposed in equations (5)–(7) for the enzymatic production of $^1O_2$ in the AR–AD system.

$$AR/AD + O_2 \rightarrow O_2^- \qquad (5)$$

$$2O_2^- \rightarrow {}^1O_2 + O_2^{2-} \qquad (6)$$

$$O_2^{2-} + 2H^+ \rightarrow H_2O_2 \qquad (7)$$

We have carried out similar quenching studies on the *direct* chemilumine-scence obtained from xanthine oxidase–xanthine. The chemiluminescence which is also interpreted in terms of the formation of the singlet oxygen dimol species and the subsequent emission of light, is inhibited by DABCO or super-oxide dismutase. Experiments to determine the spectral distribution of the chemiluminescence from the adrenodoxin reductase–adrenodoxin and the xanthine oxidase–xanthine systems are in progress.

ACKNOWLEDGEMENT

Financial support to A. P. S. from the U.S. Army Research Office—Durham and Eli Lilly and Company and to T. K. from the National Institutes of Health (AM-12713) is gratefully acknowledged.

REFERENCES

1. (a) C. S. Foote, *Accounts Chem. Res.*, **1**, 104 (1968);
   (b) D. R. Kearns, *Chem. Rev.*, **71**, 395 (1971);
   (c) E. C. Blossey, D. C. Neckers, A. L. Thayer and A. P. Schaap, *J. Amer. Chem. Soc.*, **95**, 5820 (1973) and references contained therein.
2. (a) J. D. Spikes and M. L. Knight, *Ann. N. Y. Acad. Sci.*, **171**, 149 (1970);
   (b) C. S. Foote, *Science*, **163**, 963 (1968);
   (c) T. Matsura and I. Saito, *Chem. Comm.*, 693 (1966).
3. J. Stauff, *Photochem. Photobiol.*, **4**, 1199 (1965).
4. R. M. Arneson, *Arch. Biochem. Biophys.*, **136**, 352 (1970).
5. R. M. Howes and R. H. Steele, *Res. Commun. Chem. Pathol. Pharmacol.*, **3**, 349 (1972).
6. R. C. Allen, R. L. Stjernholm and R. H. Steele, *Biochem. Biophys. Res. Comm.*, **47**, 679 (1972).
7. K. Goda, J. Chu, T. Kimura and A. P. Schaap., *Biochem. Biophys. Res. Comm.*, **52**, 1300 (1973).
8. T. Kumura, *Structure and Bonding*, **5**, 1 (1968).
9. H. M. Kalcker, *J. Biol. Chem.*, **167**, 429 (1947).
10. J. E. Vorhaben and R. H. Steele, *Biochemistry*, **6**, 1404 (1967).
11. A. U. Khan and M. Kasha, *J. Amer. Chem. Soc.*, **88**, 1574 (1966).
12. C. Ouannes and T. Wilson, *J. Amer. Chem. Soc.*, **90**, 6528 (1968).
13. I. Fridovich, *Accounts Chem. Res.*, **5**, 321 (1972).

A.2.2. Oxygen quenching of the fluorescence of pyrenebutyric acid

James A. Knopp and Ian S. Longmuir

North Carolina State University, Raleigh, North Carolina 27607, U.S.A.

SUMMARY

The quenching of fluorescence of pyrenebutyric acid by oxygen was determined in a series of aqueous salt solutions. The observed Stern–Volmer quenching constants were invariant with salt concentrations or composition, provided that the corrections were made not only for changes in viscosity but also changes in oxygen solubility. As a consequence, the extent of quenching reflects the concentration of oxygen rather than the partial pressure or activity of oxygen. This result has important implications for the use of quenching of the fluorescence of pyrenebutyric acid as a probe for oxygen concentrations in biological materials.

1 INTRODUCTION

Oxygen plays an important role in that it is a terminal electronic acceptor for many biological systems. Despite this important position in biology, very little information has been obtained concerning the oxygen concentration or oxygen gradients within biological systems. This lack of information is not due to a lack of effort but rather due to inadequate techniques to be applied in this particular problem.[1,2] Until recently, the main technique has been the Clark-type polarographic electrode which responds to P_{O_2}.[3] This electrode has two main disadvantages: (a) the biological systems tend to be disrupted

† From the Department of Biochemistry, School of Agriculture and Life Sciences, and School of Physical and Mathematical Sciences, North Carolina State University, Raleigh, North Carolina. Supported in part by grant No. GM-19358 from the National Institutes of Health Paper No. 4332 of the Journal Series of the North Carolina State University Agricultural Experiment Station, Raleigh, North Carolina 27607 (U.S.A.).

when such electrodes are inserted into tissues, organs or cells, and (b) the very compound which is being measured, i.e. oxygen, is consumed by the means of measurement in this case. The quenching of fluorescence provides an ideal way of circumventing these two problems.

The phenomenon of oxygen quenching of fluorescence was first noticed as early as 1936,[4] and many authors have observed this phenomenon in a variety of aromatic compounds. Kansky and Müller[5] apparently were the first to suggest that the quenching of oxygen might be used as a measurement of oxygen concentration. The possibility that quenching of the fluorescence of pyrenebutyric acid might be used for the measurement of oxygen in biological systems was first mentioned by Vaughan and Weber.[6] They showed that oxygen is a quencher of fluorescence of pyrenebutyric acid at normal partial pressures of oxygen. They further showed that the quenching occurs according to the Stern–Volmer relationship and that the quenching is diffusion-controlled.

In support of our measurements of intracellular oxygen concentrations by its quenching effect on pyrenebutyric acid, we wish to determine the values of the quenching constants for our systems[7] and/or for the technique of oxygen quenching of fluorescence to be useful in biological measurements of oxygen concentrations. In this way, we will be able to predict the precise quenching constants which will be used. Specifically, we are concerned with whether the concentration or the partial pressure of oxygen should be used as the amount of quencher in the Stern–Volmer relationship. While the Stern–Volmer relationship would give a linear relationship using either the partial pressure or the concentration of oxygen, the observed slopes would differ by the Henry's Law constant. We wish to determine which constant would be independent of the environment of the quenching process.

Many authors have studied the quenching of fluorescence of aromatic compounds by oxygen, either using the concentration of oxygen or the partial pressure.[8-23] However, only two of these papers[22-23] use partial pressure as a term in the Stern–Volmer relationship, while all the others employ the concentration of oxygen as determined from solubility data. To our knowledge, no one has examined whether oxygen concentration or partial pressure should be employed in the Stern–Volmer relationship. This distinction becomes important because of large variations in oxygen solubility or activity coefficient with a variety of solvents and with the introduction of high salt concentration to aqueous solutions.

2 EXPERIMENTAL

Pyrenebutyric acid was obtained from Eastman Kodak and was recrystallized several times from methanol water solution giving pale yellow to white crystals.[24] Oxygen and nitrogen were supplied by Air Products and Chemicals, Inc. as medical grade. All other chemicals were reagent grade.

The quenching experiments were performed by both of the following techniques:

(A) Using the fluorometer described previously.[25] A sample was placed in a 27 mm I.D. glass cylinder which was closed by a rubber stopper and placed at the center of the cross-beams of the fluorometer. A small glass propeller, driven by a small electric motor, rapidly mixed the solution. The partial pressure of oxygen of the solution was measured by the Instrumentation Laboratory or Yellow Springs Clark-type electrodes using a Keithley model 610C nano-ampere meter. The temperature of the solution was determined by a Yellow Springs Instrument telethermometer. The P_{O_2} electrode was situated immediately above the excitation beam. For each quenching curve determination, the P_{O_2} electrode was calibrated by purging the solution alternately with nitrogen and oxygen. Initially, fluorescence intensity, exciting at 360 nm and emitting at 390 nm, and P_{O_2} readings were taken only after five minutes of gas bubbling. However, the intermediate values were found to follow the same line as did the five-minute values, so subsequent readings were taken continuously.

(B) The appropriate solutions were bubbled with either nitrogen, air or oxygen for a period of one hour and then placed into a cuvette to bubble for an additional ten minutes. These cuvettes were immediately stoppered and placed in an Aminco–Bowman spectrofluorometer, and the intensities were measured versus a standard solution of pyrenebutyric acid. The partial pressure of oxygen was determined from the barometric pressure and the vapour pressure of the solution.

3 RESULTS

The fluorescence intensities of pyrenebutyrate solutions in 50 mM phosphate buffer, pH 7·4, were found to decrease as the partial pressure of the oxygen solution was increased. When the reciprocal of the normalized fluorescence intensity was plotted versus the partial pressure of oxygen, as measured by the Clark electrode,[3] the experimental values normally fell in a straight line as shown in Figure 1. Occasionally, deviations were found in the expected linear relationship and were most noticeable as a hysteresis in the Stern–Volmer plot. Deviations were reduced by minimizing gas phase in the measuring cylinder and by increasing the rate of stirring. The average value of 26 different determinations is given in Table 1. Despite the care in experimental technique, variations in average value can be seen to be 10% as shown by the standard deviation of the value. The temperature of these measurements was 25 ± 3 °C. The quenching constants were determined in buffer at the different wavelengths of emission which correspond to the peaks and valleys of the emission curve. The quenching constant of the values obtained agreed with each other within experimental error. The quenching constant was also found to be independent

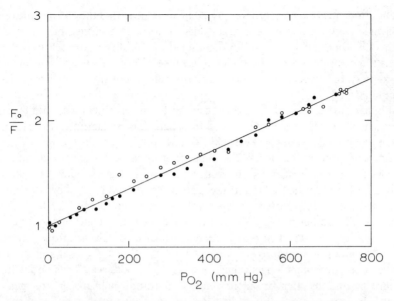

Figure 1   Quenching of fluorescence of pyrenebutyrate by oxygen. The points represent the ratio of the fluorescence intensity in the absence of oxygen to the fluorescence intensity at different values of oxygen partial pressures as measured by the Clark electrode. The fluorescence intensity was measured at 390 nm, with excitation at 366 nm of pyrenebutyrate in 0·05 M potassium phosphate buffer, pH 7·4. The partial pressure of oxygen was altered by bubbling into the solution either nitrogen ($\bigcirc$) or oxygen ($\bullet$). The temperature of the solution was $24 \pm 1\,°C$. The solid line represents the least squares analysis of the data

| | Quenching constants $\times 10^4$ | | | |
| | Experimental values | | Calculated values | |
Solvent	Procedure A	Procedure B	Based on P_{O_2}[a]	Based on $[O]_2$[b]
Buffer	$18 \pm 2^c (26)^d$	20 ± 1 (4)	—	—
1 M NaCl	14 ± 3 (10)	15 ± 1 (3)	17 ± 2	12 ± 2
2 M NaCl	10 ± 4 (11)	9 ± 1 (2)	15 ± 2	8 ± 1
3 M NaCl	—	7 (1)	13 ± 2	5 ± 1
1 M KCl	13 ± 2 (4)	17 (1)	18 ± 2	13 ± 2
2 M KCl	—	12 (1)	18 ± 2	10 ± 1
3 M KCl	—	6 (1)	18 ± 2	7 ± 1
4 M KCl	6 ± 4 (4)	—	18 ± 2	6 ± 1
1 M Na$_2$SO$_4$	6 ± 1 (5)	—	12 ± 2	6 ± 1
0·5 M K$_2$SO$_4$	—	9 (1)	15 ± 2	11 ± 1

[a] $K_{cal}(solvent) = K_{exp}(buffer) \div \eta(buffer)/\eta(solvent)$.
[b] $K_{cal}(solvent) = K_{exp}(buffer) \div \eta(buffer)/\eta(solvent) \times$ solubility of O$_2$ in solvent/solubility of O$_2$ in water.
[c] Standard deviation.
[d] The value in parenthesis represents the number of experiments.

of the concentration of pyrenebutyric acid, providing that the concentration was less than 3 μM to minimize trivial reabsorption effects. Since high concentrations of sodium and potassium salts alter the solubility of oxygen in water, we used these salts to test if the quenching process responds to the concentration or activity of oxygen in solution. The quenching constants of these salts are tabulated in Table 1. In general, the Stern–Volmer plots were linear, although nonlinear curves were obtained occasionally. This problem was minimized by increasing the pH to approximately 10. There was no significant difference in the quenching constant between pH 7·4 and 10.

For comparison purposes, the average quenching constant determined in the buffer was corrected for differences in viscosity (and hence, the diffusion constant) between the buffer and specific salt solutions. The buffer viscosity was assumed to be that of water. These values are listed in the fourth column of Table 1. The values in the fifth column were then corrected for the differences in solubility of oxygen in water versus the various salt solutions, and these derived values are given in the last column of Table 1. The salts employed in the fluorescence quenching studies were tested for possible effects on the fluorescence lifetime by measuring the relative quantum yield under nitrogen. The solutions were prepared by adding the appropriate salt to a buffered solution containing pyrenebutyric acid and then equilibrating with nitrogen by rapidly bubbling nitrogen through the solution contained in a quartz cuvette. After ten minutes of bubbling, the cells were stoppered and the fluorescence intensity was immediately measured relative to a standard solution of pyrenebutyrate saturated with air. The cells were then placed in a spectrophotometer, and the absorption spectra determined. The fluorescence intensities corrected by the percent absorption at the wavelength of excitation were found to be constant within experimental error. As a model system, the quenching of the fluorescence of phenol in water by potassium iodide was measured at concentrations of added potassium chloride. Potassium chloride is not a quencher under the concentrations employed. The iodide quenching constant was measured in the absence of potassium chloride at different temperatures and at the same temperature but in different sucrose concentrations. The observed quenching constant when plotted versus temperature over viscosity gave a linear relationship with the intercept being approximately 0, indicating that this is a diffusion-controlled process as expected. In the presence of high concentrations of potassium chloride, there is a small but definite increase in the quenching constant with increase in potassium chloride concentrations. When the logarithm of the observed quenching constants is plotted versus the square root of the ionic strength, a straight line is observed with a slope of 0·107 $\mu^{-1/2}$.

4 DISCUSSION

As seen previously,[6] the experimental quenching curves followed the expected Stern–Volmer relationship. The average quenching constant of oxygen at

$25(\pm 3)\,°C$ of phosphate buffer, pH 7·4, is $0·0018 \pm 0·0002$ (mm mercury)$^{-1}$. If Winkler's value[26] (0·0283 litre of oxygen (STP) per litre of water) for the solubility of oxygen in water at 25 °C is assumed, the quenching constant becomes $1·09 \pm 0·21$ mм$^{-1}$. This value compares favourably with $1·22 \pm 0·28$ mм$^{-1}$ ($= 139·3$ ns $\times 0·87 \pm 0·201 \times 10^{10}$ (м-s^{-1})) determined by Vaughan and Weber[6] in phosphate buffer at pH 8. The observation that the quenching constant is independent of the wavelength of emission is consistent with the concept of a single emitting species being present. This permits the selection of an emission wavelength which gives the greatest ratio of pyrenebutyrate fluorescence to background fluorescence. The selection is of concern in the *in vivo* system where NADH fluorescence is expected to be observed. The temperature dependence of the quenching constant can be determined from the data of Vaughan and Weber (Reference 6, Figure 4). For a variation of $25(\pm 3)\,°C$, quenching constants are expected to vary by no more than 8%, which is less than the standard deviation obtained from quenching constants. As a consequence, rigid temperature control is of secondary concern when using the quenching of fluorescence as a measure of oxygen concentrations. The temperature data cited above indicates that the quenching process between oxygen by pyrenebutyric acid is a diffusion-controlled one. Following the arguments of Peter Debye,[27] who extended the treatments of Smoluchowski[28] to include the ionic effects, and of Umberger and LaMer,[29] who included both charge and transient effects, it is expected for the reaction of a neutral molecule with an ion that the ionic strength would have little, if any, effect on the rate constant. As a model system, we examined the quenching of phenol by the iodide ion. Because its quenching process involves a reaction of a polar molecule with a charged ion, very little effect of ionic strength was expected. Experimental results agree with this expectation, as a change from 0 to 1 in ionic strength caused only a 20% increase in the quenching constant. This small increase is reasonable in terms of the dipole moment of 1·45 debye exhibited by phenol. As a consequence, with other factors being equal, we expect that the quenching constant would be independent of the ionic strength or salt concentration.

Examination of the quenching constant for oxygen quenching of pyrenebutyrate solutions and different salt solutions, as shown in Table 1, reveals, however, that the quenching constants do indeed vary with salt concentration and salt composition. In order to compare the empirical quenching constants determined by the plot of the reciprocal of the normalized fluorescence intensity versus the partial pressure of oxygen with different salt solutions, it is necessary to correct for the fluorescence lifetime, changes in the diffusion constant and possibly changes in the solubility of oxygen in different solvents. The solutions were corrected for the effect of the different salt concentrations on the lifetime by measuring the relative quantum yield of pyrenebutyrate solutions in the different salts. Assuming that the natural radiative lifetime (which is roughly in proportion to the extinction coefficients) is constant for all solutions, the relative quantum yield is a valid measure of the changes in the lifetime. By this criterion, there is no difference in the fluorescence lifetime and, hence, no correction necessary. Assuming the quenching mechanism is the same and the nature of the conformation of the chemical species is the same in all the salt solutions, it is unlikely then that the collision efficiency or the encounter distance would be altered by the change in salt concentrations or composition. The diffusion constant, however, should be a function of solvent composition with a specific variable being the change in the viscosity of the solvent. We have corrected for the values determined in buffer by the ratio of viscosity. These corrected values appear in column 4 of Table 1. The remaining possible correction is the conversion of partial pressure of oxygen to the concentration by the ratio of the solubilities of oxygen in the different solvents. Because the buffer is so dilute, its contribution to the solubility and the viscosity has been neglected. The expected quenching constants, as corrected by both viscosity and solubility, are given in column 5 of Table 1. The experimental quenching constants, determined in the aqueous salt solutions, agree with the values expected for oxygen concentration within experimental error. These data, then, support the conclusion that the quenching process responds to the concentration of oxygen rather than the activity of oxygen. This is in agreement with the simple picture that the quenching phenomenon measures the number of collisions between the pyrene-butyrate excited states and the oxygen molecules and that the rate at which this occurs is diffusion-controlled and is proportionate to the number of oxygen molecules within a certain distance from the pyrenebutyrate molecules.

Because of the large differences in oxygen solubility in the various parts of the cells, one would expect to find intracellular gradients of oxygen concentration even though partial pressure or activity may be the same throughout the cell. This can be arranged by putting the cells or tissues under conditions where they are not consuming oxygen. Under these conditions, there still will be differences in observed fluorescence quenching in the specimen due to the uneven distribution of oxygen concentrations arising from these known variations in the solubility of oxygen. As spatial resolution increases, this problem becomes more acute. Two alternate procedures for the generation of absolute numbers for oxygen are possible:

(A) The quenching constant is determined at each individual position using known exterior partial pressure of oxygen. This value would be strictly empirical and unpredictable and would vary with position.

(B) The quenching constant based on oxygen concentrations is assumed with the only correction required being the differences in viscosities. The fluorescence intensity or lifetime measurements would then yield concentrations of oxygen.

The first procedure would be limited by the uncertainty as to whether or not the partial pressure of oxygen is the same at the position being measured as externally varied. The second procedure is limited by the meagre knowledge of local viscosities in biological samples. These difficulties should, however, be applicable to any diffusion-controlled luminescence quenching event used as a biological probe for the measurement of oxygen. The problem presented above does not detract from the advantages inherent in the use of fluorescence quenching by oxygen as a probe of oxygen concentrations in biological systems.

ACKNOWLEDGEMENT

The authors would like to thank David Bjorkman for his technical assistance.

REFERENCES

1. *Oxygen in the Animal Organism* (ed. Dickens and Neil), Pergamon Press, London, 1963.
2. *Oxygen Transport in Blood and Tissue*, (eds. Lubbers, Luft, Thews and Witzleg) George Thiene Verlag, Stuttgart, 1969.
3. I. S. Longmuir and F. Allen, *J. Polarographic Soc.*, **8**, 63 (1961).
4. V. Gachkovskii and A. Terenin, *Bull. Acad. Sci. U.R.S.S. Classe Sci. Math. Sci. Chim.*, **1936**, 805 (1936).
5. H. Kansky and G. O. Müller, *Z. Naturforsch.*, **2a**, 167 (1947).
6. W. M. Vaughan and G. Weber, *Biochem.*, **9**, 464 (1970).
7. J. A. Knopp and I. S. Longmuir, *Biochim. Biophys. Acta*, **279**, 393 (1972).
8. E. J. Bowen and A. Norton, *Trans. Faraday Soc.*, **35**, 44 (1939).
9. E. J. Bowen and A. H. Williams, *Trans. Faraday Soc.*, **35**, 765 (1939).
10. E. J. Bowen and W. S. Metcalf, *Proc. Roy. Soc. (London)*, **A206**, 437 (1951).
11. P. J. Berry and M. Burton, *J. Chem. Phys.*, **23**, 1969 (1955).
12. C. A. Parker and W. J. Barnes, *Analyst*, **82**, 606 (1957).
13. V. Bar and A. Weinreb, *J. Chem. Phys.*, **29**, 1412 (1958).
14. I. B. Berlman and T. A. Walker, *J. Chem. Phys.*, **37**, 1888 (1962).
15. H. Ishikawa and W. A. Noyes, Jr., *J. Chem. Phys.*, **37**, 583 (1962).
16. B. Stevens and B. E. Algar, *J. Phys. Chem.*, **72**, 2582 (1968).
17. J. Yguerabide, *J. Chem. Phys.*, **49**, 1026 (1968).
18. C. S. Parmenter and J. D. Rau, *J. Chem. Phys.*, **51**, 2242 (1969).
19. A. M. Halpern and W. R. Ware, *J. Phys. Chem.*, **74**, 2413 (1970).
20. A. Morikawa and R. J. Cvetanovic, *J. Chem. Phys.*, **52**, 3237 (1970).
21. Chun Ka Luk, *Biopolymers*, **10**, 1317 (1971).
22. J. A. Miller and C. A. Baumann, *J. Amer. Chem. Soc.*, **65**, 1540 (1943).
23. A. D. Osborne and G. Porter, *Proc. Roy. Soc. (London)*, **A284**, 9 (1965).
24. J. A. Knopp and G. Weber, *J. Biol. Chem.*, **244**, 6309 (1969).
25. I. S. Longmuir and J. A. Knopp, *Biophys. J.*, submitted.
26. L. W. Winkler, *Ber.*, **24**, 3602 (1891).
27. P. Debye, *Trans. Electrochem. Soc.*, **82**, 265 (1942).
28. M. Smoluchowski, *Z. Physik. Chem.*, **92**, 129 (1917).
29. J. Q. Umberger and V. K. LaMer, *J. Amer. Chem. Soc.*, **67**, 1099 (1945).

A.2.3. Excited states of flavins in photoreception processes

J. F. McKellar and G. O. Phillips

Department of Chemistry and Applied Chemistry, University of Salford, U.K.

A. Checcucci and F. Lenci

CNR—Laboratorio per lo Studio delle Proprietà Fisiche di Biomolecole e Cellule, Pisa, Italy

SUMMARY

Green flagellates *Euglena gracilis* show oriented movements towards light; the receptive (paraflagellar body) and motor (flagellum) structures are in contact without any apparent nerve-like connections. Molecular energy transport processes are involved in photoreception which initiate subsequent biochemical reactions. As the photopigments are probably flavinic in character, their excited states should be involved in the primary molecular events of photoreception. Accordingly, the nature and the behaviour of the excited states of flavins were investigated, alone and in association with aromatic molecules and with protein which considerably modify their spectroscopic characteristics.

The nature of the excited states involved in the light-induced reactions were investigated using fluorescence spectroscopy together with continuous photolysis, microsecond and nanosecond flash techniques.

The photochemical behaviour of the individual flavins free and when bound and when integrally associated in the cell systems is discussed.

1 INTRODUCTION

Green flagellates *Euglena gracilis* exhibit directional movements toward light (phototactic activity) with an orientation mechanism based on a periodical shading of the photoreceptor by means of an absorbing organelle (stigma) constituted of carotenoids embedded in a matrix probably lipidic. An extremely interesting feature, from both the photochemical and biological aspects, is that

92

the photoreceptor (paraflagellar body) and the motor (flagellum) structures are in contact without any apparent nerve-like connections.[1]

Some authors have suggested that carotenoids could be the photoreceptor pigments,[2,3] but on the basis of several evidences this role has recently been attributed to flavin derivatives.[4-6] Actually, carotenoids do not absorb in the near ultraviolet region, which is, on the contrary, effective for phototactic activity, and the absorption spectrum of isoalloxazine derivatives are very similar to the action spectra for negative phototaxis. This resemblance is further enhanced if the phototactic action spectrum is compared with the low temperature (170 K) absorption spectrum of riboflavine.[5] On the other hand, if flavins play the role of photopigments in paraflagellar body, they are surely not free but bound to macromolecules, like proteins. Due to these bindings, important modifications to the spectroscopic properties of flavins can occur, which must be taken into account to avoid misleading correlations with the action spectra of photomovements. On the basis of these considerations we embarked on a detailed spectroscopic study of the light-initiated behaviour of the excited states of flavins alone and in association with other molecules and macromolecules, which we hope will help to clarify the complex mechanisms involved in the response of *Euglena gracilis* to light stimulation.

2 EXPERIMENTAL

The following systems have been examined: riboflavine (RF), RF + adenosine-5-diphosphate (ADP), flavin-adenine-dinucleotide (FAD), the flavoprotein D-aminoacid-oxidase (D-AAO), hot water cell extracts (HWE), ranged in an increasing complexity of the RF partner. Here we report preliminary results obtained from two distinct, but complementary, approaches to the problem. One involves the use of fluorescence spectroscopy and the other that of continuous, microsecond and nanosecond photolysis techniques.

The methods of *Euglena gracilis* cultivation are described in earlier papers from the Pisa Laboratory.[7] Pigments from cells were obtained by hot water extraction (HWE).[15-16] The riboflavin (RF), adenosine-5-diphosphate (ADP) and FAD were from commercial sources. The D-AAO (D-amino acid: oxygen oxidoreductase (deaminating), E. C. 1.4.3.3) was kindly provided to us in a highly purified form by Professor B. Curti of the University of Pavia.

The fluorescence experiments were carried out in Pisa Laboratory using a Perkin–Elmer MPF3 Spectrophotofluorometer. All samples were examined at room temperature in 50 mM (pH 7–8) phosphate buffered aqueous solution (the same buffer system as used for flavin extraction from Euglena cells) after deoxygenation by passing pure nitrogen through the solution for 30 minutes.

The photochemical experiments were carried out with continuous and flash photolysis techniques described in earlier papers from the Salford Laboratory. Materials used and methods of solution preparation were the same as employed by the Pisa Laboratory.

3 RESULTS AND DISCUSSION

3.1 Fluorescence measurements

In Figure 1 are shown the effects of both chemically bound and unbound nucleotide on the fluorescence of RF. In agreement with the work of other authors on different systems,[8] the structure and the wavelength of the emission maximum of the fluorescence spectra of RF and of RF + ADP mixtures is exactly the same at different ADP concentrations, whilst the emission intensity depends on the concentration of the nucleotide. Keeping the RF concentration at a constant value, and varying the ADP concentration, it is seen from Figure 2 that only for molarities of ADP above about 5×10^{-3} does any significant quenching due to the nucleotide become apparent.

Several workers[8-10] have extensively investigated molecular complexes between flavins and aromatic molecules, so that the reliability of these new

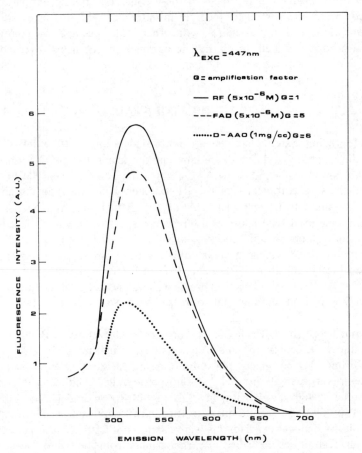

Figure 1   Fluorescence emission spectra of RF, RF + ADP, FAD and D-AAO

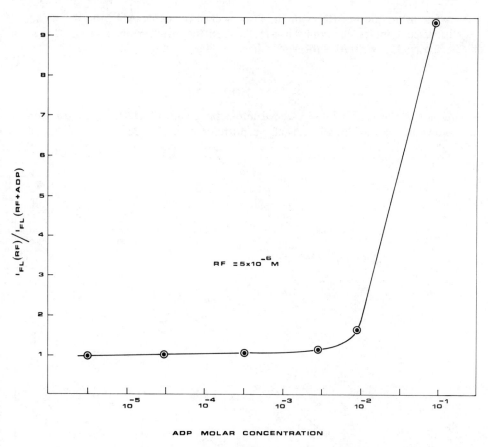

Figure 2   Quenching effect of ADP on RF fluorescence

results on the RF + ADP system can readily be compared. According to their findings the quenching of RF fluorescence that we observed in the presence of a large excess of ADP can be ascribed to the formation of a non-fluorescent molecular complex.[11] Complexes of this kind usually show only small changes in the flavin absorption spectrum and are therefore not of the charge-transfer type, wherein electronic perturbation between the complexing components is sufficient to give rise to an additional absorption band.[12] The non-fluorescent molecular complexes are stabilized by hydrogen bonding, hydrophobic interactions, London dispersion forces and other forces such as dipole–dipole and induced dipole–dipole interactions. According to Weber[11] the complexes of flavins with aromatic molecules and the flavin–nucleotide internal complex in FAD, being observed only in aqueous solutions, are to be considered as ternary complexes which include an as yet indeterminate number of water molecules.

The interaction of RF with the fluorescence quencher ADP to form a non-fluorescent molecular complex is related to the equilibrium constant for association K_{ass} by the relation:

$$K_{ass} = \frac{[RF - ADP]}{[RF][ADP]}$$

A plot of the ratio of fluorescence intensities against [ADP] concentration is shown in Figure 3, in which is also reported the value of K_{ass}.

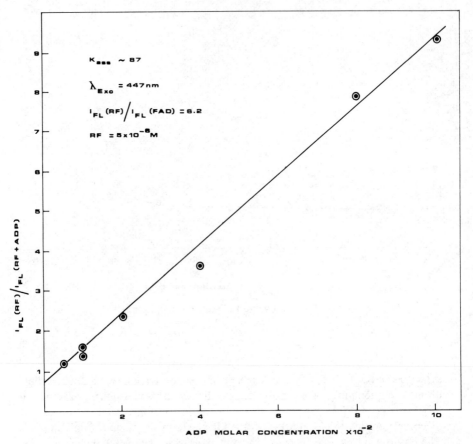

Figure 3   Plot of RF/RF + ADP fluorescence intensity against ADP concentration

From a comparison of these data with the FAD fluorescence (Figure 1) characteristics, it is found that intramolecular complexing in FAD gives rise to the same fluorescence quenching as the intermolecular complexing process in the RF + ADP system when $[ADP]/[RF] \sim 10^4$. These results, which are in agreement with those reported for RF complexing with adenine and adenosine,[10] seem to indicate the scanty importance of the diphosphate chain as regards the probability of complex formation with the flavin moiety. It is finally to be noted that the ratio between the RF fluorescence intensity and the FAD fluorescence intensity resulting from these experiments is in quite good agreement with the ratio of the respective fluorescence quantum yields reported by Bowd et al.[13]

The fluorescence intensity of FAD when bound to the D-AAO apoenzyme is about one half of the fluorescence intensity of free FAD, and its emission maximum is moreover blue shifted in comparison with that of the free dinucleotide (Figure 1). Little is known about the nature of binding of flavin prosthetic groups to proteins except that binding is usually non-covalent. Several possible mechanisms of fluorescence quenching can be considered; mainly rearrangements of the protein leading to an increased probability of radiationless energy dissipation and direct molecular overlap between aromatic side-chain groups and flavin.[14]

A comparison of the fluorescence excitation spectra for the 520 nm emission of *Euglena* hot water extract (HWE) with those of the purified D-AAO

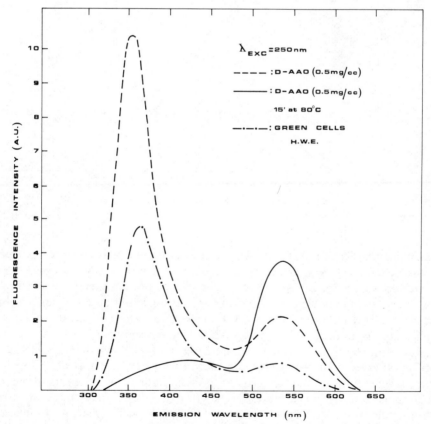

Figure 4 Fluorescence emission spectra of D-AAO and HWE from Euglena green cells ($\lambda_{excit} = 250$ nm)

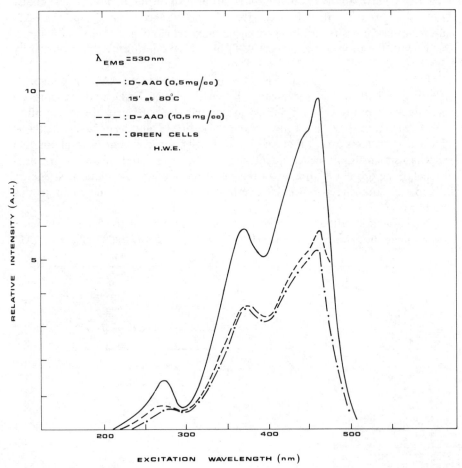

Figure 5   Fluorescence excitation spectra of D-AAO and HWE Euglena green cells (λ_{emiss} = 520 nm)

unambiguously confirms their flavinic character (Figure 5). In order to evaluate possible effects of the preparation techniques[15,16] on the extracts, fluorescence emission spectra of D-AAO were compared before and after the same heat treatment used in flavin extraction from the cells. From Figure 1 it can be seen that after heat treatment, the D-AAO fluorescence increases and the emission maximum is no longer blue-shifted, while Figure 4 shows an almost complete disappearance of the 'protein' fluorescence around 350 nm. The blue-shift effect observed with D-AAO is in agreement with that of Wu *et al.*[17] Finally, from Figure 4 it is of interest that the 'protein' fluorescence intensity at about 350 nm is relatively much greater in the green cells HWE than in the similarly heat-treated D-AAO. This observation might indicate that heat-induced denaturation is less effective in cell homogenates than in the pure protein solution. Work is in progress to clarify this point and to ascertain if and how the flavins in green cells HWE are bound to proteic structures.

3.2 Continuous photolysis

The effect of chemically bound and unbound nucleotide structures on the light-induced fading of RF is shown in Figure 6. Whether the mixtures were exposed to the light of a medium pressure Hg lamp in either quartz or pyrex reaction cells the same result was obtained: RF alone or in the presence of ADP (at the same concentration as that of RF and thus, from the fluorescence experiments, conditions under which there was no significant quenching of RF fluorescence) rapidly faded on irradiation. FAD, however, was considerably more stable to light. Evidently the fluorescence quenching effect of the nucleotide, when chemically bound to the flavin, also conveys to the flavin a much enhanced stability to photochemical reaction.

Another interesting feature of these experiments is that even when the RF/ADP mixtures were exposed to light in a quartz reaction vessel and therefore under conditions where the ADP itself was absorbing light (ADP only strongly absorbs light of wavelengths in the far ultraviolet) there was little relative difference with the experiments using a pyrex vessel. Thus light absorbed directly by ADP itself does not significantly contribute to fading of the flavin.

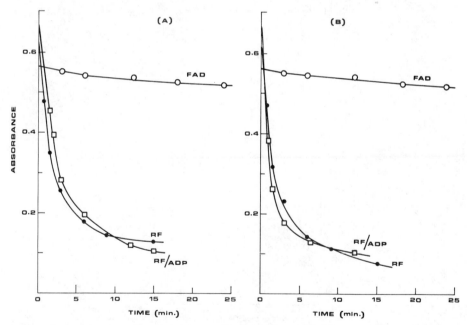

Figure 6   Continuous photolysis of RF, RF + ADP and FAD in (a) pyrex, (b) quartz reaction cells. All concentrations = 5×10^{-5} M

3.3 Microsecond flash photolysis

A comparison of RF and FAD on repetitive flash photolysis showed, like the continuous photolysis experiments, that RF is much less stable to light than

100

FAD. This is shown in Figure 7 where a marked decrease in the λ_{max} band of RF at 445 nm was observed after five photoflashes while, under the same conditions, there was no significant fading of the FAD spectrum.

A similar difference between RF and FAD was observed when they were examined by kinetic spectroscopy. This is shown in Figure 8. At 10 μs after the flash a marked decrease in the intensity of the 445 nm band of RF is observed, together with the formation of a transient species absorbing at wavelengths

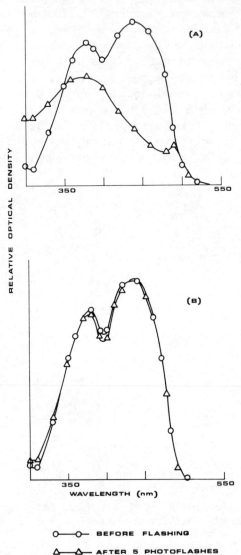

Figure 7   Repetitive flash photolysis of RF (a) and FAD (b). Both concentrations $= 1.0 \times 10^{-5}$ M

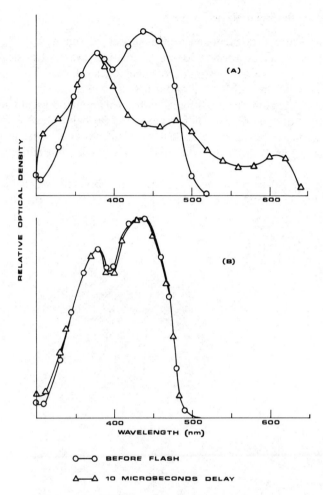

Figure 8   Microsecond flash photolysis of RF (a) and
FAD (b). Both concentrations $= 1{\cdot}0 \times 10^{-5}$ M

above about 525 nm (λ_{max} 610 nm). We have not yet studied the kinetic behaviour
of this transient but from information in the literature,[18] which is sufficient for
our purpose here, it is either a semiquinone radical or radical ion of RF. In
contrast, flash photolysis of FAD showed neither fading nor transient formation,
further evidence of its greater stability to light.

From these preliminary results we can conclude that the flavin chromophore
is much more photoactive on its own than when chemically attached to a
nucleotide structure as in the case of FAD. The presence of the nucleotide
structure in FAD thus not only reduces the fluorescence of the flavin component
as shown in Figure 1, but also reduces its photochemical activity (Figures 6, 7, 8).

3.4 Nanosecond flash photolysis

On flashing RF two transient species were observed, one with a λ_{max} at about 650 nm and the other at about 550 nm. Unfortunately, the transient spectra partially overlap, making an estimate of their λ_{max}s difficult. The spectra are shown in Figure 9. The transient at about 650 nm was quite strong in intensity and could be studied kinetically. It had a half-life of about 3·5 μs and decayed by first-order kinetics (Figure 10a). This transient is probably the RF triplet, but work is in hand to positively establish its identity. The transient at about 550 nm was too low in intensity to be studied kinetically, but had a half-life of about 20 μs.

On flashing FAD two transient species were observed which had similar spectra and lifetimes to those of RF. Figure 11 shows the spectra after flashing

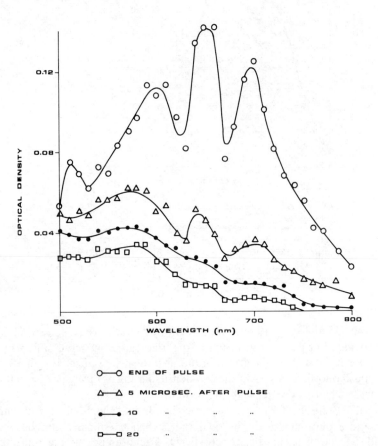

Figure 9 Nanosecond flash photolysis of RF. Concentration of RF = 3·7 × 10^{-4} M

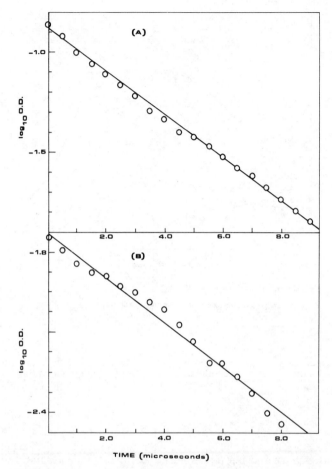

Figure 10   Kinetic first-order plots of the 650 nm transients for
RF (a) and FAD (b)

and Figure 10(b) shows the kinetics of decay of the transient at about 650 nm. Again, as with RF, the decay followed first-order kinetics.

Two features emerge from these observations. First, the intensities of the transients observed on flashing RF were much stronger than those of the corresponding FAD transients. This is in good agreement with the microsecond flash experiments which indicate a quenching effect due to the presence of the nucleotide structure in FAD. The second is that although the intensity of the 650 nm transient from FAD was significantly weaker than that of RF, its lifetime and kinetics of decay are very similar (Figure 10). If this transient is indeed the triplet, then this would indicate that the quenching effect of the nucleotide structure involves the excited singlet state of FAD, a result which would thus be in good agreement with the fluorescence experiments.

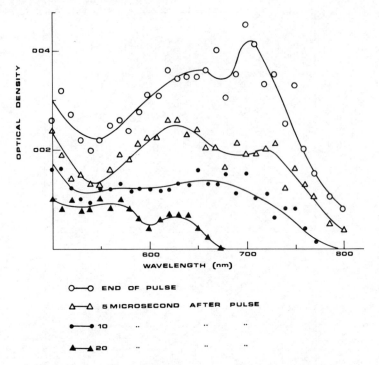

Figure 11 Nanosecond flash photolysis of FAD. Concentration
of FAD $= 4.1 \times 10^{-4}$ M

4 CONCLUSIONS

Both fluorescence and photochemical experiments indicate that the presence
of the nucleotide structure causes efficient quenching of the excited singlet of
Riboflavin in the FAD. This results in a marked increase in the photostability
of the flavin chromophore. The singlet state of the riboflavin can also be
quenched by a non-chemically-bound nucleotide, but in this case much less
efficiently.

Likewise the flavin fluorescence is quenched when the chromophore is bound
to the protein, probably because of the interaction of the isoalloxazine moiety
with the rigidly structured protein.[17,19,20]

Finally, the increase in photostability as well as the quenching of the excited
states of flavins because of the bindings and interactions with other molecules,
as it is in the actual situation of photoreceptive structure, means that absorbed
energy can be released through different channels than radiative dissipation
or photochemical degradation. This corresponds to an essential feature of the
photopigments and at the same time raises the extremely interesting problem:
how is the energy absorbed by the photoreceptor pigments transduced and
utilized in the receptor–effector system?

REFERENCES

1. A. Checcucci, in *Primary Molecular Events in Photobiology*, p. 217 (eds. A. Checcucci and R. A. Weale), Elsevier, Amsterdam, 1973, and references therein.
2. T. W. Goodwin, in *Biochemistry and Physiology of Protozoa*, Vol. 3, p. 319 (ed. S. H. Hutner), Academic Press, New York, 1964.
3. G. Tollin, in *Current Topics in Bioenergetics*, Vol. 3, p. 417 (ed. D. Rao Sanadi), Academic Press, New York, 1969.
4. B. Diehn, *Biochim. Biophys. Acta*, **177**, 136 (1969).
5. B. Diehn and B. Kint, *Physiol. Chem. Phys.*, **2**, 483 (1970).
6. G. Tollin and M. I. Robinson, *Photochem. Photobiol.*, **9**, 441 (1969).
7. A. Checcucci, G. Colombetti, G. Del Carratore, R. Ferrara and F. Lenci, *Photochem. Photobiol.*, **19**, 223 (1974).
8. G. Weber, in *Flavins and Flavoproteins*, p. 15 (ed. E. C. Slater), Elsevier, Amsterdam, 1966.
9. J. A. Roth and D. B. McCormick, *Photochem. Photobiol.*, **6**, 657 (1967).
10. D. B. McCormick, in *Molecular Associations in Biology*, p. 377 (ed. B. Pullman), Academic Press, New York, 1968.
11. G. Weber, *Biochem. J.*, **47**, 114 (1950).
12. G. R. Penzer and G. K. Radda *Quart. Rev.*, **21**, 43 (1967).
13. A. Bowd, P. Byrom, J. B. Hudson and J. H. Turnbull, *Photochem. Photobiol.*, **8**, 1 (1968).
14. J. A. D'Anna and G. Tollin, *Biochemistry*, **10**, 57 (1971).
15. K. Yagi in *Methods of Biochemical Analysis*, Vol. 10, p. 319 (ed. D. Glive), Wiley, New York, 1962.
16. J. Koziol, in *Methods in Enzymology*, Vol. 18, p. 253 (eds. D. B. McCormick and L. D. Wright), Academic Press, New York, 1971.
17. F. Y. H. Wu, S. C. Tu and D. B. McCormick, *Biochem. Biophys. Res. Comm.*, **41**, 381 (1970).
18. M. Green and G. Tollin, *Photochem. Photobiol.*, **7**, 129, 145 and 155 (1968).
19. V. Massey and B. Curti, *J. Biol. Chem.*, **241**, 3417 (1966).

A.2.4. Photochemical interaction between methylene blue and L-ascorbic acid in absence and presence of biological macromolecules

A. K. Davies, K. R. Howard, J. F. McKellar and G. O. Phillips

Department of Chemistry and Applied Chemistry, University of Salford,
Salford M5 4WT, U.K.

SUMMARY

Mixtures of methylene blue (MB) and L-ascorbic acid (AsH_2) have a greater photodegradative effect on biologically important macromolecules such as hyaluronic acid than either have separately. The existence of such a 'coupled' initiating effect between AsH_2 and a light absorbing pigment has direct implications to the problem of light-induced damage to the vitreous of the eye.

To establish the reason for this 'coupled' initiating effect in mechanistic terms, the photosensitized oxidations of AsH_2 by MB was studied and a mechanism consistent with the results obtained. Of interest was the conclusion that reaction is initiated by two distinct processes involving triplet MB:

$$MB^* + {}^3O_2 \rightarrow MB + {}^1O_2^*$$ (1)

$$MB^* + AsH_2 \rightarrow MBH^{\cdot} + AsH^{\cdot}$$ (2)

From the aspect of light-induced damage to the eye by such a 'coupled' sensitizer, we have studied the photosensitized degradation of sodium alginate, a polysaccharide closely related structurally to hyaluronic acid. These studies have shown that reaction (1) is ineffective in initiating degradation of the polymer, and that degradation is a radical-induced process triggered by reaction (2).

1 INTRODUCTION

Hyaluronic acid (HA) is a component of the extracellular fluids in connective tissue. A polymer of high molecular weight, its structure comprises the repeating unit:

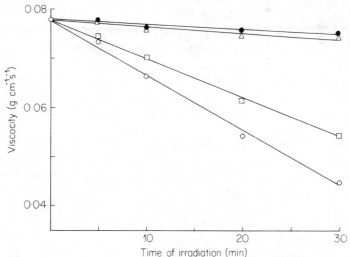

HA is susceptible to degradative depolymerization by reducing agents.[1,2] The reactions induced by these agents are termed oxidative–reductive depolymerizations (ORD), and since they occur only in the presence of oxygen the mechanism is believed to be a free radical initiated oxidative breakdown of the HA structure.[3,4]

Of particular interest to the photochemist is the fact that HA is a major component of the vitreous humour of the eye and that, in certain instances, damage to the vitreous caused by ultraviolet light has been associated with HA depolymerization.[5,6] L-ascorbic acid, another component of the vitreous humour, is also capable of depolymerizing HA. In this case the effect is accelerated by ultraviolet light.[7]

Light absorbing pigments (derived from haem) which are present in the eye, perhaps as a result of accident or surgery, also sensitize the degradation of hyaluronic acid.[5] This depolymerization probably involves the interaction of the electronically excited pigment with ascorbic acid (present at a concentration of 1×10^{-3} mol l^{-1}) to form a sensitizing 'couple'.

Recently it was found that a mixture of the dye methylene blue (MB) and L-ascorbic acid readily photosensitizes the depolymerization of HA.[8] An example of this effect is shown in Figure 1 with the closely related polymer

Figure 1   Rate of change of viscosity with time of irradiation for the systems: △ alginate; ● alginate + MB; □ alginate + AsH$_2$; ○ alginate + MB + AsH$_2$

sodium alginate. It is seen that the effect of the mixture of MB and L-ascorbic acid is significantly greater than either of the two components.

Methylene blue is therefore a useful model in the study of this phenomenon. For this reason we have examined the photochemical interaction between MB and L-ascorbic acid as a first step in understanding this photosensitized depolymerization process.

2 EXPERIMENTAL

2.1 Methods

In all experiments light from a Hanovia 220 W medium pressure mercury vapour lamp filtered through a pyrex plate (light of wavelengths > 300 nm) was employed. The intensity was varied by a series of neutral density filters.

Rates of oxygen absorption were measured using a sensitive constant pressure apparatus thermostatted by a Gallenkamp Thermostirrer at $25 \pm 0.2\,°C$. Solutions were irradiated in a 20 ml 'Pyrex' cell containing an 'oxygenation stirrer' operated at 1200 r.p.m. which ensured saturation of solutions with oxygen throughout the irradiation. The cell and stirrer were of similar design to those already described,[9] but the present cell had less 'dead space'. Rates of oxygen absorption as low as $3 \times 10^{-8}\,mol\,l^{-1}\,s^{-1}$ were easily measured.

Absorption spectra in the ultraviolet and visible regions were measured with a Pye-Unicam SP 1800 spectrophotometer coupled to a flat bed recorder type AR 25. Matched spectrosil cuvettes of path length 10 and 50 mm were employed.

The viscometer used to follow the photosensitized depolymerization of sodium alginate has been described.[10]

Determination of hydrogen peroxide was carried out as follows.[11] The irradiated solution was first decolourized by treatment with active charcoal, and filtered. The filtrate (4 ml) was mixed with a solution containing 8 ml of 2 M sulphuric acid and 10 ml of 1% Ti (IV) sulphate in 2 M sulphuric acid. The absorbance of the yellow colour produced was measured at 410 nm, and the quantity of hydrogen peroxide was obtained from a previously prepared calibration graph.

Detection of carbonyl groups in irradiated sodium alginate was carried out as follows. The polymer solution was decolourized as before and the filtered solution mixed with 2,4-dinitrophenylhydrazine solution (5 g of 2,4-DNPH in 50 ml of water, and 40 ml concentrated hydrochloric acid, diluted with 50 ml of water). The precipitated polymer was filtered and washed with 70:30 alcohol/water until the filtrate was clear and colourless. The polymeric mass which had a definite yellow colour was redissolved in a standard volume of distilled water and the absorbance at 370 nm was measured. Unirradiated sodium alginate, when treated in the same manner, was quite clear and colourless, and showed no absorbance at 370 nm.

2.2 Materials

Methylene blue (E. Gurr Ltd.) was recrystallized three times from butan-l-ol and stored in the dark. Ascorbic acid (B.D.H. Ltd.) was recrystallized twice from an ethanol/water mixture.

Furan (B.D.H.) was distilled immediately prior to use. 1,4-Diazabicyclo 2,2,2 octane (DABCO) was a gift from Imperial Chemical Industries Ltd. (Organics Division).

Water was distilled from potassium permanganate in an all-glass still. Ethanol was refluxed with 2,4-dinitrophenylhydrazine and 0.5% sulphuric acid, to remove carbonyl compounds, and fractionally distilled using an 80 cm column packed with short lengths of glass tubing. The middle fraction was collected and stored under nitrogen.

3 RESULTS AND DISCUSSION

Irradiation of mixtures of methylene blue and ascorbic acid in aqueous solution resulted in oxygen uptake as shown in Figure 2. During irradiation the absorbance of the MB (λ_{max} 665 nm) remained almost constant, while that of

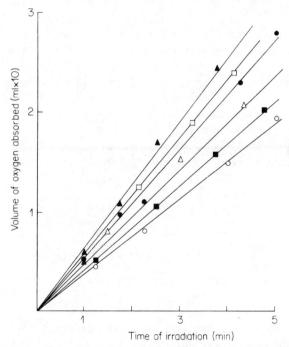

Figure 2 Typical plot of oxygen absorbed versus time of irradiation for aqueous solutions of 5×10^{-5} M MB and concentrations of AsH$_2$ of: ▲ 3.0×10^{-3} M; □ 2.5×10^{-3} M; ● 2.0×10^{-3} M; △ 1.2×10^{-3} M; ■ 1.0×10^{-3} M; ○ 0.8×10^{-4} M

the ascorbic acid (λ_{max} 265 nm) fell linearly with time up to 45 minutes. Evidently the ascorbic acid is steadily consumed while the sensitizer (MB) is regenerated.

Methylene blue has been shown to be an efficient sensitizer for the formation of excited singlet oxygen.[12] To confirm the formation of this species (under our experimental conditions) MB was irradiated in alcoholic solution in the presence and absence of furan, which is known to react efficiently with singlet oxygen according to the equation:[13]

In the absence of furan, no oxygen absorption was detected, but in its presence there was a rapid uptake. Addition of water to the solution containing furan produced only a small decrease in the rate. Since the rate constants for the processes[14]

$$^1\Delta_g O_2^* + H_2O \xrightarrow{k_1} {}^3O_2 + H_2O$$

$$^1\Sigma_g^+ O_2^* + H_2O \xrightarrow{k_2} {}^3O_2 + H_2O$$

are $k_1 = 8.4 \times 10^3 \, \text{l} \, \text{mol}^{-1} \, \text{s}^{-1}$ and $k_2 = 6.0 \times 10^8 \, \text{l} \, \text{mol}^{-1} \, \text{s}^{-1}$, this result indicates that the oxygen is in the $^1\Delta_g$ state. This conclusion is reasonable in view of the energies of the MB triplet (142 kJ mol^{-1}) and those of the excited singlet states of oxygen ($^1\Delta_g = 95$ kJ mol^{-1}, and $^1\Sigma_g^+ = 155$ kJ mol^{-1}).[15] Singlet oxygen formation was confirmed by adding DABCO (an established quencher of this species[16]) to the alcoholic solution of MB and furan. In this case there was a considerable reduction in the rate of oxygen absorption.

It has been suggested that singlet oxygen can react with ascorbic acid in a similar way to furan.[17] Thus, one possible photochemical interaction between methylene blue and ascorbic acid is (here AsH$_2$ represents ascorbic acid):

$$\text{MB} + h\nu \xrightarrow{I} {}^1\text{MB}^* \tag{1}$$

$$^1\text{MB}^* \xrightarrow{k_{isc}} {}^3\text{MB}^* \tag{2}$$

$$^3\text{MB}^* \xrightarrow{k_0} \text{MB} \tag{3}$$

$$^3\text{MB}^* + {}^3O_2 \xrightarrow{k_1} \text{MB} + {}^1O_2^* \tag{4}$$

$$^1O_2^* + \text{AsH}_2 \xrightarrow{k_2} \text{AsH}_2O_2 \tag{5}$$

$$^1O_2^* \xrightarrow{k_3} {}^3O_2 \tag{6}$$

This mechanism is consistent both with the regeneration of the dye and the gradual consumption of ascorbic acid.

The effect of the radical scavenger methylhydroquinone on the MB–ascorbic acid system was investigated. As the scavenger concentration was increased the rate of oxygen absorption decreased to a limiting value (Figure 3) which was approximately 40% of that in the absence of the scavenger. This result indicates

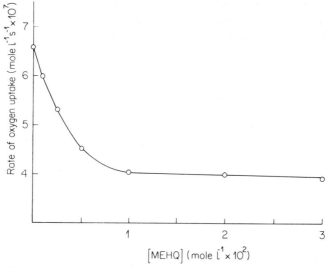

Figure 3   Effect of methyl hydroquinone on the rate of oxygen uptake by the system MB–AsH$_2$

that a free-radical generating process is involved. A radical mechanism which is consistent with the regeneration of the dye and consumption of ascorbic acid is shown below:

$$^3MB^* + AsH_2 \xrightarrow{k_4} AsH^{\cdot} + MBH^{\cdot} \tag{7}$$

$$MBH^{\cdot} + {}^3O_2 \xrightarrow{k_5} MB + HO_2^{\cdot} \tag{8}$$

$$AsH^{\cdot} + {}^3O_2 \xrightarrow{k_6} As + HO_2^{\cdot} \tag{9}$$

$$2\,HO_2^{\cdot} \xrightarrow{k_7} H_2O_2 + {}^3O_2 \tag{10}$$

The formation of hydrogen peroxide, required by this mechanism, was confirmed and the ratio: $R =$ mole H_2O_2 formed/mole O_2 absorbed was ca 0·5 at 25 °C. The value of R predicted from the free radical mechanism is 1·0. On the other hand if only the singlet oxygen mechanism were operative, R would be zero. The experimental value $R \simeq 0·5$ immediately suggested the concomitance of both mechanisms (reactions (1) to (10)).

The value $R \simeq 0·5$ can be derived from the dual mechanism on the assumption that, at 25 °C, the methylene blue triplet reacts with approximately equal efficiency by reactions (4) and (7).

Further evidence for this dual mechanism was obtained from selective quenching experiments. When DABCO (1×10^{-2} mol l^{-1}) was employed as a quencher of singlet oxygen, the rate of oxygen uptake was, as expected, much reduced but the value of R rose to $\simeq 1·18$. Under these conditions the reaction is presumably governed by the radical part of the mechanism. Conversely we interpret the remaining oxygen absorption in the presence of high concentrations of radical scavenger (Figure 3) to be due to reactions (1) to (6).

Little or no activation energy is expected for the deactivation of a triplet molecule by oxygen to produce singlet oxygen.[18] On the other hand, a finite activation energy is required for a triplet state to abstract a hydrogen atom to produce free radicals.[18] Hence a rise in temperature should favour the radical part of the mechanism, while not greatly affecting the singlet oxygen part. This should cause an increase in the ratio R. The experimental variation of R with temperature, shown in Figure 4, therefore supports the proposed dual mechanism.

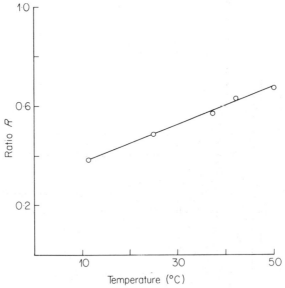

Figure 4 Variation in the ratio R with temperature for the system MB–AsH_2

A photostationary state treatment of reactions (1) to (10) gives the expression:

$$-\frac{1}{d[O_2]/dt} = \frac{1}{I}\left\{\frac{k_0}{k_1[O_2]} + 1 + \frac{k_4k_3}{k_2k_1[O_2]} + \frac{k_4[AsH_2]}{k_1[O_2]}\right\}$$
$$+ \frac{k_3}{k_2I[AsH_2]}\quad 1 + \frac{k_0}{k_1[O_2]}$$

To simplify this expression, if k_4 is assumed to be of a similar magnitude to the rate constant for the reaction of triplet thionine with allyl thiourea[19] ($k = 1.1 \times 10^7 \, l\,mol^{-1}\,s^{-1}$), then using this value together with[20] $k_1 = 3 \times 10^9 \, l\,mol^{-1}\,s^{-1}$, $[O_2] = 4.34 \times 10^{-4}\,mol\,l^{-1}$ and $[AsH_2] = 1 \times 10^{-3}$ to $5 \times 10^{-3}\,mol\,l^{-1}$, the limits for the term $k_4[AsH_2]/k_1[O_2]$ are obtained: $0.008 \le k_4[AsH_2]/k_1[O_2] \le 0.04$. The contribution of this term is thus small and for the range of ascorbic acid concentration studied, can be neglected. At a fixed concentration of oxygen the expression therefore simplifies to:

$$-\frac{1}{d[O_2]} = \frac{1}{I}\,k' + \frac{k''}{[AsH_2]}$$

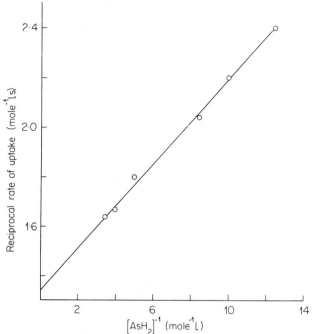

Figure 5 Reciprocal relationship between rate of oxygen uptake and ascorbic acid concentration for the system MB–AsH₂

Figure 5 shows that, as required by this expression, a plot of $(-\,d[O_2]/dt)^{-1}$ against $[AsH_2]^{-1}$ is linear and has a positive intercept on the ordinate. The rate of oxygen absorption was found to be directly proportional to the absorbed light intensity, as also required by the final expression.

3.1 Effect of the sensitizing 'couple', MB and L-ascorbic acid on sodium alginate

The important features of this interaction are:

(i) Photosensitized depolymerization of sodium alginate occurs rapidly in the presence of the couple MB and AsH_2.

(ii) Oxygen is essential for (i) since depolymerization does not occur in nitrogen-saturated solution.

(iii) Carbonyl groups are present in the degraded sodium alginate.

(iv) There is a clear correlation between the fall in viscosity of sodium alginate solutions and the consumption of ascorbic acid (Figure 6).

(v) No depolymerization occurs when MB alone is used as the sensitizer.

Excited singlet oxygen $^1O_2^*$ has been postulated as the initiating species in many biologically important photosensitized oxidations ('photodynamic action'). That methylene blue, an efficient generator of $^1O_2^*$, does not by itself

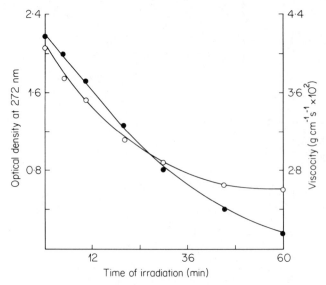

Figure 6   Change in optical density at 272 nm ○ and change in viscosity ● with time of irradiation of the system MB, AsH$_2$–alginate

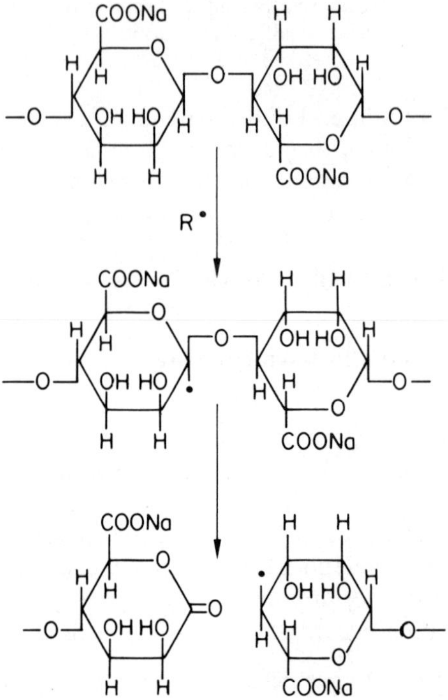

Figure 7   Proposed mechanism for the degradation of sodium alginate

sensitize the depolymerization of sodium alginate or hyaluronic acid apparently rules out $^1O_2^*$ as an effective initiating species for this type of ORD reaction. By the same token, direct chemical attack of MB* on the polymer is also ruled out.

We conclude therefore that the photosensitized degradation of sodium alginate in the presence of MB and ascorbic acid is due to the formation of radicals. By analogy with earlier work[21] we propose that the photochemically generated radicals abstract hydrogen atoms from the polymer chain leading to cleavage of the chain and formation of carbonyl end groups (Figure 7).

REFERENCES

1. C. W. Hale, *Biochem. J.*, **38**, 362 (1944).
2. B. Skanse and L. Sunblad, *Acta Physiol. Scand.*, **6**, 37 (1943).
3. W. Pigman, S. Rizvi and H. L. Holley, *Arthritis Rheumat.*, **4**, 240 (1961).
4. A. Herp, M. Harris and T. Rickards, *Carbohydrate Res.*, **16**, 395 (1971).
5. E. A. Balazs, in *The Structure of the Eye,* p. 293 (ed. G. K. Smelzer), Academic Press, New York, 1961.
6. L. Z. J. Toth, E. A. Balazs and A. F. Hawe, *Invest. Ophthalmol.*, **1**, 797 (1962).
7. E. A. Balazs and L. Sunblad, unpublished work cited in *The Amino Sugars*, Vol. 11A, p. 420 (eds. E. A. Balazs and R. W. Jeanloz), Academic Press, New York, 1965.
8. G. O. Phillips and D. Scheufel, unpublished work.
9. G. O. Phillips, A. K. Davies and J. F. McKellar, *Lab. Prac.*, **19**, 1037 (1970).
10. K. R. Howard, A. K. Davies and J. F. McKellar, *Lab. Prac.*, **22**, 577 (1973).
11. R. Belcher and A. J. Nutten, *Quantitative Inorganic Analysis*, 2nd edn. p. 326, Butterworths, London, 1967.
12. K. Gollnick, T. Franken, G. Schade and G. Dörhöfer, *Annal. N. Y. Acad. Sciences*, **171**, 89 (1970).
13. D. R. Kearns, *Chem. Rev.*, **71**, 395 (1971).
14. R. P. Wayne, *Advances in Photochemistry* (eds. J. N. Pitts, G. S. Hammond and W. A. Noyes, Jr., Interscience, New York), **6**, 311 (1969).
15. D. R. Kearns and A. U. Khan, *Photochem. Photobiol.*, **10**, 193 (1969).
16. C. Quannès and T. Wilson, *J. Amer. Chem. Soc.*, **90**, 6527 (1968).
17. P. Homann and H. Gaffron, *Photochem. Photobiol.*, **3**, 499 (1964).
18. G. O. Schenk, private communication in Reference 19.
19. H. E. A. Kramer and H. Maute, *Photochem. Photobiol.*, **15**, 7 (1972).
20. R. Nilson, P. Merkel and D. R. Kearns, *Photochem. Photobiol.*, **16**, 109 (1972).
21. G. O. Phillips, P. Barber and T. Rickards, *J. Chem. Soc.*, 3443 (1964).

A.2.5. Charge-transfer interactions in the photooxidation of phenothiazine

Hugh D. Burrows, José Dias da Silva
and Maria Isabel Ventura Batista

The Chemical Laboratory, University of Coimbra, Portugal

SUMMARY

Solutions of the pharmacologically active heterocycle phenothiazine in halogenated hydrocarbons show great photochemical sensitivity. Flash photolysis studies indicate that the reaction involves photooxidation of the phenothiazine to yield initially the radical cation, which subsequently reacts in the presence of ethanol or water to give a green coloured dimeric species. Equilibrium between the radical cation and this species, and also between radical cation and cation dimer are reported. Fluorescence and absorption spectroscopic studies indicate that phenothiazine forms contact charge-transfer complexes with halogenated hydrocarbons, and it is suggested that photooxidation proceeds via these. A comparison is made of the behaviour of phenothiazine on chemical and photochemical oxidation. The involvement of charge transfer species in the photooxidation of phenothiazine in aerated ethanol is considered, and an alternative mechanism involving singlet oxygen is suggested.

1 INTRODUCTION

The photosensitivity of several aromatic solutes in the presence of halogenated hydrocarbons has been noted for some time. However, the mechanism of the photochemical action varies considerably according to both the nature of the solute and the solvent environment. The amine N, N, N', N'-tetramethyl-p-phenylenediamine (TMPD) photoionizes in the presence of halogenated hydrocarbons through excitation of ground state contact-charge-transfer species,[1,2] while N-*iso*propylcarbazole is thought to be photooxidized by bromo- or trichloroacetic acid via its lowest excited singlet state.[3] The quenching of the fluorescence of aromatic hydrocarbons by chlorinated alkanes shows a dependence

116

on the excitation wavelength, and this has been interpreted in terms of a specific interaction in higher excited states.[4] However, this is now thought to result from the presence of ground state complexes.[5] Pulse radiolysis studies on benzene solutions of TMPD in the presence of chloracetic acid have indicated a further possible route to electron transfer, namely electron transfer from the excited triplet state.[6] In all of these cases, photooxidation occurs at photon energies considerably lower than the ionization potential of the aromatic solute.

Phenothiazine (I) (PTH) is of considerable interest, because of the pharmacological activity, and powerful antioxidant action of it and its derivatives.[7] During pulse radiolysis studies on the mechanism of the reaction of phenothiazines with alkylperoxy radicals[8] it was noticed that solutions of PTH in halogenated hydrocarbons are highly photosensitive. We have investigated the photodecomposition further, and report here the results of our study.

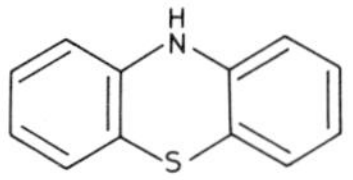

(I)

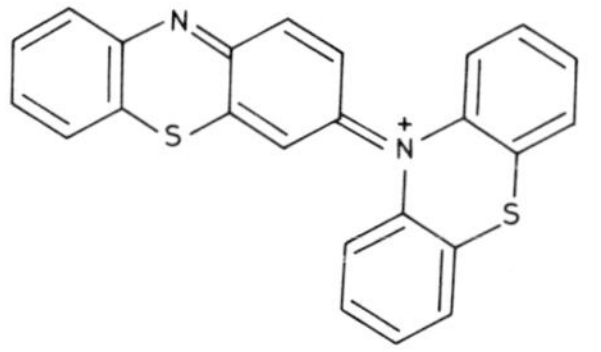

(II)

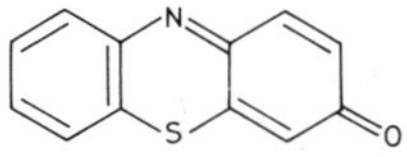

(III)

2 RESULTS AND DISCUSSION

Photolysis of either aerated or degassed solutions of PTH in carbon tetrachloride led to the formation of a pale green solution, from which a dark precipitate rapidly formed. This effect was most marked if carbon tetrachloride which had not been specially purified was used, or if traces of water were added to the solution.

Photolysis of solutions of PTH in carbon tetrachloride–ethanol (50% v/v) led to a similar colour change. However, here there was no precipitation, and the reaction was readily followed by electronic absorption spectroscopy (Figure 1 lines (a) and (b)). Following photolysis, absorptions were observed at 438, 456 (sh.), 501 (sh.), 520, 569 (sh.), and 646 nm. These absorptions decreased in intensity as a function of time over a period of a few hours, leaving a pale pink solution possessing new absorptions at 508, 557 (sh.) and 631 nm (Figure 1, line (c)). The sequence of reactions following phototolysis was also studied by

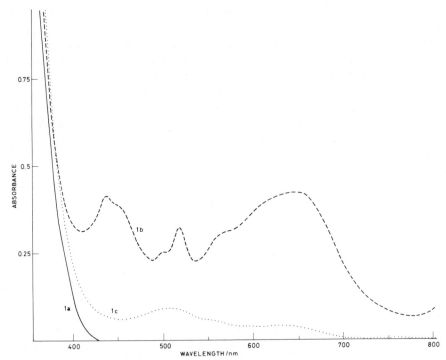

Figure 1   Absorption spectra of solutions of PTH (5×10^{-4} M) in aerated carbon tetra-chloride–ethanol (50%) (a) before photolysis, (b) immediately after photolysis, (c) 17 hours after photolysis

e.p.r. spectroscopy. The only signal which was observed following photolysis was a broad quartet (Figure 2) which decayed over a period of several hours.

Photolysis of solutions of PTH in carbon tetrachloride with acetonitrile (20% v/v) as cosolvent yielded a rather different series of spectral changes (Figure 3). Following photolysis, a golden coloured solution was formed, the spectrum of which (Figure 3, line (a)) showed absorptions at 433, 466 (sh.), 483, 503 and 520 nm, with weak bands at 635, 715 and > 790 nm. The absorptions decreased in intensity over a period of a few hours, leaving an almost colourless

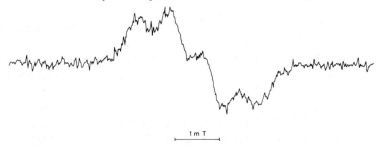

1 m T

Figure 2   E.p.r. signal observed in photolysed solution of PTH (10^{-3} M) in aerated carbon tetrachloride–ethanol (50%)

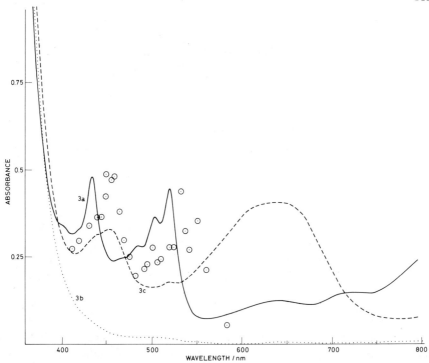

Figure 3   Absorption spectra of solutions of PTH (5×10^{-4} M) in aerated acetonitrile–carbon tetrachloride (20%) (a) immediately after photolysis, (b) 17 hours after photolysis, (c) spectrum obtained following addition of ethanol (*ca* 1:1) to solution corresponding to spectrum (a): circles, spectrum obtained immediately after flash photolysis of PTH ($1 \cdot 1 \times 10^{-4}$ M) in aerated carbon tetrachloride

solution (Figure 3, line (b)). Again, on studying the reaction by e.p.r. spectroscopy, only the quartet signal corresponding to Figure 2 was observed. This decayed within *ca* 3 hours. In the presence of traces of ethanol, however, spectral changes were observed similar to those with 50% carbon tetrachloride–ethanol. Further, if ethanol was added to the golden-orange solution corresponding to Figure 3, line (a), the solution immediately went green, yielding a solution with a similar spectrum to that with the carbon tetrachloride–ethanol solution (Figure 3, line (c)). If water was used instead of ethanol, a green colour rapidly developed at the phase boundary, and eventually diffused completely into the aqueous layer, leaving the carbon tetrachloride layer colourless.

Assignment of the species responsible for these spectral changes can be made from a consideration of literature data on oxidized species from PTH. Considering first the green coloured species observed in the presence of alcohol or water, a green coloured compound has been reported to be formed on oxidation of PTH under a variety of conditions. The nature of this species has been discussed by Hanson and Norman[9] who showed that it has the dimeric structure (II). This species is known to be rather unstable, and to eventually give the red coloured phenothiazinone λ_{max} 495 nm in CH_3CN[9] (III). We believe that the behaviour of the photolysed carbon tetrachloride–ethanol solutions can best be rationalized in terms of formation of the dimer (II) and subsequent decomposition of this to give the ketone (III) as the main reaction pathway for the photolysis in this solvent. On photolysis of carbon tetrachloride–acetonitrile solutions of PTH in the absence of ethanol or water, a species was observed with main absorptions at 433, 483, 503 and 520 nm. The spectrum is very similar to that of the phenothiazine radical cation (PTH.$^+$) formed either by dissolving the radical cation perchlorate in acetonitrile,[9] or by pulse radiolysis of PTH in chloroform.[8] To see whether this species was a precursor of the green colour in neat carbon tetrachloride, a solution of PTH in carbon tetrachloride was subjected to flash photolysis. A species was formed immediately after the flash ($\sim$ 20 μs) with absorptions similar to the carbon tetrachloride–acetonitrile solution (Figure 3, circles). The position of the absorptions was shifted slightly, possibly because of a solvent effect. This species decayed by good first-order kinetics with $k = 2.45$ ($\pm$ 1.16) s^{-1} independent of wavelength and PTH concentration over the range 2.5×10^{-5} M to 5×10^{-4} M. The decay was faster in the presence of traces of ethanol or water. However, it was not possible to give a good kinetic analysis of this behaviour.

Further information of the nature of the changes occurring can be obtained from a consideration of the e.p.r. spectral data. With both carbon tetrachloride–ethanol and carbon tetrachloride–acetonitrile solutions the same e.p.r. spectrum was observed. This was a broad quartet which is identical to the low-resolution e.p.r. spectrum of the radical cation PTH$^{.+}$ reported for electrochemical oxidation of PTH in acetonitrile.[10] This spectrum decays eventually in both cases to yield diamagnetic solutions. In the presence of ethanol, where the dimer (II) is formed, no e.p.r. spectrum was observed which could be assigned to the protonated triplet of this species reported by Tsujino.[11] Nor were we able to see any spectrum corresponding to the phenothiazine neutral radical (PT$^.$).[12,13]

It appears that for the conversion of the radical cation (I) into the dimeric species (II), either alcohol or water is necessary. Probably, as suggested by Hanson and Norman,[9] the function of the water or alcohol is to act as a base to deprotonate PTH$^{.+}$ to give the neutral radical, which subsequently dimerizes. The absence of e.p.r. spectra for PT$^.$ or Tsujino's protonated dimer may merely imply that these are present in too low concentrations to be detected in our system. Further, the fact that even in the presence of ethanol the e.p.r. signal of PTH$^{.+}$ remains long after the formation of the dimer (II) indicates that an equilibrium pathway must exist between the monomeric and dimeric species. This provides a ready explanation for the dimeric species being formed in neat carbon tetrachloride, but less so, or not at all in carbon tetrachloride–acetonitrile, as in the former system precipitation can disturb the equilibrium. Again, the results are consistent with the complex scheme proposed by Hanson and Norman.[9] The absorptions at 438 and 520 nm in the photolysed system of PTH in carbon tetrachloride–ethanol are probably also best assigned to PTH$^{.+}$.

To obtain information on the primary photochemical step leading to the formation of the cation radical, the initial optical density of PTH$^{.+}$ after flash photolysis of solutions of PTH in carbon tetrachloride was studied as a function of the light intensity. Good linear plots of the optical density at either 450 ($\pm$ 10) or 530 ($\pm$ 10) nm versus the light intensity were obtained whilst plots of the optical density against the square of the flash energy showed curvature, thus demonstrating that the primary photochemical act leading to formation of the radical cation only requires one photon, in contrast to photoionization in hydrocarbon media which is commonly a two-photon process.[14]

The fluorescence of PTH was studied in the presence of carbon tetrachloride. Addition of carbon tetrachloride to solutions of PTH in cyclohexane markedly reduced the intensity of the PTH fluorescence at 435 nm, and the quenching followed good Stern–Volmer kinetics for carbon tetrachloride concentrations in the range 0 to 0·6 M. A typical plot is shown (Figure 4). The slope of the

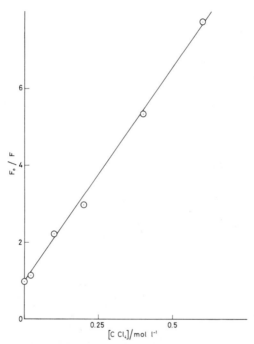

Figure 4 Stern–Volmer plot for quenching of fluorescence of a solution of PTH ($2 \cdot 2 \times 10^{-4}$ M) in cyclohexane by carbon tetrachloride. Excitation 290 nm, emission 435 nm

Stern–Volmer plot appeared to vary slightly with the excitation wavelength (Table 1). However, at certain wavelengths there was considerable scatter in the points, which was probably due to the build up of photolysis products, and the apparent wavelength variation was within the limits of experimental error.

To determine whether static or dynamic processes might be responsible for the quenching, absorption spectra were run of solutions of PTH in n-hexane in

122

Table 1 Stern–Volmer constants for the quenching of the fluorescence at $435(\pm 2\frac{1}{2})$ nm of a solution of phenothiazine $(2.2 \times 10^{-4}$ M$)$ in cyclohexane by carbon tetrachloride

Excitation wavelength nm (± 5 nm)	Quenching constant[a] $(k_Q\tau)$ 1 mol^{-1}
280[b]	9.9 ($\pm$ 1.6)
290[b]	11.1 ($\pm$ 0.4)
320	9.6 ($\pm$ 2.5)

[a] Solutions were very photosensitive, and considerable scatter was observed at certain wavelengths.
[b] With excitation at these wavelengths, new fluorescence bands were observed between 460 and 500 nm due to photolysis products.

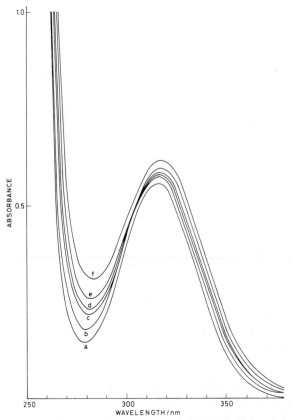

Figure 5 Absorption spectra of solutions of PTH $(1.40 \times 10^{-4}$ M$)$ in n-hexane (a) alone, and in the presence of (b) 1.01 M, (c) 2.41 M, (d) 2.99 M, (e) 3.99 M, (f) 6.35 M, carbon tetrachloride

the presence of varying concentrations of carbon tetrachloride (Figure 5). These revealed slight changes in both the intensity and position of the low-energy absorption band, strongly reminiscent of the behaviour observed with aromatic amines and halocarbons,[2] which has been assigned to contact charge-transfer. The changes in optical density as a function of concentration fitted the Benesi–Hildebrand equation,[15] with association constants 0.137 (± 0.048) $1 \, mol^{-1}$ determined at 360 nm, and 0.071 (± 0.016) $1 \, mol^{-1}$ determined at 280 nm, giving good evidence for weak complexing between PTH and carbon tetrachloride. The difference in association constants at the different wavelengths may indicate the presence of $2:1$ or higher complexes.[16] Previously, charge transfer complex formation has been demonstrated between PTH (or its derivatives) and a number of strong electron acceptors, such as quinones and nitrocompounds.[17] It is of interest, then, that PTH can also form complexes with much weaker acceptors such as carbon tetrachloride.

In the absence of fluorescence lifetime data, it is not possible to say how much, if any, of the observed fluorescence quenching results from dynamic interaction between the singlet excited state of PTH, and how much from static quenching resulting from ground state complexes. However, correction of the fluorescence quenching data for the absorbance by ground state contact species does leave a residual fluorescence quenching which may result from dynamic quenching. Moreover, it seems likely that the two processes will lead to a similar charge-transfer state. The mechanism of the primary act can be represented

$$\text{PTH} + \text{CCl}_4 \xrightarrow{h\nu} \text{PTH}^{\cdot +} + {}^{\cdot}\text{CCl}_3 + \text{Cl}^- \tag{1}$$

Electron transfer to CCl_4 is expected to lead to dissociative electron capture.

We have further investigated the oxidation of PTH in these media by using the chemical oxidant cerium (IV) ammonium nitrate. In 50% carbon tetrachloride–ethanol, ceric ammonium nitrate (1.78×10^{-4} M) very rapidly oxidized PTH (0.92×10^{-3} M) to give a transient orange-coloured solution, and then very rapidly (< 1 second) a green coloured solution possessing a similar spectrum to the photooxidized solution in this solvent, with absorptions at 643, 520 (sh.) and 452 nm. The absorptions at 438, 501 and 569 nm were not observed in this case. However, they may have been buried under the other strong absorptions. The spectrum decayed over a period of several hours to give a spectrum with absorptions at 500 and 624 nm, identical to that observed with the photo-oxidized solution, and probably best assigned to the phenothiazinone (III). On studying the reaction by e.p.r. spectroscopy, the only signal which is observed is the broad quartet corresponding to Figure 2, and the decay of this parallels the decay of the absorption at 520 nm. To provide further information on the system the kinetics of decay of the electronic spectral absorptions were analysed for both chemically and photochemically oxidized solutions of PTH in carbon tetrachloride–ethanol (Table 2). The solutions followed similar kinetic behaviour. Whilst the absorptions at 640 and 520 nm followed an overall second-order decay, the 450 nm band fitted first-order kinetics rather better. In view of this complex kinetic behaviour, assignment of the 450 nm

Table 2   Kinetics of decay of absorptions following thermal and photochemical oxidation
of phenothiazine solutions

Solution	Wavelength/ nm	Rate constant
PTH (0.46×10^{-3} M) in 50% carbon	640	k_2/ε^a 4.33×10^{-3} cm s^{-1}
tetrachloride–ethanol (aerated) following	520	k_2/ε^a 1.27×10^{-2} cm s^{-1}
photolysis	450	k_1^b 8.73×10^{-4} s^{-1}
PTH (0.92×10^{-3} M), ceric ammonium nitrate	640	k_2/ε^a 3.51×10^{-3} cm s^{-1}
(1.78×10^{-4} M) in 50% carbon tetrachloride–	520	k_2/ε^a 1.04×10^{-2} cm s^{-1}
ethanol	450	k_1^c 6.83×10^{-4} s^{-1}
PTH (1.46×10^{-3} M) in 20% acetonitrile–carbon	790	k_1^c 1.55×10^{-4} s^{-1}
tetrachloride (aertated) following photolysis	715	k_1^c 4.20×10^{-4} s^{-1}
	520	k_2/ε^a 2.72×10^{-3} cm s^{-1}
	433	k_2/ε^a 2.62×10^{-3} cm s^{-1}
PTH (1.46×10^{-3} M), ceric ammonium nitrate	640	k_2/ε^d 1.9×10^{-4} cm s^{-1}
(1.59×10^{-4}) in 20% acetonitrile–carbon	520	k_2/ε^d 3.9×10^{-4} cm s^{-1}
tetrachloride		

[a] Second-order kinetics.
[b] Complex kinetics, but first-order to 70% reaction.
[c] First-order kinetics.
[d] Complex kinetics; second-order rate constant determined for first 25% reaction.

absorption requires further data. However, the dimer (II) possesses a band at
446 nm[9], so the 450 nm band may well be a mixture of absorptions. In acetonitrile–carbon tetrachloride (20%), oxidation by ceric ammonium nitrate followed
a different course from the photochemical reaction. Immediately after mixing
PTH (1.46×10^{-3} M) and ceric ammonium nitrate (1.59×10^{-4} M) in this
medium an orange-coloured solution was formed which rapidly decayed
(< 5 s) to a green-coloured solution with absorptions at 640, 520 (sh.), 454 and
435 (sh.) nm, similar to the spectrum observed with carbon tetrachloride–
ethanol as solvent. It would appear that if ethanol and water act as bases in
converting the radical cation (I) to the dimer (II), then probably the nitrate ion
can perform this function too. The fact that photochemical and thermal oxidations lead to different species, because of further reactions of the initially formed
species with the chemical oxidant, indicates the potential use of photochemical
routes for performing oxidations under mild conditions. The green colour
decayed slowly. However, on leaving the solution for 24 hours, a species possessing new intense absorptions at 455 and 660 nm was formed; this species was
not characterized. Kinetics of decay of the thermally oxidized species, and also
of the species following photooxidation in this medium, are reported in Table 2.
It is of interest that on photooxidation of PTH in acetonitrile–carbon tetrachloride, absorptions at 715 and > 790 nm decay by a first-order process. We
believe these can be best assigned to the radical cation dimer $(PTH^{\cdot+})_2$, which
in the solid state possesses bands at 690 and 830 nm.[18]

Having demonstrated photooxidation of PTH by carbon tetrachloride,
various other halogenated hydrocarbons were investigated for their efficiency as

photooxidants. Solutions of phenothiazine and halocarbons in ethanol were subjected to intense light flashes from conventional flash tubes (to prevent secondary photolysis of the strongly absorbing products) and the optical densities of the dimer (II) at 640 nm were monitored. The yields of dimer (II) (relative to carbon tetrachloride as photooxidant) are given in Table 3. Chloroform, trichloroethene and chloracetic acid all showed some photosensitivity, although rather less than carbon tetrachloride. Solutions of PTH in the presence of carbon tetrabromide were, however, highly photosensitive. Photolysis of a solution of PTH (5×10^{-4} M) and CBr_4 (10^{-2} M) in ethanol yielded initially absorptions at 440, 500 (sh.), 518, 562, 600 (sh.) and 640 nm. Over a period of a few minutes the 440 and 518 nm absorptions decreased in intensity, while the 640 nm band showed a corresponding increase. The optical density at 640 nm was taken before this further thermal reaction had occurred. It is notable that the intensity of the absorption following photolysis is similar to that obtained from photolysing PTH with 1 M carbon tetrachloride in ethanol for the same length of time, thus demonstrating the superior photooxidizing ability of CBr_4, which conforms well to its established higher electron affinity.[4] With all of the halogenated hydrocarbons, fluorescence quenching and absorption spectral shifts similar to those with carbon tetrachloride were observed, suggesting a similar mechanism for the photooxidation.

Solutions of PTH in ethanol in the presence of oxygen have been reported to be photooxidized.[12,19,20] Following photolysis, we found that solutions of PTH in aerated ethanol did give photoreaction, but the yield of the product was very much less than in the presence of halocarbons. There is some disagreement between various workers on the nature of the products of this reaction. Shine and coworkers[19] showed that on photolysis, an e.p.r. signal was observed which they assigned to the neutral phenothiazinyl radical. Jackson and Patel[12] on the other hand have suggested that the signal observed by Shine and coworkers is probably better assigned to the nitroxide. However, they observed the neutral radical on photolysis of PTH in ethanol under slightly different conditions. Finally, Koizumi and coworkers[20] have demonstrated the importance of oxygen in this reaction and have shown that the reaction proceeds

Table 3   Relative yields of dimer (II) following photolysis of aerated ethanolic solutions of PTH (5×10^{-4} M) in the presence of halocarbons

Solute	Concentration mol l^{-1}	Relative yield[a] of dimer (II) measured as absorbance at 640 nm 30 seconds after photolysis
None	—	0·01
$ClCH_2CO_2H$	1	0·06
Trichlorethylene	1	0·13
$CHCl_3$	1	0·64
CCl_4	1	1
CBr_4	10^{-2}	1·59

[a] Yields relative to 1 M CCl_4.

via the PTH triplet state. These workers observed an e.p.r. spectrum which they assigned to the neutral radical, while following flash photolysis of solutions of PTH in aerated ethanol they observed a transient absorption at 385 nm which they assigned to an intermediate charge-transfer complex between PTH and oxygen, either in the lowest excited state or the ground state. The feasibility of such a complex is supported by our demonstration of contact charge-transfer complexes between PTH and weak acceptors such as carbon tetrachloride. However, it is noteworthy that the phenothiazinyl neutral radical (PT.) has also been shown to possess a band at 385 nm,[8,9] so that alternative assignment of their flash photolysis spectrum to this species cannot be ruled out. An attractive alternative explanation for their results is that oxidation occurs via the intermediacy of singlet oxygen

$$^3PTH^* + O_2 \rightarrow PTH + {}^1O_2 \tag{2}$$

$$^1O_2 + PTH \rightarrow HO_2{}^{\cdot} + PT^{\cdot} \tag{3}$$

and that the 385 nm transient is, in fact, the neutral PT$^{\cdot}$ radical. Singlet oxygen is known to be formed by photosensitization with the closely related methylene blue system,[21] and reaction (2) would explain the effect of oxygen on the PTH triplet state.

In conclusion, photooxidation by halocarbons provides a ready route for generation of phenothiazine radical cations and studying the complex sequence of reactions involved in their decay. It appears that many of these reactions involve equilibria between monomeric and dimeric species, such as both the equilibrium between the radical cation and the dimer (II) (probably with the intermediacy of the neutral phenothiazinyl radical[9]), and probably also an equilibrium between the radical cation and cation dimer

$$2\,PTH^{\cdot+} \rightleftharpoons (PTH^{\cdot+})_2 \tag{4}$$

We are currently investigating photooxidation of PTH through other charge-transfer species to find systems which do not involve concurrent generation of haloalkyl radicals.

ACKNOWLEDGEMENTS

The authors gratefully acknowledge financial support for this work from the Fundação Gulbenkian and the Instituto de Alta Cultura (C.E.Q.N.R. project CQ2).

REFERENCES

1. W. C. Meyer, *J. Phys. Chem.*, **74**, 2118, 2127 (1970).
2. K. M. C. Davis and M. F. Farmer, *J. Chem. Soc. (B)*, 28 (1967).
3. G. Pfister and D. J. Williams, *J. Chem. Phys.*, **59**, 2683 (1973).

4. M. S. S. C. Leite and K. Razi Naqvi, *Chem. Phys. Letters*, **4**, 35 (1969).
5. W. R. Ware and C. Lewis, *J. Chem. Phys.*, **57**, 3546 (1972).
6. D. Greatorex, T. J. Kemp and J. P. Roberts, *J. Phys. Chem.*, **73**, 1616 (1969).
7. See, for example, S. P. Massie, *Chem. Rev.*, **54**, 797 (1954).
8. H. D. Burrows, T. J. Kemp and M. J. Welbourn, *J.C.S. Perkin II*, 969 (1973).
9. P. Hanson and R. O. C. Norman, *J.C.S. Perkin II*, 264 (1973).
10. J. P. Billon, G. Cauquis and J. Combrisson, *J. Chim. Phys.*, 374 (1964).
11. Y. Tsujino, *Tetrahedron Letters*, 2545 (1968).
12. C. Jackson and N. K. D. Patel, *Tetrahedron Letters*, 2255 (1967).
13. B. C. Gilbert, P. Hanson, R. O. C. Norman and B. T. Sutcliffe, *Chem. Comm.*, 161 (1966).
14. K. D. Cadogan and A. C. Albrecht, *J. Chem. Phys.*, **43**, 2550 (1965); *J. Phys. Chem.*, **72**, 929 (1968).
15. H. A. Benesi and J. H. Hildebrand, *J. Amer. Chem. Soc.*, **71**, 2703 (1949).
16. See, however, R. Foster and I. Hormann, *J. Chem. Soc.* (*B*), 171 (1966) for an alternative interpretation,
17. R. Foster and P. Hanson, *Biochim. Biophys. Acta,* **112**, 482 (1966); R. Foster and C. A. Fyfe, *Biochim. Biophys. Acta*, **112**, 490 (1966).
18. Y. Iida, *Bull. Chem. Soc. Japan*, **44**, 663 (1971).
19. H. S. Shine, C. V. Veneziani and E. E. Mach, *J. Org. Chem.*, **31**, 3395 (1966).
20. T. Iwaoka, H. Kokubun and M. Koizumi, *Bull. Chem. Soc. Japan*, **44**, 341, 3466 (1971).
21. See, for example, T. Wilson, *J. Amer. Chem. Soc.*, **88**, 2898 (1966).

A.2.6. Cannabinols: Photochemistry and determination in biological fluids

A. Bowd, D. A. Swann, B. Towell and J. H. Turnbull

*Applied Chemistry Branch, Royal Military College of Science,
Shrivenham, Wilts., U.K.*

SUMMARY

A simple, sensitive method for the fluorimetric estimation of cannabis constituents following extraction from an ethanol/water mouthwash is described. The technique utilizes the quantitative formation of a highly fluorescent photoproduct following ultraviolet irradiation of alcoholic solutions containing cannabinols. Concentrations of cannabinol in mouthwash as low as 15 ng/cm^3 can be determined. Spectral characteristics and structure of the photoproduct are discussed.

1 INTRODUCTION

Methods currently available for the determination of cannabis constituents in biological materials are largely based upon thin-layer chromatography.[1-5] None of these procedures fully meets the requirement for a reliable, sensitive and specific method for the quantitative estimation of cannabis constituents. In this respect fluorimetry appeared most promising, since previous luminescence studies[6] have indicated that the cannabinols, whilst possessing only weak intrinsic fluorescence emission, can be converted on irradiation in ethanol to a highly fluorescent photoproduct having a characteristic structured emission spectrum. All the four major cannabis constituents, Δ^9-tetrahydrocannabinol (Δ^9-THC), Δ^8-tetrahydrocannabinol (Δ^8-THC), cannabidiol (CBD) and cannabinol (CBN), appear to be converted to the same fluorescent end product, although the rate of formation differs markedly between the four compounds.

For dilute solutions of cannabinols irradiated at 285 nm, the fluorescence intensity of the photoproduct rapidly increases, reaching a maximum after

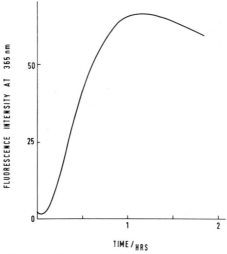

Figure 1   The increase in fluorescence intensity of the CBN photoproduct on continuous irradiation at 285 nm. (Emission recorded at 365 nm)

about 1–2 hr, and then slowly decreases, presumably due to photodecomposition of the product (Figure 1). When the maximum fluorescence intensity developed is plotted against the initial concentration of cannabinol, a linear dependence is obtained over the range 10^{-8}–10^{-5} M.

In the present paper we apply this technique to the estimation of CBN added to an ethanol/water mouthwash (10% EtOH v:v). We have used an extraction procedure similar to that of Stone and Stevens[4] to remove interfering fluorescent substances from the mouthwash. We have demonstrated that the method is extremely sensitive, gives a high percentage recovery of cannabinol, and good reproducibility.

2 EXPERIMENTAL

2.1 Instrumentation

The spectrofluorimeter used in this work has been described previously.[6,7] Light from a 2 kW xenon source is focused onto the entrance slit of the excitation monochromator with both slits set at 10 mm, giving a half-band width of 16 nm. The sample is contained in a quartz cuvette (12 mm^2) mounted in a thermostatically controlled cell compartment at 25 °C. Fluorescence is viewed at right angles by the emission monochromator, with slits of 5 mm (half-band width 8 nm). The detector is an EMI 9558 QB photomultiplier tube.

2.2 Materials

Samples of Δ^9-THC and pure crystalline CBN were supplied by the Home Office Central Research Establishment, Aldermaston. The alumina column was prepared by filling a Pasteur pipette to 4 cm with Hopkin and Williams alkaline aluminium oxide, 100–200 mesh, activity 1. All other reagents used were of analytical grade.

2.3 Construction of calibration curves

Solutions of CBN in ethanol (99·7–100 %) were prepared over a concentration range 10^{-8} M to 10^{-5} M. Each solution was irradiated at 285 nm and the fluorescence emission at 365 nm (the wavelength of a photoproduct fluorescence peak) was recorded continuously. The irradiation was stopped after the fluorescence had reached a maximum and had started to decline (1–2 hr). The increase in fluorescence signal, as indicated by the increase in the photomultiplier anode current, was plotted against concentration (Figure 3). A similar calibration curve was constructed for Δ^9-THC over the concentration range 10^{-7} M to 10^{-5} M.

2.4 Extraction of CBN from mouthwash

The mouth is washed with four successive 15 cm³ volumes of aqueous 9% ethanol. This mouthwash is added to 0·5 cm³ of ethanolic CBN solution of appropriate concentration to give a total volume of 50 cm³. An equal volume of 2% w:v sodium hydroxide solution is added to the wash and mixed. 50 cm³ of light petroleum (60–80 °C) is added, and the CBN extracted in a separating funnel by vigorous shaking. After standing, the lower aqueous layer is discarded and the petroleum extract is washed once with 50 cm³ M HCl and once with water. Any emulsion in the petroleum layer is broken by the addition of anhydrous sodium sulphate. The dried petroleum is filtered and evaporated on a rotary evaporator to 0·5 cm³. The concentrated extract is placed on an alkaline alumina column and washed through with a further 10 cm³ of light petroleum. This washing removes most of the fatty materials in the mouthwash. The CBN is then eluted with 10 cm³ of 100% ethanol, and this solution is subjected to irradiation at 285 nm in the spectrofluorimeter. The increase in the fluorescence intensity at 365 nm is measured and by reference to the calibration graph the concentration of CBN in the ethanolic eluate can be determined.

3 RESULTS AND DISCUSSION

3.1 Spectral characteristics

The fluorescence emission spectra of CBN in ethanol after various periods of irradiation at 285 nm are shown in Figure 2. The intrinsic fluorescence emission of CBN is a low intensity peak at 310 nm, but this is rather difficult to observe because of the intense photoproduct fluorescence which develops after a few minutes irradiation. Preliminary experiments have shown 285 nm to be the most effective excitation wavelength for photoproduct formation. Temperature and the presence of dissolved oxygen appear to have very little effect on the rate.

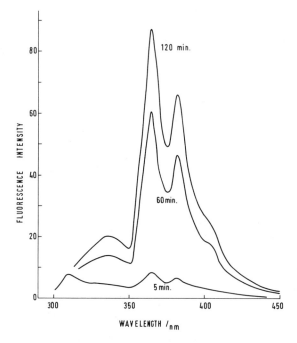

Figure 2   Fluorescence spectra of CBN in ethanol
after various periods of irradiation at 285 nm

3.2 Calibration curves

The maximum fluorescence intensity of the CBN photoproduct varies linearly with concentration over the range 10^{-8} M to 10^{-5} M (Figure 3). At concentrations below 10^{-8} M the combined background fluorescence and photomultiplier dark current signals are too great for significant fluorescence intensity differences to be observed. Above 10^{-5} M inner filter effects become significant and the calibration is no longer linear. For Δ^9-THC the calibration is linear over the concentration range 10^{-7} M to 10^{-5} M. This lower sensitivity is due to the slower rate of photoproduct formation compared to CBN, and the consequent increased significance of photoproduct decomposition.

3.3 Extractions

The efficiency of the extraction process depends upon the concentration of CBN initially present in the mouthwash. For mouthwash solution containing 10^{-6} M CBN (*ca* 300 ng/cm^3) the percentage recovery is very high, whereas at concentrations around 5×10^{-8} M (*ca* 15 ng/cm^3) less than 60% of the total CBN is recovered (Table 1). We have shown, however, that the recovery factor is reproducible for a particular concentration, and a correction factor can

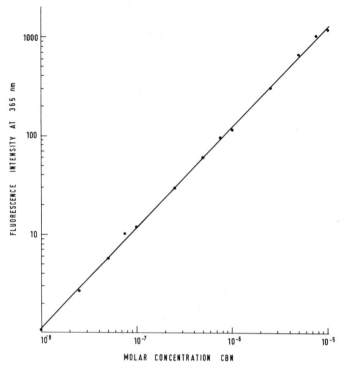

Figure 3   Calibration plot for CBN in ethanol

Table 1   Recovery of cannabinol (CBN) after extraction

Concentration of CBN in mouthwash (50 cm³)	Increase in fluorescence intensity/10^{-7} A	Concentration of CBN in eluate (10 cm³)	% Recovery
5×10^{-8} M	17·2	$1·46 \times 10^{-7}$ M	58
$7·5 \times 10^{-8}$ M	28·0	$2·36 \times 10^{-7}$ M	63
10^{-7} M	42·8	$3·6 \times 10^{-7}$ M	72
$2·5 \times 10^{-7}$ M	122·4	$1·01 \times 10^{-6}$ M	81
5×10^{-7} M	272·4	$2·2 \times 10^{-6}$ M	88
10^{-6} M	560	$4·5 \times 10^{-6}$ M	90

therefore be applied to the measured CBN concentration in the eluate in order to determine the concentration of CBN originally present.

3.4 Nature of the photoproduct

The fluorescent photoproduct was isolated and purified by GLC and thin layer chromatography. The UV absorption, fluorescence and phosphorescence spectra were examined, and showed well developed structural features characteristic of the hydroxyphenanthrene nucleus (Table 2). The mass spectrum ($M^{+}m/e$ 292) shows that the photoproduct is formed from cannabinol (1) by the elimination of one mole of water. On the basis of these results we propose that the photoproduct has the structure of a substituted 4-hydroxyphenanthrene (2).

Table 2

	UV absorption λ_{max} (nm)	Fluorescence 25 °C λ_{max} (nm)	Phosphorescence 77 K λ_{max} (nm)
Cannabinol	222, 286	310	445, 475, 507
CBN photoproduct	256, 277, 307, 343, 360	365, 382, 402 (s)	477, 515, 555
3-OH phenanthrene	251, 278, 303, 341, 357	362, 379, 400 (s)	468, 503, 545
4-OH phenanthrene[a]	250, 275, 304, 339, 357		

[a] Data from C. Djerassi, H. Bendas and P. Sengupta, *J. Org. Chem.*, **20**, 1046 (1955).

(1)   (2)

3.5 Conclusions

The fluorescence technique outlined in this paper is suitable for the estimation of trace amounts of CBN and other cannabis constituents which undergo photochemical reaction to form the fluorescent photoproduct. A detailed study of the nature of the photochemical mechanism involved is currently in progress.

ACKNOWLEDGEMENT

We would like to thank the Director, Home Office Central Research Establishment, Aldermaston, for his interest in this work.

REFERENCES

1. J. M. Andersen, E. Nielsen, J. Schon, A. Steentoft and K. Worm, *Acta Pharmacol. et Toxicol.*, **29**, 111 (1971).
2. J. Christiansen and O. J. Rafaelsen, *Psychopharmacologia (Berl.)*, **15**, 60 (1969).
3. E. J. Woodhouse, *Amer. J. Public Heath*, **62**, 1394 (1972).
4. H. M. Stone and H. M. Stevens, *J. For. Sci. Soc.*, **9**, 31 (1969).
5. R. Hackel, *Arch. Toxikol.*, **29**, 341 (1972).
6. A. Bowd, P. Byrom, J. B. Hudson and J. H. Turnbull, *Talanta,* **18**, 697 (1971).
7. A. Bowd, P. Byrom, J. B. Hudson and J. H. Turnbull, *Photochem. Photobiol.*, **8**, 1 (1968).

A.2.7. On the luminescence of ketosteroids

A. Weinreb and A. Werner

Department of Physics, Bar-Ilan University, Ramat-Gan, Israel

SUMMARY

Some features of the fluorescence of ketoandrostanes and ketoestrogens are stated. The former is highly influenced by the position in the steroid frame at which the carbonyl group is attached. The latter is strongly determined by intramolecular energy transfer from the aromatic A-ring to the carbonyl group which is attached to the D-ring. The fluorescent behaviour of 6 Ketoestradiol is described in detail. For the explanation of this behaviour, particularly the excitation spectrum, a hypothetical term scheme is proposed.

1 INTRODUCTION

The fluorescence of estrogens is an important means for their detection in medical research and clinical application. The highly fluorescing material is, however, always obtained upon denaturation (usually by the addition of a strong acid). The natural fluorescence of many of these compounds, on the other hand, is extremely low, and can thus hardly serve as a diagnostic tool. It is thus perhaps not too surprising that it has received very little attention, and practically no research on this aspect has been reported in the open literature, though other spectroscopic properties of this large important group of compounds, like IR absorption, optical rotation and dichroism, have been extensively investigated.

Since the proper excited states of these molecules may be of great importance in biological systems under conditions of irradiation, and perhaps even for the understanding of their biochemical function, we embarked several years ago upon a systematic survey of the luminescence of these systems. In this paper we

135

describe briefly some general features and concentrate on one particular compound of rather basic biological importance.

From the vantage point of fluorescence as well as biological function it is practical to divide the ketosteroids into two groups: androstanes and estrogens. In the former all four rings of the steroid frame are formed by σ-bonds, with the carbonyl group attached at any one of the positions emphasized by a circle in Figure 1. In the latter the A (or A and B) ring is aromatic, which by the very difference in bond angles causes a strain in the neighbouring saturated rings. It is clear that the fluorescent behaviour of these two groups will be quite different.

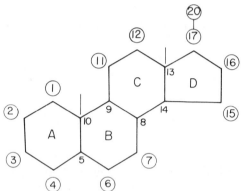

Figure 1   Stereochemical configuration and numbering system of a steroid frame

2 ANDROSTANES

We first consider the androstanes. It is found that the shape of the absorption spectrum of the C=O group when attached to the steroid frame is relatively independent of the position (except for the 16-position), while the height of the spectrum may vary by as much as a factor of two for the first absorption band, which is of $n\pi^*$ character. The fluorescence intensity, however, varies drastically with position, and differences of more than an order of magnitude are observed. This can be seen from Table 1 (the last number or numbers in the designation of the compound indicate the position of the ketogroup).

These results are interpreted to be closely related to the vibrational pattern of the molecule at the respective positions. This strongly influences the radiative process as well as radiationless transitions from the excited state of the chromophore to a high vibrational level of the ground state or to a dissociative state of different geometry.[1]

The results show also a strong interaction between non-adjacent carbonyl groups: from the table it is seen that the carbonyl group in the C-11-position causes a remarkable decrease in the fluorescence efficiency of the carbonyl group in the C-17-position (which otherwise has the greatest quantum yield). We are not aware of such interaction between carbonyl groups in any other system.

Table 1 Fluorescence of Keto-androstanes

Compound	Intensity
5α-Androstane-3-one	83
5α-Cholestane-3β-ol-6-one	310
5α-Cholestane-3α-ol-7-one	150
5β-Androstane-3α,17β-diol-11-one	<20
Desoxybilianic acid (C-12)	220
5-Androstene-3β, 17β-diol-16-one	280
5α-Androstane-17-one	430
5α-Pregnane-3β-ol-20-one	64
5α-Androstane-3α-ol-11,17-dione	47
5α-Androstane-3,11,17-trione	57

A feature common to all ketoandrostanes is the absence of any phosphorescence (with decay time longer than 2×10^{-4} s), though many aliphatic carbonyl compounds are known to phosphoresce efficiently.

3 ESTROGENS

In order to understand the fluorescent behaviour of ketoestrogens we have first to study the fluorescence of the parent compound without the carbonyl group. Consider, for example, estradiol which is a very potent steroid. The fluorescence spectrum of this compound is shown in Figure 2. Figure 3 shows the fluorescence spectrum of estrone. In this compound the carbonyl group replaces the hydroxyl group in the 17-position of estradiol. It should be noted

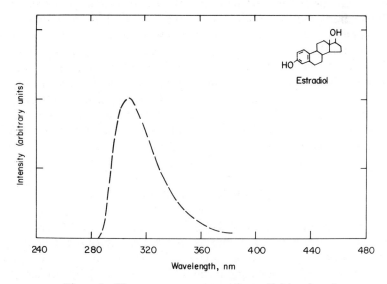

Figure 2 Fluorescence spectrum of estradiol in ethanol

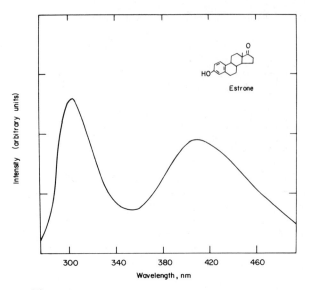

Figure 3   Fluorescence spectrum of estrone in ethanol

that the intensity of the peak at 308 nm in the spectrum of estrone is smaller by a factor of 30 than that of the corresponding peak in the spectrum of estradiol, under the same conditions of excitation. The broad band with peak at 410 nm is identical with the fluorescence spectrum of 5α-Androstane-17-one. Since the absorption coefficient of the phenolic chromophore at the wavelength of excitation (280 nm) is much higher than that of the carbonyl group, it is concluded that the fluorescence of the carbonyl group is solely due to transfer of excitation energy from the former to the latter. An exact evaluation of the transfer efficiency in terms of the overlap integral of the two chromophores reveals that the Forster radius (6 Å) comes close to the 'actual' distance within 10%. A better match cannot be expected since the dipole–dipole interaction model can hardly hold for shorter distances, while the concept of the distance itself becomes problematic. In contrast to the highly efficient energy transfer in estrone (and many other ketoestrogens of phenolic or phenylic parentage) there is total absence of the effect in equilenin (Figure 4). The fluorescence spectrum of this compound equals in shape and intensity that of the parent compound 17-β-dihydroequilenin. This fact is due to the vanishingly small overlap integral of the naphtholic and the carbonylic chromophore.

4 6-KETOESTRADIOL

A rather uncommon fluorescent behaviour is encountered in 6-Ketoestradiol (6KE$_2$). In fact, this compound constitutes a separate group of ketoestrogens, in that the carbonyl group is conjugated to the phenolic chromophore. Figure 5, curve a, shows the absorption spectrum of 6KE$_2$ in ethanol. The molar absorp-

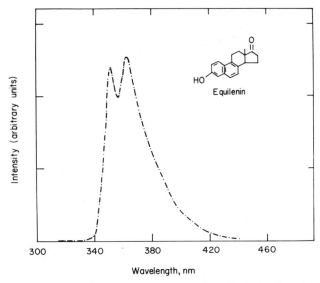

Figure 4   Fluorescence spectrum of equilenin in ethanol

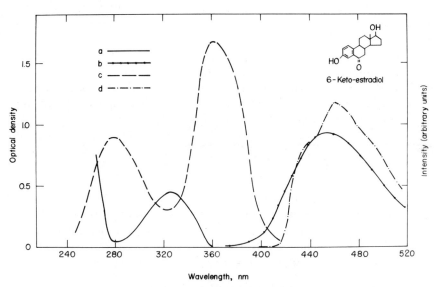

Figure 5   (a) Absorption spectrum of $6KE_2$.
(b) Fluorescence spectrum of $6KE_2$.
(c) Excitation spectrum.
(d) Phosphorescence spectrum.
The intensity scale is different for each curve

tion coefficient at the peak at 327 nm is 4575 litres $mole^{-1}$ cm^{-1}. Although the location of the peak resembles the absorption spectrum of some familiar aromatic carbonyl compounds, like benzaldehyde, acetophenone and benzophenone, the absorption coefficients for the long-wavelength peaks of these compounds are two orders of magnitude lower than that for $6KE_2$. The high value of the absorption coefficient $6KE_2$ at 327 nm indicates almost unequivocally that this transition is of $\pi\pi^*$ character as contrasted to the $n\pi^*$ character of the lowest energy peak in the other aromatic carbonyl compounds. It is, however, interesting to note that the absorption spectrum of m-hydroxy-acetophenone, whose structure resembles that of the chromophoric group of $6KE_2$, is almost identical with the spectrum of $6KE_2$. (The absorption coefficient of this compound at its first absorption peak (311 nm) is 3980 litres mole^{-1} cm^{-1}.) In m-hydroxy-acetophenone, as well as in $6KE_2$, the $\pi\pi^*$ transition hides the $n\pi^*$ transition which is well resolved in other aromatic carbonyl compounds. When the solvent is changed from alcohol to dioxane (a less polar solvent) a distinct blue shift of 70 Å in the first absorption peak of $6KE_2$ is observed. This again is characteristic for a $\pi\pi^*$ transition.

Curve b of Figure 5 shows the fluorescence spectrum of $6KE_2$ in ethanol. The spectrum is structureless with maximum at 455 nm. In EPA the maximum is shifted to the blue by 750 cm^{-1}. On the other hand, m-hydroxy-acetophenone does not exhibit any detectable fluorescence. We are thus led to the conclusion that it is a steric effect, probably the absence of rotation due to the attachment of the chromophore to the rigid steroid frame, which reduces the radiationless degradation of the excitation energy in $6KE_2$. No fluorescence is observed when $6KE_2$ is dissolved in dioxane.

The excitation spectrum of $6KE_2$ is shown in curve c of Figure 5. It is seen that the excitation spectrum is inverted with regard to the absorption spectrum, the fluorescence being strongest at excitation wavelengths at which the absorption is lowest. This feature is independent of concentration and it is still observed at concentrations for which the optical density is extremely low.

A strong phosphorescence is observed in a solid solution of $6KE_2$ in EPA. The spectrum is structureless with peak at 460 nm and a shoulder at 440 nm. The phosphorescence is efficiently excited at 330 nm, a wavelength at which the fluorescence is barely excited. The decay time of the phosphorescence is 0·3 s, indicating a $\pi\pi^*$ transition.

In order to explain the fluorescence of $6KE_2$, accounting simultaneously for its spectral distribution, excitation spectrum and phosphorescence spectrum, a rather unique term scheme has to be assumed. Such a hypothetical scheme is presented in Figure 6. The particular features of this scheme are:

(1) The vibrational ground state of the first excited singlet $\pi\pi^*$ level lies below the $S_1 n\pi^*$ level and its potential curve rises steeply for distances which correspond to the equilibrium configuration of the ground state of the molecule.

(2) Excitation into $S_1 \pi\pi^*$ and in particular into its absorption maximum (327 nm) leads to intersystem crossing to the $T \pi\pi^*$ level and is responsible for the observed phosphorescence (P).

(3) Excitation at 370 nm leads to the $S_1 n\pi^*$ level which crosses the $S_1 \pi\pi^*$ level at a point which is so low on the potential curve that intersystem crossing to $T \pi\pi^*$ is strongly inhibited. Ultimately the vibrational ground state of $S_1 \pi\pi^*$

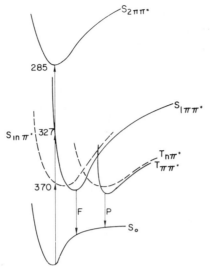

Figure 6   Hypothetical system of potential curves of $6KE_2$

is reached from which the observed fluorescence (F) is emitted. The assumed internal conversion from $S_1 n\pi^*$ to $S_1 \pi\pi^*$ may indeed be much more efficient than intersystem crossing from $S_1 n\pi^*$ to $T n\pi^*$.

(4) Excitation into $S_2 \pi\pi^*$ (285 nm) leads to fast internal conversion to $S_1 n\pi^*$ rather than to $S_1 \pi\pi^*$ because of the (assumed) better matching of Franck–Condon coefficients.

This level scheme which is of course entirely hypothetical accounts for all the observations described above. It received additional support from the fact that no fluorescence is observed when the molecule is dissolved in dioxane. We have seen that changing the solvent from alcohol to EPA causes a shift of the fluorescence peak by $750 \, cm^{-1}$. Changing to dioxane will cause a further shift. Since the vibrationless $n\pi^*$ and $\pi\pi^*$ singlet levels are rather close in energy their order can be readily reversed by changing the solvent. The singlet $\pi\pi^*$ level will be elevated in energy and the probability of intersystem crossing to the triplet state will be enhanced, causing a complete disappearance of the fluorescence in dioxane.

Though the luminescence pattern of $6KE_2$ seems rather uncommon it is not unique in this regard. Yoshihara and Kearns[2] have observed, what they term an anomalous phosphorescence of fluorenone with $0 - 0$ transition above the origin of the lowest singlet–singlet absorption band. This phosphorescence was observed only for excitation wavelengths which are shorter than 340 nm. They tend to ascribe this emission to a disruption of the C=O band under these conditions of excitation which leads to a fluorene-like or biphenyl-like phosphorescence.

Balke and Becker[3] report a rather unusual excitation spectrum of all-*trans*-Retinal, the quantum yield being greatest near the onset of absorption and decreasing steeply with increasing absorption coefficient. They explain their results by competitive photochemistry (relative to internal conversion) at higher vibrionic levels of the first excited state.

Recently a similarly uncommon excitation spectrum has been reported by Christensen and Kohler[4] for the fluorescence of visual chromophores. This is explained by an *ad hoc* term scheme of the molecule.

It seems therefore, that the conjugation of the carbonyl group to a free electron system creates in certain instances the necessary conditions for an unusual fluorescent behaviour.

REFERENCES

1. A. Weinreb and A. Werner, *Photochem. and Photobiol.*, **15**, 4 43 (1972).
2. K. Yoshihara and D. R. Kearns, *J. Chem. Phys.*, **45**, 1991 (1966).
3. D. E. Balke and R. S. Becker, *J. Amer. Chem. Soc.*, **89**, 506 (1967).
4. R. L. Christensen and B. A. Kohler, *VII International Conference on Photochemistry*, Jerusalem, 1973.

A.2.8. Physical and chemical quenching of $O_2\,^1\Delta_g$ by α-tocopherol (Vitamin E)

B. Stevens and R. D. Small, Jr.

Department of Chemistry, University of South Florida, Tampa, Florida 33620,

S. R. Perez

Xerox Corporation, J. C. Wilson Center for Technology, Rochester, New York 14644, U.S.A.

Although the need for vitamin E in human nutrition is established, there is some disagreement as to whether it serves generally as an antioxidant or as a specific cofactor.[1] The evidence indicates however that its presence in the cell is necessary to control lipid peroxidation which may proceed via a free radical mechanism[2] or involve $O_2\,^1\Delta_g$ as a reactive intermediate which is known to produce hydroperoxides with allylic groups presumably present in the unsaturated lipid fraction.[3] An ideal $O_2\,^1\Delta_g$ scavenger from the biological viewpoint is one which physically quenches the excited oxygen molecule, an example being β-carotene which, it has been suggested,[4] protects the plant from destructive photodynamic effects during photosynthesis. It is therefore of interest to examine the possibility that vitamin E can play a similar protective rôle in animals where, however, a source of $O_2\,^1\Delta_g$ has yet to be established.

The $O_2\,^1\Delta_g$ quenching efficiency of D-α-tocopherol(T) (Eastman) was estimated from its inhibition of the self-sensitized photoperoxidation of rubrene (R) at 546 nm where α-tocopherol does not absorb. The relevant processes are summarized in the following scheme which accommodates inhibition by α-tocopherol quenching of the rubrene triplet state (process 3) which is the $O_2\,^1\Delta_g$ precursor:

$$R + h\nu \rightarrow {}^1R^*$$

$${}^1R^* + O_2\,^3\Sigma \rightarrow {}^3R + O_2\,^3\Sigma \tag{1}$$

$${}^3R + O_2\,^3\Sigma \rightarrow R + O_2\,^1\Delta \tag{2}$$

$${}^3R + T \rightarrow R + T \tag{3}$$

$$R + O_2\,^1\Delta \rightarrow RO_2 \tag{4}$$

$$T + O_2\,^1\Delta \rightarrow \text{quenching} \tag{5}$$

$$O_2\,^1\Delta \rightarrow O_2\,^3\Sigma \tag{6}$$

143

Here RO_2 denotes rubrene peroxide, the sole product of the uninhibited reaction, and process 5 includes both physical quenching and chemical reaction. Under photostationary conditions processes (1)–(6) provide the expression

$$\gamma_0/\gamma = \{1 + k_3[T]/k_2[O_2]\}\{1 + k_5[T]/(k_4[R] + k_6)\} \tag{I}$$

for the reduction in quantum yield of RO_2 formation from γ_0 to γ in the presence of inhibitor at concentration $[T]$; this leads to three distinguishable alternatives:

(i) γ_0/γ is a linear function of $[T]$ but is independent of rubrene concentration if triplet state quenching (process (3) is solely responsible for inhibition;
(ii) γ_0/γ is a non-linear function of $[T]$ if the additive quenches both $O_2{}^1\Delta_g$ and the rubrene triplet state;
(iii) if the linear dependence of γ_0/γ on $[T]$ increases with a reduction in rubrene concentration inhibition may be attributed to additive quenching of $O_2{}^1\Delta_g$ only (process (5)).

Relative quantum yields of rubrene autoperoxidation were estimated as relative initial rates of rubrene consumption in air-saturated benzene or cyclohexane at 25 °C, obtained from continuous recordings of the optical density of rubrene at the actinic wavelength (546 nm) as a function of exposure time in the presence of different concentrations of α-tocopherol. This procedure is based on the relationship

$$\gamma_0/\gamma = \frac{(d[R]_0/dt)}{(d[R]_0/dt)_T} \quad \frac{(I_a)_T}{(I_a)} = \frac{(dOD_0/dt)}{(dOD_0/dt)_T}$$

where subscript T denotes the presence of inhibitor and for a constant initial rubrene concentration the absorbed light intensity $(I_a)_T = (I_a)$.

As shown in Figure 1, γ_0/γ varies directly with concentration of α-tocopherol, and the extent of inhibition $d(\gamma_0/\gamma)/d[T]$ increases with a reduction in rubrene concentration consistent with additive quenching of $O_2{}^1\Delta_g$ (process 5) as the predominant inhibiting process (alternative (iii) above); accordingly equation (I) may be rearranged to

$$1/\beta_T = \{d(\gamma_0/\gamma)/d[T]\}\{1 + [R]/\beta_R\} \tag{II}$$

where β_T $(= k_6/k_5)$ and β_R $(= k_6/k_4)$ are the respective $O_2{}^1\Delta_g$ reactivity indices for α-tocopherol and rubrene respectively. Use of equation (II) and reported values for β_R and k_6 in the solvents used provides the tabulated data for k_5.

The values of k_5 are of the order of magnitude reported for α-tocopherol quenching of $O_2{}^1\Delta_g$ in pyridine[7] $(k_5 = 2.5 \times 10^8 \text{ M}^{-1}\text{ s}^{-1})$ and in *isooctane*[8] $(1.2 \times 10^8 \text{ M}^{-1}\text{ s}^{-1})$ and are higher than those obtained for $O_2{}^1\Delta_g$ addition to rubrene, one of the most reactive $O_2{}^1\Delta_g$ acceptors, but are some two orders of magnitude below the rate constant for $O_2{}^1\Delta_g$ quenching by energy transfer to β-carotene which is virtually diffusion-limited.

The fraction α of $O_2{}^1\Delta_g$ quenching encounters (process (5)) which lead to

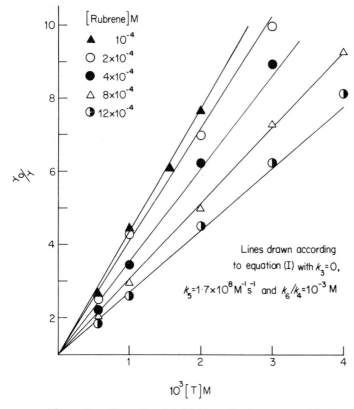

Figure 1   α-Tocopherol inhibition of rubrene peroxidation in
benzene at 25 °C

chemical change may be estimated from spectrophotometric measurements of the rate of α-tocopherol consumption relative to that of an established $O_2{}^1\Delta_g$ acceptor of similar reactivity which exhibits weak absorption in the region of 300 nm and strong absorption at longer wavelengths; 9,10-diphenyl-

Table 1   Reactivity indices β_T and rate constants k_5 for α-tocopherol quenching of $O_2{}^1\Delta_g$
at 25 °C

	Solvent	
	Benzene	Cyclohexane
$\beta_R(M)(= k_6/k_4)^5$	$1\cdot0 \times 10^{-3}$	$1\cdot4 \times 10^{-3}$
$\beta_T(M)(= k_6/k_5)$	$2\cdot7 \pm 0\cdot4 \times 10^{-4}$	$6\cdot8 \pm 0\cdot8 \times 10^{-4}$
$k_6 (s^{-1})^6$	$4\cdot2 \times 10^4$	$5\cdot9 \times 10^4$
$k_5 (M^{-1} s^{-1})$	$1\cdot7 \pm 0\cdot3 \times 10^8$	$0\cdot9 \pm 0\cdot2 \times 10^8$

anthracene(D) proved to be suitable for this purpose. Integration of the relative rate equations[9]

$$\frac{-\,d[T]/dt}{-\,d[D]/dt} = \frac{d[T]}{d[D]} = \frac{\alpha k_5}{k_4'}\frac{[T][O_2{}^1\Delta]}{[D][O_2{}^1\Delta]}$$

yields

$$\frac{\log\,([T_0]/[T])}{\log\,([D_0]/[D])} = \frac{\alpha\beta_D}{\beta_T}$$

where $\beta_D = k_6/k_4'$ and k_4' denotes the rate constant for $O_2{}^1\Delta_g$ addition to 9,10-diphenylanthracene. In practice the absorption spectrum of a solution of α-tocopherol and diphenylanthracene in cyclohexane was recorded before and after several exposures to the Hg line at 365 nm, and the concentration changes computed from changes in optical density at 300 nm (α-tocopherol) and 373 nm (diphenylanthracene) with appropriate corrections for absorption at 300 nm by diphenylanthracene and its peroxide. The mean of six determinations is expressed as

$$\alpha\beta_D/\beta_T = 0.9 \pm 0.1$$

which with $\beta_D = 0.056$ M (in cyclohexane[10]) leads to

$$\beta_T/\alpha = 0.06 \text{ M}$$

or

$$\alpha k_6 = 1.0 \times 10^6 \text{ M}^{-1}\text{ s}^{-1}$$

Accordingly $\alpha \simeq 0.01$ indicating that the α-tocopherol quenching of $O_2{}^1\Delta_g$ (process (5)) may be regarded as essentially (99%) a physical process. In contrast to quenching by β-carotene however this is unlikely to involve the energy transfer process (7)

$$T + O_2{}^1\Delta \rightarrow {}^3T + O_2{}^3\Sigma \qquad (7)$$

insofar as the α-tocopherol triplet state ${}^3T$ is not expected to lie below $O_2{}^1\Delta_g$ at 8000 cm^{-1}. On the other hand the change in spin angular momentum attending the overall process (8)

$$T + O_2{}^1\Delta \rightarrow T + O_2{}^3\Sigma \qquad (8)$$

requires the intervention of either a charge transfer or biradical intermediate; in the former case the quenching rate constant k_5 should be sensitive to solvent polarity, whereas a biradical intermediate may be susceptible to solvent scavenging leading to an increase in the extent of chemical quenching in H-donating solvents.

ACKNOWLEDGEMENT

The continued support of the National Science Foundation under Research Grant No. GP28331X is gratefully acknowledged.

REFERENCES

1. J. Green, *Ann. N.Y. Acad. Sci.*, **203**, 29 (1972).
2. F. D. Gunstone and T. P. Hilditch, *J. Chem. Soc.* (*London*), 836 (1945); *Nature,* **116**, 558 (1950).
3. H. R. Rawls and P. J. van Santen, *Ann. New York Acad. Sci.*, **171**, 135 (1970).
4. C. S. Foote, R. W. Denny, L. Weaver, Y. Chang and J. Peters, *Ann. New York Acad. Sci.*, **171**, 139 (1970).
5. B. Stevens and S. R. Perez, *Mol. Photochem.*, **6**, 1 (1974).
6. P. B. Merkel and D. R. Kearns, *J. Amer. Chem. Soc.*, **94**, 7244 (1972).
7. A. A. Lamola and A. M. Trozzolo, personal communication.
8. C. S. Foote, personal communication.
9. T. Wilson, *J. Amer. Chem. Soc.*, **88**, 2898 (1966).
10. Estimated from β_D in benzene and values of k_6 in benzene and cyclohexane.

Session B
Excited states of DNA and nucleotides

B. Excited state interactions and energy transfer processes in the photochemistry of protein–nucleic acid complexes

Claude Hélène

Centre de Biophysique Moléculaire 45045 Orléans Cédex, France

1 INTRODUCTION

Many studies have been devoted to excited states of nuclcic acids and proteins (and their constituents). Two kinds of information can be obtained from such studies:

(i) they contribute to our knowledge of short-range interactions between constituents (e.g. nucleic acid bases, aromatic amino-acids, ...) inside these macromolecules and therefore lead to a better description of their conformation in solution; and

(ii) they allow us to understand the mechanism of photochemical reactions in these biologically important macromolecules and thus to control (inhibit or sensitize) photoproduct formation.

Investigations of the excited states of oligo- and poly- nucleotides, or nucleoside aggregates and of DNAs of different base compositions have been very useful in understanding how the excitation energy absorbed by purine and pyrimidine bases migrates at the singlet and triplet levels and how it is trapped at the triplet level on thymine residues (for recent reviews see References 1–3). Excited state studies have also permitted, for example, the selective production of photosensitized lesions in nucleic acids and the investigation of their biological significance.[4]

The investigation of the excited states of aromatic amino-acids in proteins has led to a better knowledge of their environment, of their interaction with neighbouring residues, of energy transfer processes and of photochemical inactivation of, e.g., enzymes.[5] Photosensitized reactions involving, e.g., dyes or coenzymes are also helpful in introducing selective photochemical modifications[6] and in identifying neighbouring amino-acid residues.[7]

However, nucleic acids are not isolated inside the cell and are most often linked to proteins. For example, in the chromatin of eukaryotic cells, an important part of the DNA is covered by histones. In ribosomes, proteins and RNAs are associated to give a highly ordered structure. Even in prokaryotic cells, DNA is interacting with proteins such as repressor molecules, RNA polymerase, DNA polymerase, ... The photochemical behaviour of these protein–nucleic acid complexes can be expected to be different from that of the separated macromolecules. Complex formation may, of course, modify the kinetics and the quantum yield of photoproduct formation in both macro-molecules. Some reactions might be inhibited or sensitized. Moreover, new possibilities could be introduced for the production of mixed photoproducts involving constituents of the two macromolecules. Electron and proton transfers as well as migration of excitation energy could be involved in the photochemistry of nucleic acid–protein complexes. It has already been dis-covered that cross-linking of nucleic acids and proteins is responsible for the decrease in the amount of DNA that can be isolated free of proteins in ultraviolet-irradiated bacteria.[8] Covalent bonds between uracil or thymine and amino-acids are formed under ultraviolet irradiation.[9,10] Recently, evidence has been provided for the photoinduced covalent attachment of RNA polymerase to DNA[11] and of lac-repressor to BUdR-containing lac-operator.[12]

I will review here the results that have been obtained in the study of excited-state interactions between amino-acids and nucleic acid constituents. Energy transfer processes will be discussed as well as some photosensitized reactions involving either tryptophan or bases as sensitizers.

2 GROUND-STATE AND EXCITED-STATE INTERACTION BETWEEN NUCLEIC ACID BASES AND AMINO-ACIDS

2.1 Interactions in aggregates formed in frozen aqueous solutions

When an aqueous solution is frozen, solute molecules do not remain dispersed but form aggregates. Aromatic molecules form stacked arrays in which energy transfer may occur over long distances. For example in frozen aqueous solutions of adenosine at 77 K, such a transfer occurs at the triplet level over more than 100 molecules and this requires a good overlap of the electronic clouds of neighbouring molecules.[13,14] Aggregate formation is also responsible for the photochemical dimerization of thymine in frozen aqueous solutions.[15] A review of physicochemical properties of frozen aqueous solutions can be found in Reference 16.

The freezing of an aqueous solution containing two types of aromatic mole-cules induces the formation of mixed aggregates. Intermolecular complexes which are often not observed in the fluid state can thus be formed in the frozen state.[16]

When equimolar aqueous mixtures of nucleosides and aromatic amino-acids are frozen down to 77 K, the luminescence spectrum of the mixture is not the

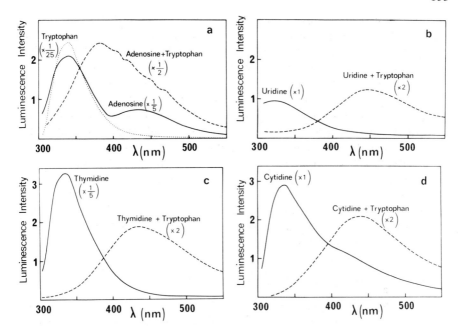

Figure 1 Luminescence spectra at 77 K of frozen aqueous solutions (5 × 10⁻³ M) of tryptophan (····), nucleosides (——) and tryptophan–nucleoside equimolar mixtures (5 × 10⁻³ M each) (----)

superposition of the spectra of the separated molecules. The most striking examples have been provided by nucleoside–tryptophan mixtures.[17] As can be seen in Figure 1, the fluorescence emissions of tryptophan and nucleosides are completely quenched in frozen aqueous mixtures and a new fluorescence band appears at longer wavelengths. The formation of aggregates may often be prevented by addition of salts or organic solvents.[18–20] In a water–propylene glycol (1v/1v) mixture at 77 K, the new fluorescence band is not observed and the fluorescence spectrum is the superposition of the emissions due to tryptophan and nucleoside.

The fluorescence of tyrosine is quenched in the presence of nucleosides.[21] The fluorescence of pyrimidine nucleosides is also quenched, while that of purine nucleosides is not markedly affected. Methylation of the hydroxyl group of tyrosine (p-methoxyphenylalanine) does not introduce marked differences as compared with tyrosine.

In phenylalanine–nucleoside mixtures at 77 K, the fluorescence spectrum of the nucleoside is not affected under neutral conditions (pH 7). A 15 nm red shift of the fluorescence spectrum of protonated cytidine (in acidic medium) is observed in the presence of phenylalanine.[22]

Histidine induces changes in the fluorescence spectra of pyridimine nucleosides in equimolar frozen aqueous solutions at 77 K.[22]

All the above results have been interpreted in terms of charge-transfer interactions in the excited states of complexes formed between aromatic amino-acids and nucleic acid bases. Indole, phenol, benzene and imidazole rings are electron donors, whereas the purine or pyrimidine bases behave as electron acceptors. Complexes are formed prior to excitation in the mixed aggregates formed in frozen aqueous solutions. These complexes result from the stacking of the aromatic molecules. In a few cases,[17] changes in absorption spectra have been observed in frozen equimolar mixtures when compared with the separated components. A new absorption appears at longer wavelengths which reflects the electron donor–acceptor interaction. It is expected that electron transfer would be increased in the excited state as compared to the ground state.[23] As a matter of fact, fluorescence spectra are much more affected than absorption spectra. The charge-transfer character of the new fluorescence band observed in equimolar mixtures is supported by the observation that the spectroscopic effects are larger with pyrimidine than with purine bases as expected from the energy of the lowest empty molecular orbitals.[24] They are also much larger with protonated bases than with neutral ones.

2.2 Interactions in oligopeptide–nucleic acid complexes

The binding of oligopeptides containing aromatic amino-acids to polynucleotides has been recently studied by different methods such as fluorescence, circular dichroism and proton magnetic resonance.[25–31]

The fluorescence quantum yield of Lys-Trp-Lys decreases in the presence of polynucleotides and DNA (taking into account the screening effect of nucleic acid bases in fluorescence measurements). The extent of quenching depends on nucleic acid and peptide concentrations. Preliminary measurements show that the fluorescence lifetime of the Trp residue is not changed, indicating that static quenching is involved. Quantitative analysis of fluorescence data[31] provided evidence for the presence of two types of complexes in equilibrium. In complex I, the fluorescence quantum yield of the Trp residue is identical to that of the free peptide. In complex II, this fluorescence quantum yield is reduced to zero. Proton magnetic resonance studies[29] showed that the indole ring and bases are stacked in complex II, whereas the Trp ring does not interact with bases in complex I. Fluorescence quenching in complex II was therefore ascribed to the stacking interaction as already observed in aggregates (see §2.1). Stacked complex II is favoured in single-stranded polynucleotides and denatured DNA as compared to double-stranded polynucleotides and native DNA. If native DNA is submitted to ultraviolet irradiation, pyrimidine dimers are formed.[32] Regions of the DNA containing these dimers are locally denatured.[33] The probability of stacking interactions involving the indole ring of Lys-Trp-Lys and bases is about 10 times higher in these denatured regions as compared to double-stranded (native) parts. Photosensitized splitting of pyrimidine dimers results from this specific interaction (see below).

Quenching of tyrosine fluorescence is observed in the complexes of the oligopeptide Lys-Tyr-Lys with poly(A) or DNA. Stacking of Tyr with adenine bases is demonstrated by PMR studies of poly(A)–Lys-Tyr-Lys complexes.[29] In

DNA complexes, PMR investigations did not provide any evidence for stacking of Tyr with bases. Stacking of tyrosine with purine and pyrimidine bases leads to a quenching of tyrosine fluorescence. In DNA–Lys-Tyr-Lys complexes, fluorescence quenching must be ascribed to another kind of interaction since no stacking exists. There are several possibilities which should be submitted to an experimental investigation:

(i) Tyrosine might transfer its excitation energy to nucleic acid bases.[34] This would lead to fluorescence quenching since DNA bases are only very weakly fluorescent in aqueous solutions at room temperature.[3]

(ii) Hydrogen-bonding involving the hydroxyl group of tyrosine might also lead to fluorescence quenching as a result of proton transfer in the excited state.[35] Phosphate groups are known to quench the fluorescence of tyrosine[36] but mono-ionized phosphates of nucleic acids are only weak quenchers. It has been recently shown that the tyrosine hydroxyl group can form hydrogen bonds with bases[37] and this hydrogen bonding interaction may result in fluorescence quenching.

In order to test the possible role of the hydroxyl group of tyrosine, the oligopeptide Lys-Tyr-OMe-Lys (where the hydroxyl group is methylated) has been synthesized. This methylation does not prevent the stacking interaction with adenine bases in poly(A) complexes as shown by PMR investigations. The fluorescence of Tyr-OMe is quenched and a new fluorescence band appears at longer wavelengths (Figure 2). This fluorescence is only observed with single-stranded poly(A) and appears to be characteristic of Tyr OMe–Adenine stacked complexes. The quenching of tyrosine fluorescence by nucleosides in mixed

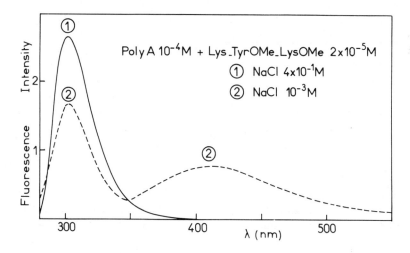

Figure 2   Fluorescence spectrum of Lys-Tyr OMe-Lys (2×10^{-5} M) in the presence of 10^{-4} M poly(A) at two ionic strengths. The complex is fully dissociated in 0·4 M NaCl. Spectra are not corrected for monochromator transmission and photomultiplier sensitivity. Excitation at 275 nm

aggregates has been ascribed to an electron donor–acceptor interaction.[21] The fluorescence band at 410 nm could be the corresponding charge-transfer fluorescence band in Tyr OMe–Adenine complexes. It must be noticed that complex formation is accompanied by small changes in absorption spectra which can be measured by differential spectrophotometry. These changes reflect both the interaction between Tyr or Tyr OMe and Adenine rings and the conformational change (base destacking) of poly(A).[30] Absorption changes induced by the binding of Lys-Tyr-Lys and Lys-Tyr OMe-Lys to poly(A) are identical indicating that the nature of interactions between aromatic rings are very similar in both cases. The new fluorescence band observed with Tyr OMe may result from a higher radiative transition probability if the Tyr OMe–Adenine complex is compared to the Tyr–Adenine complex.

3 ENERGY TRANSFER BETWEEN NUCLEIC ACID BASES AND AMINO-ACIDS

In Table 1 are reported the energies of the lowest excited singlet and triplet states of nucleotides, tyrosine and tryptophan, as determined from luminescence spectra at 77 K in a water–ethylene glycol mixture (1v/1v). Tyrosine should be able to transfer its excitation energy to nucleic acid bases both at the singlet and triplet levels. Tryptophan has the lowest triplet state and bases might transfer their triplet excitation energy to tryptophan. At the singlet level, tryptophan might be either the energy donor (with respect to T, G and C) or the energy acceptor (with respect to A and U). However, it must be kept in mind that these energy values reported in Table 1 are deduced from luminescence measurements in glasses at low temperature (77 K). They could be slightly modified in fluid medium at room temperature, especially in macromolecules where particular local environments might induce energy shifts.

Calculations of the Förster distance R_0 for singlet energy transfer between tyrosine or tryptophan and nucleic acid bases have been made by Montenay–Garestier.[34,38] At room temperature, due to the very low fluorescence quantum yield and the consequently very short lifetime of nucleic acid bases,[3] energy transfer from bases is not expected.

Förster distances for energy transfer from Trp to bases are quite short (< 10 Å), whereas they range between 10 and 17 Å for energy transfer from Tyr to bases. The conclusions are not very much different at low temperature, except that the increase of the fluorescence quantum yield of nucleotides at low temperature in rigid glasses leads to a higher probability of singlet energy transfer from bases to tryptophan ($8 < R_0 < 12$ Å). Förster distances for energy transfer at 77 K from bases to Tyr are less than 10 Å, whereas those for energy transfer from Tyr to bases range from 17 to 21 Å.

In order to provide evidence for energy transfer processes, experiments were performed both in mixed aggregates obtained in frozen aqueous solutions and in oligopeptide–nucleic acid complexes.

Table 1   Energy of the lowest excited singlet and triplet states of nucleotides and aromatic amino-acids determined in a water–ethylene glycol glass at 77 K

	Singlet energy (cm^{-1})	Triplet energy (cm^{-1})
AMP[1]	35,200	26,700
UMP[1]	34,900	—
TMP[1]	34,100	26,300
GMP[1]	34,000	27,200
CMP[1]	33,700	27,900
Tyrosine[34]	37,000	29,000
Tryptophan[39]	34,400	25,000

3.1 Energy transfer between nucleic acid bases and aromatic amino-acids in aggregates

As described above (§2.1), aggregates are formed in frozen aqueous solutions of nucleosides or amino-acids. Energy transfer processes can be conveniently studied in these aggregates by introducing small amounts of an acceptor molecule. It has been shown that stacking interactions between aromatic molecules, e.g. adenosine, favour migration of the excitation energy at the triplet level.[13,14] If small amounts of tryptophan are added to a solution of adenosine, the phosphorescence of adenosine is quenched while the phosphorescence of tryptophan is sensitized as shown in Figure 3.[39] The latter is easily characterized by its vibronic structure and its long lifetime (6.2 s as compared with 2.3 s for adenosine phosphorescence). A plot of phosphorescence intensity versus the ratio (r) of tryptophan and adenosine concentrations shows that about 200 adenosine molecules are able to transfer their excitation energy to tryptophan (Figure 3). No change can be observed in the fluorescence spectrum of adenosine for small values of r (<0.1) for which the tryptophan phosphorescence has already reached its maximum intensity. This indicates that the transfer described above (Figure 3) takes place at the triplet level. Energy migrates between adenine molecules in stacked arrays until it is trapped by a tryptophan molecule.

At higher values of $r (>0.1)$, a new phenomenon appears due to energy transfer at the singlet level from adenosine to adenosine–tryptophan complexes whose fluorescence is red-shifted (as described above (§2.1).[17] When r reaches values close to 1, a third phenomenon occurs due to the heterogeneity of distribution of molecular species in mixed aggregates. The phosphorescence spectrum is always characteristic of tryptophan, but it shifts toward longer wavelengths. This effect is even more marked in the case of 5-methoxytryptophan.[14]

The red-shift of the phosphorescence spectrum has been interpreted as resulting from the formation of short aggregates of the indole derivative intercalated between adenosine aggregates (heterogeneity of distribution).

All nucleosides have been shown to transfer their triplet excitation energy to

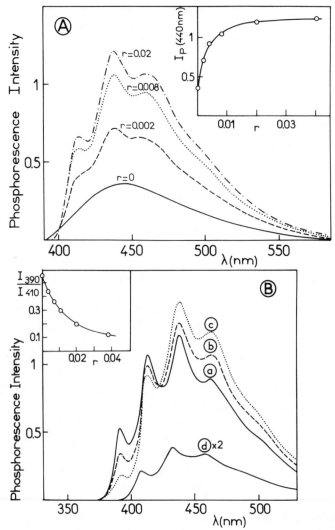

Figure 3 (from Reference 39) (A) Phosphorescence spectra of adenosine–tryptophan mixtures in frozen aqueous solution at 77 K. Adenosine concentration is constant (10^{-3} M). The ratios of tryptophan to adenosine concentrations are indicated on the spectra.

Inset: Change in phosphorescence intensity at 440 nm versus the ratio (r) of tryptophan and adenosine concentrations.

(B) Phosphorescence spectra of poly(A) –Lys-Trp-Lys mixtures in a water–ethylene glycol (1v/1v) glass at 77 K. The concentration of poly(A) is $2 \cdot 9 \times 10^{-4}$ M. The concentration of Lys-Trp-Lys is 0 (a), 2×10^{-6} M (b), $5 \cdot 7 \times 10^{-6}$ M (c). (d) is the phosphorescence spectrum of 2×10^{-6} M Lys-Trp-Lys (intensity multiplied by 2).

Inset: Change in the phosphorescence intensity ratio at two wavelengths (390 and 410 nm) versus the ratio r of Lys-Trp-Lys and poly(A) concentrations

tryptophan in mixed aggregates, although quantitative studies concerned only adenosine–tryptophan mixtures.[39]

Tyrosine is able to transfer its excitation energy both at the singlet and at the triplet level to nucleic acid bases in mixed aggregates.[34,38] Both transfers probably involve migration of the energy between neighbouring tyrosine molecules in the stacked aggregates until trapping takes place at the nucleoside site.

3.2 Energy transfer between nucleic acid bases and aromatic amino-acids in oligopeptide–nucleic acid complexes

Oligopeptides such as Lys-Trp-Lys form complexes with polynucleotides and DNA (see §2.2 and References 25, 27, 29–31). A fraction of the complexes involves a stacking of the tryptophan ring with bases. Such a stacking favours energy transfer at the triplet level which can be easily demonstrated in poly(A)–Lys-Trp-Lys complexes (see Figure 3b).[39] These studies were performed in water–ethylene glycol glasses at 77 K. The fluorescence of the tryptophyl residue is quenched as already observed in fluid medium,[29,31] whereas that of poly(A) is not affected. The phosphorescence of poly(A) is quenched by small amounts of Lys-Trp-Lys, while that of Trp is sensitized (Figure 3b). This result is therefore quite similar to that observed in mixed aggregates (§3.1). The excitation energy in poly(A) is known to be delocalized at the triplet level over approximately 60 bases.[40] One paramagnetic metal cation acting as energy trap will quench the phosphorescence of about 120 bases when bound to poly(A). A quantitative study of the phosphorescence of poly(A)–Lys-Trp-Lys complexes[39] indicates that one peptide is able to quench the phosphorescence of about 70 adenine bases. Analysis of Trp fluorescence quenching in aqueous solution at 275 K led to the conclusion that about 2/3 of the bound peptides have their Trp ring stacked with bases.[31] Thus, the results obtained with poly(A)–Lys-Trp-Lys complexes are quite comparable to those obtained with poly(A)–cation complexes.

The possibility of singlet energy transfer from nucleic acid bases to tryptophan in poly(A) and DNA complexes with Lys-Trp-Lys was investigated at room temperature by comparing the excitation spectra of nucleic acid–Lys-Trp-Lys mixtures at low ionic strength where the complex is formed and at high ionic strength where the complex is dissociated (but the screening effect of the nucleic acid is still present). It must be remembered (see §2.2) that only the bound peptide molecules whose Trp residues do not stack with bases (complex I) emit fluorescence. Our experiments[31] led to the conclusion that bases are not able to transfer their singlet excitation energy to these Trp residues. This does not exclude energy transfer to those Trp residues which are stacked with bases (complex II). Low-temperature measurements described above did provide evidence for efficient triplet energy transfer to Trp residues in complex II.

In DNA–Lys-Trp-Lys complexes, evidence for triplet energy transfer from bases to Trp is also provided by an increase in the Trp phosphorescence-to-fluorescence ratio,[39] but the results are less clear than with poly(A) complexes because the fraction of stacked complexes is much smaller[31] and the extent of energy delocalization in DNA is much smaller than in poly(A).[1]

The fluorescence and phosphorescence of Lys-Tyr-Lys are quenched in the presence of poly(A) and DNA.[41] However, accurate quantitative measurements are prevented by the high screening effect of the nucleic acid. It has been reported that the Tyr phosphorescence of histones is quenched upon binding to DNA.[42] It was concluded that triplet energy transfer from intercalated Tyr residues to DNA bases was responsible for this quenching. However, experiments on stacked Tyr–base complexes (see §2.1) have shown that stacking leads to a quenching of Tyr fluorescence by introducing a new non-radiative deactivation path for the lowest excited singlet state of Tyrosine. A concomitant decrease in the population of the triplet state is observed. Therefore it is unexpected that triplet energy transfer from Tyr to bases would be observed in stacked Tyr–Base complexes because the Tyr triplet state would not be populated. Only if a base is intercalated in stacks of tyrosine molecules can its phosphorescence be sensitized by energy transfer from the tyrosine triplet state because intersystem crossing will take place in Tyr molecules which are not base neighbours.[34]

3.3 Interactions between nucleic acid bases and disulphides

Disulphide bridges are important in maintaining the biologically active form of proteins. They have been shown to act as quenchers of the excited states of tryptophan and tyrosine in model compounds.[43] In proteins they can photochemically react either after direct ultraviolet excitation or after scavenging electrons ejected from Trp residues.[44] As a matter of fact disulphides are very efficient electron scavengers in aqueous solutions.[45] Among them, cystamine has often been used as a radioprotector.[46]

Cystamine, which has two positively charged amino groups at pH 7, binds electrostatically to polynucleotides and DNA.[47] It was therefore of interest to determine whether the disulphide group could interact with purine and pyrimidine bases, and whether energy transfer from bases to disulphide groups could be involved in the photochemistry of protein–nucleic acid complexes. The luminescence of poly(A)–cystamine complexes was therefore investigated in low-temperature glasses.[48] As shown in Figure 4, the fluorescence of poly(A) is slightly red-shifted in the presence of cystamine, but its phosphorescence is completely quenched for small values of the ratio (r) of cystamine and poly(A) concentrations, Hexamethylenediamine has two positively charged amino groups at pH 7 separated by 6 carbon atoms. This molecule binds to poly(A) as does cystamine. This binding induces the same effect as cystamine on the fluorescence spectrum of poly(A), but it does not quench the phosphorescence of adenine bases. Therefore phosphorescence quenching in poly(A)–cystamine complexes is due to the presence of the disulphide group. Similar results were obtained in adenosine aggregates in frozen aqueous solutions. Small amounts of

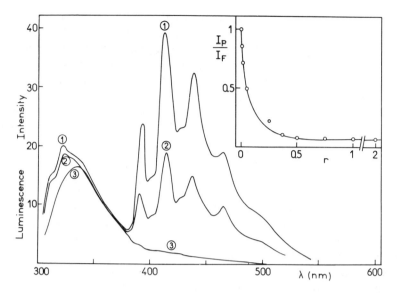

Figure 4 Total luminescence spectrum of poly(A)–cystamine mixtures in a water–ethylene glycol glass at 77 K. The concentration of poly(A) remains constant (10^{-3} M). The ratios (r) of cystamine and poly(A) concentrations are 0 (spectrum 1), 0·05 (spectrum 2) and 1 (spectrum 3).

cystamine quench the phosphorescence of adenosine aggregates, whereas the fluorescence is not affected. This quenching can be accounted for by triplet energy transfer to the disulphide group (whose triplet level is not known) or by an electron transfer mechanism. This last mechanism seems more likely since disulphide bridges are good electron acceptors, but the first mechanism cannot be disproved.

Interactions of cystamine with nucleic acid bases other than adenine has been investigated in aggregates at neutral pH.[48] The fluorescence of thymidine is quenched by cystamine. In cystamine–cytidine frozen aqueous mixtures, a new fluorescence band appears at 380 nm.

All these results show that disulphide bridges might interact with nucleic acid bases and might be involved in trapping the excitation energy absorbed by nucleic acid bases.

4 PHOTOCHEMICAL REACTIONS INVOLVING AMINO-ACIDS AND NUCLEIC ACID COMPONENTS

4.1 Biphotonic processes

Ultraviolet excitation of nucleic acid bases in frozen aqueous solutions leads to photoejection of electrons as a consequence of a biphotonic process involving the triplet state as the intermediate species absorbing the second photon.[49,50] A similar conclusion was reached in the case of aromatic amino-acids.[51,54]

These biphotonic processes might also lead to photosensitized reactions in frozen mixtures. For instance, purine derivatives photosensitize the formation of radicals in ethanol added to the solution before freezing.[51] Deamination of amino-acids (e.g. glycine) can be photosensitized in frozen aqueous solutions by adenosine as well as by tryptophan.[55] This photosensitized reaction involves a biphotonic process taking place at the adenosine molecule and results probably from electron capture by the NH_3^+ group of the amino-acid.

4.2 Photochemical addition of amino-acids to nucleic acid bases

Photochemical reactions of amino-acid side chains containing chemical functions, such as primary or secondary alcohol, carboxylic acid, amine,..., with excited nucleic acid bases could be of importance in the photochemical behaviour of protein–nucleic acid complexes. Photochemical reactions of purine bases in DNA with alcohols[56] and of caffeine with amino-acids[57] have already been demonstrated. The formation of covalent bonds between uracil or thymine and amino acids has also been shown to take place under ultraviolet irradiation.[4,5,58] Some typical photoproducts are shown in Figure 5.

4.3 Photosensitized splitting of thymine dimers by tryptophan derivatives

In frozen aqueous solutions, tryptophan can act as a photosensitizer of the splitting reaction of thymine dimers containing a cyclobutyl ring.[59] Thymine dimers are formed when DNA is submitted to ultraviolet irradiation in aqueous solutions.[32] These thymine dimers can be removed from DNA by a light-requiring enzyme, the photoreactivating enzyme, whose action spectrum extends up to 400 nm.[60] The chromophore absorbing the photoreactivating light as well as the mechanism of splitting are not known yet.

In frozen equimolar aqueous mixtures of tryptophan and thymine dimer (isolated pure cis-syn dimer), the fluorescence of tryptophan is quenched and the splitting of thymine dimers is photosensitized as shown by absorbance and

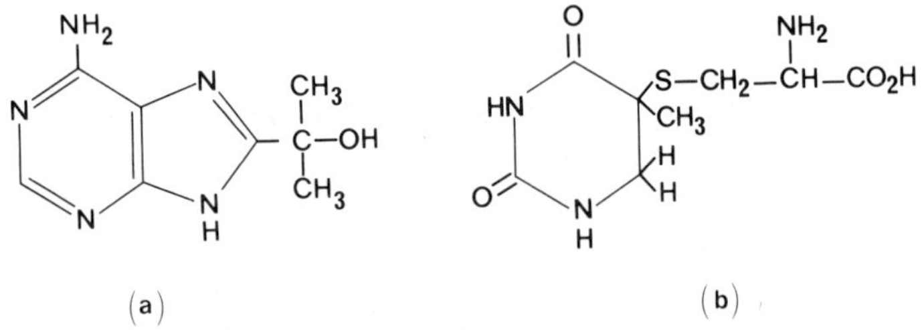

(a)   (b)

Figure 5   Photoproducts obtained by reaction of (a) adenine with isopropanol, and (b) thymine with cystein.

chromatographic measurements.[59] Fluorescence quenching and photosensitized splitting were ascribed to electron transfer from the excited tryptophan molecule in its singlet state to a neighbouring dimer in its ground state. The photosensitized splitting of thymine dimers also occurs in fluid medium and the fluorescence of the indole derivative is quenched in a diffusion-controlled reaction.[4] Electron capture by thymine dimers followed by dissociation into monomers has also been observed in pulse radiolysis experiments.[61]

As already mentioned in §2.2, the tripeptide Lys-Trp-Lys binds selectively to denatured parts of DNA. When thymine dimers are produced in DNA by ultraviolet irradiation, DNA is locally denatured in these dimer-containing regions.[33] If the complex Lys-Trp-Lys–DNA is further irradiated, thymine dimers in DNA are split and regenerate monomers[62] (Table 2). When the complex is dissociated by increasing the ionic strength, the rate of splitting strongly decreases.

Table 2   Photosensitized splitting of thymine dimers in *E. coli* DNA by Lys-Trp-Lys. Thymine dimers were first produced in DNA by ultraviolet irradiation (280 nm) until about 25% of the thymines were dimerized, then Lys-Trp-Lys was added and the mixture was further irradiated as indicated in the first column. Thymine dimer concentration was determined by paper chromatography after acid hydrolysis

Irradiation duration	Lys-Trp-Lys concentration	NaCl concentration	% Split dimers
10 min	10^{-4} M	10^{-3} M	44
10 min	10^{-4} M	10^{-1} M	10·5
2 hr	10^{-4} M	10^{-3} M	59

5 CONCLUSION

Although excited states and energy transfer processes in nucleic acids and proteins have been studied for a long time, only very recently have these problems been investigated in protein–nucleic acid complexes. The results obtained until now show that the photochemical behaviour of these complexes may be much more complicated than the already complex behaviour of each individual macromolecule. New interactions are expected to take place in the excited state as well as in the ground state at the recognition site of both molecules. It has been shown, for example, that aromatic amino-acids may form electron donor–acceptor complexes with nucleic acid bases. Excited-state processes such as electron or proton transfer as well as migration of the excitation energy, might be involved in the photochemistry of protein–nucleic acid complexes.

The formation of covalent bonds between amino-acid residues of proteins and nucleic acid bases might represent a new way of getting some insight into the origin of specific interactions between proteins and nucleic acids. The selective recognition of base sequences in nucleic acids by proteins (or enzymes) is involved at every step of genetic expression (and its regulation). This selective recognition rests upon a structural complementarity and upon specific interactions between the amino-acids of the protein and the recognized bases of the nucleic acid. The photochemical formation—direct or sensitized—of covalent links between the two macromolecules and the analysis of the chemical nature of these links might allow a better knowledge of the nature of the interacting components in the protein–nucleic acid recognition site.

ACKNOWLEDGEMENTS

I wish to thank all my collaborators who contributed to an important part of the work described in this review and especially Drs T. Montenay-Garestier, M. Charlier, F. Brun and J. J. Toulmé for their investigation of the properties of aggregates, of energy transfer processes, of photosensitized reactions and of the fluorescence of peptide–nucleic acid complexes.

Part of this work has been supported by the Délégation Générale à la Recherche Scientifique et Technique (contract n°72–7–0498) and the Commissariat à l'Energie Atomique (contract n° SA 2086).

I wish to thank the Editor of *Photochemistry and Photobiology* for the permission to reproduce Figure 3.

REFERENCES

1. J. Eisinger and A. A. Lamola, *Excited states of proteins and nucleic acids*, pp. 107–198, (eds. Steiner and Weinry), Plenum Press, N.Y. 1971.
2. C. Hélène, *Physico-chemical Properties of Nucleic Acids,* Vol. 1, pp. 119–142 (ed. J. Duchesne), Academic Press, 1973.
3. P. Vigny and M. Duquesne, Paper B1, this Volume.
4. A. A. Lamola, *Mol. Photochem.*, **4**, 107 (1972).
5. J. W. Longworth, in Reference 1, pp. 319–484.
6. L. I. Grossweiner and A. G. Kepka, *Photochem. Photobiol.*, **16**, 305–314 (1972).
7. G. Jori, G. Gennari, C. Toniolo and E. Scoffone, *J. Mol. Biol.*, **59**, 151–168 (1971).
8. K. C. Smith, *Photochem. Photobiol.*, **3**, 415–427 (1969).
9. K. C. Smith, *Biochem. Biophys. Res. Comm.*, **34**, 354–357 (1969).
10. (a) K. C. Smith and R. T. Aplin, *Biochemistry*, **5**, 2125–2130 (1966); (b) K. C. Smith, *Biochem. Biophys. Res. Com.*, **39**, 1011–1016 (1970).
11. A. Markowitz, *Biochim. Biophys. Acta*, **281**, 522–534 (1972).
12. S. Lin and A. D. Riggs, *Proc. Nat. Acad. Sci. U.S.* (in press).
13. C. Hélène and Th. Montenay-Garestier, *Chem. Phys. Lett.* **2**, 25–28 (1968).
14. Th. Montenay-Garestier and C. Hélène, *J. Chim. Phys.*, **70**, 1391–1399 (1973).
15. S. Y. Wang, *Fed. Proceed.*, **24**, S-71 (1965).
16. Th. Montenay-Garestier, M. Charlier and C. Hélène, *Photochemistry and Photobiology of Nucleic Acids* (ed. S. Y. Wang), Academic Press, N.Y. (in press).

17. T. Montenay-Garestier and C. Hélène, *Biochemistry*, **10**, 300–306 (1971).
18. C. Hélène, R. Santus and M. Ptak, *C.R. Acad. Sci. Paris*, **262C**, 1349–1352 (1966).
19. W. Kleinwächter, *Coll. Czech. Chem. Comm.*, **37**, 1622 (1972).
20. T. Montenay-Garestier, *J. Chim. Phys.*, **70**, 1379–1384 (1973).
21. C. Hélène, T. Montenay-Garestier and J. L. Dimicoli, *Biochim. Biophys. Acta*, **254**, 349 (1971).
22. T. Montenay-Garestier and C. Hélène, *J. Chim. Phys.*, **70**, 1385–1390 (1973).
23. N. Mataga and Y. Murata, *J. Am. Chem. Soc.*, **91**, 3144 (1969).
24. B. Pullman and A. Pullman, *Quantum Biochemistry*, (John Wiley, N.Y., 1963.
25. C. Hélène and J. L. Dimicoli, *FEBS Letters*, **26**, 6–10 (1972).
26. E. J. Gabbay, K. Sanford and C. S. Baxter, *Biochemistry*, **11**, 3429–3435 (1972); *Biochemistry*, **12**, 4021–4029 (1973).
27. C. Hélène, J. L. Dimicoli, H. N. Borazan, M. Durand, J. C. Maurizot and J. J. Toulme, Jerusalem Symposia on Quantum Chemistry and Biochemistry, Vol V, *Conformation of Biological Molecules and Polymers*, pp. 361–380, Jerusalem, 1973.
28. R. K. Novak and J. Donhal, *Nature N.B.*, **243**, 155–157 (1973).
29. J. L. Dimicoli and C. Hélène, *Biochemistry*, **13**, 714–730 (1974).
30. M. Durand, J. C. Maurizot, H. N. Borazan and C. Hélène, *Biochemistry*, **14**, 563–570 (1975).
31. F. Brun, J. J. Toulme and C. Hélène, *Biochemistry*, **14**, 558–563 (1975).
32. R. B. Setlow, *Photochem. Photobiol.*, **7**, 643–649 (1968).
33. R. B. Setlow and W. L. Carrier, *Photochem. Photobiol.*, **2**, 49–57 (1963).
34. T. Montenay-Garestier, Paper B5, this Volume.
35. J. Feitelson, *J. Phys. Chem.*, **68**, 391 (1964).
36. R. F. Chen and P. F. Cohen, *Arch. Biochem. Biophys.*, **114**, 514–522 (1966).
37. H. Sellini, J. C. Maurizot, J. L. Dimicoli and C. Hélène, *FEBS Letters*, **30**, 219–224 (1973).
38. T. Montenay-Garestier, *Photochem. Photobiol.* (in press).
39. C. Hélène, *Photochem. Photobiol.*, **18**, 255–262 (1973).
40. J. Eisinger and R. G. Shulman, *Proc. Nat. Acad. Sci. U.S.*, **55**, 1387–1391 (1966).
41. C. Hélène, unpublished results.
42. A. Matsuyama and C. Nagata, *Biochim. Biophys. Acta*, **224**, 588 (1970).
43. R. W. Cowgill, *Biochim. Biophys. Acta*, **140**, 37–44 (1967).
44. L. I. Grossweiner and Y. Usui, *Photochem. Photobiol.*, **11**, 53 (1969).
45. M. Z. Hoffman and E. Hayon, *J. Am. Chem. Soc.*, **94**, 7950–7959 (1972).
46. G. Kollmann, N. Castel and B. Shapiro, *Int. J. Radiat. Biol.*, **18**, 587–594 (1970).
47. E. Jellum, *Int. J. Radiat. Biol.*, **9**, 185–200 (1965).
48. T. Montenay-Garestier, F. Brun and C. Helene (to be published).
49. C. Hélène, R. Santus and P. Douzou, *Photochem. Photobiol.*, **5**, 127–133 (1966).
50. M. D. Sevilla, *J. Phys. Chem.*, **75**, 626–631 (1971).
51. R. Santus, C. Hélène and M. Ptak, *Photochem. Photobiol.*, **7**, 341–360 (1968).
52. Y. A. Vladimirov, D. I. Roshchupkin and E. E. Fesenko, *Photochem. Photobiol.*, **11**, 227 (1970).
53. (a) R. Santus, A. Hélène, C. Hélène and M. Ptak, *J. Phys. Chem.*, **74**, 550–561 (1970); (b) M. Aubailly, M. Bazin and R. Santus, *Chem. Phys. Letters*, **12**, 70 (1971).
54. J. Moan and H. B. Steen, *J. Phys. Chem.*, **75**, 2887 (1971).
55. R. Santus, C. Hélène and M. Ptak, *C.R. Acad. Sci.*, **262D**, 2077–2080 (1966).
56. R. Ben-Ishai, M. Green, E. Graff, D. Elad, H. Steinmaus and J. Salomon, *Photochem. Photobiol.*, **17**, 155–167 (1973).
57. D. Elad and I. Rosenthal, *Chem. Comm.*, 905–906 (1969).
58. A. J. Varghese, *Biochem. Biophys. Res. Comm.*, **51**, 858–862 (1963).
59. C. Hélène and M. Charlier, *Biochem. Biophys. Res. Comm.*, **43**, 252 (1971).
60. R. F. Beers, R. M. Herriott and R. C. Tilghman, *Molecular and Cellular Repair Mechanisms*, pp. 53–103, Johns Hopkins University Press, 1972.

61. R. Santus, C. Hélène, J. Ovadia and L. I. Grossweiner, *Photochem. Photobiol.*, **16**, 65 (1972).
62. M. Charlier and C. Hélène, *Photochem. Photobiol.*, **21**, 31–37 (1975).
63. H. Steinmaus, D. Elad and R. Ben-Ishai, *Biochem. Biophys. Res. Comm.*, **40**, 1021–1025 (1970).

B.1. On the fluorescence properties of nucleotides and polynucleotides at room temperature

P. Vigny and M. Duquesne

Institut du Radium—Laboratoire Curie†
11, rue Pierre et Marie-Curie—75231 Paris—Cedex 05 France

SUMMARY

The previously unmeasurable fluorescence of DNA components in neutral aqueous solutions at room temperature has been investigated using a spectrophotofluorometer built in the Laboratory. The fluorescence properties of the five common nucleotides under the above conditions are presented and compared with results obtained in ethylene glycol at 77 K. The fluorescence spectra of some di- and polynucleotides are examined and discussed in terms of interactions between monomers, leading to the conclusion that intramolecular excimers can be formed at room temperature, this process competing with singlet emission.

1 INTRODUCTION

Fluorescence spectroscopy of nucleotides and polynucleotides is of importance on account of the information that can be obtained concerning their ground state properties and because their excited electronic states may be involved in such biological functions as DNA photodamage. Until recently, except for experiments at extreme pH values, under which conditions some nucleic bases are known to exhibit fluorescence, very few data were available on the excited states of such molecules at room temperature because the systems are quenched to a high degree. In this respect, they differ from many aromatic compounds for which internal conversion from the first excited singlet state is unimportant. Most work has been carried out at 77 K in glasses where quantum yields are of the order of magnitude of 10^{-1} or 10^{-2}, which enables a

† Laboratoire Associé au C.N.R.S/N⁰ 198.

normal recording of luminescence spectra. The major question which has been constantly raised[1,2] is whether results obtained in a rigid medium at 77 K can be extrapolated to fluid aqueous solutions at room temperature. This is particularly important for the photochemistry of DNA under biological conditions and for the use of intrinsic probes to obtain structural information. Two types of approach have been used to get information on these excited states at room temperature:

(i) The first approach uses energy transfer to added quenchers. Lamola and Eisinger[3] showed that the europium ion can scavenge the excited states of nucleotides and emit in the red, but they were not able to get informations on polynucleotides (due to the complicated kinetics).

(ii) A direct approach has been chosen in this Laboratory, in trying to detect fluorescence from nucleic bases, nucleosides and nucleotides,[4] and some di- and poly-nucleotides[5] at room temperature and in neutral solutions. Daniels and Hauswirth[6] have done the same in the case of nucleic bases. As shown in Figure 1, relative fluorescence quantum yields under such conditions are of the order of magnitude of 10^{-4} or less.

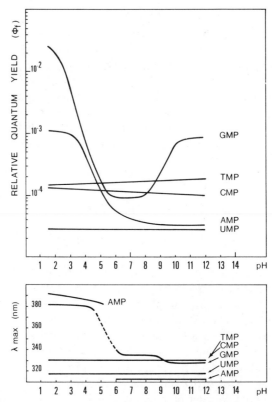

Figure 1   Room-temperature relative fluorescence quantum yields (ϕ_F) and maximum intensity wavelengths (λ max) of nucleotides as a function of pH

2 EXPERIMENTAL

Fluorescence spectra were obtained with a spectrophotofluorometer, specially built in the Laboratory for the detection of very weak fluorescences from biological molecules.[7] It uses the photon-counting method which allows an improvement of the signal-to-noise ratio by increasing the counting time.[8] Moreover the sample cell is mounted on a table which can be displaced in the X- and Y-directions, permitting the investigation of absorption and self-absorption processes within the cell. Corrected fluorescence and fluorescence excitation spectra can thus be obtained even in the case of non-dilute solutions, which appears to be of interest when quantum yields are low. As an example, Figure 2 shows the fluorescence spectrum of Adenine ($\phi_F = 2.6 \times 10^{-4}$) as it is recorded on the multiscale analyser, at concentration $C = 10^{-4}$ M. Each point

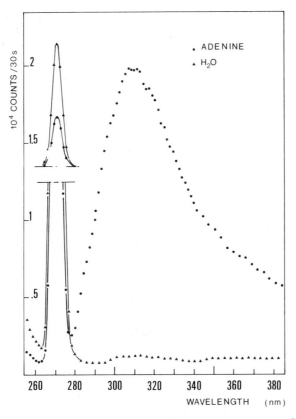

Figure 2   Room-temperature fluorescence spectrum of Adenine in water ($\phi_F = 2.6 \times 10^{-4}$). (Experimental conditions: $C = 10^{-4}$ M, $\lambda_0 = 248$ nm, $\Delta\lambda_0 = 4.2$ nm, $\Delta\lambda_i = 3.2$ nm. Counting time = 30 s at each point. The Raman scattering has been divided by a factor of ten in the left upper part of the figure)

corresponds to a number of counts during 30 seconds at a given emission wavelength. When the dilute solution approximation was not allowed, solvent noise was corrected to obtain its exact value in the presence of the fluorescent molecules, as described elsewhere.[7]

Most of the products used in this study were commercial ones. Bases, nucleosides, nucleotides and dinucleotides were purchased from Calbiochem (A grade), and polynucleotides from Miles Laboratories. Suprasil quartz cells were carefully selected, and the water used was triple distilled from $KMnO_4$ and $Ba(OH)_2$ solution. The concentration of the products was 10^{-4} to 10^{-3} M (in monomer).

3 NUCLEOTIDES

Typical room-temperature corrected fluorescence spectra are given for Thymine (Figure 3) and Adenine (Figure 4) and for their nucleosides and nucleo-

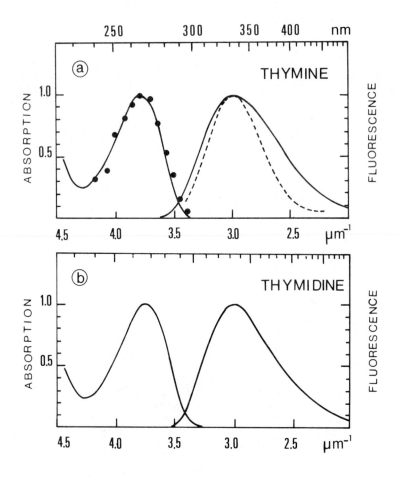

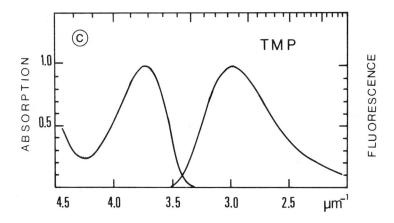

Figure 3 Absorption and fluorescence spectra of Thymine (a), Thymidine (b), TMP (c) at 300 K. (a) Dotted line (····): fluorescence spectrum at 77 K; dots (●): fluorescence excitation spectrum at 300 K. (Experimental conditions: $C = 10^{-4}$ M, $\lambda_0 = 248$ nm, $\Delta\lambda_0 = 4\cdot2$ nm, $\Delta\lambda_i = 3\cdot2$ nm)

tides. As compared to low-temperature spectra, they become broad and structureless while the fluorescence yields decrease: the TMP molecule, for example, has a room-temperature fluorescence quantum yield of $1\cdot2 \times 10^{-4}$ compared with 10^{-1} at 77 K. A surprising feature is that no important red shifts can be observed when going from rigid samples to fluid solutions. The fluorescence spectra of a given base, of its nucleoside and of its nucleotide are quite similar in profile. Their quantum yields are of the same order of magnitude for the pyrimidines, but they are smaller for the purine nucleosides and nucleotides than for the purine bases themselves.[4,9]

The fluorescence excitation spectra, which are of great interest, are difficult to determine due to the low quantum yields. Such a spectrum is shown in Figure

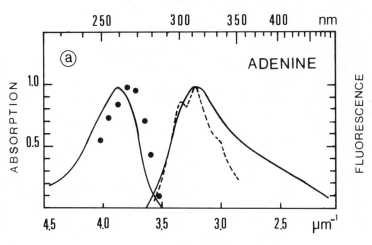

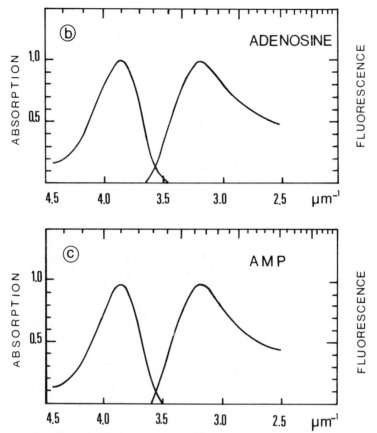

Figure 4 Absorption and fluorescence spectra of Adenine (a), Adenosine (b), AMP (c) at 300 K. (a) Dotted line ($\cdots$): fluorescence spectrum at 77 K; dots ($\bullet$): fluorescence excitation spectrum at 300 K. (Experimental conditions: $C = 10^{-4}$ M $\lambda_0 = 248$ nm, $\Delta\lambda_0 = 4\cdot2$ nm, $\Delta\lambda_i = 3\cdot2$ nm)

3(a) for Thymine (represented by dots). As already mentioned[7] it appears to be superimposable on the absorption spectrum which is not in agreement with previous results.[6] This is however not a general rule as shown in Figure 4(a) for Adenine. The observed shift between absorption and excitation spectra may be taken as an implication that there is a variation in the emission yield as a function of excitation energy.

A summary of the room-temperature fluorescence properties of nucleotides is given in Table 1, where they are compared to low-temperature data and to room-temperature data obtained by the Europium method.[3] From this table it appears that the $0 - 0$ singlet energies are in the order AMP > UMP > TMP > CMP > GMP which is different from that observed at 77 K.[10] This would mean that the relative excited states energies of nucleotides may be sufficiently altered by temperature to affect the direction of energy transfer.

Table 1 Comparison between room-temperature and low-temperature fluorescence properties of nucleotides at pH 7. (Low-temperature data are from References 10 and 2. Room-temperature 0 – 0 singlet energies have been determined by the absorption–emission intersection. Room-temperature fluorescence quantum yields have been estimated with reference to Adenine at 300 K [$\Phi_F = 2 \cdot 6 \times 10^{-4}$].[6] Calculated singlet lifetimes have been obtained by using Strickler and Berg's relation[16] and assuming that the entire low-energy absorption band is responsible for emission. They must therefore be considered as a lower limit)

Compound	λ max (nm)		0 – 0 energy (cm^{-1})		Quantum yield (Φ_F)		Calculated singlet lifetimes at 300 K (s)	
	77 K	300 K	77 K	300 K	77 K	300 K	Eu^{3+} method[3]	This work
AMP	313	312 $\pm$ 2	35,200	35,550	0·01	$0 \cdot 5 \times 10^{-4}$	$1 \cdot 3 \times 10^{-12}$	$0 \cdot 3 \times 10^{-12}$
GMP	323	340 $\pm$ 5	34,000	33,800	0·13	$0 \cdot 8 \times 10^{-4}$	$4 \cdot 4 \times 10^{-12}$	$0 \cdot 5 \times 10^{-12}$
CMP	322	330 $\pm$ 3	33,700	34,000	0·05	$1 \cdot 2 \times 10^{-4}$	$3 \cdot 6 \times 10^{-12}$	$1 \cdot 0 \times 10^{-12}$
TMP	322	330 $\pm$ 3	34,100	34,100	0·16	$1 \cdot 2 \times 10^{-4}$	$1 \cdot 8 \times 10^{-11}$	$0 \cdot 8 \times 10^{-12}$
UMP	314	320 $\pm$ 5	34,900	35,300	0·01	$0 \cdot 3 \times 10^{-4}$	$2 \cdot 3 \times 10^{-11}$	$0 \cdot 3 \times 10^{-12}$

4 POLYNUCLEOTIDES

Two questions about polynucleotides and nucleic acids are yet unresolved. The first one is to know whether there is, or is not, energy migration between neighbouring bases under physiological conditions. It would be also of interest to know if fluorescence from di- and poly-nucleotides could originate from excimers. Such excited complexes between two stacked bases which give an emission at longer wavelength, the energy being lowered by the interaction, are very often found at 77 K. It is still too early to answer the first important question. On the other hand, it is now established that the room-temperature emission from some di- and poly-nucleotides comes from excimers.[5,14] Figure 5 shows the fluorescence spectra of Adenine derivatives at pH 7 (from the base to the polynucleotide). Beside the monomer contribution, a new broad fluorescence emission appears in ApA with maximum intensity at 420 nm and also in

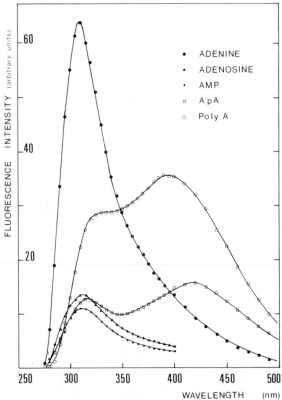

Figure 5   The emission spectra of Adenine, Adenosine, Adenylic acid (AMP) Adenylyl 3'-5' Adenosine (ApA) and polyadenylic acid (poly A) at room temperature and pH 7. (The spectra are corrected and normalized so that their relative fluorescence quantum yield is proportional to their surface)

poly A with a maximum intensity at 395 nm. According to what is known about excimer formation, such an emission should be more or less intense depending on the stacking of the two bases of the dinucleotide. At room temperature ApA is supposed to be in a stacked conformation.[11] Moreover, this stacking is very temperature dependent and becomes less important when temperature is increased.[12] Figure 6 shows that the intensity of the second fluorescence

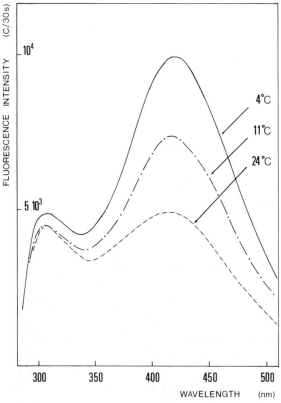

Figure 6   The emission spectra of Adenylyl 3'-5' Adenosine
(ApA) at various temperatures (uncorrected spectra)

emission is effectively temperature dependent and it is notably increased when the temperature is lowered from 24 °C to 4 °C. The fluorescence of the acidic form of poly C, already described by Favre[13] might be also interpreted in terms of excimer emission.[14] Figure 7 shows the fluorescence of the two forms of poly C. At neutral pH, poly C is known to be a random coil. Its fluorescence is quite monomer-like, with a small contribution of excimer above 400 nm. Things are quite different at pH 4 where poly C is known to form a double-stranded helical structure. Because of this structure in which the cytidine rings are stacked, excimer formation is very important leading to the increased emission at 420 nm.

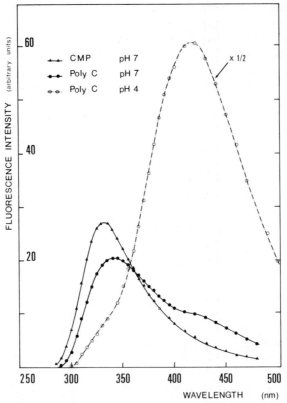

Figure 7 The emission spectra of cytidylic acid (CMP) at pH 7 and of polycytidylic acid (Poly C) at pH 7 and pH 4. (The spectra are corrected and normalized so that their relative fluorescence quantum yield is proportional to their surface)

5 CONCLUSION

Recent improvements of the fluorescence techniques have permitted direct observation of the fluorescence of nucleotides and polynucleotides at room temperature and neutral pH. The major feature is, of course, the quenching of the excited singlet states of nucleotides which leads to fluorescence quantum yields of 10^{-4} and less. At the present time, we do not have a good explanation of the room-temperature quenching. There were some indications that going to a liquid solvent might be the most important factor, rather than the increased temperature.[2] But it is surprising that the emission spectra observed in fluid solutions are not significantly red shifted compared to the spectra in rigid samples. Comparison of the $0 - 0$ singlet energies between low and room temperature shows that the energies of the excited states of mononucleotides may be altered by temperature, and that extrapolation from low-temperature measurements to processes occurring under biological conditions requires great care.

Concerning polynucleotides, it is interesting to notice that the quantum yields are not lower than those of mononucleotides. Here again, in analogy to what happens at low temperature, it was expected to find the opposite.[15] But the most interesting feature is that, in some di- and poly-nucleotides, excimer emission can compete with singlet emission and, in some cases, predominate. This emission can lead to structural information on the polynucleotide in solution, but it is also important since it is known that excimer formation represents an excitation energy trap.

ACKNOWLEDGEMENTS

The authors are indebted to J. Tanguy and A. Bernheim for their technical assistance.

REFERENCES

1. J. Eisinger, A. A. Lamola, J. W. Longworth and W. B. Gratzer, *Nature*, **226**, 113 (1970).
2. J. Eisinger and M. Gueron in *Principles of Chemistry of Nucleic Acids* ed. P. O. P. T'so (Academic Press, New York) 1974.
3. A. A. Lamola and J. Eisinger, *Biochim. Biophys. Acta*, **240**, 313 (1971).
4. P. Vigny, *C.R. Acad. Sc. Paris*, **D272**, 2247 (1971); P. Vigny, *C.R. Acad. Sc. Paris*, **D272**, 3206 (1971).
5. P. Vigny, *C.R. Acad. Sc. Paris*, **D277**, 1941 (1973).
6. W. Hauswirth and M. Daniels, *Photochem. Photobiol.*, **13**, 157 (1971); M. Daniels and W. Hauswirth, *Science*, **171**, 675 (1971).
7. P. Vigny and M. Duquesne, *Photochem. Photobiol.*, **20**, 15 (1974).
8. M. Duquesne, P. Vigny and N. Gabillat, *Photochem. Photobiol.*, **11**, 519 (1970).
9. P. Vigny (in preparation).
10. M. Gueron, J. Eisinger and R. G. Shulman, *J. Chem. Phys.*, **47**, 4077 (1967).
11. M. Warshaw and I. Tinoco, *J. Mol. Biol.*, **19**, 29 (1966).
12. J. M. Massoulie and A. M. Michelson, *C.R. Acad. Sc. Paris*, **D259**, 2923 (1964).
13. A. Favre, *F.E.B.S. Letters*, **22**, 280 (1972).
14. P. Vigny and A. Favre, *Photochem. Photobiol.*, **20**, 3457 (1974).
15. W. Kleinwächter, *Studia Biophysica*, **33**, 1 (1972).
16. S. J. Strickler and R. A. Berg, *J. Chem. Phys.*, **37**, 814 (1962).
17. P. I. Hønnas and H. B. Steen, *Photochem. Photobiol.*, **11**, 67 (1970).

B.2. Fluorescent probes of chromosome structure

Samuel A. Latt† and Scott Brodie‡

†Clinical Genetics Division, Children's Hospital Medical Center,
and Department of Pediatrics, Harvard Medical School, Boston,
Massachusetts, and ‡Department of Chemistry, Wesleyan University,
Middletown, Connecticut.

SUMMARY

Fluorescent dyes can serve as spectroscopic probes of chromosome structure. Examination of the excited states of complexes of dyes such as quinacrine and 33258 Hoechst with DNA aids understanding of the environmental features of chromosomes stained with these dyes.

Chromosomes stained with quinacrine exhibit a characteristic banded fluorescence pattern. This pattern may relate to chromosomal DNA base composition, since the fluorescence efficiency of quinacrine–DNA complexes decreases as the DNA guanine–cytosine content increases. We present evidence suggesting that guanine–cytosine base pairs in DNA quench quinacrine fluorescence in at least two ways. The major effect, evident at low dye/DNA ratios, is a quenching of quinacrine fluorescence with a less than proportionate reduction in dye fluorescence lifetime, perhaps reflecting events occurring prior to relaxation of excited dye molecules to the pre-fluorescent first excited singlet state. At higher dye/DNA ratios, energy transfer between bound quinacrine molecules apparently occurs, converting dyes bound at quenching sites into energy sinks. Both processes exhibit a marked non-linear dependence on DNA base composition, serving to highlight the fluorescence of chromosome regions containing stretches of DNA with a very high adenine–thymine content.

The fluorescence of 33258 Hoechst is quenched by incorporation of the base analogue 5-bromodeoxyuridine into DNA dye binding sites. This effect can be used for microfluorometric analysis of DNA replication in chromosomes and for fluorometric detection of process related to chromosome stability, such as sister chromatid exchanges. Characterization of the 33258 Hoechst–5-bromodeoxyuridine interaction suggests additional uses of this dye for studying chromosome structure.

1 INTRODUCTION

Fluorescent staining techniques, using the dyes quinacrine and quinacrine mustard, have demonstrated that metaphase chromosomes possess a longitudinal array of regions of differing fluorescent intensities.[1-5] Such observations have allowed unambiguous chromosome identification and revealed the existence of an elaborate underlying organization in these chromosomes. Interest in the chemical features responsible for the staining patterns has stimulated research into the manner in which the fluorescence of these dyes reflects their environment.[6,7]

An alternate cytochemical approach to chromosome analysis involves selection of a dye exhibiting fluorescence sensitive to known chemical features. For example, the fluorescence of complexes of the dye 33258 Hoechst† with DNA is reduced if the thymidine analogue BrdU is incorporated into the DNA.[8] This phenomenon has recently been used for the microfluorometric detection of DNA synthesis in metaphase chromosomes.[8]

The present report deals with fluorescence properties of quinacrine–DNA and 33258 Hoechst–DNA complexes which may relate to the fluorescent chromosome staining patterns obtained with these dyes.

2 QUINACRINE AND QUINACRINE MUSTARD

Early studies of chromosome fluorescence utilized quinacrine mustard,[1] a bifunctional alkylating agent derived from quinacrine. Subsequently, it was established that quinacrine exhibits the same staining properties.[2] The major difference between the two compounds appears to centre on the relative stabilities of their complexes with DNA.[9]

Fluorescence properties of quinacrine–DNA complexes have been studied in attempts to determine the mechanism for quinacrine or quinacrine mustard staining patterns. Upon complexing with DNA, the spectra of these dyes undergo characteristic shifts. Most notable is the bathochromic shift in the long-wavelength absorption band, also observable in the fluorescence excitation spectrum.[7] Microspectrofluorometric measurements indicate that the excitation spectrum of quinacrine mustard exhibits a shift upon binding to interphase or metaphase chromosomes similar to that following binding to DNA (Figure 1). This result supports the hypothesis that the fluorescence of chromosomes

† *Abbreviations:*
33258 Hoechst: (2-[2-(4-Hydroxyphenyl)-6-benzimidazolyl]-6-(1-methyl-4-piperazyl)-benzimidazol·3HCl)
BrdU: 5-Bromodeoxyuridine
Quinacrine: 2-Methoxy-6-chloro-9-[1-methyl-4-diethylaminobutylamino] acridine . 2HCl
Quinacrine Mustard: 2-Methoxy-6-chloro-9-[4-bis(2-chloroethyl) amino-1-methylbutylamino] acridine . 2HCl
Hepes: N-2-Hydroxyethylpiperazine-N′-2-ethane sulphonic acid

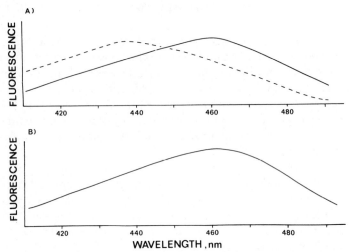

Figure 1   Microfluorometric excitation spectra of quinacrine mustard; effect of interaction with calf thymus DNA or a human leukocyte nucleus. Spectra were obtained with a Leitz MPV II microspectrofluorometer. Excitation from a 450 watt xenon source was dispersed by a grating monochromator (bandwidth 20 nm) deflected onto the sample with incident illumination employing a BG-12 bandpass filter, a TK 495 dichroic mirror, a K 495 suppression filter, and a 100 × achromat objective. Fluorescence emission collected by this objective traversed a 530 nm barrier filter before photometric detection. (A) Samples consisting of a concentrated (10^{-3} M) solution of quinacrine mustard (---) or an aggregate formed from 2×10^{-4} M quinacrine mustard and 4×10^{-3} M calf thymus DNA (——) (both in 0·01 M NaCl, 0·005 M Hepes, pH 7) were mounted on a microscopic slide under a standard glass coverslip. (B) The sample was a nucleus of a cell stained with a 10^{-4} M solution of quinacrine mustard and mounted in 0·16 M sodium phosphate, 0·04 M sodium citrate, pH 7. Spectra have not been corrected for the variation in instrumental excitation efficiency with wavelength. Amplitudes have been normalized to comparable magnitudes

interacting with the quinacrine chromophore arises from dye molecules complexed with chromosomal DNA.

The fluorescence quantum yields of quinacrine–DNA complexes increase with the adenine–thymine (A–T) content of the DNA.[6,7] This phenomenon has been invoked as the basis for quinacrine staining patterns,[6,7,10] although the exact correspondence between quinacrine fluorescence efficiency and chromosomal DNA composition is unclear. For example, the dependence of quinacrine fluorescence on DNA A–T content is distinctly non-linear (Table 1).[6,7,11] The presence of a guanine–cytosine (G–C) base pair on either side of an intercalated quinacrine molecule appears sufficient for quenching of its fluorescence.[14]

DNA base composition can in principle affect quinacrine fluorescence both before and after relaxation of excited dye molecules to their immediately pre-fluorescent state. As a first approximation, the relative efficiencies of pre-

Table 1   Fluorescence parameters of quinacrine–DNA complexes

DNA source[a]	Base composition[12] (%A–T pairs)	Quantum[b] yield(Q)	Lifetime[c] (τ)	T_2/T_1[d]
Free dye	—	0·08	3·5	—
Poly (dA-dT)	100	0·48	10·7	1·06
Cl. perfringens	70	0·13	8·6	0·54
Calf thymus	56	0·06	7·8	0·26
E. coli	51	0·05	7·5	0·24
M. lysodeikticus	28	0·005	4·5	<0·09

[a] DNA phosphate/quinacrine ratios were all in excess of 87. All solutions were 21–22°C buffered at pH 7 with 0·005 M Hepes to which 0·01 M Na Cl was added.

[b] Determined following excitation at 458 nm relative to a value of 0·51 for quinine sulphate in 0·1 N H_2SO_4[13] (the latter excited at 350 nm). Absorptivities were less than 0·01.

[c] Fluorescent lifetimes are given in nanoseconds. They were calculated as the best single-exponential fits of fluorescence decay curves, using data acquired over approximately the first 20 nanoseconds of these curves.

[d] T_2 and T_1 were determined from experiments such as those shown in Figure 3, and analysed as described in the text.

and post-relaxational transitions in different quinacrine–DNA complexes can be estimated from values of dye fluorescence quantum yield (Q) and lifetime (τ). The former parameter reflects the overall relaxation and emission efficiency, while the fluorescence lifetime is determined by post-relaxational events. Both quantities have been determined for complexes of quinacrine with DNA from four sources (with different base compositions) and with the synthetic polymer poly (dA-dT).†

One set of measurements was done at dye–DNA phosphate ratios of about 1:100 to minimize interactions between bound dyes. For all complexes examined, the average fluorescence lifetime (Table 1) was greater than that of the free dye (3·5 ns). In contrast, the fluorescence quantum yield of the dye was reduced upon binding to DNA from calf thymus, *E. coli*, or *M. lysodeikticus*, but increased upon binding to either *Cl. perfringens* DNA or poly (dA–dT). Relative to the quinacrine–poly (dA-dT) complex, fluorescence of the other complexes decreases as the G–C content of the DNA increases, as expected from previous reports.[6,7]

Quinacrine–DNA complexes with higher fluorescence efficiencies exhibit longer fluorescence lifetimes (Table 1). However, the lifetimes vary less than three-fold (4·5–10·7 ns) among complexes for which the quantum yields vary nearly one hundred-fold (0·005–0·48). One interpretation of these results is that the major impact of DNA base composition on quinacrine fluorescence occurs prior to relaxation of the excited dye, consistent with it reflecting a strong

† Fluorescence lifetime measurements utilized a single-photon nanosecond fluorometer[15] kindly made available by Professor Lubert Stryer, Yale University.

interaction in the dye–DNA complex. A frequently cited hypothetical quenching mechanism by which G–C base pairs could reduce quinacrine fluorescence prior to excited state relaxation would involve a charge-transfer interaction.[14,16–18] Heterogeneity in dye fluorescence efficiencies could also account for such a disparity between the variations in average quantum yield and lifetime.

Measurements of the transfer of electronic excitation energy from quinacrine to a second dye can be used to study dye excited state relaxation efficiency. The efficiency with which quinacrine fluorescence is quenched by energy transfer (designated at T_1) reflects only post-relaxational excited state processes. However, the average efficiency with which a quantum absorbed by quinacrine is transferred to an energy acceptor (designated as T_2) depends in addition on the relaxation efficiency of an excited quinacrine molecule. The ratio T_2/T_1 is also sensitive to heterogeneity in dye luminescence properties.

The dye ethidium was chosen as an energy acceptor for quinacrine fluorescence, since it forms a tight complex with DNA[19] and absorbs in the same spectral region in which quinacrine fluoresces (Figure 2). In addition, the

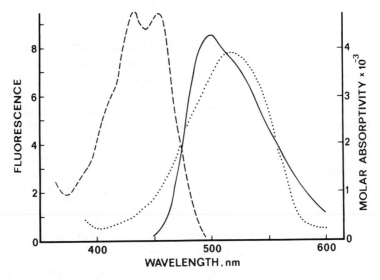

Figure 2   Spectral overlap between the fluorescence emission spectrum of quinacrine complexed with DNA (——) and the absorption of quinacrine–DNA (– – –), or ethidium–DNA (· · ·) complexes. The fluorescence emission spectrum of 3×10^{-6} M quinacrine complexed with 2×10^{-4} M calf thymus DNA was obtained with a Hitachi MPF-3 spectrofluorometer. Excitation was at 424 nm (10 nm bandwidth). Emission was scanned with a 6 nm bandwidth and corrected for variation of instrument response with wavelength. For measurement of absorption spectra, the quinacrine–DNA and ethidium–DNA complexes were both 2×10^{-5} M in dye and 2×10^{-4} M in DNA. The buffer for all solutions was 0·01 M NaCl, 0·005 M Hepes, pH 7·0

binding of ethidium to DNA is competitive with that of quinacrine,[19] suggesting that the two dyes interact similarly with DNA. The critical distance for 50% efficient resonance energy transfer (R_0) between these dyes, when bound to DNA, calculated from spectroscopic parameters according to Förster's theory,[20–22] is approximately 35Å.

Addition of increments of ethidium to a quinacrine–DNA complex results in a reduction of quinacrine fluorescence due to energy transfer to ethidium, the fluorescence of which is concomitantly enhanced. Under conditions chosen such that effectively all added dye is bound, i.e. (1) low ionic strength, (2) low dye/DNA phosphate ratios and (3) DNA concentrations in excess of the dye dissociation constant,[19,23] the fluorescence quenching increases linearly with the amount of ethidium added (Figure 3A). The amount of ethidium required to reduce quinacrine fluorescence by 50% increases linearly with the DNA concentration (Figure 3B). This quantity depends on the type of DNA used although the values determined are of the magnitude expected based on an R_0 near 35Å.

The average efficiency with which light absorbed by quinacrine is transferred to ethidium, T_2, can be calculated from measurements of ethidium fluorescence

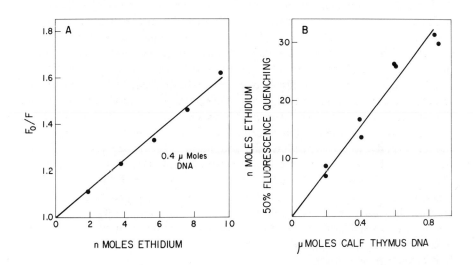

Figure 3   Energy-transfer quenching of quinacrine fluorescence by ethidium in complexes with calf thymus DNA. (A) The fluorescence of a 2 ml aliquot of a 0.7×10^{-6} M solution of quinacrine complexed with 2×10^{-4} M calf thymus DNA (in 0.01 M NaCl, 0.005 M Hepes, pH 7) was measured after successive 20 μl additions of a 1×10^{-4} M solution of ethidium bromide. Excitation was at 420 nm. Quinacrine fluorescence (F) was measured at 493 nm. F_0 is the initial fluorescence in the absence of added ethidium. The ratio F_0/F is plotted versus the amount of added ethidium. Values on the ordinate were corrected for volume changes due to ethidium addition. (B) The amount of ethidium at which quinacrine fluorescence is reduced 50%, obtained by extrapolating plots like those of part A to the amount of added ethidium at which $F_0/F = 2$, is plotted versus the amount of calf thymus DNA present in the solution. Total quinacrine concentration in all cases was 0.7×10^{-6} M

(at 577 nm) following excitation at 420 nm (a wavelength at which both quinacrine and ethidium absorb) both in the presence and absence of quinacrine.

$$T_2 = \frac{F_1 - F_2 - F_3}{F_2} \, \alpha \tag{1}$$

where F_1 is the fluorescence at 577 nm of the complex of quinacrine and ethidium with DNA (following excitation at 420 nm), F_2 is the value of F_1 determined in the absence of quinacrine, F_3 is the contribution of quinacrine fluorescence to F_1 and α is the ratio of ethidium and quinacrine absorbancies (at 420 nm) in the DNA complex.

The ratio T_2/T_1 was determined for different quinacrine–DNA complexes (Table 1). While the experimental uncertainty in these estimates is considerable, the values obtained decrease as fluorescence quantum yields decrease, as expected in view of the comparatively small variations in fluorescence lifetimes. For the complex of quinacrine with poly (dA-dT), T_2/T_1 is effectively unity. A–T base pairs thus have little effect on excited dye relaxation efficiency. However, some post-relaxational quenching apparently occurs in the poly (dA-dT) complex since the overall quantum yield is significantly less than one (Table 1).

While the above analyses were performed at very low dye/DNA phosphate ratios, the saturation by dye of the DNA in cytological chromosome preparations may be much higher. These preparations are stained with quinacrine dye concentrations ($\sim 10^{-3}$ M) orders of magnitude above the dissociation constant of the quinacrine–DNA complex ($\sim 10^{-7}$ M).[9,23] Since the absorption of quinacrine, when bound to DNA, overlaps its fluorescence (Figure 2), energy transfer between bound dye molecules should occur at moderate dye/DNA phosphate ratios. Migration of energy between quinacrine molecules until it reaches a dye with a low fluorescence efficiency would manifest itself as a reduction in dye fluorescence intensity and lifetime. This process has previously been invoked to account for the fluorescence of acriflavine–DNA complexes.[24] Dye surrounded by long stretches of non-quenching (A–T base pair) DNA binding sites would be expected to be unaffected by this process, thereby increasing the relative fluorescence intensity from these regions compared to regions with mixed base composition. Energy transfer between dyes intercalated into DNA would also be expected to result in a moderate reduction in dye fluorescence polarization.

The experimental results agree with the above expectations. At DNA phosphate/dye ratios above 50, fluorescence intensity and lifetime are independent of DNA/dye ratio. The limiting values are different for each type of DNA. Data for complexes of quinacrine with calf thymus DNA are shown in Figure 4. As DNA phosphate/dye ratios drop below 50, dye fluorescence intensity and lifetime begin to decrease,† ultimately dropping by nearly a factor of two. This effect is also marked for the complexes with *Cl. perfringens*, and *E. coli* DNA. For the

---

† In preliminary experiments, decreases in dye fluorescence polarization, expected as dye binding saturation increases, have also been observed.

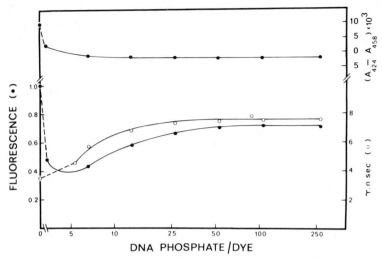

Figure 4 Saturation dependence of the fluorescence and absorption of quinacrine calf thymus DNA complexes. A constant amount of quinacrine (final concentration 2.7×10^{-6} M) was added to solutions spanning a wide range of DNA concentrations (based on a molar extinction coefficient at 260 nm of 6.8×10^3 cm^{-1} M^{-1}). All solutions were buffered at pH 7 with 0.005 M Hepes and in addition contained 0.01 M NaCl. The fluorescence intensity of quinacrine was measured at 493 nm following excitation at 458 nm. Absorption of the free dye at 458 nm was approximately 0.9 that of the dye–DNA complexes at this wavelength. Fluorescence lifetimes were the best single-exponential fits of fluorescence decay curves obtained on a single-photon nanosecond fluorometer.[15] Absorptivity measurements, employing the 0.02 absorbance scale of a Cary 118 spectrophotometer, were performed on the same samples that were used for fluorescence intensity measurements. Dashed lines connect the measurements at the lowest DNA phosphate/dye ratio employed with those on free quinacrine

complex with *M. lysodeikticus* DNA, which fluoresces with very low efficiency, changes in fluorescence intensity are small, while those in fluorescence decay time are of the same magnitude as the experimental error. Importantly, until very low phosphate/dye ratios are reached, the dye absorption spectrum remains unchanged from that of the complexes at lower saturation. Since the absorption spectra of free and DNA-bound molecules differ, these absorption measurements establish that the fluorescence changes occur in the DNA complexes, and do not merely reflect failure of some dye to bind. At phosphate/dye ratios below 5, not all added dye is bound, as indicated by the absorbance measurements, and the solution fluorescence parameters begin to approach those of the free dye.

Results with the quinacrine–poly (dA-dT) complex are in contrast to those with natural DNA samples. As the saturation of poly (dA-dT) by quinacrine increases, emission intensity and lifetime do not decrease but in fact increase slightly (Figure 5). (The increase may reflect some structural change brought about by the dye in this synthetic polymer, which has the unusual configuration of a string of base-paired helical loops.[26]† The results with this polynucleotide, apparently devoid of quenching centres, rules against a second dye binding mode as the explanation for saturation-dependent quenching in the other complexes. These data support the idea that the decreases in quinacrine fluorescence intensity and lifetime when complexed with natural DNA at high dye/phosphate ratios are due to energy transfer between bound quinacrine molecules.

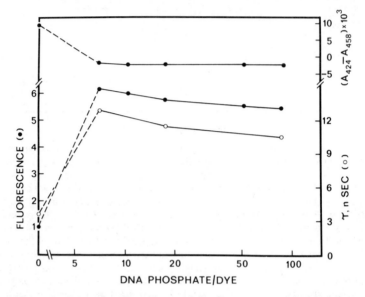

Figure 5 Saturation dependence of the fluorescence and absorption of quinacrine–poly (dA-dT) complexes. The fluorescence intensities and fluorescence lifetimes of complexes of 2.7×10^{-6} M quinacrine with poly (dA–dT) are plotted versus the final DNA phosphate/dye ratio. Also shown is the difference between solution absorptivities at 424 and 458 nm. Techniques and conditions are otherwise like those of Figure 4

Energy transfer between quinacrine molecules complexed with chromosomes at a high dye/free phosphate saturation would amplify the difference in intensity between regions of different DNA A–T content. This effect would thus depend non-linearly on DNA base composition, and would serve to reinforce the even greater differences in fluorescence efficiency intrinsic to the dye–DNA complexes at low saturation. These results suggest that much of the observed fluorescence from some chromosome regions (e.g. those exhibiting especially bright quinacrine fluorescence) could originate from a relatively small proportion of quinacrine molecules complexed to sequences of very high A–T content. The excited state interactions of quinacrine might thus act to enhance contrast in the fluorescence image of chromosomes stained with this dye. Additional

† Similar measurements indicate that the fluorescence efficiency and lifetime of complexes of quinacrine with poly(dA) . poly(dT) do not depend on a dye/phosphate ratio over a similar range.

factors such as chromosomal proteins and chromosome condensation might also influence quinacrine fluorescence.

3 BIS-BENZIMIDAZOLE DYES

While quinacrine fluorescence is sensitive to the composition of chromosomal regions, it is of limited use in studying chromosome function. For example, it is often of interest to follow the replication of chromosome regions differentiated by quinacrine staining. For this purpose, the effective resolution of auto-radiography, the traditional cytological method for analysis of DNA synthesis, is insufficient. A different approach is an optical analysis of DNA synthesis, which most directly would employ DNA base analogues with distinctive absorptive or fluorescent properties. Unfortunately, available base analogues do not make such an approach practical. An alternative method is to use the environmental sensitivity of fluorescent chromophores, employing them to report indirectly the incorporation of nucleotide analogues containing substituents, such as heavy, polarizable atoms (e.g. Br), which can perturb dye luminescence.[27]

Towards this goal, the BrdU-sensitivity of the fluorescence of a number of dyes was examined.[8] The test employed consisted of a comparison of the fluorescence of dye bound to poly (dA-dBrU) with that bound to poly (dA-dT). While the fluorescence of two commonly used dyes, quinacrine and ethidium bromide, is not very sensitive to BrdU, that of a number of *bis*-benzimidazole dyes, such as 33258 Hoechst,† is about four-fold less when bound to poly (dA-dBrU) than when bound to poly (dA-dT).[8] In contrast to quinacrine, the fluorescence of 33258 Hoechst bound to calf thymus DNA (44 % G–C) at a very low dye/phosphate ratio is only slightly less than that of the dye when complexed with poly (dA-dT).

Incorporation of BrdU into the DNA of leukocytes also reduces the fluorescence of cytological preparations of these cells stained by 33258 Hoechst.[8] Comparison of the fluorescence intensity of cells stained with 33258 Hoechst after one or two divisions in medium containing BrdU established that successive incorporation of BrdU into one or both chains of the DNA of a chromatid progressively quenches dye fluorescence. Differentiation of regions due to substitution of one polynucleotide chain allows DNA synthesis kinetics to be followed by fluorescence microscopy, using appropriate combinations of dT and BrdU pulses. Thus, the late-replicating X chromosome can easily be identified, and the kinetics of its replication followed.[28]

Differential substitution of entire polynucleotide chains of a chromatid, as after two divisions in BrdU, affords fluorometric demonstration of the semi-conservative distribution of DNA to daughter chromatids. The chromatid with the lower number of DNA chains substituted with BrdU fluoresces more brightly than its sister. Sharp reciprocal alterations in fluorescence along a chromosome

† These dyes were generous gifts of Dr H. Loewe, Hoechst A/G, Frankfurt, Germany.

signal sister chromatid exchanges. Since chromosomes can first be stained with quinacrine (revealing the standard band pattern), destained, and then restained with 33258 Hoechst, the location of these exchanges can now be determined relative to the standard band patterns.[29]

Determination of the mechanism by which BrdU quenches 33258 Hoechst fluorescence will also aid the use of this dye as a probe of protein–nucleic acid interactions in chromatin. The fluorescent lifetimes of complexes of 33258 Hoechst with poly (dA-dT) and poly (dA-dBrdU) are both 4 ns under the same conditions (low ionic strength, pH 7) in which the fluorescence efficiency of these complexes differs by about a factor of four. These results might reflect heterogeneity in dye–polynucleotide interactions within these complexes. Alternatively, they could result if quenching of 33258 Hoechst fluorescence by BrdU occurred prior to relaxation of an excited dye molecule. This in turn would indicate that the interaction between BrdU and the dye is very strong. While the precise quenching mechanism remains speculative, it may reflect a short-range interaction, perhaps involving charge or proton transfer. Proximity of the dye binding site to the bromine atom, which is located in the major groove of the DNA double helix, would make 33258 Hoechst a sensitive probe of interactions in this region in chromosomes.

ACKNOWLEDGEMENTS

The authors thank Professor Lubert Stryer for the use of his single photon nanosecond fluorometer and computational facilities and Mr James Wohlleb for diligent technical assistance. This work was supported by the National Institute of Child Health and Human Development (HD-06276—Mental Retardation and Human Development and HD-04807—Program Project), National Institute of General Medical Sciences (GM-21121), Harvard Medical School (Milton Fund), the Charles A. King Trust Fund, and Wesleyan University (S.B., partial salary support).

REFERENCES

1. T. Caspersson, S. Farber, G. E. Foley, J. Kudynowski, E. J. Modest, E. Simonsson, U. Wagh and L. Zech, *Exptl. Cell Res.*, **49**, 219 (1968).
2. T. Caspersson, L. Zech, E. J. Modest, G. E. Foley, U. Wagh and E. Simonsson, *Exptl. Cell Res.*, **58**, 141 (1969).
3. L. Zech, *Exptl. Cell Res.*, **58**, 463 (1969).
4. T. Caspersson, J. Lindsten, G. Lomakka, A. Moller and L. Zech, *Int. Rev. Exper. Pathol.*, **11**, 2 (1972).
5. O. J. Miller, D. A. Miller and D. Warburton, *Prog. Med. Genet.*, **9**, 1 (1973).
6. B. Weisblum and P. L. De Haseth, *Proc. Nat. Acad. Sci. USA*, **69**, 629 (1972).
7. U. Pachmann and R. Rigler, *Exptl. Cell Res.*, **72**, 602 (1972).
8. S. Latt, *Proc. Nat. Acad. Sci. USA*, **70**, 3395 (1973).
9. T. Caspersson, L. Zech and E. J. Modest, *Science*, **170**, 762 (1970).

10. J. R. Ellison and H. J. Barr, *Chromosoma*, **36**, 375 (1972).
11. B. Weisblum, *Nature*, **246**, 150 (1973).
12. H. A. Sober, *Handbook of Biochemistry, Selected Data for Molecular Biology*, 2nd edn., Chemical Rubber Co., 1970.
13. R. F. Chen, *Nat. Bur. Stds.* (Special Pub.), **378**, 183 (1973).
14. R. Rigler, 'Interactions between Acridines and DNA', in *Chromosome Identification*, Twenty-Third Nobel Symposium, p. 335 (eds. T. Caspersson and L. Zech), Academic Press, New York, 1973.
15. J. Yguerabide, *Meth. Enz.*, **XXVI**, 498 (1972).
16. J. Weiss, *Trans. Faraday Soc.*, **35**, 48 (1939).
17. L. E. Orgel, *Quart. Rev.*, **8**, 422 (1954).
18. H. Leonhardt and A. Weller, *Zeit. Phys. Chem.*, **29**, 277 (1961).
19. J. B. Le Pecq and C. Paoletti, *J. Molec. Biol.*, **27**, 87 (1967).
20. T. Förster, *Ann. Phys.*, **2**, 55 (1948).
21. T. Förster, in *Modern Quantum Chemistry*, Vol. III, p. 93 (ed. O. Sinanoglu), Academic Press, New York, 1965.
22. L. Stryer, *Science*, **162**, 526 (1968).
23. E. J. Modest and S. K. Sengupta, 'Chemical correlates of chromosome banding', in *Chromosome Identification*, Twenty-Third Nobel Symposium, p. 327 (eds. T. Caspersson and L. Zech), Academic Press, New York, 1973.
24. L. M. Chan and Q. Van Winkle, *J. Molec. Biol.*, **40**, 491 (1969).
25. R. P. Haugland, J. Yguerabide and L. Stryer, *Proc. Nat. Acad. Sci. USA*, **63**, 23 (1969).
26. T. M. Hickey and E. Hamori, *Biochemistry*, **11**, 2327 (1972).
27. W. C. Galley and R. M. Purkey, *Proc. Nat. Acad. Sci. USA*, **69**, 2198 (1972).
28. S. A. Latt, *Exptl. Cell Res.*, **86**, 412 (1974).
29. S. A. Latt, *Science*, **185**, 74 (1974).

B.3. 4-Thiouridine, a built-in probe for structural changes in transfer-RNA

N. Shalitin and J. Feitelson

Department of Physical Chemistry, The Hebrew University, Jerusalem, Israel

SUMMARY

The luminescence of an aqueous solution of 4-thiouridine is compared to its emission when forming part of the polynucleotide chain ot t-RNA. In both cases excitation into the last absorption band at 335 nm yields a weak emission in the 520–550 nm region. However, while in aqueous solution this emission has a lifetime of ~ 240 ns, it increases in t-RNA to $\tau \simeq 6\,\mu s$. Oxygen and Cl^- ions quench the thiouridine emission efficiently in aqueous solution, while Na^+ and Mg^{++} ions have no influence on it. On the other hand, thiouridine which forms part of a t-RNA molecule is quite insensitive to Cl^- ions and to O_2, while its emission is greatly enhanced by Na^+ and Mg^{++} ions. From these salt effects as well as from data on the temperature dependence of the emission yield and decay, it is concluded that the site of the thiouridine residue is very well protected within the tertiary structure of t-RNA, and that changes both in the tertiary and in the second structures of the polynucleotide can be observed by following the emission characteristics of its thiouridine residue.

1 INTRODUCTION

The base 4-thiouridine, thioU for short, is present in the various transfer-RNA molecules obtained from *E. Coli.*[1,2] It is found at position 8 from the 5′ end of the molecule,[3,4] and it thus occupies a strategically important location between the double helices of the CCA and the dihydroU stems of the t-RNA clover leaf. The spectroscopic properties of thioU might therefore be expected to be influenced by changes in the secondary and tertiary structures of the molecule. It has a characteristic absorption band at 335 nm and a room-temperature emission at 550 nm. When incorporated into biologically active t-RNA, thioU forms photodimers with Cytidine in position 13.[6,7] Both its absorption and emission seem to reflect the conformation of the t-RNA molecule in the vicinity of the thioU residue.[8,9,10]

The long-wavelength emission properties of thioU in solution have been described in a previous study.[11] Here we wish to present a comparison between these data and measurements performed on thioU in the intact t-RNA molecule, a comparison which shows that the emission spectrum, quantum yield and lifetime can be used to study structural changes in t-RNA.

2 EXPERIMENTAL METHODS

Partially purified t-RNA, from *E.Coli-w*, was obtained by courtesy of Mr Ehud Ziv of our Biochemistry department. Two preparations, both of which had undergone separation on BD cellulose, were used in this study. One was more than 90 % pure t-RNA$_1^{Val}$ and had a biological activity of 50 %. The other had an overall activity of 80 %, about equally divided between four t-RNA species active towards valine, methionine, alanine and glycine.

The preparations were freed from salts by dialysis against neutral EDTA (0.01 M) in the presence of 0.1 M NaCl and subsequent dialysis against triply distilled water. The absorbance ratio in water at 260 nm and 340 nm of the final preparation was $OD_{340}/OD_{260} \simeq 0.03$ for the t-RNA$_1^{Val}$ and 0.022 for the t-RNA mixture, respectively.

All measurements were carried out at a t-RNA concentration of $\sim 2 \times 10^{-5}$ M. The chemicals used were of analytical grade. Water was triply distilled and ethylene glycol was of spectroscopic grade. Emission spectra were measured in a standard fluorimeter consisting of a 250 W Xe lamp, two Bausch and Lomb 500 nm focal length monochromators and an EMI 6256 S photomultiplier. Quantum yields were measured by comparison with quinine bisulphate fluorescence.[12] Emission decay curves were determined by exciting with a (10 ns) nitrogen laser and feeding the signal directly into a Textronix 545 oscilloscope.

3 RESULTS

For t-RNA to be biologically active the presence of Na$^+$ ions (0.01–0.1 N) and of Mg^{++} ions (10^{-2}–10^{-3} M) is required. It was therefore of interest to study the influence of these ions on the emission of the thioU residue in the t-RNA molecule.

The emission quantum yield of the thioU residue in a salt-free t-RNA solution is similar to the corresponding value for free thioU ($\phi \simeq 3 \times 10^{-4}$). Its room temperature emission band lies at 540 nm, somewhat to the blue of the 550 nm band of thioU. In both cases the excitation spectrum of this emission coincides with the long-wavelength absorption band of the thioU moiety at 335 nm.

Though the spectral properties of the above two thioU systems are fairly similar, the effects of salts on these properties are dramatically different.

3.1 Effects of salts and of oxygen on the emission

In the following experiments the preparation of pure t-RNA^{Val} was used since particularly the melting curves, described in the next paragraph, might well differ for different t-RNA species.

NaCl and MgCl$_2$ quench the emission of thioU, the Stern–Volmer constant for NaCl being $k_{sv} = 125 \ M^{-1}$ which corresponds to a deactivation rate constant of $k_d = 8 \times 10^8 \ M^{-1} s^{-1}$. Since both CH$_3$COONa and (CH$_3$COO)$_2$ Mg do not affect the thioU emission (k_{sv} is close to zero), the quenching can be attributed to the Cl$^-$ ions only and it can be concluded that Na$^+$ or Mg^{++} ions do not influence the thioU emission.[13] On the other hand the emission of t-RNA is increased in 0·1 N NaCl by a factor of 2·5. A further addition of Mg^{++} ions (10^{-3} M) causes an additional increase by a factor of two to three. The final solution (0·1 M NaCl, 0·001 M MgCl$_2$) has a quantum yield about six times larger ($\phi \simeq 2 \times 10^{-3}$) than that of free thioU in solution. It must here be mentioned that the thioU emission in t-RNA is very little affected by the presence of the Cl$^-$ ions. Their quenching rate constant has a value of $k_d \simeq 2 \cdot 3 \times 10^{-5} \ M^{-1} s^{-1}$, in agreement with the data of Favre.[13] Since the above Na$^+$ and Mg^{++} ions presumably stabilize the structure of the active t-RNA molecule, all subsequent measurements were performed in the presence of either Na$^+$ ions alone or together with Mg^{++} ions.

Oxygen decreases the quantum yield of the emission of thioU in solution with a deactivation rate constant of $k_d = 6 \times 10^9 \ M^{-1} s^{-1}$. A similar value is obtained from emission decay measurements in the presence and in the absence of oxygen. For thioU incorporated in t-RNA the value of the quenching rate constant is only $k_d \leq 1 \cdot 6 \times 10^7 \ M^{-1} s^{-1}$. No measurable decrease in the decay rate is observed in the presence of oxygen as can be expected for such a small value of k_d. The actual lifetime of excited thioU in solution at room temperature (22 °C) is about $\tau = 240$ ns. The corresponding value for a thioU residue in t-RNA at a salt concentration of [Na$^+$] = 10^{-2} M and [Mg^{++}] = 3 × 10^{-3} M was found to be 25 times larger, i.e. $\tau = 6 \ \mu s$.

3.2 Effect of temperature on the emission

20 to 70 °C range. Measurements were carried out at 0·1 M NaCl and 3 × 10^{-3} M (CH$_3$COO)$_2$ Mg concentrations. Thus they can be compared with the corresponding absorbance data of Seno *et al.*[8] in 0·2 M NaCl and 10^{-2} M (CH$_3$COO)$_2$ Mg. The solution pH was adjusted by 0·01 M tris buffer of pH 7·2. Upon raising the temperature from 20 to 70 °C the emission of thioU in a solution of similar salt content decreases gradually by about 50% of its initial value. The emission of a thioU residue in t-RNA is appreciably increased at room temperature (22 °C) by the presence of 0·1 N Na$^+$ ions as described above. However, increasing the temperature causes a drop in emission at about 45 °C

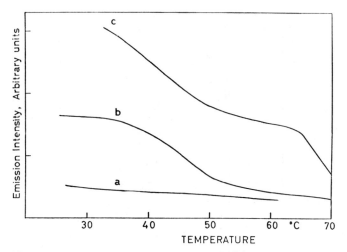

Figure 1 Temperature dependence of t-RNA$_1^{Val}$ emission.
(a) In tris buffer 0·01 N pH 7·2.
(b) In presence of 0·1 M NaCl and above buffer.
(c) In presence of 0·1 M NaCl, 10^{-3} M Mg(CH$_3$COO)$_2$ and buffer

to a level comparable to that of a salt-free solution. When both Na$^+$ (10^{-1} M) and Mg^{++} (3×10^{-3} M) ions are present, the emission decreases in two steps as shown in Figure 1. The first decrease is seen to occur between 30 and 45 °C, while a second step in the emission versus temperature curve is observed near 65 °C, after which the emission intensity approaches the value found for the salt-free solution.

$+20$ to -196 °C range. In these experiments the preparation containing four t-RNA species was used (see §2). Measurements of the emission in this range were carried out in an ethylene glycol–water (1:1) mixture in order to obtain a clear glass at low temperature. For the same reason the lower concentrations of 10^{-2} M Na(CH$_3$COO) and of 3×10^{-3} M (CH$_3$COO)$_2$ Mg$^-$ were used in these experiments.

The emission intensity of the thioU residue in t-RNA increases with decrease of temperature from a quantum yield of $\phi \simeq 2 \times 10^{-3}$ at 22 °C to about $\phi \simeq 0·2$ at 77 K (Figure 2). For free thioU in solution the quantum yield at liquid-air temperature has a similar value ($\phi \sim 0·16$), while the yield at 22 °C is by an order of magnitude smaller ($\phi \simeq 3 \times 10^{-4}$). Furthermore two distinct emissions peaking at ~ 530 and at 475 nm and having different lifetimes were observed for free thioU in the low-temperature glass. In t-RNA with decreasing temperature a single peak is observed which shifts with decreasing temperature from 535 nm at 22 °C to 480 nm at 77 K. When, however, the lifetime of this emission is measured at the short- and long-wavelength edges of this peak two distinctly different decay times can be resolved. In liquid air (77 K) at 445 nm a purely exponential decay of $\tau = 2·6 \pm 0·1$ ms was measured. At longer wavelengths at this same temperature a second short-lived component

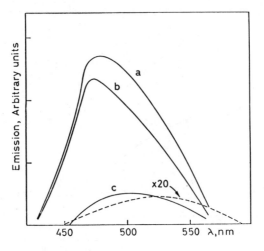

Figure 2   Emission spectrum of t-RNA excited at
335 nm in ethylene glycol–water glass at 77 K.
(a) Overall spectrum, (b) long-lived component
($\tau_1 = 2.6$ ms), (c) short-lived component ($\tau_2 = 0.6$ ms). Broken line—emission in aqueous solution
at room temperature ($\sim 20\,°C$)

appeared in the decay curve. At the long-wavelength edge of the emission peak
the short decay predominated, though some of the long-lived component still
prevailed. From the previous data in free thioU it appears that the two different
species decay independently, i.e. we are dealing with two simultaneous, and not
consecutive, processes. If this is the case also in t-RNA, the time dependence of
the emission is described by

$$I(t) = \alpha e^{-t/\tau_1} + \beta e^{-t/\tau_2} \tag{1}$$

where $I(t)$ denotes the emission intensity at time t, and α and β represent the
initial populations of the two excited species with lifetimes τ_1 and τ_2. In our
case τ_1, the lifetime at the short wavelength, is known since it could be measured
directly. At long times only the long-lived component contributes to the emission, and hence in the long-time limit $\log I(t)$ should decrease linearly, with
slope $1/\tau_1$ as a function of time. A straight line with the *known* slope of $1/\tau_1$ was
fitted to the long-time edge of the log $I(t)$ decay curve and the value of log α was
determined from its intercept at time zero. By subtracting $\alpha e^{-t/\tau_1}$ from $I(t)$ and
plotting $\log[I(t) - \alpha e^{-t/\tau_1}]$ against t, τ_2 was obtained from the slope and log β
from the intercept of the straight line. By repeating the procedure for different
wavelengths the same value of τ_2 should always be obtained, if indeed the decay
curve is described by two exponentials. This was found to be the case within the
experimental error range, and a value of $\tau_2 = 0.6 \pm 0.06$ ms at 77 K was
obtained. The ratio of the initial intensities, i.e. the populations of the two
states emitting with different lifetimes, is given by the difference of the two
intercepts $\log \alpha - \log \beta = \log(\alpha/\beta)$ in the above plot (see Figure 3). On the

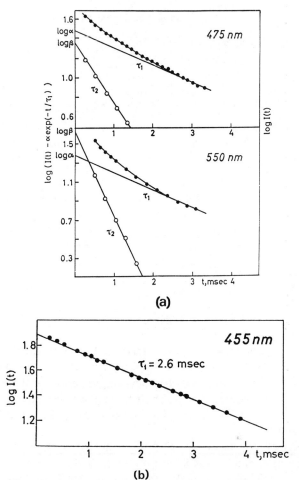

(a)

(b)

Figure 3   Separation of components from lifetime measurements. Experimental value of log $I(t)$ is shown by full circles. $\log[I(t) - \alpha \exp(-t/\tau_1)]$ refers to the open circles only. The slope of the straight line (τ_1) is taken from measurements at $\lambda = 455$ nm where $\beta = 0$

other hand integration from time zero to infinity of equation (1),

$$\int_0^\infty I_\lambda(t)\, dt = \alpha\tau_1 + \beta\tau_2 = F(\lambda) \tag{2}$$

yields the intensity in a certain wavelength interval (λ to ($\lambda + \Delta\lambda$)), a quantity which is also directly obtained from the steady state emission spectrum, $F(\lambda)$. A knowledge of the quantities, α/β, τ_1 and τ_2, as well as $\alpha\tau_1 + \beta\tau_2$, as obtained from the emission spectrum, allows us therefore to separate this spectrum into its two components (see Figure 2).

The various emission properties of thioU in solution and for a thioU residue in the nucleotide chain of t-RNA are summarized in Table 1.

Table 1   Emission properties of thioU

	In aqueous solution	As a constituent of the t-RNA chain
Quenching by Cl⁻	$k_{Cl} = 7 \times 10^8 \, M^{-1} s^{-1}$	$k_{Cl} = 2.3 \times 10^5 \, M^{-1} s^{-1}$
by O_2	$k_{O_2} = 6 \times 10^9 \, M^{-1} s^{-1}$	$k_{O_2} \leq 1.6 \times 10^7 \, M^{-1} s^{-1}$
Influence on quantum yield of		
Na^+ 0.1 M	None	Increase × 2.5
of addit. $Mg^{++} 10^{-3}$ M	None	Further increase × 2–3
Lifetimes τ		
20 °C	240 ns	6 μs
−196 °C	0.4 ms, 1.0 ms	0.6 ms, 2.6 ms
Quantum yield		
at 20 °C	3×10^{-4}	2×10^{-3}
at −196 °C	0.16 (in ethylene glycol water)	0.20 (in ethylene glycol water)

4 DISCUSSION

Lately the highly sensitive NMR method has been developed to probe specifically the various base pairs for structural changes in their vicinity.[14] The luminescence emission described here might supplement this information in that it reflects changes in the neighbourhood of a base which is not hydrogen-bonded, but is located between two base-paired regions—the CCA and the dihydroU stems.

From the 'melting curves' of t-RNA it is seen that in the presence of Na^+ ions the base-paired regions of t-RNA form double helical structures. On the other hand, since for t-RNA to be biologically active a small amount of Mg^{++} ions is required, it seems that its tertiary structure is not complete in the absence of the latter.

The temperature dependence of the thioU residue emission in 0.1 N NaCl shows that the t-RNA structure of which thioU is a part is disrupted at about 45 °C. The measurements by Seno et al.[8] of the hypochromic effect at 335 nm, which disappears at ∼50 °C show that the above structure involves stacking of the thioU base. The hypochromicity at 260 nm, on the other hand, which reflects the overall helical structure of the nucleotides, shows a gradual decrease in stacked regions until above 70 °C the whole secondary structure of t-RNA is disrupted.

In solution containing Na^+ and Mg^{++} ions the absorption data of the above authors both at 260 nm and at 335 nm show a much higher melting temperature of ∼75 °C. This can be taken to mean that the double helical structure is stabilized by Mg^{++} ions. The large increase in emission upon excitation at 335 nm, when Mg^{++} ions are added, can therefore hardly be due to any additional stacking but must be attributed to interaction of the already stacked thioU within its neighbourhood residues. This increase, we think, might reflect changes in the tertiary structure in the thioU vicinity. The temperature dependence of the emission therefore can be taken to indicate that the first decrease in emission intensity between 30 and 45 °C (see Figure 1) is due to such structural changes near the thioU residue leaving the stacks intact, while in the second decrease at ~ 70 °C the secondary structure is also disrupted.†

That the thioU luminescence reflects structural features of t-RNA can also be deduced from a comparison between the quantum yield and decay of thioU in solution and in the t-RNA molecule. The low-temperature emission dependence and the resolution of the spectrum at 77 K (Figure 2) shows that in both cases are we dealing with similar triplet emissions. We had previously proposed that the short-wave length emission ($\lambda = 475$ nm) represents the triplet of thioU while the long-wavelength emission ($\lambda \sim 560$ nm at room temperature) is a triplet which derives by a fast first-order process either directly from the singlet or from the thioU triplet. This first-order process was thought to involve either an intramolecular process or a rearrangement in the neighbourhood of the excited chromophore. The great increase in the triplet lifetime of the thioU residue in t-RNA as compared to the free molecule indicates first and foremost that it is very well protected within the t-RNA structure from outside quenching effects (see Table 1). Moreover, the quantity ϕ/τ is a measure of the initial population of the excited state. In t-RNA at room temperature ϕ/τ is appreciably smaller than for thioU in solution (see Table 1). This indicates that within the rigid structure of t-RNA the above first-order process is hampered—an effect which can also be observed in a solution of free thioU in a low-temperature glass. It seems therefore that both τ and ϕ/τ might be well suited to yield information about structural changes in the vicinity of the thioU residue in t-RNA.

† The slightly lower melting temperatures in our work when compared to Seno et al.[8] are probably due to the lower overall salt concentrations used by us (0.1 vs 0.2 N NaCl, 3×10^{-3} vs 10^{-2} M $Mg(CH_3COO)_2$).

REFERENCES

1. M. N. Lipsett, *J. Biol. Chem.*, **240**, 3975 (1965).
2. M. N. Lipsett, *Biochem. Biophys. Res. Comm.*, **20**, 224 (1965).
3. H. G. Zachau, *Angew. Chem. Internat. Edit.*, **8**, 711 (1969).
4. M. Yaniv and B. G. Barrell, *Nature*, **222**, 278 (1969).
5. F. Pochon, C. Balny, K. H. Scheit and A. M. Michelson, *Biochem. Biophys. Acta,* **228**, 49 (1971).
6. A. Favre, A. M. Michelson and M. Yaniv, *J. Mol. Biol.*, **58**, 367 (1971).
7. A. Favre, B. Roques and J. L. Fourrey, *FEBS Lett.*, **24**, 209 (1972).
8. T. Seno, M. Kobayashi and S. Nishimura, *Biochem. Biophys. Acta*, **174**, 71 (1969).
9. F. Pochon and S. S. Cohen, *Biochem. Biophys. Res. Comm.*, **47**, 720 (1972).
10. M. Dourlent, M. Yaniv and C. Helene, *Eur. J. Biochem.*, **19**, 108 (1971).
11. N. Shalitin and J. Feitelson, *J. Chem. Phys.*, **59**, 1045 (1973).

12. J. G. Calvert and J. N. Pitts, *Photochemistry*, p. 799, Wiley, New York, 1966.
13. A. Favre, *Photochem. Photobiol.*, **19**, 15 (1974).
14. (a) Y. P. Wong, B. R. Reid and D. R. Kearns, *Proc. Nat. Acad. Sci. US.*, **70**, 2193 (1973);
    (b) R. G. Schulman, C. W. Hilbers, D. R. Kearns, B. R. Reid and Y. P. Wong, *J. Mol. Biol.*, **78**, 57 (1973).

B.4. The triplet state of benzo[a]pyrene as a probe of its microenvironment in DNA complexes

N. E. Geacintov, W. Moller and E. Zager

Chemistry Department and Radiation and Solid State Laboratory
New York University, New York, New York 10003, U.S.A.

SUMMARY

Physical complexes of benzo[a]pyrene (BP) and DNA in aqueous solutions were prepared by the methods of Boyland and Green and a transient absorption was observed in a flash photolysis experiment. This transient absorption is attributed to the triplet state of benzo[a]pyrene. In degassed solutions the triplet lifetime is ~ 2–5 ms and decreases in the presence of molecular oxygen. The Stern–Volmer oxygen quenching constant is $\sim 1.3 \times 10^8$ M^{-1}. The triplet lifetime of BP complexed with DNA is 20–80 times longer than in fluid benzene solution, while the quenching constant is ~ 20 times smaller. These results indicate that the BP molecule intercalated in the DNA is located in a relatively viscous environment and is less accessible to other molecules, even if they are as small as the oxygen molecule.

With pyrene–DNA complexes the triplet could not be detected, which indicates that its lifetime is less than 50 microseconds. These preliminary results indicate that the triplet lifetime of aromatic hydrocarbons complexed with DNA is a sensitive probe of the microenvironment and is a function of structural parameters which determine the mode of interaction of these molecules (many of which are carcinogenic) with nucleic acids.

1 INTRODUCTION

In 1962 Boyland and Green[1] and Liquori *et al.*[2] reported that several aromatic hydrocarbons including benzo[a]pyrene, or 3,4-benzopyrene (BP), are more soluble in aqueous solutions of DNA than in water. The polycyclic aromatic hydrocarbons are solubilized by the hydrophobic core of stacked bases in DNA. Some doubts were initially raised about this effect,[3] but the earlier work was extended and the results of Boyland and Green, and Liquori *et al.* were later confirmed.[4-6]

Nagata *et al.*[7] used the flow method[8] to demonstrate a flow-induced orientation and dichroism of suspensions of DNA–hydrocarbon and DNA–dye complexes. The observed dichroism is indeed consistent with a model in which BP, pyrene and phenanthrene are intercalated in the double helix with their planes perpendicular to the axis of the helix. The flow dichroism was distinctly different for another set of molecules including coronene, tetracene and 20-methyl cholanthrene. Isenberg and coworkers have shown that the degree of complexing with DNA of this group of molecules is very small and have indicated that this is due to their large size.[9,10] Green and McCarter[11] extended the flow-induced orientation method and observed a rather strong flow-induced polarization of fluorescence of BP complexed with DNA in aqueous solution. Their results were consistent with those of Nagata and coworkers in showing that the planes of the aromatic molecules are perpendicular to the axis of the helix, thus providing further evidence for the intercalation model. It has been proposed that hydrophobic interactions are significant in these aromatic hydrocarbon–DNA complexes.[12-14]

The formation of these complexes is accompanied by a red shift of about 10 nm in the first absorption band of BP ($S_0 \rightarrow S_1$). Furthermore, there is a considerable quenching of the fluorescence of BP upon complex formation.

In this paper we present preliminary observations on the triplet lifetime of BP in aqueous DNA complexes measured by Porter's flash photolysis technique[15,16] and on the quenching of the triplet by molecular oxygen. The triplet state of BP complexed with DNA has a relatively long lifetime (~ 2–5 ms) and should therefore be a sensitive probe of its microenvironment. Molecular oxygen is a well-known quencher of the triplet state[17] by a diffusion-controlled bimolecular process which depends on the viscosity of the medium, the local oxygen concentration, and on the accessibility to oxygen of the complexed BP molecule.

In principle, both the triplet lifetime and the quenching of the triplet state by different quenchers may provide valuable information on the mode of binding and the reactivity of aromatic hydrocarbons complexed with DNA. Using these techniques it should be possible to investigate the modes of interaction of different polycyclic aromatic carcinogens and structurally similar noncarcinogens with nucleic acids and to determine if there are any significant differences.

2 EXPERIMENTAL

Benzo[a]pyrene (BP) (K and K Laboratories) was purified by thin layer chromatography. Calf thymus DNA (sodium salt) was purchased from Sigma Chemical Company. The BP–DNA complexes were prepared by the method of Boyland and Green.[1] The DNA concentration in salt-free water was 0·05% by weight and the complexed BP concentration was of the order of $\sim 10^{-6}$ M.

The flash photolysis apparatus was of the type described by Porter[15] using microsecond flash lamps for photoexcitation and was a commercially available unit (FP–2R) supplied by Northern Precision Co., Ltd., London. Any transient absorption with a lifetime shorter than 50 μs cannot be detected with this apparatus. The lifetimes of the transient absorption of photoexcited BP in the nucleic acid solutions were greater than ~ 0.4 ms. Any short-lived (< 50 μs) components which may or may not have been present in our experiments were thus not detectable.

The aqueous solutions were degassed by four to five successive freeze–pump–thaw cycles. The solutions were frozen in either liquid nitrogen or in dry ice–acetone solutions. The lifetime of the transient absorption of photoflash excited BP–DNA complexes is shortened if molecular oxygen is present. The lifetime as a function of the oxygen concentration was determined by varying the partial pressure of oxygen above the aqueous solutions. The oxygen pressure was first determined in a vessel of known volume V_1 isolated by a stopcock from the vessel (also of known volume V_2) containing the solution. The stopcock was then opened and the partial pressure of oxygen was calculated from the values of V_1 and V_2. The oxygen concentration in the aqueous solution was calculated using Henry's law. All measurements were performed at room temperature (24 °C).

3 RESULTS

An absorption spectrum ($S_0 \rightarrow S_1$) of the BP–DNA solutions used in these experiments is shown in Figure 1. The main peak is at 394 nm which agrees well with the spectra previously reported.[1,2,4,14]

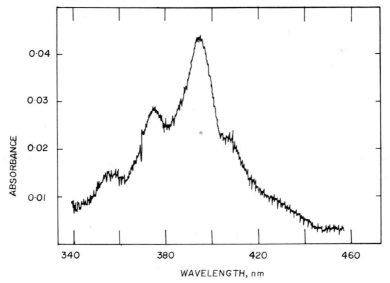

Figure 1 Absorption spectrum of a benzo[a]pyrene–DNA (calf thymus) complex in water (0.05 % by weight)

The decay of the transient absorption was exponential, within experimental error, with both degassed and oxygen-containing solutions. This is shown in the semi-logarithmic plots in Figure 2. The experimental points were taken directly from polaroid photographs of oscilloscope traces. The quenching was reversible and reproducible.

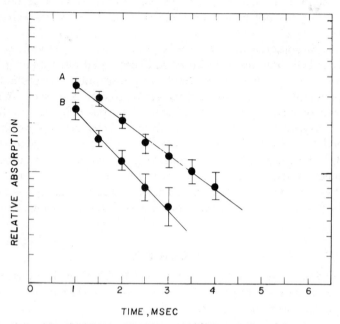

Figure 2 Decay of transient absorption of flash-photoexcited benzo[a]pyrene–DNA complex in aqueous solution. (A) Degassed solution, (B) partial pressure of oxygen is 1·5 mm (24 °C)

The wavelength dependence of the transient absorption was determined using a degassed solution and is shown in Figure 3. The different wavelengths were selected with a monochromator, and a photograph of the transient absorption trace was taken at each wavelength. The points were taken at a point on the time scale corresponding to the peak in the absorption curve.[15,16] The triplet–triplet absorption spectrum of BP in degassed benzene solution as determined by Slifkin and Walmsley[18] is also shown for comparison in Figure 3. The absorption in the DNA is broader and somewhat red-shifted in comparison to the triplet–triplet absorption in benzene.

4 DISCUSSION

The identification of a transient absorption in a flash photolysis experiment as a triplet–triplet absorption ($T_1 \rightarrow T_n$) is not necessarily a straightforward matter. Other intermediate products such as ions, radicals and isomers may

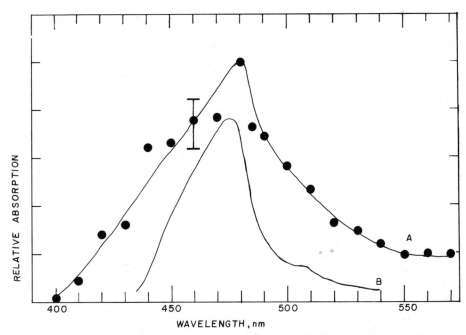

Figure 3   (A) Wavelength dependence of transient absorption of flash-photoexcited benzo-[a]pyrene–DNA complex in aqueous solution. (B) Triplet–triplet absorption spectrum of benzo[a]pyrene in degassed benzene solution.[18] (The error bar indicates the maximum observed deviation in these measurements)

also give rise to a transient absorption.[19] A positive identification is usually made upon correlation with the phosphorescence and/or ESR spectra deter-mined under conditions similar to those of the triplet–triplet absorption. Because of the low concentrations of BP in our nucleic acid solutions and the short triplet lifetimes, these experiments do not appear to be feasible. Although such a positive identification has not been made, the identification of the transient absorption observed with BP solutions in benzene[18] as the triplet–triplet absorption of BP appears to be the most reasonable one. On the basis of the similarity of the two spectra and the near-coincidence of the maxima shown in Figure 3 we conclude that we are also most likely observing the absorption of the triplet of BP in DNA complexes.

The triplet excited states of aromatic hydrocarbons are relatively long lived. In rigid media the triplet lifetimes can be as long as 10–20 s, depending on the molecule. In fluid solutions at room temperature, on the other hand, the triplet lifetimes are usually less than 100 μs, even in degassed solutions. These life-times depend on the viscosity of the medium.[16]

In thoroughly degassed BP–DNA solutions the triplet lifetime was repro-ducible from experiment to experiment ($\pm 10\%$) with the same sample, but differed with different samples for reasons which are not yet known. The lifetimes in the absence of oxygen were in the range of 2–5 ms. Slifkin and Walmsley found that the triplet lifetime of BP in degassed benzene solution was 0·084 ms, or about 20–80 times shorter than in the BP–DNA complexes. Since the triplet lifetime is a function of the viscosity of the medium in which the aromatic molecule is dissolved, it appears that BP intercalated in the DNA macromolecules is located in a region of significantly higher effective viscosity than that of benzene (0·63 cP). Triplets of aromatic hydrocarbons are reversibly quenched by molecular oxygen.[17] The measured decay times τ decrease with increasing oxygen concentration $[O_2]$ according to the well-known Stern–Volmer equation:

$$1/\tau = 1/\tau_0 + K_q[O_2] \tag{1}$$

where τ_0 is the triplet lifetime in oxygen-free solutions. The quenching constant K_q may be written as[20]

$$K_q = \tfrac{1}{9}K_dP \tag{2}$$

where $\tfrac{1}{9}$ is a spin statistical factor, K_d is the bimolecular encounter rate and P is the probability of quenching per encounter. For BP in both polar and non-polar fluid solvents at room temperature $P = 1.$[20] Furthermore, it has been shown that the P value for a given aromatic hydrocarbon dissolved in both fluid[20] and rigid[21] solvents at room temperature is approximately the same in both solvents despite the large differences in the viscosity. It is therefore likely that the oxygen quenching of the BP triplet in DNA complexes depends only

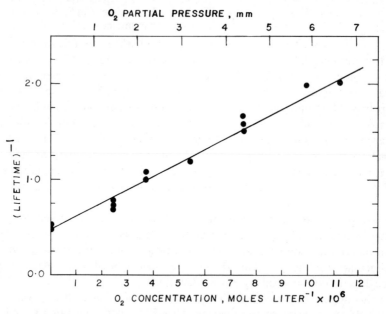

Figure 4   Stern–Volmer plot of the reciprocal triplet lifetime as a function of oxygen concentration (benzo[a]pyrene–DNA complex in aqueous solution). The individual experimental points refer to separate experiments with the same solution

on the rate of encounter of the two reacting molecules, and thus on the micro-viscosity and accessibility of the BP molecule to oxygen.

The reciprocal triplet lifetime of BP in aqueous BP–DNA solutions as a function of oxygen concentration is shown in Figure 4. As predicted by the Stern–Volmer equation, a straight line is obtained which shows that the quenching is dynamic in nature, i.e. depends on the rate of encounter of an oxygen molecule with a BP triplet. The Stern–Volmer quenching constant K_q calculated from the slope in Figure 4 is $\sim 1.3 \times 10^8 \text{ s}^{-1} \text{ M}^{-1}$. Gijzeman et al.[20] find that K_q is $2.8 \times 10^9 \text{ s}^{-1} \text{ M}^{-1}$ in benzene, about 23 times larger than in the BP–DNA complex. These results indicate that (i) the effective oxygen concentration at the site where the intercalated BP molecule is located may be less than in the aqueous phase (which is calculated from Henry's law), or that (ii) the rate of encounter as embodied in K_d is decreased, perhaps because of a higher effective viscosity at the site of the BP molecules. Both effects may also be operative simultaneously.

Paramagnetic metal ions are also known to be quenchers of aromatic hydrocarbon triplets;[16] however, they are less effective than oxygen or nitric oxide because of their larger effective radius in aqueous solutions. Quenching experiments with these metallic species should therefore give a direct measure of the accessibility of the complexed aromatic hydrocarbon molecules.

Similar experiments were also performed with pyrene–DNA complexes. However, no transient absorption was detected within the limits of the resolution of the flash photolysis apparatus employed. This indicates that in degassed pyrene–DNA complexes the triplet lifetime is less than $\sim 50 \, \mu s$. In rigid media, on the other hand, the pyrene triplet lifetime is longer than that of BP.[22] These preliminary results indicate that the interactions in the aromatic hydrocarbon–DNA complexes are strongly dependent on the nature of the aromatic molecule. It therefore appears feasible to probe the physico-chemical interactions of different structurally related aromatic hydrocarbons by flash photolysis and to gain a deeper insight into the mode of interaction between polycyclic aromatic molecules with nucleic acids.

ACKNOWLEDGEMENTS

This investigation was supported by Public Health Service Research Grant No. CA 14980 from the National Cancer Institute, and in part by a Health Research Council City of New York Grant No. U-2312. We are indebted to Professor D. I. Schuster for permission to use his flash photolysis apparatus.

REFERENCES

1. E. Boyland and B. Green, *Brit. J. Cancer*, **16**, 507 (1962).
2. A. M. Liquori, B. DeLerma, F. Ascoli, C. Botré and M. Trasciatti, *J. Mol. Biol.*, **5**, 521 (1962).

3. B. C. Giovanella, L. E. McKinney and C. Heidelberger, *J. Mol. Biol.*, **8**, 20 (1964).
4. E. Boyland and B. Green, *J. Mol. Biol.*, **9**, 589 (1964).
5. E. Boyland, B. Green and S. L. Liu, *Biochim. Biophys. Acta*, **87**, 653 (1964).
6. J. K. Ball, J. A. McCarter and M. F. Smith, *Biochim. Biophys. Acta*, **103**, 275 (1965).
7. C. Nagata, M. Kodama, Y. Tagashira and A. Imamura, *Biopolymers*, **4**, 409 (1966).
8. L. S. Lerman, *Proc. Nat. Acad. Sci.*, **49**, 94 (1963).
9. I. Isenberg, S. L. Baird, Jr. and R. Bersohn, *Biopolymers*, **5**, 477 (1967).
10. M. Craig and I. Isenberg, *Biopolymers*, **9**, 689 (1970).
11. B. Green and J. A. McCarter, *J. Mol. Biol.*, **29**, 447 (1967).
12. P. O. P. Ts'o and P. Lu, *Proc. Nat. Acad. Sci.*, **51**, 17 (1964).
13. P. O. P. Ts'o, *Ann. N.Y. Acad. Sci.*, **153**, 785 (1969).
14. I. Isenberg, S. L. Baird, Jr. and R. Bersohn, *Ann. N.Y. Acad. Sci.*, **153**, 780 (1969).
15. G. Porter, in *Techniques of Organic Chemistry*, 2nd edn., Vol 8 (ed. A. Weissberger), Wiley–Interscience, New York, 1963.
16. G. Porter and M. R. Wright, *Disc. Farad. Soc.*, **27**, 18 (1959).
17. J. B. Birks, *Photophysics of Aromatic Molecules*, Wiley–Interscience, London and New York, 1970.
18. M. A. Slifkin and R. H. Walmsley, *Photochem. Photobiol.*, **13**, 57 (1971).
19. H. Labhart and W. Heinzelmann, *Organic Molecular Photophysics,* Vol. 1, p. 297 (ed. J. B. Birks), John Wiley, London and New York, 1973.
20. O. L. J. Gijzeman, F. Kaufman and G. Porter, *J. Chem. Soc. Farad. Trans. II*, **69**, 708 (1973).
21. R. Benson and N. E. Geacintov, *J. Chem. Phys.*, **60**, 3251 (1974).
22. S. P. McGlynn, T. Azumi and M. Kinoshita, *Molecular Spectroscopy of the Triplet State*, p. 161, Prentice Hall, Englewood Cliffs, New Jersey, 1969.

Note added in proof: Recent experiments with carefully deproteinized DNA have yielded benzo[a]pyrene triplet lifetimes as long as 35 ms in rigorously degassed solutions.

B.5. Interaction and energy transfer between tyrosine and nucleic acid bases

T. Montenay-Garestier

Museum National d'Histoire Naturelle, Laboratoire de Biophysique,
61 Rue buffon 75005, Paris, France

SUMMARY

Luminescence measurements at 77 K provide evidence for efficient energy transfer at the singlet and triplet levels from tyrosine to nucleic acid bases. These results have been obtained in frozen aqueous solutions where solute molecules form aggregates. The excitation energy is shown to migrate between tyrosine molecules in tyrosine aggregates until it is trapped by a nucleic acid base or a tyrosine–base complex. In tyrosine aggregates the energy migration involves about 70 tyrosine molecules at the singlet level and several hundred molecules at the triplet level.

The critical Förster distances for energy transfer at the singlet level between tyrosine and nucleic acid bases have been calculated in fluid aqueous solutions at room temperature and in organic glasses at 77 K. These calculations show that an efficient energy transfer may occur from tyrosine to nucleic acid bases, Förster distances ranging from 11 Å to 24 Å, depending on the nature of the base.

These results as well as previous reports on tyrosine–base complex formation are discussed with respect to the role that could be played by energy transfer processes in the photochemical behaviour of protein–nucleic acid complexes.

1 INTRODUCTION

The photochemical behaviour of nucleic acid–protein complexes can be expected to be different from those of the separated macromolecules. For example, energy and electron transfer processes between nucleic acid bases and aromatic amino acids could be involved.[1-2] We are presently investigating these possibilities and the present report deals with excited state interactions

207

and energy transfer processes between tyrosine and nucleic acid bases. It has been previously reported that the rapid freezing of dilute aqueous solutions induces the formation of aggregates of solute molecules.[3] Energy transfer processes can be conveniently studied in these aggregates.[4-5] Molecular interactions due to the close proximity of molecules in these structures may also allow us to observe complex formation between two different molecular species.[6] Using this method interactions and complex formation have been already demonstrated between nucleic acid bases and aromatic amino acids such as tryptophan,[7] tyrosine,[8] and phenylalanine and histidine.[6] It has also been shown that the nature of interactions induced in these aggregates is identical to that occurring in concentrated fluid solutions.[9]

Complex formation between tyrosine or tyramine and nucleosides or nucleotides has been previously demonstrated[8] by fluorescence studies of frozen aqueous solutions at 77 K. Nuclear magnetic resonance and circular dichroism studies in aqueous solutions at room temperature showed that tyramine molecules were able to stack with adenine bases in poly(A). In frozen equimolar aqueous mixtures of tyrosine or tyramine with nucleosides, the fluorescence emission of tyrosine is quenched. The fluorescence of pyrimidine nucleosides is also quenched, whereas that of purine nucleosides is much less affected. Methylation of the hydroxyl group of tyrosine leads to the appearance of a new fluorescence at longer wavelengths in the presence of pyrimidine nucleosides. These results have been ascribed to electron transfer from the excited phenol ring to the purine or pyrimidine rings as proposed previously to explain the quenching of tryptophan fluorescence by nucleic acid bases.[7] The different behaviour of pyrimidine and purine bases is consistent with the fact that pyrimidines are better electron acceptors than purines.

The energies of the lowest singlet and triplet states of tyrosine as determined from the blue edges of the fluorescence and phosphorescence spectra in water–ethylene glycol solutions at 77 K are 37,000 cm^{-1} and 29,000 cm^{-1} respectively. These values can be compared to those of nucleotides[10] and it can be noticed that the singlet and triplet states of tyrosine have a higher energy than those of nucleic acid bases (Table 1). Thus, singlet or triplet energy transfer from tyrosine to nucleic acid bases can be expected.

Table 1   Energy of the lowest excited singlet and triplet states of nucleotides determined in a water–ethylene glycol glass at 77 K[10] and tyrosine under the same conditions

	Singlet energy (cm^{-1})	Triplet energy (cm^{-1})
Tyrosine	37,000	29,000
AMP	35,200	26,700
UMP	34,900	—
TMP	34,100	26,300
GMP	34,000	27,200
CMP	33,700	27,900

2 ENERGY TRANSFER FROM TYROSINE TO NUCLEIC ACID BASES IN FROZEN AQUEOUS SOLUTIONS

2.1 Singlet energy transfer

When small amounts of nucleosides were added to an aqueous solution of tyrosine (10^{-3} M or 2×10^{-3} M) before freezing, the fluorescence emission at 300 nm was quenched whatever the nature of the nucleoside (Figure 1). To show

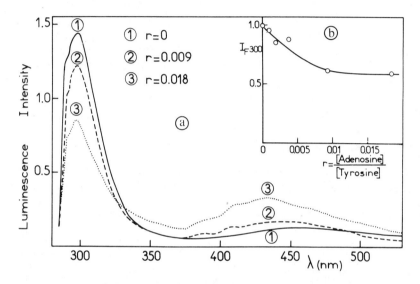

Figure 1 (a) Luminescence spectra of adenosine–tyrosine mixtures in frozen aqueous solutions at 77 K. The ratio (r) of adenosine to tyrosine concentrations is indicated on the spectra. Tyrosine concentration is constant (10^{-3} M). Excitation wavelength: 270 nm. (b) Fluorescence intensity of adenosine–tyrosine mixtures at 300 nm versus the ratio of adenosine to tyrosine concentration (see text)

that this quenching was not due to the screening effect of nucleosides at the excitation wavelength (270 nm), we have investigated tyrosine–nucleoside mixtures in a water–propylene glycol glass at 77 K. In this latter case, the molecules remained dispersed and neither energy transfer nor complex formation was expected to take place. At the concentration ratios used in the studies of aggregates in frozen aqueous solutions, we did not observe any significant decrease in tyrosine fluorescence intensity. Analysis of the fluorescence quenching-curves led to the conclusion that a single nucleoside molecule is able to quench the fluorescence of about 70 tyrosine molecules stacked in aggregates. At higher ratios of nucleoside to tyrosine concentrations, the measured fluorescence intensity at 300 nm did not vanish (Figure 1b) because the contribution of the fluorescence emission of the nucleoside became important. This is especially true in the case of purine nucleosides whose fluorescence quantum yield is not affected by interaction with tyrosine (see above). Although the fluorescence quantum yield of thymidine and cytidine is considerably reduced by interaction with tyrosine, it still contributes to the fluorescence intensity at 300 nm. In the equimolar mixture of tyrosine and thymidine, the thymidine emission is ten-fold reduced as compared to that of thymidine aggregates. Although the excited state properties of nucleoside–tyrosine complexes did not allow us to observe sensitized fluorescence of the nucleosides in tyrosine aggregates, it appears likely that the quenching of tyrosine fluorescence observed at low nucleoside–tyrosine ratios ($r < 0.1$) is due to energy transfer from tyrosine to nucleosides. One nucleoside molecule is able to quench the fluorescence of about seventy tyrosine molecules independent of the nature of the nucleoside. However the Förster critical distances for singlet energy transfer are quire different from one nucleoside to the other (see after). Therefore it appears reasonable to assume that singlet excitation energy is migrating in tyrosine aggregates from tyrosine to tyrosine until it is trapped by a nucleoside or a tyrosine–nucleoside complex. Such a singlet excitation migration was already observed in poly L–tyrosine.[11]

2.2 Triplet energy transfer

Beside tyrosine fluorescence quenching in mixed aggregates a sensitized phosphorescence of nucleosides could be observed. This sensitization appeared at lower ratios (r) of nucleoside to tyrosine concentrations and seemed to be a very efficient process, especially in the case of purine nuclosides.

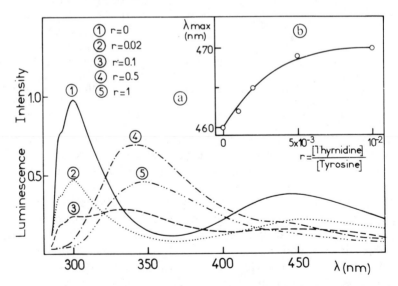

Figure 2 (a) Luminescence spectra and (b) phosphorescence maximum wavelength, as a function of the ratios of thymidine to tyrosine concentrations of thymidine–tyrosine mixtures in frozen aqueous mixtures at 77 K. The ratio (r) of thymidine to tyrosine concentrations is indicated on the spectra. Tyrosine concentration is constant (10^{-3} M). Excitation wavelength: 270 nm

For very small amounts of nucleosides added in tyrosine aqueous solutions ($r = 2 \times 10^{-3}$), the sensitized phosphorescence of the nucleoside appeared, which could be identified by its vibronic structure (in the cases of adenosine and guanosine), by its maximum wavelength (thymidine) and by its lifetime.

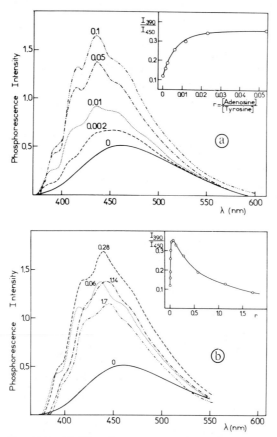

Figure 3 (a) Phosphorescence spectra of tyrosine–adenosine mixtures in frozen aqueous solutions at 77 K at low values of the ratio (r) of adenosine to tyrosine concentrations (indicated on the spectra). Tyrosine concentration is constant (10^{-3} M). Excitation at 270 nm.

Inset: Change in the phosphorescence intensity ratio at 390 and 450 nm versus r

(b) Phosphorescence spectra of adenosine–tyrosine mixtures at high values of the ratio r (indicated on the spectra). A red shift of the blue edge of the spectra is observed.

Inset: The phosphorescence intensity ratio at 390 nm and 450 nm decreases at high ratios of adenosine to tyrosine concentrations: stacks of adenosine are formed in tyrosine aggregates (see text)

The half-life of tyrosine phosphorescence in aggregates was 0·4 s (non-exponential decay).[12] Frozen aqueous solutions of tyrosine containing 1 % of adenosine or thymidine exhibited phosphorescence emission with 2·5 s and 0·26 s lifetimes, respectively. At the concentrations of nucleosides used in these experiments (low values of r), the phosphorescence intensity of the unsensitized nucleoside would be quite weak. Quantitative studies of the ratio of the phosphorescence intensities at two different wavelengths and of the maximum phosphorescence wavelength shift as a function of nucleoside concentration demonstrated that several hundreds of tyrosine molecules were involved in triplet excitation energy transfer to nucleosides. For a concentration ratio of 10^{-2} (Nu/Tyr) the sensitized phosphorescence reached a plateau (Figures 2a, 3a, 4b) from which it could be calculated that triplet energy migrating over 170, 200 and 700 tyrosine molecules was trapped by a thymidine, adenosine and guanosine molecule, respectively. This very efficient sensitization must be attributed to a triplet–triplet energy transfer from tyrosine to nucleosides. Since triplet–triplet energy transfer requires a good overlap of the electronic clouds of the donor and acceptor molecules,[13] the results obtained here led to

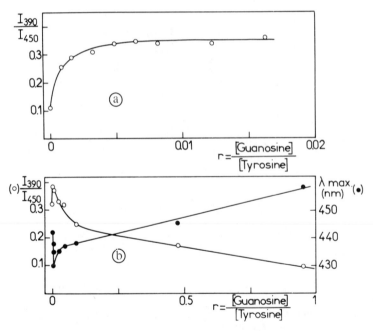

Figure 4 (a) The phosphorescence intensity ratio at 390 nm and 450 nm guanosine–tyrosine mixtures in frozen aqueous solutions at 77 K versus ratio of guanosine to tyrosine concentrations.
(b) Change of the phosphorescence intensity ratio (○) and shift of the phosphorescence maximum wavelength (●) at high ratios of guanosine to tyrosine concentrations. Stacks of guanosine are formed in tyrosine aggregates (see text). Tyrosine concentration is constant $(10^{-3}$ M)

the conclusion that triplet excitation energy migrates in tyrosine aggregates (hopping model) until it is trapped by a nucleoside molecule whose phosphorescence is sensitized.

When the ratio of nucleoside and tyrosine concentrations increases from 0·01 to 1 (or more), a new phenomenon appeared due to an heterogeneity of solute distribution in the aggregates. An important red shift and an intensity decrease of the sensitized phosphorescence of the purine nucleosides occurred (Figures 2b, 3b). This behaviour has been ascribed to the formation of short stacks of each of the solute molecules rather than that of a regularly alternated structure. These phenomena have already been observed in mixed aggregates of tryptophan derivatives and nucleosides.[5] In this work, no evidence for sensitization of cytidine or uridine phosphorescence could be provided, probably because tyrosine and cytidine phosphorescence spectra are too similar and the phosphorescence quantum yield of uridine is too weak.

3 CALCULATION OF ENERGY TRANSFER PROBABILITY

3.1 Förster critical distances

In order to understand these experimental results, we undertook some theoretical calculations. We have evaluated the Förster critical distances R_0 for singlet energy transfer between tyrosine and nucleosides at room (300 K) and low (77 K) temperatures (Tables 2 and 3).

Table 2 Förster critical distance R_0 for singlet energy transfer *from tyrosine to nucleosides* in aqueous solutions at room temperature. (Φ_D of tyrosine is taken equal to 0·21)[15]

Nucleoside	$J(10^{-28}\,\mathrm{M}^{-1}\,\mathrm{m}^6)$	R_0 (Å)
Guanosine	5·7	16·8
Cytidine	4·9	16·3
Thymidine	3·75	15·6
Uridine	1·2	12·9
Adenosine	0·45	10·9

Table 3 Förster critical distance R_0 for singlet energy transfer *from tyrosine to nucleosides* in ethanolic solutions at 77 K. (Φ_D of tyrosine is taken equal to 0·425)[15]

Nucleoside	$J(10^{-27}\,\mathrm{M}^{-1}\,\mathrm{m}^6)$	R_0 (Å)
Guanosine	2·3	23·6
Cytidine	1·7	22·2
Thymidine	1·3	21·2
Uridine	0·95	20·1
Adenosine	0·9	20

The following conditions are assumed:

(1) The dipolar coupling between donor and acceptor is supposed to be small compared with the unresolved absorption band of the acceptor, that is to say that we are in the 'very weak coupling' case of Förster theory.[14]

(2) The dipole approximation is valid, which means that donor–acceptor distances are large compared with the intramolecular distances so that higher multipoles do not need to be taken into consideration and exchange interactions are negligible.

(3) Vibrational relaxation to the lowest excited state of the donor is fast compared with the energy transfer rate (after-relaxation transfer[10]). This approximation will be verified below.

The distance R_0 at which the rate of energy transfer from the donor is equal to the sum of rates of all other modes of deactivation is given by:

$$R_0^6 = 8.8 \times 10^{-25} \frac{\kappa^2}{n^4} \Phi_D J$$

where J is the overlap integral

$$\int F(\bar{v}) \cdot \varepsilon(\bar{v}) \cdot \frac{d\bar{v}}{\bar{v}^4}$$

$F(\bar{v})$ is the fluorescence quantum spectrum normalized to unity on a wave-number scale (m^{-1}), $\varepsilon(\bar{v})$ is the molar decadic extinction coefficient, n is the index of refraction of the medium intervening between the donor and acceptor, κ is the dipole–dipole orientation factor and Φ_D is the donor emission quantum yield.

The index of refraction is taken equal to 1·333 at room temperature for aqueous solutions and 1·359 for ethanolic glasses at liquid-nitrogen temperature. In all calculations an average value of κ^2 equal to $\frac{2}{3}$ has been used (see ref. 21). The value of the tyrosine fluorescence quantum yield is still the subject of discussions.[15] The value of 0·21 used in the above calculations is that measured by Teale and Weber.[16]

We cannot evaluate transfer distances in the opposite direction from nucleosides to tyrosine at 300 K since the room-temperature fluorescence quantum yield of nucleosides is negligible (Φ_D are in the range of 10^{-5}–10^{-6} [17]).

At low temperature, the overlap between the fluorescence spectrum of tyrosine and the absorption spectra of nucleosides is more important than at room temperature and Förster distances are increased accordingly (see Table 3). These values calculated above for energy transfer distances from tyrosine to nucleosides at room temperature as well as at low temperature are slightly higher than those calculated for energy transfer from tyrosine to tryptophan (14·7 Å and 17 Å at 300 and 77 K respectively).[18] Calculations relative to the transfer in the opposite direction (from bases to tyrosine) lead to non-negligible values of critical distances (Table 4). These distances are within a factor of one-half of the distances of transfer from tyrosine to bases. Since the probability of transfer increases as the sixth power of the critical distance, transfer from tyrosine to nucleic acid bases will be much more efficient than the transfer in the opposite direction (see ref. 21).

Table 4 Förster critical distance R_0 for singlet energy transfer *from nucleosides to tyrosine* in ethanolic glasses at 77 K. (The fluorescence quantum yields of nucleotides used are those given in Reference 10)

Nucleoside	$J(10^{-29} \text{ M}^{-1} \text{ m}^6)$	R_0 (Å)
Guanosine	0·0475	4·65
Cytidine	4·5	8·5
Thymidine	1·62	8·8
Adenosine	2·75	6·0

This efficient singlet energy transfer from tyrosine to nucleosides has not yet been demonstrated at room temperature. We are presently investigating this possibility.

At liquid-nitrogen temperature energy transfer at the singlet level involves 70 molecules of tyrosine whatever the nature of the nucleoside present in the solution (see above). This result seems to demonstrate that the transfer between tyrosine residues is predominant over the tyrosine–nucleoside transfer which can occur at the final stage only. This transfer between tyrosine molecules in aggregates may occur by two different mechanisms (exchange and Förster mechanisms). The R_0 distance for singlet energy transfer between two tyrosine molecules in ethylene glycol–water glass has been calculated to be 11 Å at 77 K.[18]

3.2 Triplet energy transfer

At the triplet level energy transfer occurs by an exchange mechanism the probability of which increases with the overlap of the phosphorescence spectrum of the donor and the singlet–triplet absorption spectrum of the acceptor.[13] Although singlet–triplet absorption spectra of nucleosides are not known, the overlap of the phosphorescence spectrum of tyrosine obtained in a glass at 77 K and the mirror-image of the phosphorescence spectra of nucleosides under the same conditions seems to be correlated with the number of tyrosine molecules whose phosphorescence is quenched by a nucleoside molecule (it decreases in the order G > A > T).

4 CONCLUSION

The different energy transfer processes described here might be of importance in the photochemistry of nucleic acid–protein complexes. The aromatic amino-acids, tyrosine and tryptophan, may participate in complex formation by a direct stacking interaction with the bases.[19,20] Photochemical studies of the complexes should take into account the possibility of energy transfer processes as well as the change in excited-state properties of the complexes compared with separated molecules. Even without stacking interactions, the eventuality of energy transfer mechanisms at relatively long distances should be taken into account in the photochemical behaviour of nucleic acid–protein complexes.

ACKNOWLEDGEMENTS

I wish to thank Dr C. Hélène for helpful discussions and criticism of the manuscript.

REFERENCES

1. C. Hélène, *Photochem. Photobiol.*, **18**, 255–262 (1973).
2. C. Hélène and M. Charlier, *Biochem. Biophys. Res. Comm.*, **43**, 252–257 (1971).
3. C. Hélène, *Biochem. Biophys. Res. Comm.*, **22**, 237–242 (1966).
4. C. Hélène and Th. Montenay-Garestier, *Chem. Phys. Letters*, **2**, 25–28 (1968).
5. Th. Montenay-Garestier and C. Hélène, *J. Chim. Phys.*, **10**, 1391–1399 (1973).
6. Th. Montenay-Garestier and C. Hélène, *J. Chim. Phys.*, **10**, 1385–1390 (1973).
7. Th. Montenay-Garestier and C. Hélène, *Biochemistry*, **10**, 300–306 (1971).
8. C. Hélène, Th. Montenay-Garestier and J. L. Dimicoli, *Biochim. Biophys. Acta*, **254**, 359–365 (1971).
9. J. L. Dimicoli and C. Hélène, *Biochimie*, **53**, 331–345 (1971).
10. M. Guéron, J. Eisinger and R. G. Shulman, *J. Chem. Phys.*, **47**, 4077–4091 (1967).
11. J. W. Longworth and R. O. Rahn, *Biochim. Biophys. Acta*, **147**, 526–535 (1967).
12. J. Nag-Chaudhuri and L. Augenstein, *Biopolymers Symp.*, **1**, 441–452 (1964).
13. D. L. Dexter, *J. Chem. Phys.*, **21**, 836–850 (1953).
14. Th. Förster, in *Modern Quantum Chemistry*, Vol III, pp. 93–137 (ed. O. Sinanoǧlu), Academic Press, N.Y. and London, 1965.
15. J. W. Longworth, in *Excited States of Proteins and Nucleic Acids*, pp. 319–484 (eds. R. F. Steiner and I. Weinryb), Plenum Press, N.Y., 1971.
16. F. W. J. Teale and G. Weber, *Biochem. J.*, **65**, 476–482 (1957).
17. P. Vigny, *C.R. Acad. Sc. Paris*, **D272**, 3206–3209 (1971).
18. J. Eisinger, B. Feuer and A. A. Lamola, *Biochemistry*, **8**, 3908–3915 (1969).
19. C. Hélène, J. L. Dimicoli and F. Brun, *Biochemistry*, **10**, 3802–3809 (1971).
20. C. Hélène and J. L. Dimicoli, *F.E.B.S. Letters*, **26**, 6–10 (1972).
21. Th. Montenay-Garestier, *Photochem. Photobiol.*, **21**, in press (1975).

B.6. Photolysis of 5-bromouracil in solution

J. M. Campbell, C. von Sonntag, and D. Schulte-Frohlinde

*Institut für Strahlenchemie im Max-Planck-Institut für Kohlenforschung,
4330 Mülheim a.d. Ruhr, Stiftstraße 34–36, Germany*

5-Bromouracil is able to replace thymine in the DNA molecule of bacteria and phages. In this way the cells and the phages are made more sensitive to radiation damage and therefore can be deactivated by significantly smaller radiation doses.[1]

It is presently assumed that absorption of UV light by 5-bromouracil (BU) results in a homolytic splitting of the C—Br bond to give a uracilyl radical and a Br atom (reaction 1).

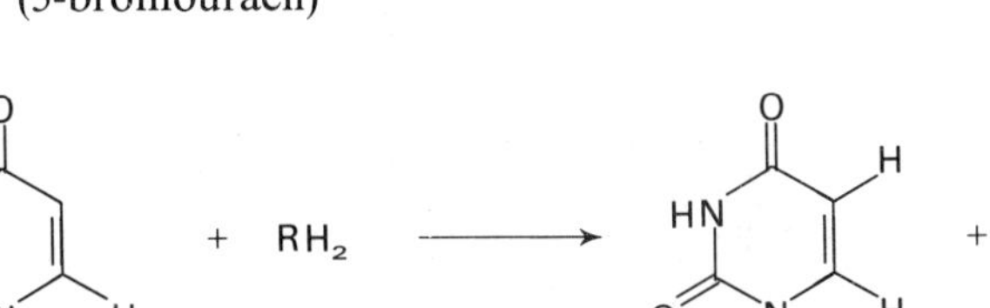

$$\tag{1}$$

(5-bromouracil)

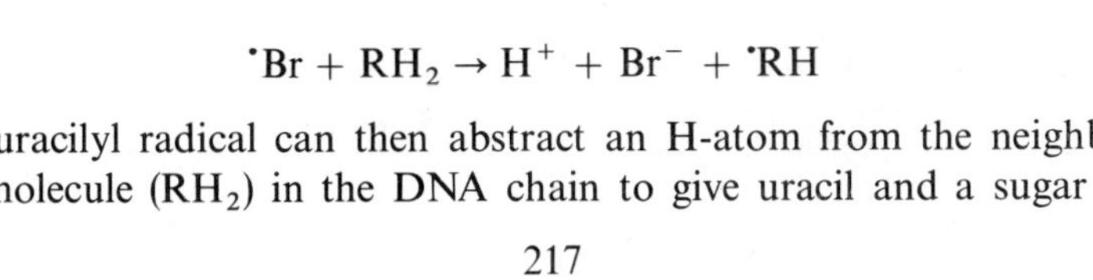

$$\tag{2}$$

(uracil)

$$\cdot Br + RH_2 \rightarrow H^+ + Br^- + \cdot RH \tag{3}$$

The uracilyl radical can then abstract an H-atom from the neighbouring sugar molecule (RH_2) in the DNA chain to give uracil and a sugar radical (reaction 2). This resulting sugar radical leads to a break in the DNA chain at the phosphate ester bond.[2] This chain break prevents reproduction of the DNA and is thus lethal if the break cannot be repaired by the cell. If this interpretation is correct, the quantum yield of BU destruction should compare with the quantum yield of chain breaks in the BU-substituted DNA (BU–DNA). The quantum yield of single strand breaks in BU–DNA in aqueous solutions has been shown to be over a factor of 10 times larger than the quantum yield of BU decomposition also in an aqueous solution.[2,3] In order to determine the reason for this discrepancy, the photolysis of BU in solution has been investigated in more detail.

In an oxygen-free aqueous solution in the presence of t-butanol as H-atom donor (RH_2 in reactions (2) and (3)), the photolysis of BU leads to uracil and

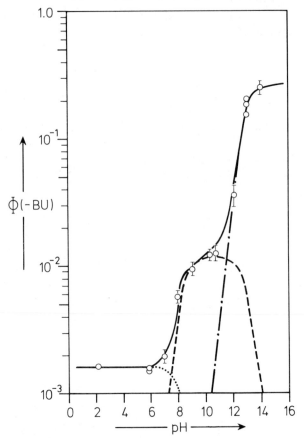

Figure 1   Photolysis of 5-bromouracil (3–6×10^{-4} M) in deoxygenated aqueous solution containing t-butanol (1.6×10^{-1} M) at 254 nm. Quantum yield for the loss of BU: ○, experimental; ——, calculated ϕ (—BU); ....., neutral form; – – – –, monoanion; —·—·—, dianon

HBr (reactions (2) and (3)). In methanol as a solvent the primary stable products are also uracil and HBr.

The formation of uracil, H^+ and Br^- corresponds to the loss of bromouracil. At higher conversions of BU, uracil is destroyed by secondary photolysis reactions to give predominantly the uracil-5, 6-hydrate. In the region of pH 2 to 11 the results are in agreement with a homolytic splitting of the C—Br bond as the primary photolytical step. Photolysis of the dianion in strongly alkaline solution gives rise to a further UV absorbing product. The solvated electron has also been detected using N_2O in the pH > 12 region.[4] The quantum yield of BU destruction at 254 nm increases stepwise from a value of 1.6×10^{-3} at low pH to 0·28 at pH 14 (Figure 1). The inflection points correspond to the two known pK values of BU in the ground state[5] indicating a dissociation of the C—Br bond prior to the establishment of the ionic equilibrium in the excited state.

The quantum yield of bromouracil destruction has also been studied as a function of wavelength (Table 1), alcohol concentration (Figure 2) and solvent composition (Table 2). In aqueous solution at low concentration of H-donors

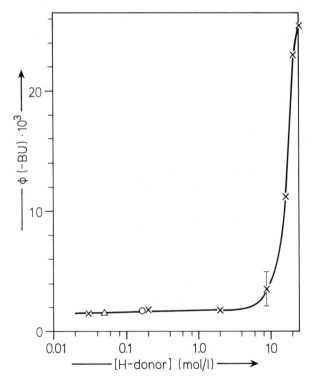

Figure 2   Quantum yield of 5-bromouracil (2×10^{-4} M) disappearance at 254 nm and 20 °C in aqueous solution as a function of the concentration of radical scavengers: ×, methanol; △, *iso*propanol; ○, *t*-tutanol

the same quantum yield of BU decomposition is found independent of the nature of the H-donor (Figure 2). In pure H-donor solvents the quantum yield of BU consumption is much higher and depends on the nature of the solvent (Table 2).

Table 1 Photolysis of 5-bromouracil in deoxygenated aqueous solution containing t-butanol ($1\cdot6 \times 10^{-1}$ M) at 20 °C. Quantum yield for the loss of BU in different ionic forms

	282 nm	254 nm	214 nm
Bromouracil, neutral form pK = $8\cdot05^{5}$	$2\cdot9 \times 10^{-4}$	$1\cdot6 \times 10^{-3}$	$4\cdot0 \times 10^{-2}$
Bromouracil, monoanion pK = 13^{5}	$1\cdot6 \times 10^{-3}$	$1\cdot4 \times 10^{-2}$	
Bromouracil, dianion	$2\cdot2 \times 10^{-1}$	$2\cdot8 \times 10^{-1}$	

Table 2 Quantum yields for the consumption of 5-bromouracil ($\lambda = 254$ nm)

Solvent	Quantum yield $\times 10^{3}$
0, 16 mol/1 t-Butanol in H_2O	1·6
t-Butanol	5·5
iso-Propanol	18
Methanol	25
Tetrahydrofuran	45
BU-DNA in water[3]	~28

The increase in quantum yield of BU consumption in aqueous solution with increasing H-donor concentration occurs only at high concentrations of donor. The quantum yield rises from $1\cdot6 \times 10^{-3}$ in less than 40 % volume methanol in water to $2\cdot5 \times 10^{-2}$ in pure methanol. These results are in agreement with the assumption that the H-donors in high concentration react directly with the initially formed radicals within the solvent cage. It follows from this that the higher quantum yield observed for chain breaks in the photolysis of BU incorporated in DNA compared to that for photolysis of BU in aqueous solution is due to the high *local* concentration of H-atom donors (sugar molecules) within the DNA molecule even if the DNA molecule as a whole is present in aqueous solution. A sugar molecule in the backbone of the DNA is in direct contact with the Br atom of the bromouracil in the BU–DNA and acts as a scavenger of the initially formed uracilyl and bromine radicals.[6]

REFERENCES

1. B. Djordjevic and W. Szybalsky, *J. Exptl. Med.*, **112**, 509 (1960).
2. F. Hutchinson, *Quart. Rev. Biophys.*, **6**, 201 (1973).

3. Lion, private communication. Lion calculates the quantum yield of strand breaks as the number of breaks observed per number of photons absorbed in the BU nucleotides, assuming the BU nucleotides absorb only the fraction of light corresponding to the mole fraction of BU in the DNA. This disregards the number of photons absorbed in the other nucleotides and assumes negligible energy transfer to the BU moiety.

4. M. E. Langmuir and E. Hayon, *J. Chem. Phys.*, **51**, 4893 (1969).

5. K. Berens and D. Shugar, *Acta Biochemica Polonica*, **10**, 25 (1963).

6. For more complete data see: J. M. Campbell, C. von Sonntag and D. Schulte-Frohlinde, *Z. Naturforsch*, **29b**, 750 (1974).

B.7. Luminescence lifetimes of thymine and uracil containing DNA†

J. W. Longworth, V. L. Koenig‡ and W. L. Carrier

Biology Division, Oak Ridge National Laboratory, Oak Ridge, TN 37830, U.S.A.
‡Division of Biochemistry, University of Texas
Medical Branch, Galveston, TX 77550, U.S.A.

ABSTRACT

PBS-2, a phage of *B. subtilis* has a DNA which lacks thymine but contains uracil in its place. The luminescence features of this DNA were compared with calf thymus DNA, and were in general similar in character. The phosphorescence of thymine DNA in its native conformation is characterized by a structureless spectrum associated with thymine residues. PBS-2 DNA possesses a similar structureless spectrum. A comparable phosphorescence is noted from the synthetic polynucleotide poly r(A–U) · (A–U)$_n$. Uracil in polynucleotides has a triplet energy comparable to thymine.

Solution of DNAs in a neutral organic solvent, ethylene glycol, creates a random coil conformation. The phosphorescence is now predominantly from adenine residues for both DNA samples. Adenine is readily identified by its fine structure spectrum and decay constant.

The phosphorescence decay of DNA is composed of two components in glycol:water mixtures, 0·3 s and 2 s. For thymine DNA it is the shorter component which predominates, uracil DNA the longer. Phosphorescence risetimes appear homogeneous and intermediate between the two components. A multi-exponential decay, though faster is observed in ice matrices. There is clearly considerable complexity in the sites of phosphorescence in native DNA.

The fluorescence lifetime of both thymine and uracil DNA was determined at 77 K, and is *c.* 10 ns. Denaturation with acid, alkali or neutral organic solvent considerably shortened the lifetime (2–4 ns) and shifted the peak from 370 nm to 335 nm. We suggest this arises from the exciplex nature of fluorescence of native DNA.

† Research sponsored by the U.S. Atomic Energy Commission under contract with the Union Carbide Corporation.

B.9. Studies on cellular fluorescence excited by a nanosecond-pulsed tunable laser†

S. Cova

Istituto di Fisica del Politecnico, Milan, Italy

G. Prenna

Centro di Studio per l'Istochimica del C.N.R. Istituto di Anatomia Comparata dell'Università, Pavia, Italy

C. A. Sacchi and O. Svelto

Laboratorio de Fisica del Plasma ed Elettronica Quantistica del C.N.R., Istituto di Fisica del Politecnico, Milan, Italy

SUMMARY

An apparatus is described that is suitable for measurements of fluorescent emission from single cells. The apparatus is based on the use of a nanosecond-pulsed tunable dye laser and of pulse-electronic techniques. The apparatus allows high-sensitivity fluorometry and measurements of the main parameters of the fluorescent transition. Consideration is given, in particular, to measurements of the decay time, of the quantum yield and of the absorption cross-section, which are illustrated by experimental examples.

1 INTRODUCTION

A current trend in fluorescence microscopy is to study biophysical and bio-chemical processes in single cells, eventually in the living state. The aim of such studies is to obtain information on the connections between the structure of biological macromolecules and the functional activities, the differentiation and the ambient of the cells. Typical examples of this trend are (i) biophysical studies of the secondary structure of nucleic acids, using acridine derivatives as a

† Work supported by the Italian Consiglio Nazionale delle Ricerche.

fluorescent molecular probe;[1,2,3] (ii) biochemical studies of enzyme reactions, which either result in a spontaneous fluorescence emission[4] or take place on a fluorogenic substrate.[5,6] A motivation for this trend is the fact that information of great biological interest has already been obtained by fluorescence studies performed *in vitro*, i.e. on suitable solutions containing the molecules (proteins, nucleic acids, etc....) to be studied, on an ensemble of subcellular particles, isolated membranes or their artificial models. For instance, data have been obtained on the structure of nucleic acids and proteins and on structural and functional properties of membranes etc.[7,8,9,10]

The fluorescence emissions to be measured in biological samples may be due either to particular constituents of the biological substance (*primary fluorescence*) or to suitable fluorescent dyes, which are attached to specific positions of the biological macromolecules (*secondary fluorescence*). The emission characteristics of these dyes are in general dependent on the interactions with the macromolecule. Both in primary and secondary fluorescences the following properties of the emitted light are of interest: (a) the emission and excitation spectra, (b) the polarization, (c) the decay time and (d) the quantum yield.

Apart from the histochemical problems involved in the selection of suitable staining techniques, in fluorescence studies on single cells problems are encountered concerning the measurement and instrumentation, as many difficulties are caused by the complexity, the inhomogeneity and the small dimensions of the cells. Several instruments have so far been developed;[2,11,12,13] however, they are basically able only to measure the spectral parameters (a) above. Furthermore, since they use incoherent lamps as excitation sources (e.g. mercury or xenon lamps), they suffer from a few limitations, namely: (i) the lamp brightness is somewhat limited, and measurements on low concentrations of the fluorescent material, are either not possible, or require a long time; (ii) the use of a continuous light source often represents a serious limitation, because of photodecomposition.

The present paper, which follows a preliminary one on the same subject,[15] reports more extensive results obtained by using a pulsed, tunable dye laser as the excitation source in a fluorescence microscope. Although our apparatus can in principle measure all the above parameters (a)–(d), the present work will deal only with measurements of parameters (c) and (d). In particular, due to the high brightness and short duration (~ 2 ns) of our laser pulses, reliable decay time measurements of the light emitted by a single cell have been made possible for the first time. Other fluorescence quantities of biological interest have also been measured directly, as discussed below. The experimental cases to be presented have been selected to illustrate the potentialities of our system rather than on the basis of their direct biological relevance.

2 EXPERIMENTAL APPARATUS

The experimental apparatus is shown in the block diagram in Figure 1, and is basically the same as that described previously.[15] The dye laser uses a solution

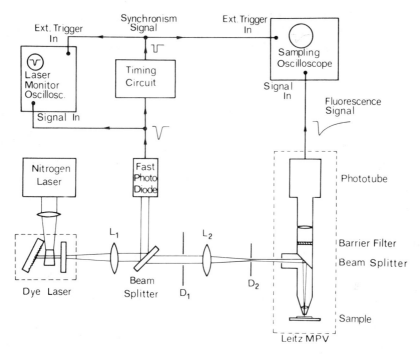

Figure 1　Block diagram of the experimental apparatus

of Rhodamine 6G in methanol and it is pumped by the light emitted at 337·1 nm by a pulsed nitrogen laser. It gives a continuous train of light pulses with a repetition rate of up to 100 pps. Each pulse has a duration of ~ 2 ns and a peak power of up to ~ 50 kw. The optical bandwidth is limited to ~ 1 nm by a diffraction grating, and its centre can be tuned from 560 to 630 nm. The same laser is potentially tunable, however, from 410 to 670 nm by using other laser dyes. The laser light is sent into a Leitz MPV cytofluorometer (resolving power of the microscope $\sim 0·3\ \mu m$). The laser spot size on the plane of the sample can simply be changed by changing the lenses L_1 and/or L_2. As the beam is practically diffraction-limited, it can be concentrated down to the resolving spot of the microscope, as experimentally verified. In the experiment, a spot size of $\sim 40\ \mu m$ was, however, used so as to obtain uniform excitation of the cell nucleus.

Two different fast detectors were used for the incoming and the fluorescence light signals. To monitor the incoming laser light, a beam-splitter was placed after L_1 and a fraction of the beam was thus sent to a fast photodiode (Hadron TRG, mod 105C, risetime $<0·3$ ns). To monitor the fluorescence light a fast photomultiplier tube (PMT) (RCA 70045D, risetime $\sim 0·7$ ns) was mounted at the top of the Leitz MPV instrument. The electrical signal of the fast photodiode served the dual purpose of monitoring the laser power and providing a suitable signal for a precision timing circuit. The high PMT amplification also enables single photons to be detected. On the other hand, when multi-photon fluorescence pulses are measured, the peak intensity must be limited in order to keep the PMT output current within the linear range. Optimum operating conditions (i.e. total voltage and voltage divider) are different for single-photon and multi-photon pulse measurements. As a great flexibility was required during this stage of the work, compromise conditions were chosen. The electrical PMT pulses were sent to a fast oscilloscope, and the output pulse from the timing circuit was used as an external trigger signal. Synchronous measurements were thus obtained down to single-photon levels. In most measurements, a sampling oscilloscope with a storage tube (Tektronix 564) was employed, thus making use of its versatile internal facilities (see § 3). Conventional, fast oscilloscopes were also used in cases where single pulses, e.g. as required at single photon levels, had to be observed.

3 EXPERIMENTAL INVESTIGATIONS

The experimental investigations were performed in three main directions: (a) fluorometry with high sensitivity, i.e. on cellular samples in which the fluorescent substance was of low concentration and/or low quantum yield; (b) fluorescence waveforms and related physical parameters (lifetimes etc.); and (c) saturation behaviour of the fluorescence and related parameters (quantum yield etc.).

(a) High-sensitivity fluorometry

In fluorometry with conventional stationary excitation sources (e.g. xenon or mercury lamps) problems are met with weakly fluorescent samples. The detected intensity is often so low, that the output of the PMT is a random sequence of single-photon pulses: their average rate may even be lower than the rate of the *background pulses uncorrelated to excitation*. The latter pulses originate from spurious electron emission in the PMT (dark current) and from detection of stray light. Only a fraction of them can be discriminated by a pulse-height selection;[17] their average rate p_B varies with the experimental conditions (usually $p_B = 10^2$ to a few 10^4 pps) and is often affected by *drift*. In order to obtain reliable measurements it is therefore often necessary to use long averaging times, modulation of the fluorescence light (choppers etc.) and synchronous measurements (lock-in detectors or synchronous reversible single-photon counting).[18]

A remarkable increase in the stationary fluorescence intensity may be obtained by replacing conventional sources with a continuous-wave tunable dye laser, in order to have a high-intensity, narrow band excitation. However, photodecomposition of the fluorescent substance[14] and saturation of the fluorescence (see (c) below) set limits to the improvement. In particular, at high excitation intensity, significant photodecomposition is observed after absorption of small energy quantities, i.e. in short times (< 1 s).

The use of a pulsed dye laser and of pulse-measurement techniques in practice avoids the limitation due to photodecomposition. A high-excitation intensity ($\sim 10^5$ times that of conventional lamps) is obtained over a well-defined and very short time-interval T_L (a few nanoseconds). The average number $n_B = p_B T_L$ of *uncorrelated* background pulses falling within this interval is very low (in the range of 10^{-6} to 10^{-4}). The average number n_f of detected fluorescence photons in the same interval can therefore be much higher, while the absorbed energy per pulse is so limited, that no significant photodecomposition is observed after many pulses.[15] By using a synchronous measurement over this interval, a high value of the ratio of fluorescence to uncorrelated background can thus be obtained, even at single-photon levels (i.e. at $n_f < 1$) if averaging is used. This has been experimentally verified, by using a conventional fast oscilloscope triggered by the synchronism signal. The reference time position corresponding to the fluorescence was preliminarily determined by observing intense fluorescences. Weakly fluorescent cells, whose measurement was unpractical by conventional stationary excitation, were then observed. The rate of pulses falling within 10 ns around the reference position and in a displaced position was observed, and the former was always found to be much higher.

With suitable procedures for accurate subtraction of the uncorrelated background,[19] even measurements of fluorescences that give n_f comparable to n_B may be envisaged. However, the lower limit to the measurement of weak fluorescences is probably due to spurious fluorescences, which constitute a *background correlated to the excitation.*

(b) Fluorescence waveform

A basic objective of the waveform measurements in pulsed fluorescence is to measure the fluorescence decay time τ. This parameter depends on intramolecular and intermolecular interactions, as binding between macromolecules and energy transfer between absorbing and emitting chromophores. Structural information about biological systems may be thus obtained by measuring the τs and their variations due to biological processes. Different fluorescences can be identified by their τ-values. In cases where more than one fluorescence is excited by the laser pulse, it may be possible to measure the different components separately, by a time-domain analysis, if the waveform is measured with high precision.

In this stage of our work, the simple method of measuring PMT current waveforms with a sampling oscilloscope (see §2) was used, to obtain measurements of τ with a simple, fast procedure, although not with high precision and accuracy. The laser intensity was adjusted to give fluorescence pulses above the single-photon level, but within the linear range of the PMT and well below the saturation of the fluorescence (see (c) below). The oscilloscope was used in single-sweep mode, and various measurements were taken alternately (to avoid drift effects in the oscilloscope) of the fluorescence pulse and of the laser pulse, reflected towards the PMT by a glass slide in the object plane. In this

228

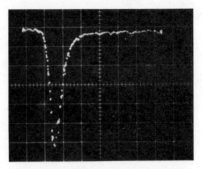

Figure 2 Typical laser pulse, as detected by the fast photodiode (Hadron TRG, mod. 105 C, risetime <0.3 ns) and displayed on the sampling oscilloscope (Tektronix 564 with S2 sampling head, 2 ns/div.)

last case, the barrier filter indicated in Figure 1 had been removed. The laser pulse (Figure 2) gives the resolution function of the apparatus for this time-domain analysis (e.g. see Reference 17). An experimental example is shown in

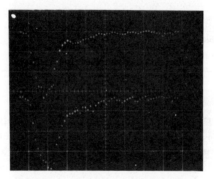

Figure 3 Typical PMT current waveforms, as displayed in single-sweep operation of the sampling oscilloscope, (2 ns/div.) triggered by the synchronism signal (see Figure 1).
(a) Laser pulse. The laser light is reflected toward the PMT by a glass slide in the object plane; the barrier filter has been removed.
(b) Fluorescence pulse given by a single cell-nucleus of frog's erythrocytes stained with a water solution (0.3%) of Rhodamine 3GO (pH 1.2)

The relative displacement of the barycentres of the two pulses is a measure of the fluorescence decay time ($\tau = 0.85 \pm 0.08$ ns, see text)

Figure 3. Average waveforms were then computed by using 5 to 10 photographs to smooth the fluctuations due to the statistics of the low number of detected photons and of the PMT amplification (practically no contribution was due to fluctuations of the laser intensity, see § 2). Values of τ were computed by a first-moment analysis, i.e. by the shift in the barycentre of the fluorescence pulse in respect of that of the laser pulse. Figure 3 refers to a sample made of frogs' erythrocytes, stained with a water solution (0·3%) of Rhodamine 3 GO (pH = 1·2). In this case, the fluorescence decay time is evaluated as $\tau = 0·85 \pm 0·08$ ns. It is worth noting that the same dye was measured to give a different value of $\tau = 2·8 \pm 0·2$ ns in an *in vitro* solution.

The measurement made with the present technique on the frogs' erythrocytes stained by the conventional Feulgen reaction for DNA demonstration, confirmed the value of $\tau = 0·2 \pm 0·05$ ns previously found in a different way.[15]

(c) Saturation of the fluorescence emission

To illustrate the saturation phenomenon, let us first consider, for the sake of simplicity, the case of continuous laser excitation. In this case, if we plot the fluorescent power *vs* the laser intensity, we find that the plot is linear only at low laser intensities. At high laser intensities, indeed, the fluorescence increases less than linearly and tends to reach a limiting value (saturation). This saturation is due to the fact that, at high laser intensities, an appreciable number of molecules will be raised to the upper fluorescent level. The absorption coefficient of the material (which is proportional to the difference between the numbers of molecules in the lower and upper levels) will thus decrease, and this will in turn produce a decrease in fluorescent intensity.

For a time-varying excitation (as in our case), the phenomenon is more complicated to describe since account should also be taken of the finite delay between excitation and fluorescent emission. The problem is, however, amenable to a simple analytical formulation. In fact it can be shown[20] that the total fluorescent power† P_f of a sample of volume V is related to the incident laser intensity I_L by the following differential equation:

$$\frac{dP_f}{dt} + \frac{P_f}{\tau}\left(1 + \frac{I_L}{I_s}\right) = k\,I_L, \tag{1}$$

in which τ is the upper state lifetime and $k = \alpha V \omega_f/\tau_{sp}\omega_L$. Here α is the absorption coefficient, τ_{sp} is the spontaneous emission lifetime, and ω_f and ω_L are the peak fluorescence and laser excitation frequencies, respectively. The quantity I_s, which appears in the brackets, is a characteristic intensity (called the saturation intensity) given by the expression:

$$I_s = \hbar\omega_L/2\sigma\tau \tag{2}$$

in which σ is the absorption cross-section of the material transition, calculated at the frequency ω_L. Note that the quantity I_s depends only on the given

† Of course, only a fraction of P_f will be collected by the microscope objective and sent to the PMT.

transition and on the frequency ω_L of the laser excitation. Its physical meaning is easily obtained by considering a steady-state regime, wherein $dP_f/dt = 0$. In this case, from (1), we get

$$P_f = (k\tau I_s) \frac{(I_L/I_s)}{1 + (I_L/I_s)} \tag{3}$$

which shows that for $I_L = I_s$, the fluorescent power P_f is $(1/2)$ of that which would correspond to the linear regime.

It is useful to measure the quantity I_s since its value leads directly, through (2), to the measurement of $\sigma\tau$. This makes possible an evaluation (done directly in a single cell) of the quantum yield η_q of the fluorescent transition. In fact it can be shown[20] that η_q is related to $\sigma\tau$ by the equation

$$\sigma\tau = \left(\frac{\lambda}{2\eta}\right)^2 g_t(\omega_L)\eta_q \tag{4}$$

in which λ is the wavelength of the excitation light, η is the refractive index of the medium and $g_t(\omega_L)$ is the value of the normalized lineshape absorption spectrum at the excitation frequency ω_L. The lineshape spectrum $g_t(\omega)$ is intended to be normalized so that

$$\int_0^\infty g_t(\omega)\,d\omega = 1.$$

Furthermore, since τ can be independently obtained as previously shown,† the measurement of I_s makes possible the measurement of σ directly in the cell.

In our case we are dealing with a time-varying excitation: therefore, in order to measure I_s it is convenient to consider the time at which the maximum of the fluorescent emission occurs. Here $dP_f/dt = 0$, and from (1) we get again equation (3). This means that (3) remains valid, also for time-varying excitations, provided that P_f and I_L are respectively the peak of the fluorescent power and the corresponding laser intensity, measured at the same time instant.

The validity of (3) has been verified and the corresponding values of I_s have been measured for our biological samples. The measurement can be performed both in the steady state (i.e. in the case where τ is much shorter than the laser pulse duration) and in the transient regime by using the sampling oscilloscope triggered by the synchronism signal.

Figure 4 shows the plot of the experimental points obtained in the case of frogs' erythrocytes, stained by the conventional Feulgen reaction for DNA demonstration. The solid line is the theoretical behaviour described by (3) with $I_s = 2.8 \times 10^6$ W/cm². The theoretical curve fits the experimental points well, thus indicating that the saturation model is a good description of the physical process. Since the value of τ for this case, as measured with our system, was $\tau \simeq 0.2$ ns,[15] from knowledge of I_s and τ we obtain from (2) that

---

† It is worth noticing here that the previous decay time measurements were made at a laser intensity I_L, much smaller than I_s. In this case, the bracketed term of (1) can be approximated to unity and the equation can be reduced to the well-known linear one.

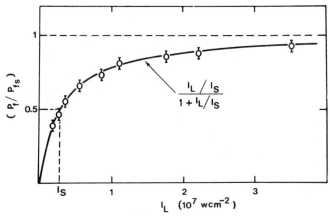

Figure 4   Plot of the fluorescence power P_f versus the laser excitation intensity I_L, showing the saturation of the fluorescent transition. The sample was a frog's erythocyte cell, stained by the conventional Feulgen reaction for DNA demonstration. The experimental points and the theoretical curve of equation (3) are shown

$\sigma = 3 \cdot 07 \times 10^{-16}\,\text{cm}^2$ at the laser wavelength $\lambda = 566\,\text{nm}$. This value is in acceptable agreement with that we obtained by absorption measurement on an *in vitro* solution of the same dye ($\sigma = 4 \cdot 2 \times 10^{-16}\,\text{cm}^2$). To determine the quantum yield η_q we used (4) with $\eta = 1 \cdot 55$ and $g_t = 2 \cdot 1 \times 10^{-15}\,\text{s}$ at $\lambda = 566\,\text{nm}$, as evaluated from the shape of the absorption spectrum. We found $\eta_q \simeq 8 \cdot 9\%$.

The saturation behaviour has been experimentally verified also with cells stained by a water solution of Rhodamine 3GO. The saturation intensity has been found to be $I_s = 42 \cdot 5 \times 10^6\,\text{W cm}^{-2}$, which gives, at $\lambda = 566\,\text{nm}$, $\sigma\tau = 4 \cdot 13 \times 10^{-27}\,\text{cm}^2\,\text{s}$. Since $\tau = 0 \cdot 85\,\text{ns}$, we obtain $\sigma = 4 \cdot 86 \times 10^{-18}\,\text{cm}^2$, in agreement with the value found from an *in vitro* measurement of the absorption coefficient *vs* the concentration. From the value of $\sigma\tau$ we obtain, for the quantum yield, $\eta_q = 2 \cdot 13\%$ (with $\eta = 1 \cdot 55$ and $g_t = 5 \cdot 8 \times 10^{-16}\,\text{s}$ at $\lambda = 566\,\text{nm}$).

Before ending this section on the saturation of the fluorescent emission, it is appropriate to discuss the relevance of the measurements that have been made and to point out further possibilities. We have seen that the measurement of I_s leads to the determination of the quantum yield. We believe that this quantity is of great biological relevance, since its magnitude is certainly affected by the interaction of the fluorescent probe with the substrate and the cellular ambient. The measurement of σ (which is derived when τ is also known) makes possible the measurement of the concentration N of fluorescent molecules in the cell, if the measurement of the absorption coefficient $\alpha = \sigma N$ is also possible (through an absorption measurement).

4 FURTHER EXPERIMENTAL POSSIBILITIES

As already outlined, apart from the type of measurement described in §3, other applications of our experimental apparatus are possible. They require only minor modifications and/or the use of more sophisticated electronic measuring instrumentation. We shall briefly mention some of the more significant ones. It is possible to measure excitation spectra directly, and emission spectra by introducing monochromators in the fluorescence light path. In both these cases the advantages over conventional apparatus would be considerable. The time-domain analysis of the waveforms should make possible a time-resolved spectroscopy, i.e. the measurement of the spectra of fluorescent components with different lifetimes. The time-domain analysis may also be useful to discriminate part of the correlated background, i.e. unwanted fluorescences with lifetimes different from those to be studied. Polarization studies can easily be made.

5 CONCLUSION

Our work shows that, by using nanosecond-pulsed tunable lasers and fast pulse-measurement techniques, it is possible to measure, even in single cells, the main parameters of fluorescence emission transitions. These fluorescence techniques can therefore be employed in many different biological studies, as outlined in the Introduction. In our laboratory, work is in progress on chromosome fluorescent bands for studies on the DNA-base-sequences.

REFERENCES

1. R. Rigler, Jr., *Acta Physiol. Scand.*, **67**, Suppl. 267, 1 (1966).
2. S. S. West, *Physical Techniques in Biological Research*, 2nd edn., p. 253 (ed. A. W. Pollister), Academic Press, New York and London, 253, 1969.
3. J. W. McInness and M. McClintock, *Biopolymers*, **9**, 1407 (1970).
4. E. Kohen, C. Kohen and B. Thorell, *Biochim. Biophys. Acta*, **234**, 531 (1971).
5. G. Prenna, to be published.
6. M. Sernetz, *Fluorescence Techniques in Cell Biology*, (eds. A. A. Thaer and M. Sernetz), Springer Verlag, Berlin–Heidelberg–New York, 1973.
7. S. V. Konev, *Fluorescence and Phosphorescence of Proteins and Nucleic Acids*, Plenum Press, New York, 1967.
8. J. B. Le Pecq and C. Paoletti, *J. Molec. Biol.*, **27**, 87 (1967).
9. G. K. Radda and J. Vanderkooi, *Biochim. Biophys. Acta*, **265**, 509 (1972).
10. L. Stryer, *Science*, **162**, 526 (1968).
11. T. Caspersson, G. Lomakka and R. Rigler, Jr., *Acta. Histochem.*, Suppl. VI, 123 (1965).
12. F. W. D. Rost and A. G. E. Pearse, *J. Microsc.*, **94**, 93 (1971).
13. S. Cova, G. Prenna and G. Mazzini, *Histochem. J.*, **6**, 279 (1974).
14. G. Prenna, G. Mazzini and S. Cova, *Histochem. J.*, **6**, 259 (1974).
15. C. A. Sacchi, O. Svelto and G. Prenna, *Histochem. J.*, **251** (1974).
16. G. Amsel and R. Bosshard, *Rev. Sci. Instrum.*, **41**, 503 (1970).
17. S. Cova, M. Bertolaccini and C. Bussolati, *Phys. Stat. Sol. (a)*, **18**, 11 (1973).
18. F. T. Arecchi, E. Gatti and A. Sona, *Rev. Sci. Instrum.*, **37**, 942 (1966).
19. S. Cova and M. Bertolaccini, *Rev. Sci. Instrum.*, **41**, 1153 (1970).
20. O. Svelto, *Principi dei Laser*, Tamburini, Milano, 1972. (English edition to be published.)
21. S. Udenfried, *Fluorescence Assay in Biology and Medicine*, Vol. 2, p. 165, Academic Press, New York and London, 1969.

B.11. Exciton-like splitting: a diagnostic for acridine dye–nucleic acid complexes

M. Nastasi, R. W. Yip

Division of Chemistry

V. L. Seligy, A. G. Szabo, and R. E. Williams

Division of Biological Sciences
National Research Council of Canada
Ottawa, Ontario K1A 0R6 Canada

The complexes between cationic acridine dyes and nucleic acids are of considerable interest because of (1) their potential as a regioselective label in the staining of chromosomal DNA in eukaryotes;[1-8] (2) the wide variety of biological effects in which the interaction between the dye and nucleic acids have been deeply implicated, e.g. antimalarial, carcinogenic and bacteriostatic activity, induction of mutations, photodynamic action;[8-10] and (3) the methods of investigation and interpretations are relevant in the more general field of small molecule/macromolecule interactions.

Based on the X-ray study on fibres of the proflavine/DNA complex, together with hydrodynamic measurements in solution, Lerman[11-13] concluded that the dye molecules were intercalated between adjacent base pair layers by extension and lengthening of the sugar-phosphate backbone. A variety of recent experiments[8,14-19] have been directed towards substantiating the intercalation model or modifications thereof. Most of the information concerning the nature of the complex in solution has been derived from studies of the complicated equilibria between free and bound dye in the form of binding curves[8] (Scatchard plots). Despite this effort, Lerman's original experimental evidence remains the most persuasive for the intercalation model, namely that the mass per unit contour length of the DNA is diminished on complexing with the dye.[12]

Recently several workers[1-3,14,15,20,21] have studied the complexes formed between the anti-malarial drug and chromosomal staining agent, quinacrine

Issued as N.R.C.C. No. 13997

(QAC), and nucleic acids. Correlations of the effect of the type of nucleic acid used on QAC fluorescence have been developed, but the molecular nature of the complex remains unanswered.

To gain further information on the structure of the complexes between acridine dyes and nucleic acids, we have carried out ultraviolet absorption experiments in order to detect and measure possible exciton interaction in complexes formed from the acridine dye quinacrine (QAC) with DNA, poly(dAT)

Quinacrine

and poly(dG). poly(dC). In parallel with these experiments, we have measured the rate constants for quenching of the fluorescence of the complexed dye by iodide ions which provided us with an independent set of data. Coupled with theoretical calculations we believe that the spectroscopic data could be quite important in unravelling the structures of acridine dye–nucleic acid and similar complexes as they occur in solution.

It is well known that the interaction of excited states of weakly coupled composite systems can be treated by the molecular exciton model.[22] Examples are molecular crystals, dimers, aggregates and polynucleotides. The interaction between dyes and nucleic acids in the strong binding case (at high phosphate: dye $(P:D)$ ratios) can similarly be treated.[23] In the simple theory, where only the transition dipole–dipole potential term is considered, significant interaction occurs if (a) there is a small energy difference between the interacting states and (b) if the interacting transition dipoles have large transition moments. The interaction between two transition dipoles with identical energies (resonance condition) takes on a particularly simple form. The excitation exchange matrix element v_{ab}^{ij}, which is the energy of interaction of the transition dipole moment $\mathbf{M}_a^i$ of the ith transition of a with the transition dipole moment, $\mathbf{M}_b^j$ of the jth transition of b, is given by:[22]

$$V_{ab}^{ij} = r_{ab}^{-3}|\mathbf{M}_a^{i\cdot}\mathbf{M}_b^j - 3(\mathbf{M}_a^{i\cdot}r_{ab})(\mathbf{M}_b^{j\cdot}r_{ab})r_{ab}^{-2}|$$

Thus the interaction falls off with distance with a $1/r^3$ dependence.

The absorption spectrum of QAC has three absorption maxima, with two weak bands in the visible (S_1 and S_2) at $22{,}470\ cm^{-1}$ ($445\ nm$, $f = 0.18$) and $29{,}070\ cm^{-1}$ ($344\ nm$, $f = 0.08$), respectively and a very strong absorption maximum (S_3) at $35{,}710\ cm^{-1}$ ($280\ nm$, $f = 1.15$) in aqueous solution at pH 7. Therefore on the basis of the magnitude of the transition moment and the energy of the transition, the S_3 absorption band of QAC is the only band which could possibly show significant exciton-like interaction (in addition to dispersion and other solvent shifts) with the polynucleotide transition at 38,460 cm^{-1} (260 nm, $f = 0.10$). The other two bands (S_2 and S_1) at lower energies ought to show dispersion and possibly other shifts, but not exciton-like interaction.

To ensure complete binding and that only strong primary binding is occurring,[8] we worked with nucleotide phosphate: dye ratios of at least 6:1 and in most cases 20:1. As a consequence, the strong 260 nm absorption in the spectrum of the polynucleotide which is unperturbed by bound dye, completely masks any small changes in the dye transition in the same region. It was therefore necessary to record the difference spectra of various dye–polynucleotide solutions placed in the sample beam against identical polynucleotide samples in the reference beam of our Cary 14 Spectrophotometer.

Figure 1 shows the spectra of free QAC and the difference spectra of QAC bound to DNA, poly(dAT) and poly(dG).poly(dC).

If we examine the difference spectrum of QAC bound to DNA (Figure 1a) we see that the 445 nm and 344 nm bands in the spectrum of the free QAC are observed at 455 nm (21,980 cm^{-1}) and 350 nm (28,570 cm^{-1}) in the spectrum of the bound dye. As expected, the two lowest energy transitions show dispersion shifts only. In each case the vibronic structure and Franck–Condon envelope of the band are not altered and the entire bands exhibit shifts of 490 cm^{-1} and 500 cm^{-1}, respectively. However, in the 260–300 nm region two overlapping bands are now observed with peaks at 290 nm (34,480 cm^{-1}) and 270 nm (37,040 cm^{-1}). If these two bands are decomposed into their components, we obtain two bands with maxima at 34,480 cm^{-1} and 37,540 cm^{-1}. The energy difference between these peaks (Δ) is 3060 cm^{-1}. Thus Δ is larger than the energy difference between the unperturbed interacting states, ΔE_{31} (38, 460–35, 710 = 2750 cm^{-1}) by 310 cm^{-1}.

Relative to an area of 1·15 for the oscillator strength of the $S_0 \rightarrow S_3$ transition of the free dye, we observed areas of 0·40 and 0·30 for the 290 nm and 270 nm components, respectively, for the spectrum of the complex. A similar band pattern is observed with the QAC/poly(dAT) complex (Figure 1b). Decomposing the two overlapping bands as before we obtained two bands at 37,315 cm^{-1} and 34,015 cm^{-1}, giving $\Delta = 3300$ cm^{-1}, again larger (by 550 cm^{-1}) than $\Delta E_{31} = 2750$ cm^{-1}.

In contrast to the results with DNA and poly(dAT), the difference spectrum of QAC bound to poly(dG).poly(dC) (Figure 1c) contains only a single broad band in the region of the 280 nm absorption of the spectrum of free QAC.

When we varied the DNA phosphate to dye ratio from 42:1 to 5:1, we observed no noticeable change in the difference spectra. However, when the $P:D$ ratio was further reduced from 5:1 to 1:1, the doublet rapidly gave way to a single broadened maximum at 280 nm. Since the low $P:D$ region corresponds to that in which weak binding is known to occur,[24] the absorption at 280 nm can be attributed to weakly bound or free dye.

The dramatic difference in the band pattern in the 280 nm region of the spectra of free QAC and of QAC bound to DNA and poly(dAT) is good evidence

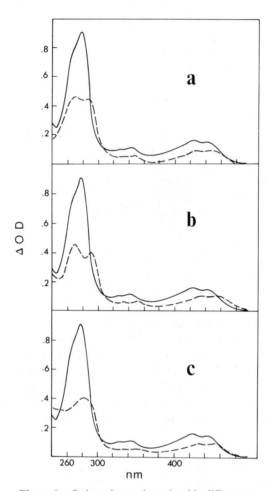

Figure 1 Quinacrine–polynucleotide difference spectra. Full lines; free quinacrine 1.55×10^{-5} M. Dotted lines: quinacrine 1.55×10^{-5} M, polynucleotide 3.1×10^{-4} M.

(a) Calf thymus DNA, (b) poly(dAT), (c) poly-(dG) . poly(dC).

Difference spectra were measured on a Cary 14 spectrophotometer (dispersion = 20 Å/mm) using 1 cm cells at 23 °C. The reference cell contained a concentration of polynucleotide equal to that in the sample cell which contained the quinacrine–polynucleotide solution. Identical maxima were observed for reference optical densities of 0.6 (slits, 1 mm) and 1.85 (slits, 2.5 mm)

that the interaction between the S_3 state of QAC and its environment is not due to a simple dispersion shift.

Since the interaction energy $(310\,\mathrm{cm}^{-1})$ is smaller than the initial level separation $(2750\,\mathrm{cm}^{-1})$ by almost an order of magnitude, the exciton model is not strictly applicable either. However, the transition moments are associated with broad bands rather than sharp lines and these bands overlap considerably (see Figure 1). Therefore, resonance type interactions do occur as in Förster energy transfer and this may account for the exciton-like behaviour, *viz.* the large intensity redistribution and pronounced band shifts and splitting. It therefore seems reasonable to attribute the two maxima which we have observed from the difference spectra as being due to two states formed from the dipole–dipole interaction between the S_3 state of QAC and the S_1 state of DNA with intensity being stolen from the intense $S_0 \rightarrow S_3$ transition of QAC. In the case of dipole–dipole interactions the interaction energy varies as r^{-3}. We conclude therefore that the binding of QAC to poly(dG).poly(dC) and to the other two polynucleotides is such that the distance of approach between QAC and the G–C base pairs is much larger than that between QAC and the A–T base pairs. In this sense, the observed splitting in DNA and poly(dAT) and the limiting $P:D$ ratio of 5:1 at which the splitting is observed is fully consistent with the intercalation model.

Since the more tightly wound G–C helix is less likely to unwind than the other two polynucleotides in order to accommodate the intercalating dye molecules, the different difference spectra observed for poly(dAT) and DNA dye complexes on the one hand, and the poly(dG).poly(dC)–dye complex on the other supports the intercalation model in the former two polynucleotides.

In the absence of intensity redistribution, the weakly absorbing perturbed state of DNA (or poly(dAT)) would not be readily apparent from the subtraction of DNA absorption from the total spectrum. In that event, we would have seen only a single peak due to the perturbed state of the dye.

The observed splitting cannot be due to a change on binding in the extent of protonation of QAC. The spectrum of the conjugate base of QAC at pH 9 includes only a broad structureless absorption maximum at 418 nm in the visible region, whereas the visible spectrum of the bound form includes only a dispersion shifted band at 455 nm.

The difference spectra obtained in the P/D range of 5:1 to 42:1 when analysed in the following manner provide three additional significant conclusions. If OD_T is the optical density of the total spectrum and OD_Δ the optical density of the difference,

$$OD_T = E_Q C_Q + E'_p C'_p + E_p (C_p - C'_p)$$

$$OD_\Delta = OD_T - E_p C_p = E_Q C_Q + C'_p (E'_p - E_p)$$

where E_Q, E_p and E'_p are the extinction coefficients of bound QAC, unperturbed DNA bases, and perturbed DNA bases, respectively. C_Q is the QAC concentration, C_p the initial DNA concentration $(3 \times 10^{-4}$ in this case) and C'_p the concentration of DNA bases perturbed by the binding. There is a linear relationship between C'_p and C_Q: $C'_p = K C_Q$ where K is a constant depending on the number of DNA bases involved in the binding of one QAC molecule, and hence,

$$OD_\Delta = C_Q[E_Q + K(E'_p - E_p)]$$

The plots (Figure 2) of OD_Δ vs C_Q at different wavelengths between 260 and 290 nm were linear for all the P/D ratios used from 5:1 to 42:1. This shows:

(1) *K is constant*, which precludes the existence of more than one binding site in this range of dye: polynucleotide ratios.

(2) E'_p *is constant*, for all ratios and hence the dye–base interactions responsible for the observed splitting are those between dye molecules and their four nearest neighbours.

(3) E_Q *is constant*, which suggests a negligible dye–dye interaction.

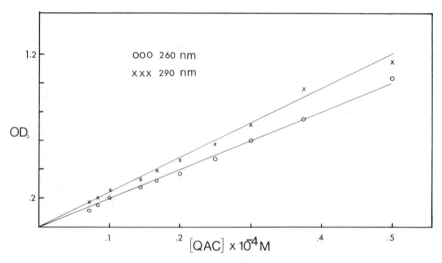

Figure 2   Optical density of OD_Δ of difference spectra of DNA–Quinacrine complexes against quinacrine concentration [QAC]: [DNA] $= 3.0 \times 10^{-4}$ M; 290 nm, $\times$; 260 nm, $\bigcirc$. See Figure 1 for conditions. All samples were at pH 7, cacodylate buffer [0.01 M]

The contour lengthening of DNA by QAC has not been studied. However, in their work on contour lengthening of DNA of different base composition when bound to proflavine, Ramstein and Leng[25] concluded that proflavine was bound more externally to G–C rich regions of DNA. Their result is in agreement with our conclusions that the distance between QAC and the G–C base pairs is larger than QAC and A–T base pairs.

The foregoing conclusion that QAC molecules are bound closer to A–T base pairs than to G–C base pairs suggests that the dye molecules are more sterically hindered and hence are more shielded from collisions with molecules in solution when bound to poly(dAT) than to poly(dG). poly(dC). One way to test this variation in shielding is to examine the quenching of the excited state of the complexed QAC molecule by species such as iodide ions, a process which is known to be collisional,[26] i.e. quenching is diffusion-controlled. We would therefore expect differences in the fluorescence quenching rate constant (k_q) for QAC/poly(dAT) and QAC/poly(dG).poly(dC) complexes. Fluorescence intensity values for iodide quenching[27] themselves are less physically significant than k_q, in that intensity studies do not take into account the significant variation in fluorescence lifetime of the dye in the different environments (Table 1).

In the three cases studied, linear Stern–Volmer plots were observed. The results recorded in Table 1 show a significant variation in k_q depending on the particular complex. We applied a correction[27] to the observed k_q values in order to ensure that the differences observed were not merely due to a small amount of free dye which may still be present even at high $P:D$ ratios. Free dye would have a larger contribution to the emission of the poly(dG).poly(dC)/ QAC complex, since the dye fluorescence is quenched on binding in this case, while it is enhanced in the case of the poly(dAT)/QAC complex. We estimated (from optical density changes at 440 nm vs various $P:D$ values) the free dye to be present to a maximum extent of 5% of the total dye concentration. There was no significant difference between the corrected and uncorrected values of k_q. Only a gross underestimation of the amount of free dye present would have invalidated our conclusions.

Table 1   Fluorescence quenching of quinacrine and quinacrine/nucleic acid complexes by iodide ions

| | Φ_F^f/Φ_F^{f+b} | τ(ns) | $k_q \times 10^{-9}$ M^{-1} s^{-1} | |
			corrected	uncorrected
QAC	1	4	7·8	—
QAC + poly(dAT)	0·4	20	0·19	0·20
QAC + poly(dG)˙poly(dC)	7	6	2·3	3·0
QAC + DNA	2·1	10	1·1	1·2

[QAC] = 2×10^{-5}M; $P:D$ = 20:1. The solution was buffered at pH 7 with 0·01 M cacodylate. The ionic strength was maintained constant at 0·1 M by the addition of appropriate amounts of NaCl such that $[Cl^-] + [I^-]$ = 0·1 M. The polynucleotides were purified by dialysis for 24 hrs. Calf thymus DNA was purchased from Worthington Biochemicals; poly(dG).poly(dC) and poly(dAT) were purchased from Miles Laboratories. Fluorescence intensities were measured using a Farrand Spectrofluorimeter exciting at 365 nm (bandwith 5 nm) and monitoring at 490 nm. Lifetimes were measured by the TRW Model 75A fluorescence lifetime apparatus. Rate constants were obtained from Stern–Volmer plots[28] of fluorescence intensity vs iodide concentration. Φ_F^f/Φ_F^{f+b} ($= A$) is the ratio of the quantum yield of the free dye to that of free + bound dye. The corrected rate constants were derived from values of I_b^0/I_b obtained according to:[27]

$$I_b^0/I_b = [1 - (1 - \alpha)A]/[(I_{f+b}^0/I_{f+b})^{-1} - (1 - \alpha)A(I_f^0/I_f)^{-1}]$$

where I is the fluorescence intensity, the superscript zero designates the fluorescence at zero iodide concentration; the subscripts f, b, f + b designate free dye, bound dye, and free plus bound dye respectively; α is the fraction of bound dye (0·95). I_{f+b} are the observed values of the fluorescence intensity at various iodide concentrations.

The value of k_q for the free dye is in agreement with the calculated diffusion-controlled value in water.[29] The lowest k_q was observed for the poly(dAT)/QAC complex, and the highest for the poly(dG).poly(dC)/QAC complex. These observations confirm our interpretation of the spectroscopic experiments described above; that QAC is bound considerably closer (intercalated) to A–T than to G–C base pairs in the polynucleotides.

The difference absorption spectrum (Figure 3) of the complex of QAC and heat-denatured shock-cooled DNA is an additional demonstration of the sensitivity and selectivity of exciton-like interactions. This difference spectrum includes only a dispersion shifted band at 290 nm with a small shoulder at 270 nm. This shoulder is likely due to the small amount of highly repetitive renatured regions of this DNA sample. Significantly the visible absorption bands of the QAC/denatured DNA complex are superimposable on that of the QAC/ native DNA complex.

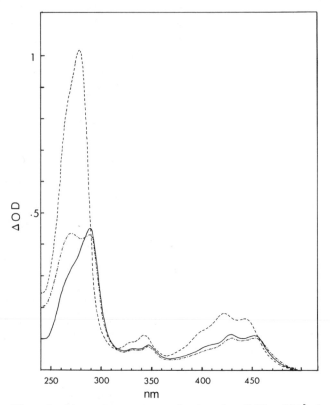

Figure 3   Absorption spectrum of quinacrine, $[1.76 \times 10^{-5} \text{ M}]$ — — —. Difference absorption spectra of quinacrine, $[1.76 \times 10^{-5} \text{ M}]$ with native DNA, $[3.0 \times 10^{-4} \text{ M}]$ — · — · —; and with denatured DNA, $[3.0 \times 10^{-4} \text{ M}]$ ———; against the respective polynucleotide solutions. See Figure 1 for conditions. All samples were at pH 7, cacodylate buffer $[0.01 \text{ M}]$

These results indicate that the observation of exciton-like splitting in the difference absorption spectra of QAC/polynucleotide complexes requires intact, double stranded, helical polynucleotides. These observations correlate with the strong primary binding of QAC with nucleic acids and are diagnostic for helix–coil transitions rather than base sequence alterations, whereas the dispersion shift of the visible absorption bands does not allow one to make these distinctions.

In summary: (1) we have observed exciton-like splitting between the S_3 state of QAC and the lowest excited state of DNA and poly(dAT) with concomitant intensity redistribution, but we have not observed this effect with QAC and poly(dG). poly(dC) or denatured DNA; (2) we conclude that QAC is bound closer to A–T base pairs (possibly by intercalation) than to G–C base pairs in polynucleotides; (3) the most important interactions responsible for the splitting are nearest-neighbour interactions; (4) dye–dye interactions are negligible for QAC–nucleic acid complexes; (5) exciton-like splitting requires intact double-stranded nucleic acids and is diagnostic for helix–coil transitions.

Coupled with theoretical calculations we believe that the exciton-like interaction could be quite important in unravelling the structures of acridine dye–nucleic acid and similar complexes as they occur in solution.

We thank Drs W. Siebrand, D. F. Williams, J. M. Morris and Professor G. Strauss (School of Chemistry, Rutgers University) for suggestions and discussion and Mr P. Tolg for technical assistance. M.N. is an N.R.C.C. postdoctoral fellow 1972–74.

REFERENCES

1. T. Caspersson, L. Zech, E. J. Modest, G. E. Foley, U. Wagh and E. Simonsson, *Expt. Cell. Res.*, **58**, 128 (1969).
2. T. Caspersson, L. Zech, U. Wagh, E. J. Modest, G. E. Foley and R. Simonsson, *Expt. Cell. Res.*, **58**, 141 (1969).
3. T. Caspersson, *Expt. Cell. Res.*, **58**, 451 (1969).
4. C. L. Y. Lee, J. P. Welch and S. H. S. Lee, *Nature, New Biology*, **241**, 142 (1973).
5. F. T. Bosman and A. Schaberg, *Nature, New Biology*, **241**, 216 (1973).
6. W. D. Peterson, Jr., W. F. Simpson, P. S. Ecklund and C. S. Stulberg, *Nature, New Biology*, **242**, 22 (1973).
7. R. E. Williams, P. F. Lurquin and V. L. Seligy, *Eur. J. Biochem.*, **29**, 426 (1972).
8. A. R. Peacocke, in *Acridines*, p. 723 (ed. R. M. Acheson), Wiley, New York, 1973.
9. L. Sankaran and B. M. Pogell, *Nature, New Biology*, **245**, 257 (1973).
10. J. B. Le Pecq, P. Jeanteur, R. Emanoil-Ravicovitch and C. Paoletti, *Biochim. Biophys. Acta*, **119**, 442 (1966).
11. L. S. Lerman, *J. Molec. Biol.*, **3**, 18 (1961).
12. L. S. Lerman, *J. Molec. Biol.*, **10**, 367 (1964).
13. L. S. Lerman, *Proc. U.S. Nat. Acad. Sci.*, **49**, 94 (1963).
14. B. Weisblum and P. L. de Haseth, *Proc. U.S. Nat. Acad. Sci.*, **69**, 629 (1972).
15. U. Pachmann and R. Rigler, *Expt. Cell Res.*, **72**, 602 (1972).
16. J. Ramstein, M. Dourlent and M. Leng, *Biochem. Biophys. Res. Comm.*, **47**, 874 (1972).
17. E. Fredericq and C. Houssier, *Biopolymers*, **11**, 2281 (1972).
18. D. G. Dalgleish, E. Dings yr and A. R. Peacocke, *Biopolymers*, **12**, 445 (1973).
19. H. J. Li and D. M. Crothers, *J. Molec. Biol.*, **39**, 461 (1969).
20. R. K. Selander and A. De la Chapelle, *Nature, New Biology*, **245**, 240 (1973).
21. B. Weisblum, *Nature*, **246**, 150 (1973).

22. E. G. McRae and M. Kasha, in *Physical Processes in Radiation Biology*, p. 23 (eds. L. Augenstein, R. Mason and B. Rosenberg), Academic Press, New York, 1964.

23. M. R. Philpott, *J. Chem. Phys.*, **53**, 968 (1970).

24. D. S. Drummond, V. F. W. Simpson-Gildemeister and A. R. Peacocke, *Biopolymers*, **3**, 135 (1965).

25. J. Ramstein and M. Leng, *Biochim. Biophys. Acta*, **281**, 18 (1972).

26. S. S. Lehrer, *Biochem.*, **10**, 3254 (1971).

27. V. I. Korunskii and Y. I. Naberukhin, *Molec. Biol. Moscow*, **6**, 594 (1972).

28. N. J. Turro, *Molecular Photochemistry*, p. 94 Benjamin, New York, 1965.

29. J. G. Calvert and J. N. Pitts, *Photochemistry*, p. 627, Wiley, New York, 1967.

Session C
Excited states of photosynthetic pigments

C. Primary actions of excited molecules in the functional membrane of photosynthesis

H. T. Witt

Max-Volmer-Institut für Physikalische Chemie und Molekularbiologie
Technische Universität Berlin, 1 Berlin 12, Germany

(Dedicated to Dr. Christoph Wolff, 25-10-1936 to 22-7-1975)

ABBREVIATIONS

ADP	Adenosine diphosphate
ATP	Adenosine triphosphate
BV	Benzylviologen
Car	Carotenoid
Chl	Chlorophyll
Cyt	Cytochrome
GmcD	Gramicidin D
$NADP^+$	Nicotinamide adenine dinucleotide phosphate
PQ	Plastoquinone
Vmc	Valinomycin
X-320	Plastosemiquinone anion

SUMMARY

(1) Light energy absorbed in the pigment system of green plants travels by energy migration in two ways: (a) at low intensities towards the reaction centres of the electron-transfer chain, and (b) at high intensities out of the bulk of Chl thus preventing Chl from destruction.

(2) The electron transfer in the chain from H_2O to $NADP^+$ is driven by two reaction centres in series, Chl a_I (P 700) and Chl a_{II} (P 690). A pool of PQ is the link between the two reaction centres. The electron transfer from H_2O to $NADP^+$ converts one part of the light energy into the form of the reducing power of NADPH.

(3) The transfer of the electrons is a vectorial transfer with a component perpendicular to the thylakoid membrane. In this way the membrane is charged

and there is an electrical field across the membrane. The electrical energy of the charged membrane is an additional state into which light energy is converted.

(4) The electronic charges on the outside and inside of the thylakoid membrane are replaced by OH^- and H^+, respectively, through protolytic reactions coupled with the redox processes. This corresponds to a pumping of protons into the inner space of the thylakoid. Two field generators and two proton pumps have been recognized. This indicates in a different way the existence of the two light reaction centres. The results are explained by a vectorial zigzag scheme.

(5) The discharging of the electrically energized membrane by H^+ efflux is coupled with the formation of ATP. With the help of NADPH and ATP, absorbed CO_2 can be reduced to sugar and all other energy-rich natural compounds.

1 INTRODUCTION

According to textbooks of biochemistry a biological molecule is a molecule with the best possible structure for its biological function. In this sense chlorophyll is the best molecule for photosynthesis, rhodopsin the best for vision, cytochromes the best for respiration, etc. However, it is evident that such a biological molecule can achieve its function only when it is incorporated in a highly specific organization. For instance, chorophyll *per se* behaves with respect to its possible reaction sequences, so far we know, like any other dye molecule. When, however, chlorophyll is embedded in a specific organization, excited chlorophyll creates reaction patterns of extraordinary attributes, specificities and efficiencies. Several different biological molecules use for their various functions a common principle of organization, the vesicle. A vesicle consists of a closed membrane of lipoprotein. Some of its characteristics can be recognized if we discuss the presumable pathway of its evolution.

Billions of years ago chemistry occurred in a 3-dimensional space, in a gaseous atmosphere and later in water, respectively. In 3-dimensional space the reactants are relatively diluted, intermediates, e.g. triplet states, with long lifetimes are useful, and the reactions can be relatively slow. Later, with the 'invention' of membranes which consist of fluid bilayers of lipids a '2-dimensional-like reaction vessel' was made available for chemistry. This has the advantage that the reactive substances can be much more concentrated. Intermediates, e.g. singlet states, with short lifetimes are useful, and the reactions can be relatively fast. Furthermore, the membranes can be equipped anisotropically with biological molecules in such a way that the excited biological molecules can react only in one specific way. Thereby, excited biological molecules are prevented from wasting their energy in meaningless side-reactions. For instance, an excited dye molecule can react in 3-dimensional space by many different pathways. The excited state can be used for photooxidations, reduction, dissociation, protolytic reactions, rearrangements, fluorescence, phosphorescence, etc. By the specific arrangement in a lipid membrane, excited chlorophyll can be forced, however, to react with high efficiency in only one way, so that, for example, it is only photooxidized. This is the paradoxical reason why in some respects studies of the photochemistry of chlorophyll *in vivo* are much simpler than its analysis *in vitro*. In a further step of evolution, membranes were used for compartmentation to realize a closed vesicle. Under these conditions of an interior and exterior system, the anisotropical arrangement in the membrane can induce vectorial reactions from one side to the other. This can lead to an accumulation of substances in the inner space. In this way the system can have additionally the function of a storage device and new types of reactions are possible. Such 'interior–exterior' systems are used as an operational board for biological molecules in photosynthesis, vision, respiration, etc. In photosynthesis the vesicles are called *thylakoids*.

In the overall reaction of photosynthesis absorbed light performs a transfer of an electron from H_2O to the electron acceptor $NADP^+$.[1] About 200 electron-transfer chains are embedded in one thylakoid. Each chain is surrounded by about 500 light energy conducting pigment molecules (chlorophyll *a* and *b* and carotenoids), i.e. each thylakoid contains about 10^5 pigment molecules, see Figure 1.[1]

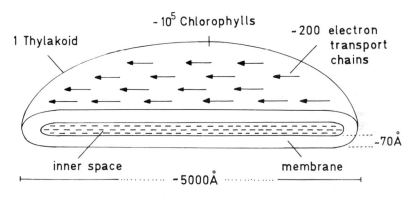

Figure 1 Scheme of the functional membrane of photosynthesis (thylakoid). The membrane is built up by lipids, proteins and pigments (chlorophylls and carotenoids)

Simultaneously with the electron transfer the formation of energy-rich ATP takes place from $ADP + P$.[2] With reduced NADPH and ATP, absorbed CO_2 can be reduced to sugar and other energy-rich organic compounds.[3]

In this lecture I want to give evidence that the specific organization of the pigments creates 5 different types of reactions which characterize certain points of the bioenergetic concept of photosynthesis (see Figure 2):

(1) Light energy is channelled by energy migration towards specific reaction centres. Superfluous light at high intensities is probably dissipated by energy

248

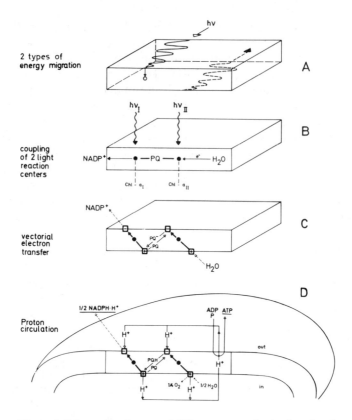

Figure 2 Schematic schemes of different events in the functional membrane of photosynthesis. (A) Two types of energy migration; *left*: quanta collection at lower intensities; *right*: quanta dissipation at high intensities. (B) In series coupling of Chl a_I and Chl a_{II} in one electron chain between the ultimate electron donor H_2O and acceptor $NADP^+$. Intermediate electron carriers have been omitted. (C) Vectorial shift of electrons across the thylakoid membrane from inside to outside at Chl a_I and Chl a_{II}. (D) Protolytic reactions with the charged membrane lead to H^+ uptake at the outside and H^+ release at the inside, i.e. to inward H^+ translocation. Subsequent field-driven efflux of protons is coupled with the formation of ATP from ADP and P

migration along a different 'pipeline' out of the bulk of chlorophyll *a* thus preventing the chlorophylls from destruction.

(2) Two special Chl molecules cooperate within the electron-transfer chain in series between H_2O and $NADP^+$.

(3) The electron-flow is not random but vectorial from inside of the membrane to outside at each active Chl. Thereby, the membrane is charged and an electrical field is set up across the membrane.

(4) The charged membrane induces through subsequent redox reactions—coupled with protolytic processes—a vectorial proton flux into the inner space.

(5) The discharging of the electrically energized membrane by the H^+ efflux through specific membrane channels is coupled with the formation of ATP from ADP + P.

2 TECHNIQUES

For the *characterization* of these 5 events the changes of optical properties are used.[4] For *excitation* of single turnovers, i.e. for the transfer of only one electron from H_2O to $NADP^+$ without recycling, normal flashes, ultrashort flashes or laser giant pulses have been applied.[5] For *registration* of the small signals within short times the repetitive pulse spectroscopic method is most suitable. The extremely high sensitivity of this technique enables the detection of very small signals. The high time resolution of this technique has extended the analysis of biological events down to 20 ns. Such techniques are described in detail elsewhere.[6]

3 TWO TYPES OF ENERGY MIGRATION

In photosynthesis the molecular machinery must be adaptable to dim light as well as to bright light. At low intensity the very few quanta nevertheless excite the small number of reaction centres with a high probability because singlet energy transfer and migration in the bulk pigment channels the quanta via the singlet states from Car to Chl and from Chl to Chl. In this way also the photoactive centres are 'visited' whereby the energy is trapped.[1] This mechanism is long known and described in detail in the paper (G.1) of Dr Pearlstein. There is another new type of energy migration in photosynthesis. At high intensities superfluous energy stays in the excited Chl states. The energy must be dissipated quickly, because it is known that excited Chl is irreversibly destroyed by photooxidation with the omnipresent O_2. Energy dissipation may occur by reemission as fluorescence or by non-radiative transitions, but these two ways cannot be utilized efficiently. An additional 'valve reaction' is probably energy transfer and migration after intersystem crossing via triplet states in the opposite direction from Chl to Car[7] (see Figure 3). Because excited Car is not irreversibly destroyed by photooxidation with O_2 in contrast to excited Chl, this process can be used to drain off harmful energy from Chl. This triplet–triplet energy transfer and migration is obviously 'invented' by nature for high intensities, and it can be regarded as the counterpart to the S–S energy transfer and migration 'invented' for lower intensities. The existence of this 'triplet valve' was demonstrated by an analysis of the absorption change during the formation of the triplet states of Carotenoids (Car^T). These absorption changes occur in less than 20 ns[8] and

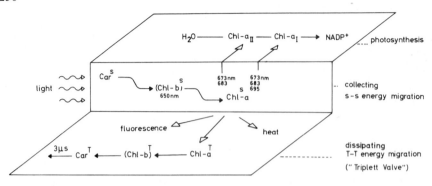

Figure 3 Schematic drawing of energy migration in the functional membrane of photosynthesis. *Centre*: Absorption of light and channelling of quanta within the bulk pigments to the photoactive centres Chl a_{I} and Chl a_{II} (see Top). *Bottom*: Draining off superfluous quanta from Chl a (a) by non-radiative transfer into heat, (b) by fluorescence and (c) by T–T energy migration and transfer to Car[7]

decay within 3 μs.[7] Three bleachings in the absorption bands of Car at 430, 460 and 490 nm and no bleaching in the red, where Chl a absorbs, indicate a Car reaction.[7,9] The simultaneous rise of an absorption at 520 nm at the position, where the absorption of Car^T *in vitro* was observed,[10] indicates the generation of Car^T *in vivo*. This interpretation is supported by the fact that this photosynthesis reaction can also be observed reversibly at $-160\,^\circ$C and by the fact that the changes, e.g. at 520 nm, can be completely quenched by paramagnetic gases such as O_2 and restored in diamagnetic gases such as N_2.[11] The quenching by paramagnetic gases is due to an accelerated decay of Car^T.[9] Because the spectrum change can be induced with red light which excites only Chl a, energy transfer and migration from Chl to Car must take place.[7] That this pathway from Chl to Car is only used for superfluous light energy follows from the observation that with increasing light intensity the formation of Car^T starts only when photosynthesis begins to become saturated.[7] From these and other experiments it is concluded that a specific organization of Car and Chl in the thylakoid membrane enables in bright light a dissipation of quanta by T–T transfer and migration. Besides the energy dissipation as heat and fluorescence, at least 20% of the energy is dissipated via this 'triplet valve'.[12]

4 COUPLING OF TWO CHLOROPHYLL REACTIONS

Two different pigment systems were tentatively postulated by Emerson in 1958[13] from the O_2 evolution as a function of the wavelength of light, a system excitable at long wavelengths (< 730 nm) and a second system excitable at shorter wavelengths (< 700 nm). Hill and Bendall postulated in 1960 that the two suggested pigment systems may be coupled in series and that the coupling is effected by cytochromes.[14] An in-series-coupling was also proposed by Kautsky in 1960.[15]

The first experimental evidence for a coupling of two systems was given independently by three spectroscopic phenomena observed by Kok, Duysens and Witt and their coworkers. In all three experiments it was shown that an intermediate reaction of photosynthesis, followed by absorption changes, is 'switched' from one direction to the contrary course of reaction when photosynthesis is triggered alternatively with <700 nm and 730 nm light.

Kok and Hoch measured the intermediate reaction of a pigment P 700.[16] Duysens et al. measured a reaction of a cytochrome.[17] Witt et al. analysed a cytochrome together with a reaction indicated at 515 nm.[18] Kok and Duysens observed an oxidation of their compounds in far-red light and a reduction in red light. We observed that in far-red light the extent of the 515 nm reaction was strongly decreased; the reduction of Cyt f$^+$ was not decreased, however, but it was slow (1 s). In red light the extent of the 515 nm reaction was increased, the extent of reduction of Cyt f$^+$ was the same, but it became fast (10^{-2} s). Further spectroscopic evidence for a coupling of two pigment systems was given in the following years. In far-red light PQ is oxidized. In red light it is reduced.[19] In far-red light the intensity of Chl fluorescence is decreased. In red light, however, it is increased.[20]

The photoactive chlorophylls within the two coupled systems have been observed directly by absorption changes (see Figure 4). The spectrum of the absorption change of Chl a_I(P 700) has been discovered by Kok[21] with a maximal bleaching at 700 nm; a further in vivo bleaching was detected at 438 nm.[22] The spectrum of Chl a_{II} (P 690) was observed by Döring et al.[23,24] and by Gläser et al.[25] with a maximal bleaching at 435 and 690 nm.

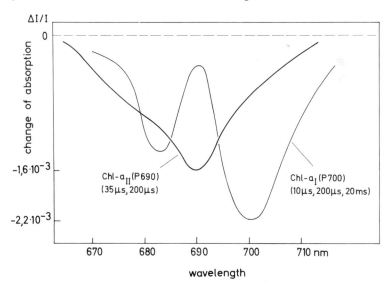

Figure 4 Transient difference spectra in the red region of the reaction of Chl a_I (P 700)[23] and Chl a_{II} (P 690).[24,25] Rise time: $\leqslant 20$ ns. Decay times are indicated in the brackets. Sample: Chloroplasts of spinach

Chl a_I (P 700). The spectrum of Chl a_I was trapped at low temperature indicating that the Chl a_I reaction represents a primary process.[26] This was supported by the measured rise time of the spectrum which is faster than 20 ns.[8] It was shown that Chl a_I is photooxidized in the light.[27,22] The redox potential has been estimated to be $+0.4$ volt.[27,28] The Chl a_I cation was detected by ESR measurements.[29] Chl a_I^+ decays in three phases: 10 μs, 200 μs and 20 ms. These times depend on the redox state of the electron donors (PC, Cyt-f, PQ) of Chl a_I^+.[30,31] A refined analysis of the spectrum of Chl a_I indicates in the red region a splitting in a double band with positions at 682 and 700 nm (see Figure 4) from which we concluded that two or more Chl a_I molecules are combined in one reaction centre with an oblique structure.[23]

Chl a_{II} (P 690). The rise time of the Chl a_{II} reaction was estimated to be faster than 20 ns, indicating that the Chl a_{II} reaction also represents very probably a primary event.[4] The reduced primary electron-acceptor of Chl a_{II}, called X-320, is a plastosemiquinone anion (absorption band at 320 nm)[32,33] and Chl a_{II} is probably photooxidized in the light.[66] The half lifetime of the absorption change

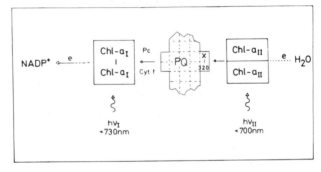

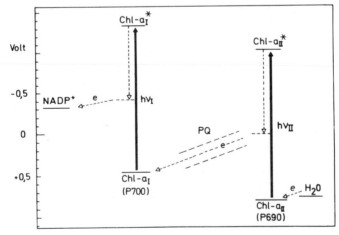

Figure 5 *Top*: Simplified electron transfer scheme from H_2O to $NADP^+$. *Bottom*: Free energy diagram. The scale indicates the midpoint potentials

of Chl a_{II} was found to be 200 μs.[24] But recently we observed with a refined technique a new large absorption change in system II also with a peak at 690 nm, but with a relaxation time of 35μs.[25] Probably, both components represent the biphasic decay of one and the same Chl a_{II}. Very likely only the cooperation of the redox reaction of two Chl a_{II} can oxidize water.[34]

The two spectra of Chl a_I and Chl a_{II} correspond exactly to what was speculated on for the reaction centres of system I ($<$730 nm) and II ($<$700 nm) (see above). According to Figure 4 Chl a_I has the characteristics of the long-wavelength system I ($<$730 nm) and Chl a_{II} that of the shorter-wavelength system II ($<$700 nm). Assuming approximately the same extinction coefficient for Chl a_I and Chl a_{II} it follows from the extent of the absorption change of both substances in Figure 4 that the ratio within one electron chain is about Chl a_I:Chl $a_{II} \approx$ 1:1,[25] a necessary condition when both are active centres and coupled in series between NADP$^+$ and H$_2$O. The coupling of the two systems of photosynthesis (see above) has been demonstrated directly by the coupling of the reaction of Chl a_I with Chla_{II}, reviewed in Reference 4. As a link between both chlorophylls a pool of plastoquinone ($\sim$ 5 molecules) has been identified.[19,32] The pools of the different electron-chains are combined with each other in the form of a strand.[34]

On the basis of the redox potentials of the components there results an energy diagram which is depicted in Figure 5. Excited Chl a_{II} lifts the electron from H$_2$O uphill to PQ. Excited Chl a_I lifts the electron a second time to NADP$^+$. It is obvious that at Chl a_I and Chl a_{II} much more energy is absorbed than is used for the electron transfer. In the next section it is shown that part of this energy is used for a vectorial redox reaction which needs additional energy for charging the membrane.

5 VECTORIAL ELECTRON TRANSFER. TWO FIELD GENERATORS

The electron transfer is not random but vectorial from the inside of the membrane to the outside, and this at each active Chl. This electron displacement must create an electrical field across the membrane. This has been measured as follows: if during photosynthesis a potential of say 100 mV is set up across the thylakoid membrane (about 100 Å thickness), this corresponds to 10^5 V/cm. In such high fields the absorption bands of all pigments are shifted by the order of one Å, depending on the optical properties of the pigments. This is called *electrochromism*. The electrochromism causes absorption changes by which the field can be measured.[35] The spectrum of the absorption change in photosynthesis which is attributed to electrical fields is depicted in Figure 6.[36] The spectrum is very complicated due to the shift of the absorption bands of Chl a, Chl b and Cars. The spectrum is generated in $<$20 ns.[37] This means that the field is set up as fast as the reaction of the photoactive Chls (see above). The decay in the dark occurs in $\sim$20 ms[38] depending on the permeability of the membrane. The extent of the changes has been calibrated in volts, and the slope of the time

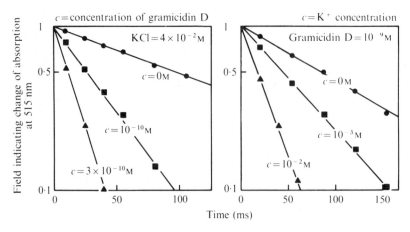

Figure 6 Time course of the field-indicating absorption change at 515 nm (log plot) in dependence of the concentration of gramicidin D (left) and of K^+ (right). Sample: Chloroplasts of spinach[35]

course of the changes in amperes.[39] The extent indicates a potential difference of ~ 100 mV. The slope of the rise corresponds to a current of $\gtrsim 1$ A/cm^2, and the slope of the decay to a current of $\sim 10\mu$A/cm^2 across the membrane. The potential generation is caused by a shift of one electron across the membrane, at each photoactive chlorophyll;[39] the decay occurs through H^+ efflux from the inner to the outer space of the thylakoid membrane.[40,54,69]

Kinetic evidence. A consequence of this interpretation is that when one increases the H^+ concentration in the inner space the potential decay must be accelerated. A 20-fold acceleration has been shown during the change from pH_{in} 8 to 5.[40,54] If on the other hand one produces artificial bypasses across the membrane, e.g. with gramicidin which makes holes in lipid membranes specifically for alkali ions, the decay also should be accelerated. The potential decay is indeed accelerated with increasing GmcD concentration as well as with increasing concentration of K^+ (see Figure 6).[35]

Spectroscopic Evidence. If the field indication is caused by band shifts of Chl *a*, Chl *b* and Car it must be possible to reproduce this effect artificially. Multilayers of Chl *a*, Chl *b* and Car were made according to the Langmuir technique[41] and the refined technique developed by Kuhn *et al.*[42] On these layers in the dark a repetitive field was applied and the absorption change due to the field was observed. The electrically induced difference spectrum at artificial multilayers is in good agreement with the light-induced difference spectrum on chloroplasts (see Figure 7).[43,44] The concept of field-indicating absorbance changes has been extended by Jackson and Crofts to photosynthetic bacteria which contain only one type of pigment in the visible spectral range, namely Carotenoids.[45] The light-induced absorption changes attributed to a field effect in bacteria are also in good agreement with the electrically induced difference spectrum on multilayers of Car (Lutein)[44] (see Figure 7).

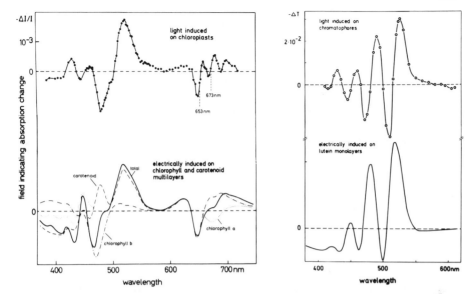

Figure 7 *Top, left*: Transient spectrum of field -indicating absorption changes induced by light pulses in chloroplast of spinach.[36] Rise time: ≤20 ns.[37] Decay times: ∼20 ms.[38] *Top, right*: The same in chromatophores of bacteria.[45] *Bottom, left*: Transient spectrum of Chl a, Chl b and Car multilayers in a microcapacitor induced by an electrical field of 10^6 V/cm.[43,44] *Bottom, right*: Transient spectrum of Car (Lutein) multilayers induced by an electrical field of 10^6 V/cm.[44]

Electrical Evidence. That the electron displacement is perpendicular to the thylakoid membrane and not only in the plane (which would also explain the above cited spectroscopic results) follows from the quenching of the field by the channel forming GmcD (see above) and it is supported by the following experiment. We compared the field-indicating absorption changes with electrical effects measured by macroscopic electrodes (the electrodes respond only to charge translocations perpendicular to the membrane).

If a non-saturating flash is fired from the top of the cuvette the upper part of each thylakoid membrane is a bit more charged than the lower part because by light absorption in the upper part the intensity is a bit reduced at the lower part. The small difference between the potential difference $\Delta\phi$ at the upper and lower part gives rise to a small net voltage $\Delta\Delta\phi$ which is detected by the electrodes (see Figure 8). In this way Fowler and Kok have measured electrical signals[46] and observed that they are diminished by one half when Chl a_I or Chl a_{II} is blocked. This confirms our results (see below) that each active Chl contributes one half to the potential.[39] However, these signals decay in about 10 μs, i.e. 10^4 times faster than the time of the field-indicating absorption change which indicates a trans-membrane flux ($\tau_\perp$). This discrepancy is due to the fast decay of $\Delta\Delta\phi$ by the equilibration of the slightly asymmetrically charged membrane via ion fluxes in the water phase parallel to the plane of the membrane ($\tau_\parallel$), i.e. $\tau_\parallel \ll \tau_\perp$.[47] To arrange conditions under which the decay of $\Delta\Delta\phi$ (measurable by electrodes) is determined by the ion flux perpendicular to the membrane ($\tau_\perp$),

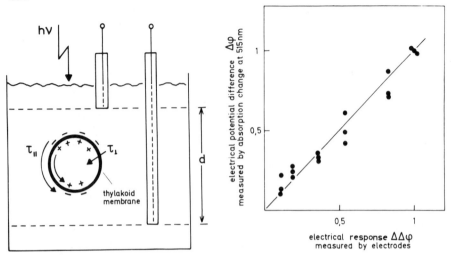

Figure 8 *Right*: Relative electrical potential difference measured by field indicating absorption changes at 515 nm as a function of the relative electrical response measured by electrodes[48] (also Witt and Zickler, unpublished). Sample: Chloroplasts of spinach. *Left*: Scheme of the measuring technique,[46,47] for details see text

we decreased the planar flux ($\tau_{||}$) by changing the viscosity of water from 1 to 10^3 centipoise by addition of sucrose and increased the transmembrane flux by addition of GmcD or Vmc (see above), so that $\tau_{||} \gg \tau_{\perp}$. Under these special conditions we observed a fair agreement between the time course of the optical signal $\Delta\phi(t)$ and that measured by electrodes $\Delta\Delta\phi(t)$.[47] Also the magnitudes of both signals are proportional to each other[48] (see Figure 8). These results support the conclusions from other experiments[35,39] that the absorption changes discussed here are due to *trans*membrane potential changes and respond *linearly* to the potential.

Summarizing we can say that by the kinetic, spectroscopic and electrical results evidence is given that the discussed absorption change represents an *intrinsic, prompt* and *linear molecular* voltmeter and ammeter which indicates potentials and currents *perpendicular* to the thylakoid membrane.

With this technique valuable insights into the electrical behaviour of the thylakoid membrane and the molecular mechanism in general have been obtained.[4,49] For instance, one of the questions already raised (see above) was whether one vectorial electron-transfer takes place at each photoactive Chl a. If we block one Chl reaction—Chl a_I or Chl a_{II}—the potential $\Delta\phi$ should be reduced to one half. This has been demonstrated[39] and indicates the existence of two field generators which are symbolized in the diagram of Figure 2 D by two heavy arrows.

6 PROTON TRANSLOCATION. TWO PROTON PUMPS

Coming to the next point I want to demonstrate that the two photoactive chlorophylls with their two vectorial electron-shifts automatically induce a vectorial H$^+$ translocation into the inner space (see Figure 2 D). We have shown in detail elsewhere that the primary electron acceptor of Chl a_{II} is a special plastoquinone out of a PQ pool.[32,33] With the acceptance of an electron by PQ a proton uptake is coupled. This is also true for the ultimate artificial electron-acceptor, in our case Benzylviogene. The positive hole produced by Chl a_{II} at the right inside of the membrane (on the unknown electron-donor of Chl a_{II}) is finally reduced by H_2O releasing $\frac{1}{4}O_2$ and $1 H^+$ to the inside. The positive hole produced by Chl a_I at the left inside of the membrane (on the electron-donor of Chl a_I : PC and Cyt-f) is reduced by the reduced quinone releasing also $1 H^+$ to the inside. In this way in a single turnover at each light reaction one proton should be translocated into the inner space of the thylakoid. This type of H^+ translocation was outlined in a hypothesis of Mitchell.[50] This proton transloca-tion was analysed by the repetitive spectroscopic technique with the addition of artificial pH indicators. The indicators could detect (a) the H^+ uptake at the membrane outside[39] and (b) the H^+ release at the inside.[51,71] The results indicate that, when in a single turnover flash one electron is transferred through the electron chain, an uptake of $2 H^+$ at the outside as well as a release of $2 H^+$ at the inside is observed. If Chl a_I or Chl a_{II} is blocked only one proton is trans-located. This indicates the existence of two proton pumps.

7 THE MOLECULAR CONCEPT

The experimental evidence (a) for 2 chlorophyll reactions, (b) for vectorial electron transfers and (c) for one H^+ translocation at each light reaction lead to a mechanism which is summarized in the vectorial zigzag scheme of Figure 2 D. This scheme is furthermore supported by the following two relations: (a) In a single turnover flash at Chl a_{II} one $PQ^{2-}/2$ is produced and a potential difference of $\Delta\phi/2$ is set up followed by a translocation of one H^+ (see above). If, however, in a flash of longer duration $n \cdot PQ^{2-}/2$ are reduced, the potential as well as the proton uptake increases proportionally to the number of n

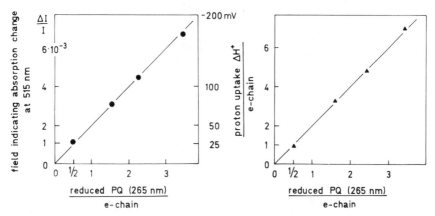

Figure 9 *Left*: Extent of field-indicating absorption changes at 515 nm induced by Chl a_{II} as a function of the amount of reduced PQ.[52,53] *Right*: Amount of H^+ uptake per electron chain induced by Chl a_{II} as a function of the amount of reduced PQ.[53] Sample: Chloroplasts of spinach

electrons injected into the pool of PQ (see Figure 9).[52,53] Both results substantiate in respect to Chl a_{II} the proposed transmembrane shift of electrons from H_2O (inside) to PQ (outside) and the H^+ uptake at PQ (outside) and H^+ release from H_2O (inside).

One can try to draw conclusions on the topography of the membrane from the functional molecular concept developed above. For instance, the vectorial trans-membrane translocation of an electron + H^+ (i.e. H-translocation) via the plastoquinone pool makes it necessary that the pool is located from the outside through the membrane to the inside. Furthermore, the observed photoinduced negative charging of the membrane at the outside, and positive charging at the inside, at both light reaction centres (in < 20 ns) and the fact that both Chl a_I[27] and Chl a_{II}[66] are photo*oxidized* (in < 20 ns) leads to the conclusion that the porphyrin rings of both photoactive chlorophylls must be located at the inside of the membrane and the primary electron acceptors, ferredoxin and X-320 (plastoquinone) at the outside. The cleavage of H_2O was established at the membrane inside (see §6 and §7). A preliminary model of the topography, which corresponds to these considerations and to the zigzag scheme outlined in Figure 1, is depicted in Figure 10. The picture also includes the electron carrier plastocyanin[70] and cytochrome-f.[17,72] The indicated ATPase[67] is probably the centre at which the proton efflux and phosphorylation are coupled.

8 PHOSPHORYLATION

According to a hypothesis of Mitchell[50] the energy which is released during the discharging of the electrically energized membrane by the efflux of protons

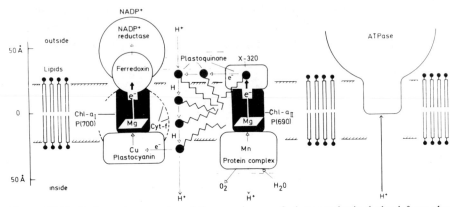

Figure 10 Preliminary topography of the membrane of photosynthesis derived from the molecular functional concept (see text). The two black 'trunks' symbolize the two photoactive centres, consisting of Chl a_I and Chl a_{II} which probably are complexed with proteins. The porphyrin rings are located at the inside of the membrane. X-320 is in its reduced form a plastosemiquinone anion. A Mn–protein complex is the centre of the cleavage of H_2O. Cyt-f and Plastocyanin are electron carriers between Chl a_I and PQ, Ferredoxin and $NADP^+$– reductase are electron carriers between Chl a_I and $NADP^+$. The proton efflux and phosphorylation are coupled in a special enzyme, the ATPase

should be coupled with the generation of ATP. The molecular voltmeter and ammeter described above was an ideal instrument to test the proposed relationships as follows. It was shown *1*. The field decay is accelerated during phosphorylation.[55,56,59] *2*. The acceleration disappears with the removal of the ATPase and reappears with its recondensation[69] (shown in chromatophores). *3*. The amount of charges (H^+) translocated via the ATPase pathway is proportional to the amount of synthesized ATP.[40,54] *4*. Charge translocation via other pathways competitively inhibits ATP generation.[35,40,49,56,69] In this respect it is of interest which minimum number of e.g. GmcD molecules act as a bypass for the intrinsic H^+ efflux (see above) and thereby—according to the coupling of H^+ flux and phosphorylation just outlined—as a deactivator of ATP formation. Figure 11 indicates that even one GmcD molecule per 10^5 chlorophylls decreases $\tau(\Delta\phi)$[35] as well as phosphorylation[40] by up to 30%. The electron transport from H_2O to NADP (measured by O_2 evolution) is, however, not disturbed at all. (O_2 evolution operates in full even at 100 times higher concentrations of GmcD[57]). About 10^5 chlorophylls are located on an area which corresponds to the size of one thylakoid. We regard the result of this 'titration experiment' as first evidence for the assumption that the functional unit of the electrical events as well as of phosphorylation—and thereby of whole photosynthesis—is a thylakoid (see Introduction).

As a consecution of the primary generated electrical potential in continuous light large amounts of H^+ ions are translocated into the inner thylakoid space[58,59] by the extrusion of an equivalent number of counter ions (e.g. K^+).[60,61] Thereby, the pH_{in} value decreases and electrical energy is in part transformed into a pH gradient. A maximum pH gradient of 3,3 was observed across the

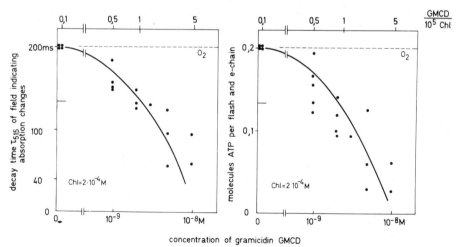

concentration of gramicidin GMCD

Figure 11 *Left*: Decay time τ of the field-indicating absorption changes at 515 nm as a function of the concentration of GmcD.[35,40] *Right*: ATP molecules per single turnover flash and electron chain as a function of the concentration of GmcD.[40] The relative activity of the electron transfer (measured by O_2 evolution) is also indicated.[57] Sample: Chloroplasts of spinach.

membrane.[62,63] Evidence for a coupling of phosphorylation with the H^+ efflux driven by the energy stored in a pH gradient was first demonstrated by Jagendorf and Uribe,[64] Rumberg[65] and others, and reviewed in Reference 65.

5. The field energy together with the subsequently generated H^+ gradient satisfies the energetic requirement for ATP synthesis (reviewed in 49).

Details are described elsewhere in References 4 and 49.

This work was supported by grants of the Deutsche Forschungsgemeinschaft. I should like to express my gratitude to all members of the Max-Volmer-Institut who contributed to the results of this chapter.

REFERENCES

1. *Bioenergetics of Photosynthesis* (ed. Govindjee), Academic Press, New York, (1975).
2. D. I. Arnon, M. B. Allen and F. R. Whatley, *Nature*, **174**, 394 (1954).
3. M. Calvin, *Angew. Chem.*, (Intern. Ed.) **74**, 165 (1962).
4. H. T. Witt, *Quart. Rev. Biophysics*, **4**, 365 (1971).
5. H. T. Witt, *Nobel Symp.*, **5**, 81 (1967).
6. H. Rüppel and H. T. Witt, *Methods Enzymol.*, **16**, 316 (1970).
7. Ch. Wolff and H. T. Witt, *Z. Naturforsch.*, **24b**, 1031 (1969).
8. K. Witt and Ch. Wolff, *Z. Naturforsch.*, **25b**, 387 (1970).
9. P. Mathis, *Progress in Photosynthesis Research*, p. 818 (ed. Metzner), (1969); P. Mathis and J. M. Galmiche, *C. R. Acad. Sc. Paris*, **264**, 1903 (1967).
10. M. Chessin, R. Livingston and T. G. Truscott, *Trans. Faraday Soc.*, **62**, 1519 (1966).
11. G. Ziegler, A. Müller and H. T. Witt, *Z. Phys. Chem.*, **29**, 13 (1961).
12. Ch. Wolff (1974), unpublished.
13. R. Emerson, *Ann. Rev. Plant Physiol.*, **9**, 1 (1958).
14. R. Hill and D. S. Bendall, *Nature*, **186**, 136 (1960).
15. H. Kautsky, W. Appel and H. Amann, *Biochem. Z.*, **332**, 277 (1960).
16. B, Kok and G. Hoch, *Light and Life*, p. 397 (ed. W. D. McElroy and B. Glass), Johns Hopkins Press, Baltimore, 1961.
17. L. N. M. Duysens, J. Amesz and B. M. Kamp, *Nature*, **190**, 510 (1961).
18. H. T. Witt, A. Müller and B. Rumberg, *Nature*, **191**, 194 (1961).
19. B. Rumberg, P. Schmidt-Mende, J. Weikard and H. T. Witt, *Photosynthetic Mechanism of Green Plants*, p. 18, Natl. Acad. Sci. Res. Council, Washington, Publ. 1145, 1963.
20. L. N. M. Duysens and H. E. Sweers, *Studies of Microalgae and Photosynthetic Bacteria*, p. 353, Univ. of Tokyo Press, Tokyo, 1963.
21. B. Kok, *Acta Bot. Neerl.*, **6**, 316 (1957).
22. B. Rumberg and H. T. Witt, *Z. Naturforsch.*, **19b**, 693 (1964).
23. G. Döring, J. L. Bailey, W. Kreutz and H. T. Witt, *Naturwiss.*, **55**, 219 (1968).
24. G. Döring, G. Renger, J. Vater and H. T. Witt, *Z. Naturforsch.*, **24b**, 1139 (1969).
25. M. Gläser, Ch. Wolff, H.-E. Buchwald and H. T. Witt, *FEBS Lett.*, **42**, 81 (1974).
26. H. T. Witt, A. Müller and B. Rumberg, *Nature*, **192**, 967 (1961).
27. B. Kok, *Biochim. Biophys. Acta*, **48**, 527 (1961).
28. B. Rumberg, *Z. Naturforsch.*, **19b**, 707 (1964).
29. H. Beinert and B. Kok, *Photosynthetic Mechanism of Green Plants*, p. 131, Natl. Acad. Sci. Res. Council, Washington, Publ. 1145, 1963.
30. W. Haehnel, G. Döring and H. T. Witt, *Z. Naturforsch.*, **26b**, 1171 (1971).
31. W. Haehnel, *Biochem. Biophys. Acta*, **305**, 618 (1973).
32. H. H. Stiehl and H. T. Witt, *Z. Naturforsch.*, **24b**, 1588 (1969).
33. K. Witt, *FEBS Lett.*, **38**, 116 (1973).
34. U. Siggel, G. Renger, H. H. Stiehl and B. Rumberg, *Biochim. Biophys. Acta*, **256**, 328 (1972); *Proc. 2nd Congr. Photosynthesis Res. Stresa 1971*, p. 753 (ed. Forti *et al*), 1972.

35. W. Junge and H. T. Witt, *Z. Naturforsch.*, **23b**, 244 (1968).
36. H. M. Emrich, W. Junge and H. T. Witt, *Z. Naturforsch.*, **24b**, 1144 (1969).
37. Ch. Wolff, H.-E. Buchwald, H. Rüppel, K. Witt and H. T. Witt, *Z. Naturforsch.*, **24b**, 1038 (1969).
38. H. T. Witt, *Naturwiss.*, **42**, 72 (1955).
39. H. Schliephake, W. Junge and H. T. Witt, *Z. Naturforsch.*, **23b**, 1571 (1968).
40. M. Boeck and H. T. Witt, *Proceedings 2nd Intern. Congr. Photosythesis. Res. Stresa 1971*, p. 903 (ed. G. Forti, M. Avron, A. Melandri, Dr W. Junk Publ., The Hague, 1972.
41. K. B. Blodgett and I. Langmuir, *Phys. Rev.*, **51**, 964 (1937).
42. H. Kuhn, D. Möbius and H. Bücher, *Techniques of Chemistry*, p. 577 (ed. A. Weissberger), Wiley–Interscience, 1972.
43. S. Schmidt, R. Reich and H. T. Witt, *Naturwiss.*, **58**, 414 (1971).
44. S. Schmidt, R. Reich and H. T. Witt, *Proceedings 2nd Intern. Congr. Photosynthesis Res. Stresa 1971*, p. 1087 (ed. G. Forti, M. Avron, A. Melandri), Dr W. Junk Publ., The Hague, 1972.
45. J. B. Jackson and A. R. Crofts, *FEBS Lett.*, **4**, 185 (1969).
46. B. Kok, *Proc. 6th Intern. Congr. Photobiol., Bochum 1972*, (ed. Schenck) paper No. 031.
47. H. T. Witt and A. Zickler, *FEBS Lett.,* **37**, 307 (1973).
48. H. T. Witt and A. Zickler, *FEBS Lett.*, **29**, 205 (1974).
49. H. T. Witt, *Bioenergetics of Photosynthesis*, p. 493 (ed. Govindjee), Academic Press, New York, 1975.
50. P. Mitchell, *Biol. Rev.*, **41**, 445 (1966).
51. W. Junge and W. Ausländer, *Biochim. Biophys. Acta*, **333**, 59 (1974).
52. H. T. Witt, B. Rumberg, P. Schmidt-Mende, U. Siggel and B. Skerra, *Angew. Chem. (Intern. Ed.)*, **4**, 799 (1965).
53. E. Reinwald, H. H. Stiehl and B. Rumberg, *Z. Naturforsch.*, **23b**, 1616 (1968).
54. P. Gräber and H. T. Witt, *Biochim. Biophys. Acta*, (1975), in press.
55. B. Rumberg and U. Siggel, *Z. Naturforsch.*, **23b**, 239 (1968).
56. W. Junge, B. Rumberg and H. Schröder, *Eur. J. Biochem.*, **14**, 575 (1970).
57. H. T. Witt, B. Rumberg and W. Junge, *19th Colloq. Ges. Biol. Chem., Mosbach*, p. 262, 1968.
58. A. T. Jagendorf and G. Hind, *Photosynthetic Mechanism of Green Plants*, p. 599, Natl. Acad. Sci. Res. Council, Washington, Publ. 1145, 1963.
59. J. Neumann and A. T. Jagendorf, *Arch. Biochem. Biophys.*, **109** (1964).
60. R. A. Dilley and L. P. Vernon, *Arch. Biochem. Biophys.*, **111**, 365 (1965).
61. H. Schröder, H. Muhle and B. Rumberg, *Proc. 2nd Intern. Congr. Photosynthesis Res. Strese 1971*, p. 919 (eds. G. Forti, M. Avron, A. Melandri), Dr W. Junk Publ., The Hague, 1972.
62. B. Rumberg and U. Siggel, *Naturwiss.*, **56**, 130 (1969).
63. H. Rottenberg, T. Grünwald and M. Avron, *Eur. J. Biochem.*, **25**, 54 (1972).
64. A. T. Jagendorf and E. Uribe, *Brookhaven Symp. Biol.*, **14**, 215 (1966).
65. B. Rumberg, *Proc. 6th Intern. Congr. Photobiol., Bochum 1972* (ed. Schenck) paper No. 036.
66. Ch. Wolff, M. Gläser and H. T. Witt, *Proc. 3rd. Intern. Congr. on Photosynthesis, Rehovot/Israel*, (1974), p. 295 (ed. Avron) Elsevier Publ.
67. E. Racker, in *Membranes of Mytochondria in Chloroplasts*, p. 127 (ed. Racker), 1970.
68. S. Saphon, J. B. Jackson, V. Lerbs and H. T. Witt, *Biochim. Biophys. Acta*, **408**, 58 (1975).
69. S. Saphon, J. B. Jackson and H. T. Witt, *Biochim. Biophys. Acta*, **408**, 67 1975).
70. G. A. Hauska, R. E. McCarthy, R. C. Berzborn and E. Racker, *J. Biol. Chem.* **240**, 3524 (1971).
71. P. Gräber and H. T. Witt, *FEBS Lett.*, in press (1975).
72. H. T. Witt, *Nature* **191**, 194 (1961).

C.1. Luminescence and phototautomerism of phytochrome and flavins

Pill-Soon Song, Quae Chae and Ming Sun

Department of Chemistry, Texas Tech University, Lubbock, Texas 79409, U.S.A.

SUMMARY

Highly purified, 'large' phytochromes (P_r and P_{fr}) are not fluorescent at room temperature. However, at reduced temperatures, 14 K $\sim$ 250 K, we observed fluorescence (Pr) with $\lambda_f \sim 675$ nm and $\Phi_f \simeq 0.04$ at 14 K and $\Phi_f \simeq 0.016$ at 250 K. The estimated rate constant for the primary photoprocess of the $P_r \rightarrow \rightarrow \rightarrow P_{fr}$ transformation responsible for the effective disposition of the excitation energy at room temperature is $\gtrsim 10^{12}$ s^{-1}. Several possibilities exist for the mechanism of the primary process: e.g. (a) proton transfer as a prerequisite step toward the $P_r \rightarrow P_{fr}$ phototautomerism, (b) transient conformational change of the chromophore, (c) transient conformational change of the apoprotein which is delicately coupled to the chromophore oscillator and (d) the combination of (b) and (c). These possibilities have been examined, and proton transfer (a) in particular, has been examined by means of a model system (phototautomeric flavins), which demonstrates proton transfer even at 100 K or below.

The model involves phototautomerism of lumichrome and alloxazine. Lumichrome (and alloxazine) fluoresces with $\lambda_f \sim 440$ (band I) and 530 nm (band II) in dioxane–pyridine and acetic acid–ethanol. Isoemissive points of fluorescence for lumichrome in these mixed solvents are at 476 and 488 nm, respectively. Band II fluorescence was shown to be due to the 6,7-dimethyliso-alloxazine, which was formed via phototautomerism during the lifetime of the excited state of lumichrome. No isoalloxazine fluorescence can be detected when the N_1 position of lumichrome or alloxazine is substituted with a methyl group. Studies of polarization characteristics and other flavin analogues further confirm our conclusion that in the singlet excited state the N_1-proton of lumichrome as well as alloxazines is shifted to N_{10} and the resulting tautomeric form is identical with that of isoalloxazines.

The aforementioned model is also operative even in the temperature range of 100–150 K or below. However, the phototautomeric proton transfer is temperature dependent, with an enthalpy change of 1·06 kcal/mole. On the other hand, the fluorescence quantum yield of P_r is essentially temperature independent in the range of 14–200 K. Thus, the primary photoprocess of P_r is not likely to be the proton transfer, although the phototautomeric proton transfer may well be a rate-determining step subsequent to the fast primary photoprocess.

1 PHYTOCHROME

Purification of phytochrome and model bilinoid pigments such as biliverdin and the high-resolution spectrofluorometer, equipped with a photon counter, have been described elsewhere.[1]

Figure 1 shows the fluorescence spectrum of phytochrome at 14 K. In addition to the 'normal' fluorescence maximum at 674·6 nm, there appears an

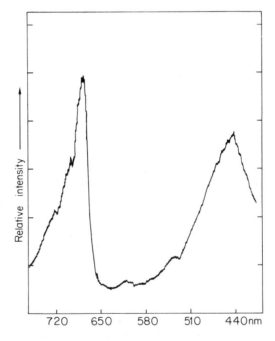

Figure 1 The fluorescence spectrum of phytochrome in glycerol–water (1:9, v/v) at 14 K. The excitation wavelength and bandpass were 380 nm and 6·4 nm, respectively, and the emission was recorded at 0·8 nm resolution

unexpected fluorescence at 440 nm, when the molecule is excited in the region of its second absorption band with 380 nm light. Both fluorescence bands are readily observable in the temperature range of 14–250 K, although their relative intensities vary with temperature. No fluorescence was observed at room temperature. (The detectability of the high-resolution spectrometer is $\Phi \geq 10^{-6}$ where Φ is the emission quantum yield.) Figure 2 shows the excitation spectrum of phytochrome, and it resembles the absorption spectrum, as expected.

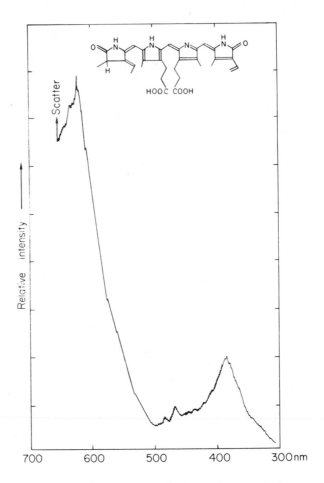

Figure 2 The fluorescence excitation spectrum of phytochrome in glycerol–water (1 : 9, v/v) with emission monitored at 580 nm. The emission bandpass was 1·6 nm, and the excitation spectrum was recorded at 0·16 nm resolution. The chromophore structure is inserted. The protein moiety is presumably bound to one of the carboxyl groups

In the fluorescence spectra of biliverdin at 77 K and of biliverdin dimethyl ester, again a two-fluorescence band system is observed, as in the case of phytochrome.

The fluorescence quantum yields of P_r in water (10% glycerol) are approximately 0·014 at 14 K and 0·005 at 250 K ($\lambda_{ex} \sim 380$ nm) and 0·038 at 14 K and 0·016 at 250 K ($\lambda_{ex} \sim 630$ nm). In contrast to earlier reports, highly purified, 'large' phytochrome P_r, used in the present work, are essentially non-fluorescent at room temperature ($\Phi_f < 10^{-4}$ under the experimental conditions employed). Fluorescence of phytochrome in solution has been reported by Hendricks et al[2] and Correll et al.[3] The latter workers obtained the fluorescence maximum of purified rye phytochrome in solution at 672 nm. More recently, however, highly purified and proteolytically undegraded phytochrome from rye showed no fluorescence in solution at room temperature.[4] The lack of fluorescence at room temperature has been confirmed in the present work.

The radiative fluorescence lifetime of P_r is 11 ns. Thus, the mean lifetime of P_r at room temperature is $11 \times 10^{-9} \times (<10^{-4}) \lesssim 10^{-13}$ s, thus the rate constant for the primary photoprocess is estimated to be $\gtrsim 10^{12}$ s^{-1}. Thus, the primary radiationless process of P_r is of comparable magnitude as in the rhodopsin → bathorhodopsin phototransformation.[5] Several possibilities exist for the mechanism of the primary process: namely (a) proton transfer as a prerequisite step toward the $P_r \rightarrow P_{fr}$ phototautomerism and (b) transient conformational change of the chromophore and/or apoprotein which is delicately coupled to the oscillator of the former.

Molecular orbital analysis of the electronic spectra (transition energies, oscillator strengths and polarization) of P_r and P_{fr} chromophores indicates that the conformations of both forms are essentially identical. The most likely conformation for the P_r chromophore is shown in Figure 3. Whether the fast primary photoprocess mentioned above involves a phototautomeric proton transfer in P_r is not established. However, in view of the fact that the proton transfer in a phototautomeric flavin model occurs even at 100 K or below and is temperature-dependent in the range of 100–200 K, proton transfer of P_r is not

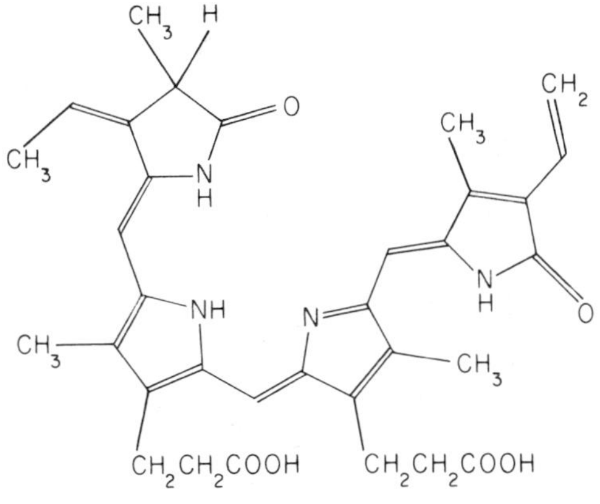

Figure 3 The most likely conformation of phytochrome arrived at from SCF MO CI (Pariser–Parr–Pople) calculations

favourable as a primary process. The next section describes the phototauto-merism of flavins.

2 PHOTOTAUTOMERISM OF FLAVINS

Figure 4 shows the fluorescence spectra of lumichrome as a function of pyri-dine concentration in dioxane. The intensity of red-shifted fluorescence (II) is increased with the pyridine concentration. A clear isoemissive point is obtained (476 nm), indicating two excited species (phototautomers) in equilibrium.

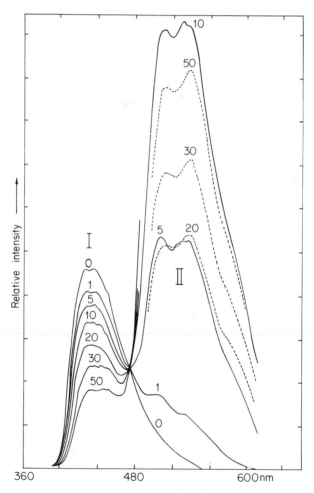

Figure 4 The fluorescence spectra of lumichrome (8.5×10^{-5} M) in dioxane at 298 K as a function of % pyridine. The dotted-line spectra are reduced by 1/3 from their actual intensity

Similar curves were obtained as a function of acetic acid concentration in ethanol. Again, an isoemissive point (488 nm) is obtained. It should be noted that the excitation spectra with respect to the two emission bands are identical in both dioxane–pyridine and ethanol–acetic acid mixtures. The excitation spectra are also identical with those of lumichrome in pure organic solvents where the two-band system is not observed, except for the usual solvent shift. This indicates that the two emission bands observed are not due to different species such as lumichrome and lumichrome–pyridine or acetic acid complexes in the ground state.

Figure 5 shows the effect of equimolar concentration of methyl pyridines (lutidines). It can be seen that 3,4-lutidine is most effective in enhancing band II, apparently reflecting the greater basicity (pK_a = 6.46) of the pyridinyl nitrogen relative to pyridine (pK_a = 5·25). On the other hand, 2,6-lutidine (pK_a = 6·6) is less effective than pyridine, possibly due to the steric requirement around the pyridinyl nitrogen. The effectiveness of 2,4-lutidine (pK_a = 6·77) nicely reflects the basicity and steric factor in catalysing the phototautomerism.

Band II (e.g., curves in pyridine in Figure 4) can be identified by superimposing it with independently measured fluorescence spectra of 10-methylisoalloxazine

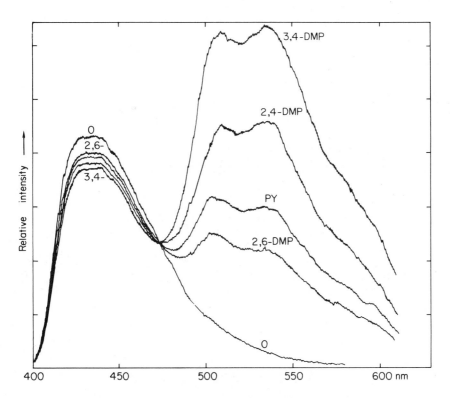

Figure 5 The fluorescence spectra of lumichrome ($\sim 8 \times 10^{-5}$ M) in dioxane at 298 K in the presence of 3×10^{-1} M pyridine and lutidines (dimethylpyridine)

(equivalent to the alloxazine tautomer) and lumiflavin (equivalent to the lumi-chrome tautomer). Furthermore, there is no phototautomerism in the presence of 10% or more pyridine in dioxane, when the N_1-proton is substituted by methyl group. N_3-Methylation does not negate the phototautomerism. This indicates that the N_1-methyl analogues do not equilibrate with the species emitting at the long wavelength (II) in their excited states.

From fluorescence polarization results, it was found that the band II intensity was suppressed with gradual increase in the viscosity of the medium. In addition, a decrease in temperature also suppresses its intensity, while the band I intensity proportionately increases. Thus, the phototautomerism involves a step which is diffusion-controlled.

Figure 6 shows a van't Hoff plot for the temperature dependence data, after correcting for the quantum yield variation for lumichrome (band I) and lumi-

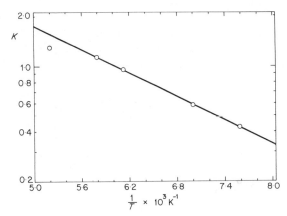

Figure 6 The temperature dependence of the quasi-equilibrium constant (K) in ether–pyridine (10%) where

$$K = \frac{I_f^{II}}{I_f^{I}} \times \frac{\Phi_f^{LC}}{\Phi_f^{LF}}$$

and I_f and Φ_f are corrected intensity of bands I and II and quantum yield of lumichrome (LC) and lumiflavin (LF), respectively

flavin (band II). Thermodynamic values for the quasiequilibrium phototautomerism of lumichrome to 7,8-dimethylisoalloxazine in the $^1(\pi, \pi^*)$ state are as follows:

$$\Delta H^{\circ *} = 1 \cdot 06 \text{ kcal/mole}$$

$$\Delta S^{\circ *} = -6 \cdot 4 \text{ e.u.}$$

$$\Delta G^{\circ *} = 2 \cdot 97 \text{ kcal/mole}$$

Assuming that the entropy change is about the same for the ground state tautomerism and knowing the transition energies of lumichrome (75 kcal/mole) and lumiflavin (64·25 kcal/mole), thermodynamic parameters for the tautomerism between alloxazine and isoalloxazine structures can be estimated, yielding

$$\Delta H° = 11\cdot 8 \text{ kcal/mole}$$

$$\Delta S° \simeq -6\cdot 4 \text{ e.u.}$$

$$\Delta G° = 13\cdot 7 \text{ kcal/mole}$$

All of the fluorescence data described above can be accommodated in terms of tautomerism between alloxazine and isoalloxazine forms in their excited states. It should be emphasized that what has been observed is not due to the excited state ionic equilibria, for example assigning band II to alloxazine or lumichrome anions. This is because the solvents used do not favour the formation of anions. Exciplex formation is not important in the present system. This is apparent, from Figure 4, since the enhancement of band II is about 8-fold or more for a given unit of decrease in the band I intensity. This is expected for the phototautomerism, since fluorescence quantum yields of isolloxazines are several times greater than those of alloxazines (for example, $\Phi_{f,\,\text{lumichrome}}/\Phi_{f,\,\text{lumiflavin}} \simeq \frac{1}{6}$).

Before illustrating possible mechanisms for the phototautomerism, it is pertinent to point out that the predicted acidity of the proton at N_1 in lumichrome increases upon excitation to the $^1(\pi, \pi^*)$ state, while the basicity of the N_{10} increases at the same time. Such a charge redistribution pattern is consistent with the acid–base catalysed phototautomerism, as shown in Figure 7. Pyridine and acetic acid act as general base and bifunctional acid–base catalysts, respectively, for the phototautomerism. The steric effect by 2,6-lutidine can be explained accordingly. The most convincing evidence in support of the phototautomerism is provided by two observations:

(a) N_1-Methyl alloxazines do not show band II and
(b) the fluorescence spectra of isoalloxazines coincide with the band II spectra.

In addition, when band II is not clearly resolved, due to viscosity and water in which resolution of the two-band system is less than that in dioxane, the enhancement in polarization degree of the band II region is characteristic of the emission from the isoalloxazine chromophore.

There was no observable solvent (D_2O) isotope effects on the phototautomeric equilibria in dioxane–pyridine and ethanol–acetic acid mixtures, when the N_1-proton was allowed to exchange with a deuteron in D_2O prior to the fluorescence measurement. Lack of the solvent isotope effect is expected for the proposed phototautomerism, particularly because $\Delta H°^*$ is small and the N_1-proton (or deuteron) is merely transferred to N_{10}.

3 CONCLUDING REMARKS

The significance of the present work is two-fold. First, the flavin system described can serve as a simple model for phototautomerism and for a dye laser.

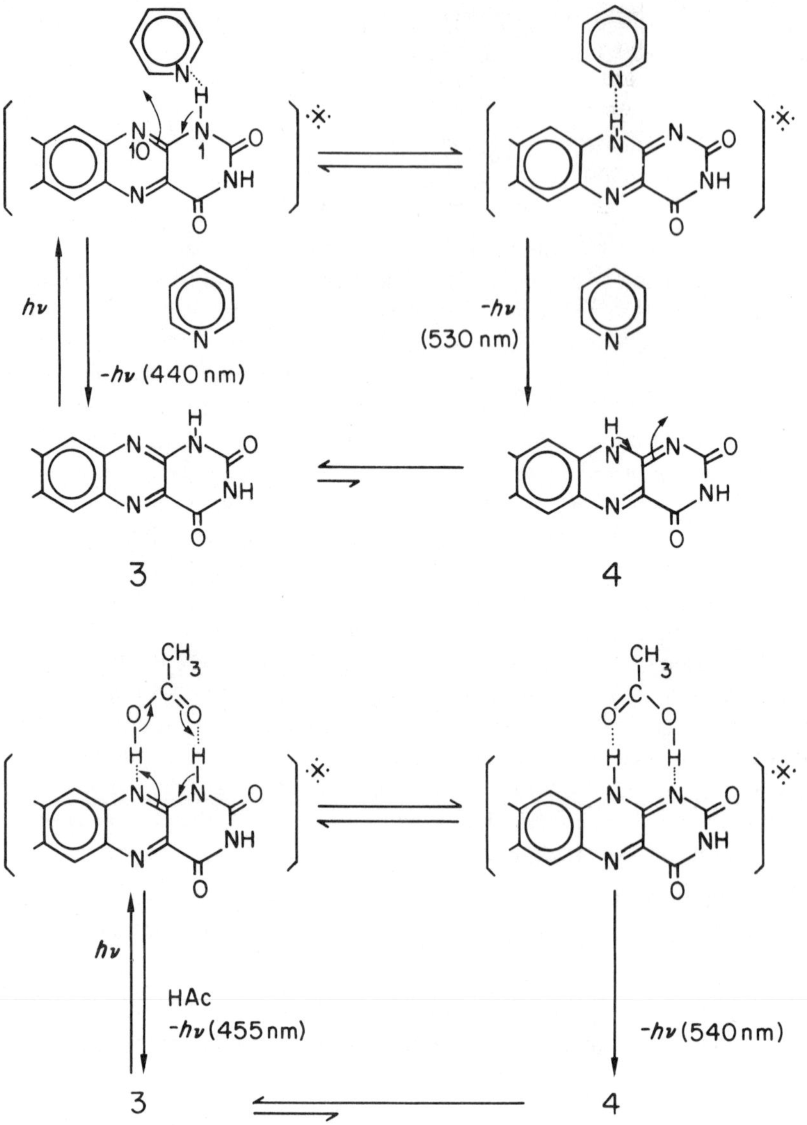

Figure 7 Possible mechanisms for the phototautomerism between lumichrome (3) and 7,8-dimethylisoalloxazine (4) catalysed by pyridine and acetic acid. The emission maxima are indicated by '−hv'

Second, the fact that the phototautomeric proton transfer in lumichrome and alloxazine occur even in rigid glass (cf. Figure 6) can be used against possible involvement of the proton transfer in the primary photoprocess of phytochrome (P$_r$). The calculated electronic structure of the excited state of P$_r$ also predicts negligible redistribution of charges so that proton transfer in the direction of phototautomerism of P_r to P_{fr} is not favourable as a primary process. The proton transfer could occur in a subsequent step, possibly as the rate-determining process.

ACKNOWLEDGEMENTS

We thank Professors Winslow Briggs and David Lightner for supplying us with samples used in this work. The present work was supported by the National Science Foundation, GB-21266, and Robert A. Welch Foundation, D-182.

REFERENCES

1. P. S. Song, Q. Chae, D. A. Lightner, W. R. Briggs and D. Hopkins, *J. Am. Chem. Soc.*, **95**, 7892 (1973).
2. S. B. Hendricks, W. L. Butler and H. W. Siegelman, *J. Phys. Chem.*, **66**, 2550 (1962).
3. D. L. Correll, E. Steers, Jr., K. M. Towe and W. Shropshire, Jr., *Biochim. Biophys. Acta*, **1968**, 46 (1968).
4. E. Tobin and W. Briggs, *Photochem. Photobiol.*, **18**, 487 (1973).
5. G. E. Busch, M. L. Applebury, A. A. Lamola and R. M. Rentzepis, *Proc. Natl. Acad. Sci. U.S.*, **69**, 2802 (1972).

C.2. Absorption, fluorescence and ESR spectra of preilluminated Chlorella

D. Frackowiak, E. Hans, A. Skowron and W. Froncisz

Institute of Physics, Poznań Technical University, Piotrowo 3, 61–138 Poznań, Poland

SUMMARY

The slow changes in optical properties of algae under illumination with various wavelengths of monochromatic light were investigated.

Changes, caused by illumination, of the absorption, emission and ESR spectra of algae suspended in various media with or without benzoquinone addition were compared. The influence of light on benzoquinone-containing samples is less than that on the samples without benzoquinone.

The yield and kinetics of all these types of photochanges depend strongly on the wavelength of the actinic light and on the type of solvent. Since these dependences are different in the case of fluorescence, ESR and absorption spectra, it seems that no simple correlation exists between free radical formation, the photoreactions responsible for absorption changes and the processes causing 'slow fluorescence induction'.

Fluorescence induction is probably related not only to the reaction undergone by the pigment but also to the structure of its environment.

The changes in fluorescence intensity subsequent to the illumination of algae (the so-called 'fluorescence induction') are complex phenomena.[1,2] The purpose of this study is to obtain more information about the molecular mechanisms responsible for the slow changes in optical properties of algae due to continuous illumination with monochromatic light.

Light intensities of order 10^{-5} watt cm^{-2} were applied: the 'light' samples were in 'range T', the 'dark' ones in the 'S phase' of fluorescence induction.[3,1,2] Changes in the absorption spectra, fluorescence intensity and ESR signals were measured. Algae were suspended in water, glycerine and ethylene glycol with or without addition of the typical fluorescence quencher—benzoquinone (bzq). The results of illumination with and without quencher were compared.

The following spectral regions were used in preillumination: 475 nm (strongly absorbed by Chl *b* belonging predominantly to PS II), 405 nm and 436 nm (absorbed by pigments from both photosystems PS I and PS II). The spectral width of the actinic beam did not exceed 10 nm.

In all solvents the influence of light in bzq-containing samples is weaker than on samples without bzq. Quenching by bzq of preilluminated *Chlorella* is less efficient than that of dark-adapted.[4] The changes in fluorescence intensity are solvent dependent. For glycol suspension, the ratio of fluorescence intensities of the sample with bzq and of that without quencher even increases as a result of illumination.[4]

ESR spectra were obtained with a Varian E-3 spectrometer. The samples were illuminated within the cavity. By proper calibration of the actinic beam and normalization of the results, the relative yield of unpaired spin formation by quanta of various spectral regions was calculated (Table, column 5).

As seen from Table 1, glycol modifies all properties of the ESR signal such as the amplitude, width and dependence on the spectrum of the actinic beam. In water and glycerine, the unpaired spin density increases more strongly with shortwave than with longwave illumination, whereas in glycol suspension long-wave illumination is more effective. In all the solvents, addition of bzq causes a decrease in the ratio of 'light' to 'dark' signal amplitudes. The addition of bzq to glycol suspension of algae leads to the disappearance of the typical *Chlorella* signal and only five lines of the bzq signal are observed.

The growth time of the signal is always shorter than its decay time (Figure 1). The decay of the signal induced by 475 nm illumination is slower than that generated with shorter wavelenths (Figure 1).

The medium in which the algae are suspended also influences the light-induced changes in absorption. As it follows from the difference (light minus dark) transmittance spectra, in glycol predominantly Chl *a* 670 (predominantly belonging to PS II form[5]) undergoes the photoreaction, whereas in glycerine (Figure 2) Chl *a* 683 and 695, which are two forms belonging to PS I,[5] are more photosensitive.

The absorption changes in glycerine are 'light triggered'[2] (increasing during some minutes after interruption of illumination and later decreasing), whereas in glycol the changes are 'light dependent'—increasing only during illumination and decreasing in the dark. In both solvents, the changes are to some extent reversible in the dark. No measurable absorption changes of suspensions in culture medium were found.

The yield of all three types of photochanges depends strongly on the wavelength of the actinic light and on the type of solvent. Since these dependences are different in the case of fluorescence, ESR and absorption spectra, it seems that no simple correlation exists between free radical formation, the photoreactions responsible for absorption changes and the processes causing 'slow fluorescence induction'.

The decrease in relative yield of light action on the pigments *in vivo* caused by the presence of bzq may be related to the difference in photoreactivity of

Table 1   ESR spectra of *Chlorella* suspensions

Medium	Type of signal	Position of signal (gauss)	Line width ΔH_{max} (gauss)	Relative yield of unpaired spin formation η_λ/η_{475}	Absorption coefficient of actinic beam in sample	Structure of signal
1	2	3	4	5	6	7
Water	Dark	3377·1	17·5	6·3	0·57	Without structure
	Light (405 nm)	3379.8	9·25			
	Dark	3374·5	17·5	9·8	0·73	Without structure
	Light (436 nm)	3377·3	9·25			
	Dark	3384·2	17·0	1·0	0·56	Without structure
	Light (475 nm)	3379·5	9·2			
Water with bzq	Dark	3374·5	6·25	0·1	0·68	Two components
	Light (405 nm)	3374·5	6·25			
	Dark	3374·25	6·25	0·18	0·85	Two components
	Light (436 nm)	3374·25	6·25			
	Dark	3374·7	6·25	1·0	0·63	Two components
	Light (475 nm)	3374·25	6·25			
Glycol	Dark	3369·0	7·5	0·16	0·60	Without structure
	Light (436 nm)	3370·0	7·5			
	Dark	3369·0	5·8	1·0	0·52	Without structure
	Light (475 nm)	3371·0	5·8			
Glycol with bzq	Dark	3375·8	10·5	bzq signal only	0·80	Five components
	Light (436 nm)	3375·8	10·5			

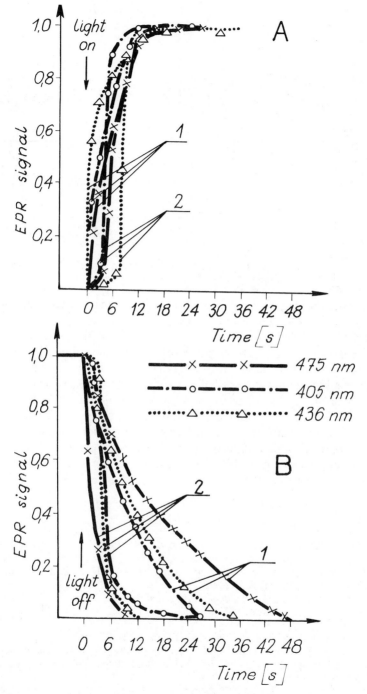

Figure 1 The kinetics of the rise A and decay B of ESR signals of
Chlorella in water: (1) without bzq, (2) with bzq

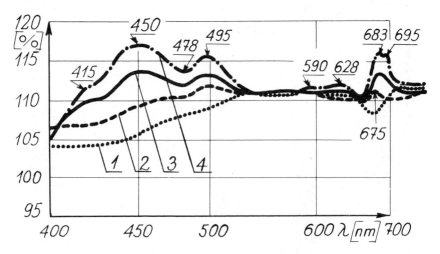

Figure 2 Difference transmittance spectra of *Chlorella* in glycerine: (1) unilluminated sample, (2) illuminated with 475 nm, (3) with 405 nm, (4) with 436 nm

various forms of the pigments. Probably, more photosensitive forms are more easily reached by the chemical quencher of fluorescence.

Similarly, the yield of chemical quenching decreases when most of the sensitive molecules are already changed by the photoreaction. The special role of glycol can be explained on the basis of the results of Inoué and Nishimura[6,7] showing that glycol influences PS II more strongly than PS I and enhances the reactivity of that system. Therefore, in glycol, PS II is modified and becomes more sensitive to chemical and light treatment. The change in ESR signal width in glycol may be due to a decrease in pigment aggregation.[8] The present results suggest that the slow light-induced change in fluorescence intensity is caused by a number of processes probably related not only to the reaction undergone by the pigment, but also to the structure of its environment.[9]

REFERENCES

1. G. Papageorgiu and Govindjee, *Biophys. J.*, **8**, 1299 (1968).
2. G. Papageorgiu and Govindjee, *Biophys. J.*, **8**, 1316 (1968).
3. J. Lavorel, in *Currents in Photosynthesis*, pp. 39–47 (eds. J. B. Thomas and J. C. Goedheer), Ad. Donker, Rotterdam, 1965.
4. D. Frackowiak and E. Hans, *Photosynthetica*, **6**, 298 (1972).
5. R. Devlin and A. V. Barker, *Photosynthesis*, Van Nostrand Reinhold Company, New York, 1971.
6. H. Inoue and M. Nishimura, *Plant Cell. Physiol.*, **12**, 137 (1971).
7. H. Inoue, K. Wakamatsu and M. Nishimura, in *Energy Transdution in Respiration and Photosynthesis*, pp. 565–578 (eds. S. Quagiariello, S. Papa and C. S. Rossi), Adriatica Editrice, Bari, 1971.
8. J. R. Norris, J. J. Katz and R. A. Uphaus, *Abstracts VI International Congress on Photobiology, Bochum*, p. 227, 1972.
9. P. Mohanty, B. Zilinskas Braun and Govindjee, *Biochim. Biophys. Acta*, **292**, 459 (1973).

C.3. Photosensitivity in bilayer lipid membranes

Donald S. Berns,* Asher Ilani** and Marc Mangel**

*Division of Laboratories and Research, New York State Department of Health, Albany, New York, U.S.A.

**Department of Physiology, Hebrew University-Hadassah Hospital Medical School, Jerusalem, Israel

SUMMARY

The phenomenon of photosensitivity of artificial bileaflet membranes made of photosynthetic pigments is described and analysed. The main feature of the phenomenon is the establishment upon illumination of an electric current across the bilayer membrane interposed between aqueous solutions of different redox potential. Other aspects of the response fit with the idea that the absorption of photons by the chlorophyll in the bilayer membrane allows electron transport across the membrane. The presence of carotenoids in the bileaflet structure is essential for the photosensitivity. The photosensitivity of the bilayer membrane is compared with the primary process in the natural photosynthetic membrane. It is argued that the presence of aqueous modifiers which alter the electronic energy barrier at the membrane–water interface and the formation of chlorophyll assemblies are the two major factors which are responsible for the different response to light of the natural and artificial membrane systems.

1 INTRODUCTION

For the past few years we have studied various aspects of the photosensitivity of artificial bilayer membranes. Such membranes may be used as a model for the much more complex photosynthetic membranes.

In this paper, we would like to present some of the major features of the experimental and theoretical studies which have been undertaken.

2 TECHNIQUE FOR THE FORMATION OF BILAYER LIPID MEMBRANES

In the early 1960s, the description of a technique for the formation of bilayer lipid membranes (BLM) by Mueller *et al.*[1] opened up new vistas in membrane research. The method of formation consists of brushing a small amount of lipid dissolved in suitable solvent across a hole in a hydrophobic partition separating two aqueous phases. When first placed on the hole, the lipid solution is globular in form. It then spontaneously thins into a bilayer. As it thins, the electrical capacity between the two aqueous phases increases and the originally coloured glob turns black (earning the alternate name of black lipid membrane for BLM). A number of different experiments have verified that the final structure is actually a bilayer.[2] Through the addition of various modifiers, BLM have been made to mimic some of the functions of biological membranes.[2]

3 PHOTOSENSITIVITY OF CHLOROPLAST EXTRACT BLM

Tien and Verma[3] were the first to show that a photovoltage response could be obtained from a BLM made of chloroplasts lipid extract. This result was verified by Ilani and Berns,[4] who also extended the studies of chloroplast extract membranes.

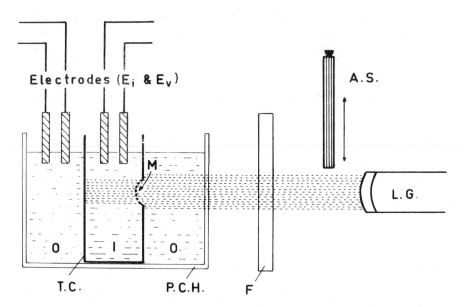

Figure 1 Scheme of experimental set-up. L. G., light guide; A. S., aluminium shutter; F, light filter; P. C. H., perspex cell holder; O, outside; I, inside; T. C., Teflon cell; M., membrane. One pair of electrodes (E_v) used for measuring potential and capacitance; second pair (E_i) used for applying a current

Figure 1 shows the set-up of their experimental apparatus. One pair of electrodes was connected to an electrometer of high input impedance and was used to measure the potential difference across the membrane. The second pair was connected to a voltage source and resistance, so that a current could be applied through the membrane. Illumination was provided by a light source and light guide. The essential features of the response to illumination are shown in Figure 2.

It can be seen that when the solutions on the two sides of the membrane were identical (I in Figure 2), the photoresponse was observed only if a current was passed through the membrane (a photoconductance effect). If a redox potential gradient was produced (II in Figure 2), a photovoltage response was observed even at zero external current. The response was modified by imposing a membrane potential. The relationship between photovoltage response and membrane potential in each case was linear; the slopes were approximately the same (lower part of Figure 2). These results are consistent with the concept that upon illumination of a BLM made of chloroplast extract electron conductive channels are opened within the membrane.

Other properties of the photoresponse of chloroplast extract BLM which support the idea that transmembrane electron movement is responsible for the developed photopotential are:

(1) The photovoltage response at zero current was dependent upon the redox gradient across the membrane and independent of the absolute value of the redox potential.[4] For example, a redox potential of about 100 mV around the Fe^{+2}/Fe^{+3} level or around the Ce^{+3}/Ce^{+4} level gave rise to similar photoresponses, even though the redox potential levels of the two couples are separated by more than 500 mV.

(2) The current produced upon illumination of the membrane was independent of temperature from 6–40 °C. Over the same range, the dark conductivity of the membrane varied by more than 300%.[4] It is plausible that the dark conductivity is ionic in nature.[2]

It was first demonstrated by Tien's group[3] and confirmed by Ilani and Berns[4] that the action spectrum of the photoresponse followed the absorption spectrum of chlorophyll. It is clear therefore that chlorophyll is an essential component of the photosensitive BLM.

4 COMPARISON BETWEEN PHOTOSENSITIVE BLM AND PHOTOSYNTHETIC MEMBRANES

Within the present understanding of photosynthesis, the primary process in the photosynthetic membrane consists of the transfer of an excited electron to a species which belongs to a redox pair of relatively high electronic energy level. Concomitantly, there is the uptake of an electron from a species belonging to a redox pair with a relatively low electronic energy level. Therefore, the primary

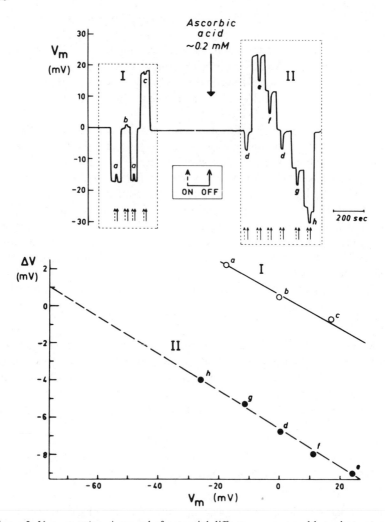

Figure 2 *Upper portion*: A record of potential difference across a chloroplast extract BLM exposed on both sides to a 0·1 M potassium acetate buffer pH 5·0 containing 1 mM $FeCl_3$ and 0·5 mM $FeCl_2$. The membrane potential was varied by applying various currents to the membrane through another pair of electrodes. The effect of light irradiation is shown at the point of the dashed and solid arrows which indicate light on and light off, respectively. Light was filtered through a 3–66 Corning filter which excluded wavelengths below about 570 nm (i.e. the incident light did not contain quanta which are absorbed by the Fe^{+++} in the aqueous solution). The intensity of the light at the level of the membrane was about 70 mW/cm^2. Ascorbic acid was added to the outside solution at a time indicated by the arrow to a concentration of about 0·2 mM. Polarity refers to the inside solution. *Lower portion*: Analysis of the photoresponse (ΔV) as a function of membrane potential (V_m) before (curve I) and after (curve II) addition of ascorbic acid to outside solution (ΔV defined as the steady state membrane potential during illumination minus membrane potential after stopping the irradiation). The points in the lower portion of the figure were calculated from the record shown in the upper portion. The labels a, b, c, etc. indicate corresponding points and record traces shown in the lower and upper portion of the figure, respectively

photosynthetic process consists of establishing or increasing a redox potential difference through the use of the photon energy.[5,6]

On the other hand, the photoresponse in BLM represents a system in which a pre-existing redox potential difference is allowed to dissipate upon illumination. Thus, the two phenomena are conceptually antagonistic. This antagonism is shown schematically in Figure 3.

There is also a profound quantitative difference between the two types of membranes. In photosynthetic membranes, the quantum efficiency (number of

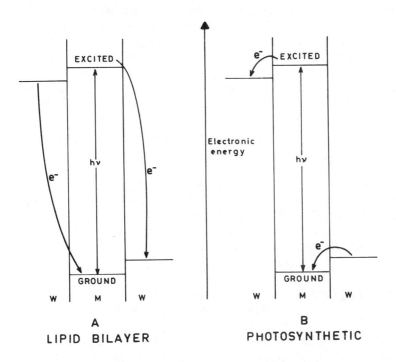

Figure 3 Schematic models of the photosensitive BLM (A) and the photosynthetic membranes (B) upon illumination. Both membranes are shown separating two redox pairs. The absorption of a photon in the BLM results in the transfer of an electron from the pair which has higher electronic energy to the membrane with a simultaneous transfer of an electron from the membrane to the pair which has lower electronic energy. Energetically, the sum of this process is the dissipation of the photon energy and of the pre-existing redox gradient. On the other hand, illumination of the photosynthetic membrane results in the movement of an excited electron to the redox pair of higher electronic energy with the concomitant movement of an electron from the pair with lower electronic energy into the membrane. Energetically, the sum of this process is the increase of the existing redox gradient and the storage of part of the photon energy. We have depicted the membranes separating the two redox couples, as indeed may be the case. However, there is no conclusive evidence against the possibility that the two redox pairs are present on the same side of the photosynthetic membrane

electrons transported/number of photons absorbed) is between 0·1 and 0·5.[5,6] In BLM, the quantum efficiency is about 10^{-6}.†

It is reasonable to assume that the quantitative difference between the two membranes results from the presence of various aqueous components (enzymes and the like) which specifically direct the electron flow in photosynthesis. According to this hypothesis, the process in the photosensitive BLM (Figure 3A) represents the spontaneous behaviour of the system, whereas the process in the photosynthetic membranes (Figure 3B) results from the alteration of barriers to electron movement brought about by the various modifiers.

Furthermore, it is possible that the quantitative difference between the two membranes is also related to the presence or absence of modifiers. These modifiers may facilitate electron movement across the membrane–solution interface and/or across the bulk membrane.

However, as will be discussed in the next section, the quantitative difference may be related to the nature of the lipidic membrane itself. It is plausible that assemblies of chlorophyll are the rule in the natural membranes whereas such aggregates may not exist in BLM, due to the presence of organic solvent. It was argued that an aggregate of chlorophyll could enhance electron exchange at the membrane–water interface.[7]

Phycocyanin, a protein isolated from blue-green algae, when placed on the more oxidized side of a photosensitive BLM could enhance the photocurrent 2–3 times.[4] The effect of phycocyanin was unidirectional. It may be that phycocyanin is indeed one of the specific electron-directing agents in the photosynthetic apparatus of blue-green algae.

5 A THEORETICAL MODEL FOR ELECTRON TRANSPORT ACROSS THE MEMBRANE

In order to explain electron movement across the membrane, Ilani and Berns[7] divided the transport process into two parts: (1) electron exchange at the membrane–water interface, (2) electron movement across the bulk membrane.

Concepts of electron tunneling used to describe other electrode processes were applied to the first stage. The main emphasis in these ideas is concerned with the requirement of overlap of the electronic energy levels of the aqueous donors and acceptors and the chromophores in the membrane.[8,9]

The second process, electron transport through the membrane, was attributed to the presence of carotenoids, which are needed to reduce the energy barrier to electron movement between the chromophores on either side of the

† A rough estimate of the quantum efficiency in BLM is: the maximum currents were about 10^{-9} A/cm² or 10^9 electrons/cm² s. The intensity of light at the membrane surface was 150 mW/cm², corresponding to approximately 3×10^{19} photons/cm² s. The membrane forming solution had an average OD of about 200. Since the membrane was 5×10^{-7} cm thick, the membrane OD was about 10^{-4}. Thus, the number of photons absorbed was around 10^{16}. Consequently, the quantum efficiency was roughly 10^{-6}

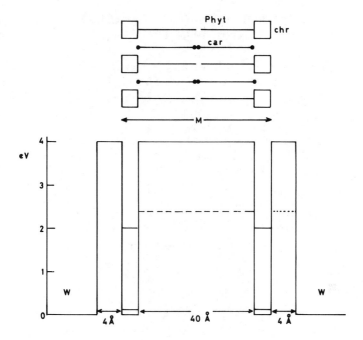

Figure 4 *Upper portion*: The scheme of a possible molecular arrangement of a BLM containing chlorophyll and carotene. Chr, chromophore portion of chlorophyll; Phyt, phytol chain of chlorophyll; Car, carotenoids. *Lower portion*: The corresponding energy barrier diagram for the membrane (M)–water (W) system. The broken line in the membrane represents the reduced energy barrier level due to carotene. The solid lines in the potential wells represent the ground and first excited states of chlorophyll. The dotted line in the right-hand side of the membrane–solution energy barrier represents the lowering of the energy barrier due to excited phycocyanin. The 4 Å width is the distance of closest approach between the redox species in water and the chromophore of the membrane

membrane. The carotenoids reduce the potential barrier due to the system of alternating double bonds.[10,11]

The schematic representation of the electronic energy levels across the membrane is shown in Figure 4. The broken line across the 40 Å barrier represents the lowered barrier when carotenoids are present. The solid lines in the small potential wells represent the ground and first excited states of chlorophyll. The model predicts that an excited electron from chlorophtll will have enough energy to traverse the membrane, if carotenoids are present. The dotted line represents a hypothetical lowering of the solution membrane energy barrier by excited phycocyanin.

6 STUDIES ON BLM CONTAINING CHLOROPHYLL AND CAROTENE[12]

One of the problems with the chloroplast extract was that its exact composition was unknown. The extraction procedure was crude and many pigments and lipids were included in the extract. Thus, it was impossible to test some of the

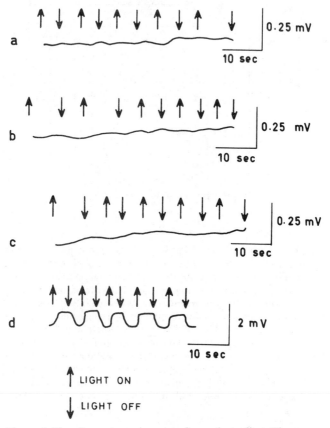

Figure 5 The pigment requirements for a photoeffect. The concentration of phosphatidyl choline was constant, 3 mg/ml throughout the series of experiments. The solvent was a mixture of t-butanol in hexane (3 : 1). (a) Phosphatidyl choline BLM; (b) phosphatidyl choline and chlorophyll (4·45 mg/ml); (c) phosphatidyl choline and carotene (0·19 mg/ml) and (d) phosphatidyl choline, chlorophyll (1·11 mg/ml) and carotene (0·16 mg/ml) BLM. Note that only when both chlorophyll and carotene were present in the membrane-forming solution was there a clear photoresponse. The basic aqueous solution was 0·1 M KCl. The inside solution also contained 16 mM $FeCl_3$ + 1 mM $FeCl_2$. The outside solution contained 1 mM $FeCl_3$ + 1 mM $FeCl_2$. The inside cup became negative with respect to the outside cup upon illumination

ideas embodied in the suggested theoretical model. For instance, the model predicted that both chlorophyll and carotene are necessary for the observation of a photoresponse in BLM. In order to test this prediction, a more synthetic approach was taken.

The pigment requirements for a photoeffect were determined as follows: a basic membrane-forming solution of phosphatidyl choline was used. Various pigments were separately added to the membrane-forming solution and the photosensitivity was tested. The results are shown in Figure 5. A membrane made of phospholipid (a), lipid and chlorophyll (b) or lipid and β-carotene (c) did not show any observable photosensitivity. However, a BLM made of lipid, chlorophyll and β-carotene (d) did exhibit photosensitivity. Thus, it was demonstrated that both chlorophyll and carotene were needed for the photosensitivity.

Figure 6 shows the results of illumination through different filters. Filters absorbing in the red range eliminated the photoresponse, while filters absorbing in the blue range did not. Thus, it was determined that though β-carotene was needed, the absorbance of light by carotene was not needed for the photosensitivity.

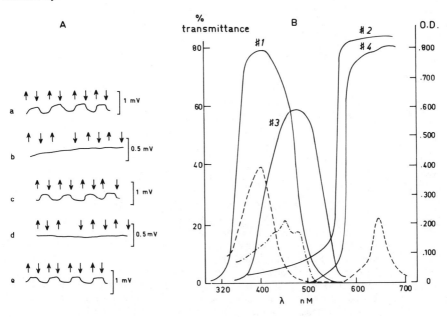

Figure 6 (A) Photoresponse as a result of illumination through different filters. BLM made of phosphatidyl choline (3 mg/ml), chlorophyll (1·74 mg/ml) and carotene (0·16 mg/ml). (a) No filter; (b) filter #1; (c) filter #2; (d) filter #3 and (e) filter #4. The transmittance curves (full lines) of the filters are shown in (B). The absorbance of chlorophyll (2·2 × 10^{-3} mg/ml; dashed line) and β-carotene (0·6 × 10^{-3} mg/ml; dotted line) solutions in t-butanol are also included. Note that the elimination of the part of the spectrum which excites carotene (filters #2 and #4) did not interfere with the photoresponse, whereas the elimination of the 680 nm range (filters #1 and #3) which excites chlorophyll abolished the response. Other details as in Figure 5

Figure 7 shows the dependence of the photocurrent on the chlorophyll concentration (when the carotene concentration was held constant) in the membrane-forming solution. The curve is quadratic in the chlorophyll concentration. This may indicate that the photosensitivity of BLM was not dependent upon the presence of large aggregates of chlorophyll.

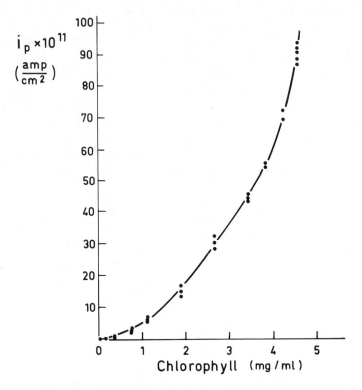

Figure 7 The dependence of photocurrent on chlorophyll concentration in the membrane-forming solution. For all membrane-forming solutions, the concentrations of β-carotene and phosphatidyl choline were 0·15 mg/ml and 3 mg/ml, respectively. Other details as in Figure 5. Each point represents the results of experiments performed on one membrane. For each membrane, a minimum of three responses were recorded. The photovoltage response was constant to within 10% of the average. The curve corresponds roughly to linear dependence of the photocurrent on the second power of chlorophyll concentration

The photocurrents observed with chloroplast extract or with chlorophyll, β-carotene containing BLM were comparable at comparable chlorophyll concentrations. That there were no large differences indicates that chlorophyll and β-carotene are ostensibly the primary functional pigments in chloroplast extract BLM.

The extremely low quantum efficiency of both types of BLM may be related to the absence of chlorophyll aggregates in the membrane. It is known that BLM contain significant concentrations of organic solvent.[13] The solvent is likely to prohibit the formation of such aggregates.

Preliminary studies of the photosensitivity of bilayer structures devoid of solvent have revealed that: (1) the quantum efficiency is much higher than the quantum efficiency of BLM with solvent, (2) the optical absorption spectra of the bilayer structures indicates the appearance of a long-wave (720–770 nm) absorption peak. This peak has been attributed to the presence of chlorophyll aggregates.

7 CONCLUSION

The photosensitive bilayer membrane represents a primitive model of the photosynthetic membrane. The presence of chlorophyll and carotenoids in such structures allows electron flow across the membrane when light is absorbed by the chlorophyll chromophores. Aqueous modifiers can alter the electronic energy barriers as demonstrated by the action of phycocyanin on the sensitive BLM.

The major factors which may transform the photosensitive bilayer into a photosynthetic one are: (a) aqueous modifiers such as enzymes which affect the electron flow through the interface and (b) the formation of chlorophyll aggregates within the membrane structure, which hopefully may be brought about by the elimination of organic solvents from the bilayer membrane.

REFERENCES

1. P. Mueller, D. O. Rudin, H. T. Tien and W. C. Wescott, *J. Phys. Chem.*, **67**, 534 (1963).
2. H. T. Tien, *The Chemistry of Biosurfaces*, Chapter 6 (ed. M. Hair), Marcel Dekker, New York, 1971.
3. H. T. Tien and S. P. Verma, *Nature*, **227**, 1232 (1970).
4. A. Ilani and D. S. Berns, *J. Mem. Biol.*, **8**, 333 (1972).
5. M. D. Kamen, *Primary Processes in Photosynthesis*, Academic Press, New York and London, 1963.
6. R. K. Clayton, *Molecular Physics in Photosynthesis*, Blaisdell, New York, 1965.
7. A. Ilani and D. S. Berns, *Biophysik*, **9**, 209 (1973).
8. R. W. Gurney, *Ions in Solution*, Cambridge University Press, Cambridge, 1936.
9. H. Gerischer, *Adv. Electrochem. Electrochem. Engin.*, **1**, 139 (1961).
10. J. R. Platt, *Science*, **129**, 372 (1959).
11. H. Kuhn, *J. Chem. Phys.*, **17**, 1198 (1946).
12. M. Mangel, D. S. Berns and A. Ilani, *J. Mem. Biol.*, **20**, 191 (1975).
13. R. E. Pagano, S. M. Ruysschaert and I. R. Miller, *J. Mem. Biol.*, **10**, 11 (1972).

C.4. Selective light scattering from oriented photosynthetic membranes

C. E. Swenberg†

Institute for Fundamental Studies, Department of Physics and Astronomy,
University of Rochester, Rochester, New York 14627, U.S.A.

N. E. Geacintov

Department of Chemistry, New York University,
New York, New York 10003, U.S.A.

SUMMARY

The anisotropy and wavelength dependence of light scattered at 90° from magnetically oriented *Chlorella* cells in suspension are interpreted using form factors calculated in the Rayleigh–Gans approximation, and are discussed in terms of the orientation of chlorophyll molecules in the photosynthetic membranes. The wavelength dependence of the scattered light for different configurations of incident and scattered light beams is constructed theoretically. The effect of shape, size, polarizability anisotropy and the chlorophyll *a* concentration *in vivo* are used in these calculations. This method yields an effective degree of orientation of chlorophyll *a in vivo*.

1 INTRODUCTION

The optical properties of biological particulate matter such as whole cells, chloroplasts and membrane fragments in suspension can provide valuable information about the structure of cell components and the orientation of pigment molecules. However, the interpretation of experiments such as absorption, linear dichroism, circular dichroism and fluorescence polarization are complicated by light-scattering effects. In particular, the scattering properties of

†Current Address: Atomic, Molecular and Polymer Physics Group, Department of Physics, University of Manchester, Manchester, England.

288

highly ordered stacks of lamellar membranes of chloroplasts have been shown to distort the absorption spectra,[1] linear dichroism[2] and circular dichroism[3] spectra of whole algal cells such as *Chlorella pyrenoidosa*. Latimer and Rabinowitch[4] have shown that the 90° scattering from *Chlorella* cells in suspension is a function of wavelength and exhibits selective scattering in the region of the absorption bands. At wavelengths below the absorption maximum there is a minimum in the scattering intensity, while at wavelengths above this absorption peak there is a maximum. In *Chlorella* the chloroplast has a hemisphere-like shape and the cells are nearly spherical with a diameter of about 3 μm. Latimer and Rabinowitch were able to account for their experimental scattering curves by a calculation based on the Mie scattering theory[5] in which the cell is replaced by a homogeneous sphere characterized by an *isotropic* index of refraction.

It has been recently shown that whole cells and chloroplasts in suspension can be oriented by magnetic fields of 10 kilogauss or more.[6] This effect can be used to study the optical properties of oriented membranes under physiological conditions. Linear dichroism[7,8] and fluorescence polarization[8] studies indicate that the chlorophyll molecules *in vivo* are highly oriented. For example the Q_y transition moment vectors which are responsible for the 678 nm absorption band in *Chlorella* and in spinach chloroplasts tend to lie in the plane or close to the plane of the lamellae. The membranes tend to orient with their planes perpendicular to the magnetic field which provides the laboratory reference axis for these optical studies.

Van Nostrand[10] has measured the wavelength (λ) dependent scattering at 90° of both oriented, $S(\lambda, H)$, and unoriented, $S(\lambda, 0)$, *Chlorella* cell suspensions, where $S(\lambda, 0)$ and $S(\lambda, H)$ denote the relative 90° scattering intensities in the absence and presence of the magnetic field respectively. The scattering ratios $S(\lambda, H)/S(\lambda, 0)$ exhibit a strong wavelength dependence, particularly in the region of the absorption bands (Figure 1). The scattering was measured either parallel or perpendicular to the membrane planes which we define as the largest projection of the cup-shaped lamellae perpendicular to the magnetic field. The scattered light was polarized perpendicular to the scattering plane.

In this paper, Van Nostrand's scattering curves are analysed semiquantitatively primarily in the region from 640–740 nm, i.e. in the region of the 'red' chlorophyll absorption band. The strong structure displayed by these scattering curves is related to an anisotropy in the polarizability of the lamellar structures and to the shape of the scattering particles. While the cell walls and organelles in living cells contribute in general to the observed scattering intensity, it should be stressed that the strong variations in wavelength are due to scattering from those cell components which are pigmented, which in our case are the membranes in the chloroplasts.

It is shown here that the structure observed in Figure 1 can be explained in terms of the intrinsic orientation of the pigment molecules which gives rise to a significant anistropy in the polarizability of the lamellar chloroplast structure. Thus, 90° scattering curves determined with oriented membranes can provide

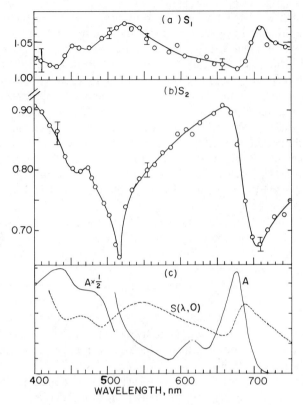

Figure 1 90° scattering from intact *Chlorella* cells in aqueous suspensions. $S(\lambda, 0)$ refers to scattering from unoriented cells (no magnetic field, H), while S_1 and S_2 are the scattering ratios $S(\lambda, H)/S(\lambda, 0)$ in the presence of a magnetic field of 10 kgauss. The lamellae tend to orient with their planes perpendicular to H. S_1 and S_2 are defined by equations (1) and (2) (see Figure 2). In (c), the absorbance (A) and scattering $S(\lambda, 0)$ are given in relative units. The experimental data are taken from the dissertation of F. Van Nostrand[10]

information about pigment orientation. This technique may prove to be particularly useful in cases when there is no fluorescence and when linear dichroism studies are complicated by the flattening effect or are not feasible for other reasons.

The lower curve (Figure 1c), $S(\lambda, 0)$, is the 90° scattering curve of an unoriented suspension of *Chlorella* cells, and is in good agreement with results of Latimer and Rabinowitch. The two upper curves (Figures 1a, b) obtained with oriented cells on the other hand, cannot be explained in terms of their theory.

In fact it will be shown that if the polarizability (and therefore the pigment orientation) is isotropic, the strong wavelength variations displayed in Figures 1(a) and 1(b) should *not* be observed.

The differential scattering ratios

$$S_1 = S_{\parallel}(\lambda, H)/S_{\parallel}(\lambda, 0) \tag{1}$$

and

$$S_2 = S_{\perp}(\lambda, H)/S_{\perp}(\lambda, 0) \tag{2}$$

are defined in terms of the orientation of the membrane planes and the directions of polarization and viewing of the scattered light. $S_{\parallel}(\lambda, H)$ and $S_{\perp}(\lambda, H)$ refer to scattered light polarized parallel and perpendicular respectively to the membrane planes. The geometry of the measurement is depicted in Figure 2. The scattering ratio parallel to the membrane planes S_1 is always $>1\cdot0$, while $S_2 < 1\cdot0$.

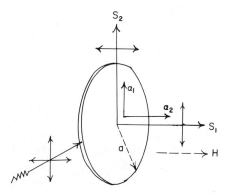

Figure 2 Scattering model and directions and polarization of the incident and scattered light S_1, S_2. The membrane segments are replaced by a thin disc of radius a and polarizabilities α_1 within the plane and α_2 normal to the plane. The assumed direction of the magnetic field H is also shown

In the next section we discuss our model (Rayleigh–Gans approximation[5]) which is used in the calculations of S_1 and S_2 and derive the appropriate theoretical equations. In the following section we discuss semiquantitatively the effects of size of the scattering particle, the polarizability anisotropy and the pigment orientation.

2 SELECTIVE LIGHT SCATTERING MODEL

Chlorella cells are approximately spherical with a radius of about 3 μm with a cup-shaped chloroplast. The chloroplast consists of membranes, or lamellar structures, and is attached to the cell wall. We define a 'partial' symmetry axis which is a vector extending from the open top of the 'cup' (which is practically devoid of membranes) to the bottom of the 'cup'. Davidovich and Knox[11] have recently analysed enlarged electron micrographs of single *Chlorella* cells to determine the membrane distribution function. They subdivided each lamellar structure in the *Chlorella* electron micrograph into very small membrane segments and calculated the relative number of these segments whose normals make an angle θ with the 'partial' symmetry axis defined above. This relative number is the membrane distribution function. They found that this function has a maximum for $\theta \cong 90°$, i.e. more membrane fragments are aligned with their planes parallel to the 'partial' symmetry axis than perpendicular to it. Thus, since the membranes tend to orient with their planes perpendicular to the magnetic field,[9] the 'partial' symmetry axis is also perpendicular to the magnetic field. In fact, the results of Davidovich and Knox explain how it is possible for the nearly spherically shaped *Chlorella* cells to orient in a magnetic field.

In our calculations we shall replace the actual whole cell by a single membrane segment which has a well defined orientation with its normal along the magnetic field direction (Figure 2). To facilitate the calculation the membrane segment is replaced by a disk of radius a and thickness l, with $a \gg l$. In our Rayleigh–Gans calculation of the scattering ratios S_1 and S_2, the disk radius is an adjustable parameter and defines the 'geometric extent' of the coherence of the scattered radiation.

The scattering ratios S_1 and S_2 are calculated for single-particle scattering only, since multiple interparticle scattering events were ruled out experimentally for the curves shown in Figure 1.

The wavelength dependence in the scattering curves displayed in Figure 1 can be traced to the chlorophyll molecules,[4] with the colourless pigments contributing a background scattering. In calculating S_1 and S_2, we shall assume that the polarizability of the chlorophyll molecules is anisotropic, and in the coordinate system of the planar porphyrin ring system of chlorophyll, the polarizability tensor per unit volume is described by $\alpha_p(\mathbf{r})$. If $\mathbf{r}$ is the distance from the axis of symmetry of the disc, then the induced dipole moment for a volume element dV in the laboratory axis system is:

$$\mathbf{P}_L = \alpha_L(\mathbf{r}) . E_L \, dV = \mathbf{R}^{-1}\mathbf{S}^{-1}\alpha_p(\mathbf{r})\mathbf{S}\mathbf{R} . \mathbf{E}_L \, dV \tag{3}$$

Here $\mathbf{S}$ denotes the transformation matrix from the molecular coordinate system to the membrane (disc) axis system and $\mathbf{R}$ represents the transformation from the laboratory axis system to the coordinate system of the disc. $\mathbf{E}_L$ is a unit vector corresponding to the state of polarization of the incident beam which, for the experimental results given in Figure 1, was unpolarized. The component of the scattered electric field in a direction $\hat{\mathbf{n}}$ ($\hat{\mathbf{n}}$ being a unit vector perpendicular to the scattered beam) is just $k^2\hat{\mathbf{n}} . \mathbf{P}_L$ to within a phase factor, and $k = 2\pi/\lambda$ is the wavevector of the incident light. In our case, all analysers were oriented perpendicular to the scattering plane (vertical configuration).

A proper treatment of the coherence effects arising from shape and form presents virtually insurmountable mathematical difficulties which are beyond the scope of this paper. To a first approximation, we adopt the Rayleigh–Gans approximation (first Born approximation of scattering theory) which treats each volume element as a Rayleigh scattering centre which experiences the *same* incident field. Although local field effects are neglected, this approximation nevertheless properly accounts for the phase differences between the various volume elements, which are denoted by $\delta(\mathbf{k}, \mathbf{r}, \hat{\mathbf{n}})$ with $\delta = 0$ at the centre of symmetry. In the context of these approximations the general formula for the scattering ratios is

$$\frac{S(\lambda, H)}{S(\lambda, 0)} = \frac{\left| \int_v \hat{\mathbf{n}} \cdot \langle \mathbf{S}^{-1} \boldsymbol{\alpha}_p(\mathbf{r})\mathbf{S} \rangle_\phi \cdot E_L \, e^{i\delta(\mathbf{k}, \mathbf{r}, \hat{\mathbf{n}})} \, d^3\mathbf{r} \right|^2}{\left\langle \left| \int_v \hat{\mathbf{n}} \cdot \langle \mathbf{R}^{-1}\mathbf{S}^{-1} \boldsymbol{\alpha}_p(\mathbf{r})\mathbf{S}\mathbf{R} \rangle_\phi \cdot E_L \, e^{i\delta(\mathbf{k}, \mathbf{r}, \hat{\mathbf{n}})} \, d^3\mathbf{r} \right|^2 \right\rangle_R} \tag{4}$$

The brackets denote the ensemble average over $\mathbf{R}$ for all orientations of the disk, i.e. the case of random orientation. The integration is carried out over the volume of the disk which reduces to a surface integral in the case of an infinitely thin disk with $l \to 0$. In general, the numerator and denominator in (4) reduce to terms which are squared products of a form factor (P) and polarizability (α) terms. The form factors may be written

$$P = A \times L \tag{5}$$

where A is the form factor which depends on the radius of the disk $r = a$, and L is a form factor which depends on the thickness l of the disk. Now

$$L(kl) = \left(\frac{\pi}{2kl} \right)^{1/2} J_{1/2}(kl) \tag{6}$$

where $J_{1/2}(kl)$ is a Bessel function. If $kl \ll 1$, then $L(kl) \approx 1$. This requires that the thickness of the disk be small enough to satisfy the relations

$$kl \ll 1 \to 2\pi l \ll \lambda \tag{7}$$

Thus, for the wavelength range for which the analysis is carried out (600–740 nm), l should be less than $\lambda/2\pi \approx 10^{-5}$ cm. Physically, this is a reasonable limit for the thickness of the lamellar stacks considered here and we therefore make the simplifying assumption that the integrals in (4) can be reduced to surface integrals.

It is known from dichroism and fluorescence polarization studies that the red chlorophyll transition moment vector Q_y (678 nm) in *Chlorella* lies primarily *in* or *near* the lamellae plane. We shall furthermore assume that the porphyrin ring is oriented at a fixed angle β with respect to the membrane plane, $\mathbf{n}_i \cdot \mathbf{m} = \sin \beta$, where $\mathbf{n}_i$ and $\mathbf{m}$ are unit vectors which are normal to the plane of the porphyrin ring, and the membrane plane respectively.

The Q_y vectors appear to be randomly oriented[7,8,12] with respect to the membrane normal $\mathbf{m}$. This randomness is represented by the bracketed average over ϕ in equation (4) and thereby characterizes the in-plane response with an isotropic polarizability which is denoted by α_1. The out-of-plane polarizability is denoted by $\alpha_2 (\neq \alpha_1)$.

For the experimental conditions of scattering perpendicular to the scattering plane (vertical analysers), the evaluation of the integrals and the ensemble average in equation (4) is tedious but straightforward[13] and gives

$$S_\parallel(\lambda, H) = C\frac{J_1^2(\omega)}{\omega^2}|\alpha_1|^2 \quad (\omega = ka) \tag{8}$$

$$S_\perp(\lambda, H) = C\frac{J_1^2(\omega)}{\omega^2}|\alpha_2|^2 \quad (\omega = \sqrt{2}ka) \tag{9}$$

and

$$S(\lambda, 0) = \frac{2C'}{\omega^2}\{-|\alpha_2|^2(I_1 - I_2) + |\alpha_2 - \alpha_1|^2(\tfrac{5}{4}I_1 - I_2)$$

$$+ \tfrac{1}{2}(\alpha_1^*\alpha_2 + \alpha_1\alpha_2^*)(2I_1 - I_2)\} \quad (\omega = \sqrt{2}ka) \tag{10}$$

C and C' are a collection of constant terms, $J_m(X)$ is the Bessel function of mth order, the asterisks represent complex conjugates, and the quantities I_1 and I_2 are defined by

$$I_1 = 1 - J_1(2\omega)/\omega \tag{11}$$

$$I_2 = \frac{2}{\omega}\sum_{n=0}^{\infty} J_{3+2n}(2\omega) \tag{12}$$

In order to calculate the wavelength dependence of the intensities of the scattered light given by (8)–(10), a reasonable model for the wavelength dependence of the polarizabilities α_1 and α_2 is necessary. The calculation is confined to the 600–740 nm region which includes the prominent 678 nm Q_y absorption band which is a π–π^* transition and which is oriented along the y-molecular axis in the plane of the porphyrin ring.[14,15] We let α_x, α_y and α_z denote the principal axis polarizabilities per unit volume in the molecular axis system, with α_y referring to the Q_y transition direction, the x direction is orthogonal to y and also in the plane of the porphyrin ring, and z is normal to this plane. We assume that in the 600–740 nm region, only α_y exhibits a strong wavelength dependence, particularly in the wavelength region of the 678 nm y-polarized absorption band, while α_x and α_z vary only slightly as a function of wavelength.

We employ the Lorentzian damped oscillator model to describe the wavelength dependence of α_y:

$$\alpha_y = \frac{C_y\gamma\omega_0}{\omega^2 - \omega_0^2 + i\gamma\omega} \tag{13}$$

where $\omega = 2\pi c/\lambda$, c is the velocity of light, and C_y is a constant which is proportional to the number of pigment molecules per unit volume and the oscillator strength of the transition. ω_0 is the resonance frequency and γ is the half-width of the 678 nm band. For our model calculations we take α_x and α_z to be wavelength independent.

Fluorescence polarization and dichroism experiments with intact *Chlorella* cells and spinach chloroplasts indicates that the Q_y transition has a component both in the plane of the membrane ($S'_\parallel$) and perpendicular to the plane ($S'_\perp$) with $S'_\parallel > S'_\perp$. The parameter $f = S'_\parallel/S'_\perp$ is therefore a measure of the degree of orientation of Q_y transition moment with respect to the lamellae plane.

Although the polarizabilities α_1 and α_2 can be expressed in terms of the molecular polarizabilities, $S'_\parallel$, $S'_\perp$ and the angle β, we find it convenient to write

$$\alpha_1 = C_\parallel + S'_\parallel \alpha_y = C_\parallel + S'_\perp f \alpha_y \tag{14}$$

$$\alpha_2 = C_\perp + S'_\perp \alpha_y \tag{15}$$

where $C_\parallel$ and $C_\perp$ are functions of β, α_x, and α_z taken to be independent of λ. Specifically

$$C_\parallel = \tfrac{1}{2}(\alpha_x + \alpha_z \sin^{-2} \beta) \tag{16}$$

$$C_\perp = \alpha_z \cos^2 \beta \tag{17}$$

and

$$f = 0.5 \cot^2 \beta \tag{18}$$

In the following section some results of the calculations are summarized.

3 RESULTS

Because of the many adjustable parameters ($f = S'_\parallel/S'_\perp$, γ, C_y, a) and the simplicity of the model, it would be unrealistic to try to fit the experimental data shown in Figure 1. Our goal is, instead, to reproduce the general functional behaviour of S_1, S_2 and $S(\lambda, 0)$ and to draw what conclusions are possible about the pigment orientation. Using the scattering theory and assuming that α is isotropic and that *Chlorella* cells can be represented by homogeneous spheres, Latimer and Rabinowitch[4] were able to reproduce the wavelength dependence of $S(\lambda, 0)$. Using their approach, however, it can easily be shown that the wavelength dependence of S_1 and S_2 shown in Figure 1 cannot be reproduced. The reason is that the polarizability terms factor out in both the numerator and denominator of (1) and (2) and thus cancel if α is isotropic (random pigment orientation). Thus, if the polarizability tensor is a scalar, then the scattering ratios reduce to ratios of form factors, i.e. a form factor for the oriented configuration divided by the ensemble (equally weighted) of all such form factors.

3.1 Effect of radius on S_1 and S_2

Assuming that α is isotropic, maxima and minima in λ can still be obtained if $r = a$, the radius of the disk is selected such that constructive (or destructive) interference occurs. This can occur if the diameter of the disk is of approximately the same size as the wavelength of the light. However, in these cases the wavelength dependence of the experimentally determined S_1, S_2 and $S(\lambda, 0)$ curves is not obtained if α is assumed to be isotropic. For example, the calculated S_1 (and S_2) curves are monotonically decreasing (or increasing) functions of wavelength when $a = 10^{-5}$ cm, whereas for $a = 6 \times 10^{-5}$ cm, S_1 exhibits a broad maximum at ~ 710 nm, which bears no resemblance to the experimental curve. In general, the higher a, i.e. the larger the dimensions over which a high degree of regular order exists, the longer the wavelengths at which peaks in S_1, S_2 and $S(\lambda, 0)$ occur when α is isotropic.

3.2 Anisotropy in α and pigment orientation

It has been shown that the scattered light is partially depolarized,[8] which indicates that α is anisotropic. While there are cell components other than chlorophyll which are ordered and are thus expected to display an anisotropic α, the scattering curve in Figure 1(a) has the appearance of the anomalous dispersion curve which is characteristic of chlorophyll. We therefore propose that the sharp structure in the S_1 and S_2 curves in the 600–740 nm region is due to an anisotropy of the α which can be traced to a preferred orientation of the chlorophyll molecules in the lamellae.[6–10]

The shapes of the calculated S_1 and S_2 curves for various values of the chlorophyll orientation parameter $f = S'_{\parallel}/S'_{\perp}$ are plotted in Figure 3. A value of $f = 0.8$ gives not only the correct order of magnitudes of S_1 and S_2, but it also displays the pronounced structure consistent with the experimental data.

This value of f, using eq. (18), implies that the $Q_y(678)$ transition moment is oriented at approximately 38° with respect to the lamellar plane. This should be viewed as an upper bound on β since our calculations underestimate the value of f due to the simplicity of the model and the neglected contributions to scattering from chlorophyll b. This conclusion is in qualitative agreement with results obtained from dichroism[7,8,12] and fluorescence polarization[12] studies which indicate that the angle between Q_y and the membrane plane is less than 30° ($f = 1.4$).

Despite the prominent absorption band at ~ 430 nm in the Soret region, the scattering peak in S_1 and S_2 at ~ 450 nm is rather small, compared to the prominent peaks at ~ 695 nm corresponding to the red ~ 678 nm transition. It is known that the Soret band transition is polarized in the plane of the porphyrin ring plane (the xy plane of the molecular coordinate system). Such a degenerate transition tends to impart a *lower* degree of anisotropy to the polarizabilities α_1 and α_2 (in the plane and perpendicular to the membrane plane), and thus should manifest itself in a less pronounced structure in S_1 and S_2. The qualitative predictions of the theory are thus in good agreement with the results in S_1 and S_2 below 460 nm.

The rather pronounced maxima and minima in S_1 and S_2 at ~ 515 nm are most likely due to the carotenoid accessory pigments, which display a broad

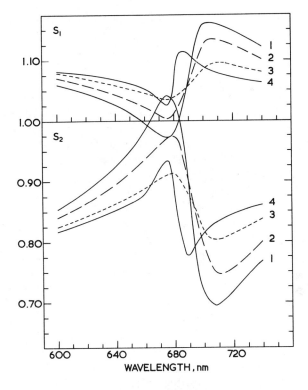

Figure 3 Scattering ratios calculated according to equation (4). The vertical scale is in absolute units. The parameters are (1) $f = 0.7$, (2) $f = 0.8$ and (3) $f = 0.9$ with $\omega_0 = 678$ nm, $\gamma = 30$ nm, $a = 10^{-5}$ cm and $C_{\parallel}/S'_{\perp} C_y = C_{\perp}/S'_{\perp} C_y = 1.0$. (4) $f = 0.8$, $\gamma = 10$ nm and $C_{\parallel}/S'_{\perp} C_y = C_{\perp}/S'_{\perp} C_y = 0.7$. For $f = \infty$ all of the Q_y, 678 nm oscillators lie in the membrane plane, whereas for $f = 0$ they are all perpendicular to this plane

absorption maximum at about ~ 480 nm. From dichroism studies[7,8,12] it is known that the transition moment vectors of the carotenoid molecules are oriented similar to Q_y, i.e. in the plane or close to the plane of the lamellae. Our scattering model also predicts qualitatively that the transition moment vectors and thus the long axes of the carotenoid molecules have a strong degree of orientation and tend to lie in the plane of the membranes.

3.3 Effect of varying C_y equation (13)

C_y is a measure of the oscillator strength of the transition and the number of pigment molecules per unit volume. Our calculations show that as C_y is increased, the wavelength-dependent part of the polarizability of course increases, which leads to a shift of S_1 (S_2), i.e. the maxima (minima), to longer wavelengths, and also enhances the magnitudes of these maxima and minima (see Figure 3, curve 4).

3.4 Effect of half-width γ

A value of $\gamma = 30$ nm was chosen in these calculations and is defined as the full width at half-maximum of the 678 nm peak. Increasing γ tends to broaden and shift the maximum in S_1 and the minimum in S_2 to longer wavelengths.

C. S. French and his coworkers have proposed that the *in vivo* 678 nm absorption band[16] is comprised of different pigment forms, each absorbing at a somewhat different wavelength, thus giving rise to a broadening of the red chlorophyll *in vivo* absorption band. Because of the crudity of the scattering model employed here, we chose to ignore the possibility of several absorption bands, and have treated the 678 nm band as arising from a single transition. However, if all these spectroscopic forms of chlorophyll have the same orientation, the results would remain qualitatively similar. Experimentally there are indications that the shorter wavelength forms possess a lower degree of orientation than the higher wavelength forms.[7,8,12] In this case the shorter wavelength forms with a lower degree of orientation would lead to a lowering of the polarizability anisotropy, reduced maxima and minima in S_1 and S_2, and a shift to longer wavelengths (of these maxima and minima).

In a preliminary calculation,[17] S_1 and S_2 were calculated without inclusion of the form factors. By using all of the spectroscopic forms of chlorophyll *in vivo* proposed by French, it was possible to quantitatively fit the experimental S_1 and S_2 curves neglecting coherence effects. However, this fit was possible because of the many parameters used and, most important, because the dominant contributions to the structure of S_1 and S_2 arise from the polarizability anisotropy, and not from the form factors.

4 VALIDITY OF THE APPROXIMATIONS

The fundamental approximation in the Rayleigh–Gans (Rayleigh–Debye[5b]) approximation is that

$$2ka|m - 1| \ll 1 \tag{19}$$

where a is the radius of the disc and $m - 1$ is the relative refractive index of the scattering medium. Not enough is known about the refractive indices of the membranes and their environment to reach any conclusions about the relative refractive indices. Kerker has summarized some comparisons between exact calculations and the Rayleigh–Gans approximation; it is shown that the range of validity of the Rayleigh–Gans approximation extends somewhat outside of the limits dictated by (19), at least for spheres and cylinders.

Since our goal is a qualitative understanding of the wavelength dependence of S_1 and S_2 which appears to be dictated mainly by the wavelength dependence of the polarizability rather than the form factors, we are not concerned with exact calculations of the form factors as such. We note that experimentally $S_1 > 1 \cdot 0$, while $S_2 < 1 \cdot 0$. These overall increases and decreases in the scattering intensities S_1 and S_2 relative to $S(\lambda, 0)$ are attributable to form factors and are correctly predicted by our Rayleigh–Gans calculations. Such increases and decreases in the scattered light intensity have been observed with flow-oriented platelets whose polarizability is isotropic (Reference 5b, p. 597).

We have considered the scattering only from a single membrane. Actually a stacking of these membranes occurs *in vivo*, and such structures are difficult to treat. If the scattering of each layer is independent of the other, then only the attenuation of the incident light has to be considered as it passes through the individual layers. It is then most likely that the overall wavelength dependence of the calculated S_1 and S_2 curves would not be too different.

5 CONCLUSIONS

The strong structure in the wavelength dependence of the scattering curves S_1 and S_2 is mainly attributable to an *anisotropy* of the polarizability α of the chlorophyll-containing chloroplast membranes. This wavelength-dependent anisotropy in α arises because of the intrinsic orientation of the chlorophyll molecules *in vivo*. The Q_y 678 nm transition moment vectors appear to lie close to the lamellar planes, according to the results of the calculation. This conclusion is in agreement with the results of dichroism and fluorescence polarization studies.

The approach to the interpretation of scattering curves developed in this paper needs to be tested with other oriented biological membranes, preferably with a known orientation of chromophores. If successful, a qualitative examination of scattering curves such as S_1 and S_2 obtained with membranes oriented by magnetic fields or other methods, should reveal the preferred orientation of the chromophores.

ACKNOWLEDGEMENT

One of the authors (C.E.S.) wishes to thank the hospitality and support of the Institute for Fundamental Studies, University of Rochester, where a part of this work was accomplished. The Institute is partially supported by the National Science Foundation Grant No. (GU-4040).

We are grateful to F. Van Nostrand for providing the data in Figure 1 and to J. Becker for helpful discussions. We thank M. Davidovich and R. S. Knox for providing us with their unpublished results on the membrane distribution function in *Chlorella*.

REFERENCES

1. K. Shibata, A. A. Bluson and M. Calvin, *Biochim. Biophys. Acta*, **15**, 461 (1954).
2. N. E. Geacintov, F. Van Nostrand and J. F. Becker, *Proc. 2nd Int. Congr. Photosynthesis Res., Stresa*, Vol. 1 pp. 283–290 (ed. G. Forti, M. Arnon and A. Melandri), Dr W. Junk N.V., The Hague, 1972.
3. K. D. Philipson and K. Sauer, *Biochemistry*, **12**, 3454 (1973).
4. P. Latimer and E. Rabinowitch, *Arch. Biochem. Biophys.*, **84**, 428 (1959).
5. (a) H. C. Van de Hulst, *Light Scattering by Small Particles*, John Wiley and Sons, Inc., New York, 1957; (b) M. Kerker, *The Scattering of Light*, Academic Press, New York, 1969.
6. N. E. Geacintov, F. Van Nostrand, J. F. Becker and J. B. Tinkel, *Biochim. Biophys. Acta*, **267**, 65 (1972).
7. J. Breton, M. Michel-Villaz and G. Paillotin, *Biochim. Biophys. Acta*, **314**, 42 (1973).
8. N. E. Geacintov, F. Van Nostrand and J. F. Becker, *Biochim. Biophys. Acta*, **347**, 443 (1974).
9. J. F. Becker, N. E. Geacintov, F. Van Nostrand and R. Van Metter, *Biochem. Biophys. Res. Commun.*, **51**, 597 (1973).
10. F. Van Nostrand, Ph.D. Thesis, New York University (1972).
11. M. Davidovich and R. S. Knox, private communication.
12. J. Breton and E. Roux, *Biochem. Biophys. Res. Commun.*, **45**, 557 (1971).
13. C. Picot, G. Weill and H. Benoit, *J. Colloid Interface Sci.*, **27**, 360 (1968).
14. M. Gouterman and L. Stryer, *J. Chem. Phys.*, **37**, 2260 (1962).
15. M. Gouterman, G. H. Wagviere and L. C. Snyder, *J. Mol. Spectr.*, **11**, 108 (1963).
16. C. S. French, *Proc. Natl. Acad. Sci. U.S.*, **68**, 2893 (1971); J. S. Brown, *Annu. Rev. Plant Physiol.*, **23**, 73 (1972).
17. R. Van Metter, C. E. Swenberg and N. E. Geacintov, *Abstracts of the First Annual Meeting of the American Society for Photobiology, Sarasota*, 1973.

Note added in proof. Which of the pigment forms are more oriented is more readily apparent by plotting S_1/S_2 from the experimental data. S_1/S_2 exhibits strong peaks at 460, 520 and 710 nm with the carotenoid molecules as strongly oriented as the chlorophyll aQ_y transition moments. The calculation of the differential scattering ratios S_1 and S_2 reported herein did not include the possibility of $C_{\parallel} \neq C_{\perp}$. These effects are discussed more fully in an article which will be published elsewhere.

C.5. Polarization of the 515 nm effect on chloroplast membranes oriented by a magnetic field

Jacques Breton and Paul Mathis

Service de Biophysique, Département de Biologie, Centre d'Etudes Nucléaires de Saclay, B.P. no. 2—91190—Gif-sur-Yvette, France

SUMMARY

The polarization of the transient absorbance changes induced by short saturating flashes of light has been studied in the spectral range 450–540 nm, on spinach chloroplasts oriented by a 10 kG magnetic field. The flashes are unpolarized and the electric vector of the analysing beam is polarized either $\parallel$ or $\perp$ to the plane of the oriented photosynthetic membranes (which are perpendicular to the magnetic field). It is shown that the observed transients represent the so-called '515 nm' or 'field' effect and that the only influence of the magnetic field is via the orientation of the lamellae.

The kinetics of the transients are identical for both polarizations, indicating that a unique phenomenon is involved. The magnitudes (ΔA) are different for the two polarizations. The degree of polarization

$$p = \frac{\Delta A_{\parallel} - \Delta A_{\perp}}{\Delta A_{\parallel} + \Delta A_{\perp}}$$

is high: $+12.5\%$ at 515 nm, $+25\%$ at 475 nm; it becomes negative in the region 482–502 nm.

These results are discussed in relation to the intrinsic orientation of the pigments within the photosynthetic membranes and to the relative orientation of the pigments with respect to the hypothetical transient electrical field.

1 INTRODUCTION

Many studies have been conducted on the transient light-induced absorbance change occurring in chloroplast suspensions and referred as the '515 nm effect' (see Reference 1, where this effect is named 'field-indicating absorption change').

It has been proposed that this effect is an electrochromism, due to a light-induced electrical field applied to the chloroplast pigments included in a membrane behaving like the dielectric of a charged capacitor.[2-4] Recent studies[5-8] have shown that these pigments are partly oriented with respect to the normal to the membrane plane. It is specially important that the chloroplast lamellae, when they are stacked in grana, can be oriented by a magnetic field under physiological conditions.[6-7]

We thought that it was possible to get more information on the electrochromic effect, and about the orientation of the pigments within the membrane, from a study of the polarization of the 515 nm effect on oriented chloroplast membranes. Our results indicate an important polarization of the 515 nm effect, with a complex wavelength dependence. However a detailed interpretation proves to be very difficult, due to many uncertainties on the effect of an electric field on the photosynthetic pigments.

2 MATERIAL AND METHODS

2.1 Biological material

Chloroplasts were prepared by grinding spinach leaves for 10 s with a Waring blender in 0·4 M sucrose–0·02 M sodium pyrophosphate at pH 8·0. The juice was filtered through 16 layers of cheese-cloth and centrifuged at 2000 g for 2 minutes. The pellet was rehomogenized in 0·4 M sucrose–0·2 M Tricine pH 8·0–ficoll 6%. These chloroplasts will be referred to as 'unbroken' in the text.

For the experiments in the presence of gramicidin D (Schuchardt), the grinding buffer was of the low-salt type (0·75 M NaCl–0·02 M Tricine pH 8·0) and the pellet was resuspended in the same buffer plus 10^{-7} M gramicidin. After a second centrifugation, the pellet was resuspended in 0·8 M sucrose–1 mM $MgCl_2$–0·01 M KCl–0·02 M Tricine pH 8·0. This treatment gives chloroplasts that can be oriented by a magnetic field, but are devoid of their outer envelope.

2.2 Measurement of absorbance changes

A square cuvette (8 × 8 mm) containing the chloroplast suspension (about 20 μg of chlorophyll per ml) was inserted in a horizontal magnetic field H (10 kG), as shown in Figure 1. The lamellae were oriented perpendicular to H.[6-8] The suspension was excited vertically, through a light guide, by unpolarized light from a xenon flash tube (10 μs) filtered by a Balzers (Calfex) interference filter and a Schott RG 630–3 mm filter. The excitation was saturating for the 515 nm effect. The measuring beam was polarized with a two-position Polaroid polarizer HN 38, either parallel (measurement of $\Delta A_{\parallel}$) or perpendicular (measurement of $\Delta A_{\perp}$) to the membrane plane. The intensity of the measuring beam ($\Delta\lambda = 2$ nm) was 2·0 erg cm^{-2} s^{-1} at 515 nm; its effect on ΔA was found to be negligible. The light going through the cuvette was collected by a 20 cm

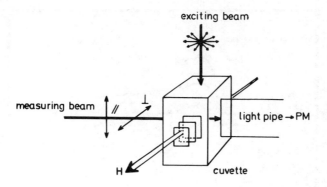

Figure 1 A schematic representation of the spatial position of the cuvette, of the exciting and measuring beams and of the direction of the magnetic field *H*. The small rectangles represent the stacked photosynthetic lamellae

light guide (10 × 20 mm, placed against the cuvette wall) and measured by a shielded photomultiplier operating at constant voltage and output. The intensity of the measuring beam, was slightly varied for other wavelengths. The signal/noise ratio was improved with a multichannel analyser (further details are given in Reference 9).

In a typical experiment the suspension was allowed to stay for 5 min in the cuvette (thermostated at 3 °C), with the magnetic field on and with the exciting flashes working (0·2 Hz); 20 signals were recorded for $\Delta A_\parallel$, then 40 signals for $\Delta A_\perp$ and 20 signals for $\Delta A_\parallel$ (or the reverse series). A measurement at a given wavelength was always inserted between two measurements at 515 nm. Any series indicating an evolution with time was discarded.

The orientation was checked by transferring the cuvette, the cuvette holder and the magnet in a spectropolarimeter that records the linear dichroism spectrum.[5]

3 RESULTS

3.1 Polarization of absorbance changes at 515 nm

A typical set of data is presented in Figure 2, reproducing the transient absorbance increase at 515 nm, for both ∥ and ⊥ polarizations of the measuring beam. The bottom trace represents the difference between the signals obtained for the two polarizations, amplified four times. The three curves presented have the same decay kinetics. The decay is slower than the decay reported by Junge and Witt[2], a difference that can be mostly attributed to the lower temperature in our measurements. The rise-time is instrument-limited (about 100 ms in Figure 2); in experiments performed with a shorter time response (5 ms) the kinetics of decay for the three curves remain identical.

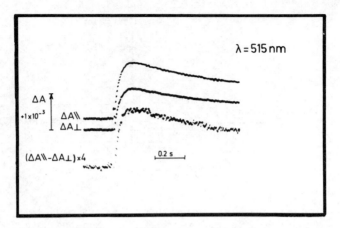

Figure 2 Oscilloscope display of the absorbance changes ΔA at 515 nm for chloroplasts oriented in a 10 kG magnetic field. The subscripts $\parallel$ and $\perp$ refer to the electric vector of the measuring beam (see Figure 1). The rise-time is instrument-limited. Chlorophyll concentration: $2\cdot2 \times 10^{-5}$ M. Optical path: $0\cdot8$ cm. Temperature: 3 °C. 40 transients were averaged

In a large series of experiments, the degree of polarization defined as

$$p = \frac{\Delta A_{\parallel} - \Delta A_{\perp}}{\Delta A_{\parallel} + \Delta A_{\perp}}$$

is $12\cdot5 \pm 2\cdot5\%$ at 515 nm. The variations are from preparation to preparation. Similar variations were observed in the linear dichroism of the pigments, measured with the spectropolarimeter.

3.2 Dependence on wavelength

The absorbance changes have been studied in the spectral range 450–540 nm. The kinetic parameters are the same for all the wavelengths studied and for both polarizations of the measuring beam. The degree of polarization varies widely in the studied spectral range. It is about 25% at 475 nm and it is negative in the range 480–500 nm (Figure 3). The values of $\Delta A_{\parallel}$ and $\Delta A_{\perp}$ are plotted versus wavelength in Figure 4; the values are normalized for a degree of polarization of $12\cdot5\%$ at 515 nm. Crossing points of the two curves are reproducibly observed around 480, 490 and 500 nm.

3.3 Control experiments

Our measurements of the degree of polarization of the absorbance change are subject to the same potential artifacts as the measurement of linear dichroism. These have been considered in Reference 7. A few control experiments were devised in order to check the nature of the observed absorbance changes and the effect of the magnetic field.

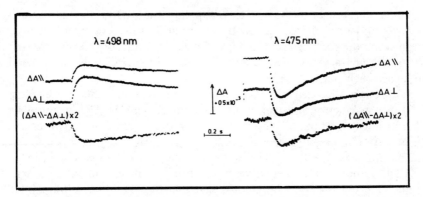

Figure 3 Oscilloscope displays of the absorbance changes at 498 and 475 nm. The conditions are as for Figure 2. Number of averaged transients: 80 at 498 nm, 40 at 475 nm. The differences in the kinetics of decay are observed from preparation to preparation

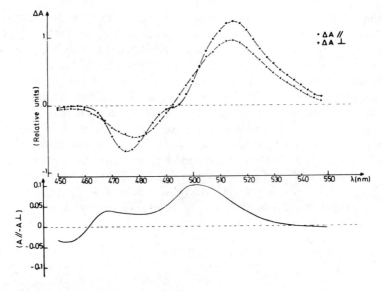

Figure 4 Wavelength dependence of:
Upper curves: Absorbance changes for the two polarizations of the measuring beam. The conditions are as for Figure 1. In order to correct for small day-to-day variations, the magnitude of ΔA was normalized to a constant value at 515 nm, for both polarizations.
Lower curve: Linear dichroism, measured in the same experimental conditions with a spectropolarimeter.

A possible contribution of light scattering was checked by varying the solid angle of collection of the measuring light after the cuvette (distance varying from 0·1 to 25 cm). We found no change in the magnitude of ΔA (in the absence of H) and in the degree of polarization (with H).

In the absence of magnetic field no polarization effect was detected. Any contribution of a photoselection by the exciting beam or by the measuring beam can thus be neglected. Such an effect was improbable in view of the saturating character of the exciting flash and of the negligible photochemical influence of the measuring beam.

With a non-polarized measuring beam we found no influence of the magnetic field on the kinetics of decay of the absorbance changes. The magnitude was slightly affected (about 5% less at 515 nm) in the presence of H. This effect can be mostly attributed to a decrease of the optical density of the suspension when membranes are oriented parallel to the direction of the measuring beam.[6]

All the above experiments were performed with 'unbroken' chloroplasts. With chloroplasts that had been broken (either by isolation in a low-salt buffer or by osmotic shock treatment) we found no influence of the magnetic field on the amplitude and kinetics of decay of signals of absorbance change. This absence of influence is true with and without polarization of the measuring beam. Similarly no linear dichroism could be detected with these preparations by the spectropolarimeter technique.

The influence of gramicidin was checked with chloroplasts isolated in a low-salt buffer (we found that 'unbroken' chloroplasts do not react easily with low concentrations of gramicidin). In order to achieve an orientation in the magnetic field the chloroplasts were resuspended in 0·8 M sucrose, probably inducing a restacking of the thylakoids. The orientation, measured by the linear dichroism of the pigments or by the degree of polarization of the 515 nm absorbance change, is about 80% of the control 'unbroken' chloroplasts. We found that treatment with 10^{-7} M gramicidin causes an acceleration of the decay of absorbance changes by an order of magnitude, in agreement with Junge and Witt.[2] The degree of polarization of the field effect, accelerated by gramicidin, is not affected by that treatment. A very slowly decaying phase,[3] not accelerated by gramicidin, seems to present a degree of polarization different to the degree of polarization of the field effect. Due to its very small amplitude, that slow phase has not been systematically studied.

4 DISCUSSION

As shown by the control experiments, the only effect of the magnetic field on the transient absorbance change that we studied is via the orientation of the membranes. The wavelength dependence, the kinetics and the action of gramicidin allow us to identify these transients with the 515 nm effect, as described by Witt and coworkers.[1-3] If this effect results from different phenomena, it was expected to discriminate between them on the basis of their degree of polarization. The identity of the kinetics of decay $\Delta A_{\parallel}$ and $\Delta A_{\perp}$ lends further support to the hypothesis that the 515 nm effect is a unique phenomenon.

The strong polarization effects reported in this study, though higher than those observed for the linear dichroism of the pigments in the covered spectral range

(Figure 4, lower curve), are of the same order of magnitude. So far the best documented hypothesis accounting for the 515 nm effect is to consider a cross-membrane delocalized electrical field that induces a Stark effect on the photosynthetic pigments.[1,2] So, the degree of polarization that we measured would reflect the polarization of the Stark effect and it would be on the dependence of the pigment orientation in the membrane with respect to the acting electrical field (roughly perpendicular to the membrane). A few observations are relevant to this point:

(1) Inversions in the sign of the degree of polarization are observed in the 482–502 nm region (Figure 4). This might reflect the difference in orientation found for the Soret band of chlorophyll *a* (polarized out of the membrane plane) and for carotenoids (an excess of carotenoids being oriented parallel to the membrane plane).[7]

(2) Schmidt et al.[10] have been nearly able to mimic the *in vivo* spectrum of the 515 nm effect by applying an electrical field to multilayers of photosynthetic pigments. However the pigment composition of the layers was markedly different from the pigment composition in the chloroplasts. This could be due to a difference in the orientation of the pigments in the two systems, resulting in a different interaction with the electrical field.

(3) Kleuser and Bücher[11] studied the electrochromism of chlorophylls *a* and *b* in monolayers. They found that the Stark effect was mainly polarized parallel to the electrical field. This is different from what we observed in the 500–540 nm region where we find a polarization of the 515 nm effect oriented parallel to the membrane plane, i.e. perpendicular to the expected gross electrical field. It is however possible that the cross-membrane electrical field has an important in-plane component, that might act on the carotenoids which are known to be mostly oriented parallel to the membrane plane.[7]

Further studies are clearly needed in order to interpret our data in more detail and to derive more information on the orientation of the pigments in the photosynthetic membrane: theoretical and experimental work on the effect of an electrical field on the absorption of photosynthetic pigments having a known orientation relatively to the field. In that respect the use of oriented monolayers might be very fruitful.

ACKNOWLEDGEMENTS

We wish to thank Drs N. E. Geacintov and W. Junge for helpful discussions and A. Peronnard for technical assistance.

REFERENCES

1. H. T. Witt, *Quarterly Rev. Biophys.*, **4**, 365 (1971).
2. W. Junge and H. T. Witt, *Z. Naturforschg.*, **23b**, 244 (1968).

3. H. M. Emrich, W. Junge and H. T. Witt, *Z. Naturforschg.*, **24b**, 1144 (1969).
4. H. T. Witt and A. Zickler, *F.E.B.S. Lett.*, **37** (1973).
5. J. Breton and E. Roux, *Biochem. Biophys. Res. Comm.*, **45**, 557 (1971).
6. N. E. Geacintov, F. Van Nostrand, J. F. Becker and J. B. Tinkel, *Biochim. Biophys. Acta*, **267**, 65 (1972).
7. J. Breton, M. Michel-Villaz and G. Paillotin, *Biochim. Biophys. Acta*, **314**, 42 (1973).
8. J. F. Becker, N. E. Geacintov, F. Van Nostrand and R. Van Metter, *Biochem. Biophys. Res. Comm.*, **51**, 597 (1973).
9. P. Mathis, M. Michel-Villaz and A. Vermeglio, *Biochem. Biophys. Res. Comm.*, **56**, 682 (1974).
10. S. Schmidt, R. Reich and H. T. Witt, *Naturwiss*, **58**, 414 (1971).
11. D. Kleuser and H. Bucher, *Z. Naturforschg.*, **24b**, 1371 (1969).

C.6. Optical detection of zero field magnetic resonance in the triplet state of chlorophyll

Richard H. Clarke and Robert H. Hofeldt

Department of Chemistry, Boston University, Boston, Massachusetts 02215, U.S.A.

SUMMARY

The zero field EPR transitions for the triplet state of chlorophyll *b* in *n*-octane solutions at 2 K have been observed by optical detection methods. The zero-field transitions were observed as both microwave-induced changes in the triplet–triplet absorption intensity and in the fluorescence intensity. An argon ion laser was used as the photoexcitation source. The observed transitions corresponded as an increase in the triplet–triplet absorption intensity and to a decrease in the fluorescence intensity. These intensity changes can be used to determine the triplet state intersystem crossing rate constants, and show that the two in-plane spin levels dominate the intersystem crossing into and out of the lowest triplet state of chlorophyll *b*.

1 INTRODUCTION

Optical detection of magnetic resonance in the triplet state of organic molecules has been highly successful in investigating the electronic structure and excited state dynamics of a wide variety of organic systems.[1-5] The measurements of rates of intersystem crossing are of central importance in determining mechanisms of spin–orbit and spin–vibronic coupling in organic molecules and, in general, in providing data for understanding paths of radiative and non-radiative relaxation in photoexcited organic systems. Both conventional high-field EPR[3,4] and phosphorescence-microwave double resonance[1,2] have proved valuable for investigating dynamics of the individual spin levels of the triplet state. In our laboratory we have recently demonstrated that triplet state EPR transitions at zero field may be detected by microwave-induced changes in triplet–triplet absorption intensity for molecules with low phosphorescence quantum yields and short triplet state lifetimes.[6,7] It has also been suggested that microwave-induced effects should be observed in the fluorescence spectra of such organic molecules.[8] In the present work we have applied optical detection of magnetic resonance techniques to an investigation of the chlorophylls in an attempt to understand the triplet state intersystem crossing mechanism in these molecules.

2 RESULTS AND DISCUSSION

We have observed zero-field magnetic resonance transitions of the lowest triplet state of chlorophyll b in n-octane solution at 2 K by triplet absorption detection of magnetic resonance (TADMR) and by fluorescence detection of magnetic resonance (FDMR). Chlorophyll b exhibits very weak phosphorescence,[9] but strong triplet–triplet (T–T) absorption maximizing at about 599 nm.[10] Further, we have observed that at temperatures below 77 K n-octane serves as a Shpolskii-type matrix for chlorophyll b.[11] Chlorophyll b exhibits a sharp line, highly structured fluorescence at 2 K, using a Spectra-Physics 4 watt argon ion laser as the photoexcitation source.

The zero-field magnetic resonance transitions occurred at 870, 1000 and 1100 MHz (Figures 1 and 2). The 1000 MHz transition could be observed

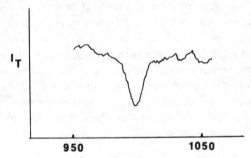

Figure 1 Zero field magnetic resonance signal of the triplet state of chlorophyll b in n-octane at 2 K detected by the microwave-induced change in the triplet–triplet absorption intensity. The signal corresponds to an increase in triplet–triplet absorption intensity (decrease in transmitted light, I_T) at 488 nm. The frequency scale is in Mhz

only under highest microwave power (incident power ~ 1 watt) in high-concentration solutions and may correspond to aggregated chlorophyll centres. Under less severe power conditions (10–20 mW) using concentrations of $\sim 10^{-5\,m}$ only the 870 and 100 MHz transitions were observed. The transitions in the TADMR spectrum correspond to an increase in the T–T absorption intensity, and to a decrease in fluorescence intensity in the FDMR spectrum. The signals disappeared as the sample was warmed above 4 K.

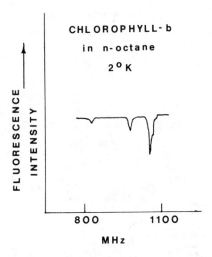

Figure 2 Zero field magnetic resonance signals of the triplet state of chlorophyll *b* in *n*-octane at 2 K detected by the microwave-induced change in the fluorescence intensity. The signal corresponds to a decrease in the fluorescence intensity, I_F, at 647 nm. The frequency scale is in MHz

From the variation in frequency of the zero-field transition with detection wavelength in the FDMR experiments it is clear that the chlorophyll molecules exist in several non-equivalent sites in the *n*-octane matrix. As shown in Figure 3 the highest-energy zero-field EPR transition varies from 1069 MHz to 1103 MHz, as the wavelength is varied among the sharp-line bands of the fluorescence spectrum. The intensity of the FDMR signals also varies with wavelength, with the strongest magnetic resonance peaks observed at 6440 Å, corresponding to ~1% decrease in the fluorescence intensity.

Both triplet absorption detection and fluorescence detection of magnetic resonance arise from the same effect—a change in the overall steady state population in the triplet state by microwave saturation of the zero-field transitions, since the intersystem crossing rates into and out of the triplet state can be quite different in each of the triplet spin levels.[1,2] An increase in the overall triplet state population will cause an increase in the T–T absorption intensity and a decrease in the fluorescence intensity. The intersystem crossing rate constants can be measured directly for the triplet spin sublevels by observing the changes in either fluorescence intensity or T–T absorption intensity as a function of time after turning on a saturating microwave field for each of the three zero-field transitions.[12] Results of such experiments show that the fastest decaying spin levels of the chlorophyll *b* triplet state are the two in-plane levels (x and y) with depopulation rates $k_x \approx k_y$. The out-of-plane (z) spin level is the longest lived with $k_z \approx 0.1\, k_x$. The relative populating rates are in approximately the same order as the depopulating rates. Further, the intersystem crossing rates appear to be independent of the site examined, showing little variation in rate constants (beyond experimental error) as the fluorescence detection wavelength is varied.

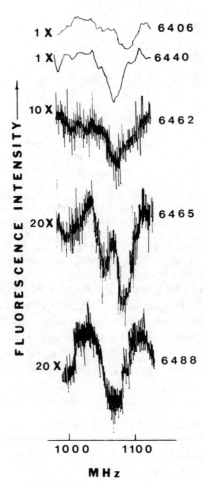

Figure 3 Variation of the fluorescence-detected magnetic resonance transition at ~ 1100 MHz with detection wavelength (indicated to the right of each trace) for chlorophyll b in n-octane at 2 K. The relative amount of signal averaging required to produce the signal intensities shown is indicated to the left of each trace

ACKNOWLEDGEMENT

Support for this research from the U.S. Army Research Office (Durham), the Research Corporation and the Petroleum Research Fund, administered by the American Chemical Society, is gratefully acknowledged. R. H. C. has also benefitted from a fellowship from the Alfred P. Sloan Foundation.

REFERENCES

1. J. Schmidt, D. A. Antheunis and J. H. van der Waals, *Mol. Phys.*, **22**, 1 (1971).
2. M. A. El Sayed, *J. Chem. Phys.*, **54**, 680 (1971).
3. M. Schwoerer and H. Sixl, *Z. Naturforsch.*, **24a**, 952 (1969).
4. R. H. Clarke, *Chem. Phys. Letters*, **6**, 413 (1970).
5. M. J. Buckley and C. B. Harris, *J. Chem. Phys.*, **56**, 137 (1972).
6. R. H. Clarke and J. M. Hayes, *J. Chem. Phys.*, **57**, 569 (1972).
7. R. H. Clarke and J. M. Hayes, *J. Chem. Phys.*, **59**, 3113 (1973).
8. C. B. Harris and R. J. Hoover, *J. Chem. Phys.*, **56**, 2199 (1972).
9. R. L. Amster, *Photochem. Photobiol.*, **9**, 331 (1969).
10. R. Livingston, *J. Amer. Chem. Soc.*, **77**, 2179 (1955).
11. F. F. Litvin and R. I. Personov, *Dokl. Akad. SSSR.* (translation), **188**, 118 (1969).
12. R. H. Clarke, J. M. Hayes and R. H. Hofeldt, *J. Mag. Resonance*, **13**, 68 (1974).

C.7. Spin polarization in the lowest triplet state of some photosynthetic pigments

J. F. Kleibeuker and T. J. Schaafsma

Laboratory of Molecular Physics, Agricultural University, De Dreijen 6, Wageningen, The Netherlands

SUMMARY

ESR studies at and above 77 K of several photosynthetic pigments reveal the existence of spin polarization in their lowest triplet state. In the presence of a central metal ion or sidegroups, e.g. in chlorophyll a/b, it is found that spin relaxation between the magnetic sublevels of this triplet state depends on the type of solvent. In the unsubstituted porphin free base no such effect is observed.

Data are presented on the zero-field parameters of free base porphin, and both chlorophyll a and b in different solvents.

From the kinetic behaviour of the light-induced ESR signals it is concluded that population and depopulation mainly involves the spin level $|y\rangle$, describing a spin moving in a plane perpendicular to the molecular plane. The spin–lattice relaxation time is found to be anisotropic, as expected for this type of molecule.

1 INTRODUCTION

The rôle of the triplet state of chlorophyll in photosynthesis has not been established to a satisfactory extent. Although it has been postulated for some time as an essential intermediate,[1,2] objections, mainly based on the long lifetime of the triplet state, relative to the period in which 'everything happens', have not been silenced by decisive experiments. A recent study of the triplet kinetics of free base porphin[3] has revealed an intimate coupling between energy transport in the singlet system and transitions between the magnetic sublevels of the lowest electronic triplet state. Time-dependent processes within the triplet manifold of photosynthetic pigments cannot be considered as a small perturbation on the optical pumping cycle, since compounds of this type have rather high quantum yields of triplet formation, even at room temperature.[4] Whereas spectroscopy at very low temperature provides valuable data on molecular properties of the triplet state, its possible rôle in photosynthesis at ambient temperature can only be clarified by kinetic studies over an extended temperature range. Even in the case that the triplet state is not an essential intermediate in biological energy conversion, it is attractive to employ its unique magnetic properties to probe the significant events in photosynthetic model compounds and in the *in vivo* unit. These are two important reasons which have prompted us to study in some detail the kinetic behaviour of electron spin resonance signals of the photoexcited lowest triplet state of some pigments, which are important in photosynthesis. Results are presented for free base porphin and both chlorophyll *a* and *b* at and above 77 K.

2 SPIN POLARIZATION WITH CONTINUOUS ILLUMINATION

In zero magnetic field the lowest electronic triplet state of a molecule with less than threefold symmetry is split into three non-degenerate levels, characterized by their spin functions $|x\rangle$, $|y\rangle$, and $|z\rangle$; x and y refer to two mutually orthogonal and non-equivalent in-plane axes, z to the out-of-plane axis of the molecule, in our case assumed to be planar. A spin function $|i\rangle$ describes a spin moving in the plane $i = 0$. The location of the x- and y-axes in the molecular plane is undefined.

The lowest triplet state of chlorophyll-*a* and -*b* and of free base porphin is assumed to have $\pi\pi^*$ character in the rather polar solvents used in our experiments.[5] For such states the level-ordering has been shown to be E_x, $E_y > E_z$. We arbitrarily choose $E_x > E_y$. The $\Delta m = \pm 1$ ESR spectrum of a randomly oriented collection of triplet molecules consists of six lines centred around H_0 and located at

$$H_z^-, H_x^-, H_y^-, H_y^+, H_x^+, H_z^+,$$

with the magnetic field H increasing from left to right. First-order field positions are related to the spin–spin Hamiltonian

$$\mathscr{H} = D(S_z^2 - \tfrac{1}{3}S^2) + E(S_x^2 - S_y^2) \tag{1}$$

by

$$H_x^\pm = H_0 \pm \tfrac{1}{2}(D + 3E)$$
$$H_y^\pm = H_0 \pm \tfrac{1}{2}(D - 3E) \tag{2}$$
$$H_z^\pm = H_0 \pm D$$

High and low field positions and canonical orientations are labeled $\pm$, x, y and z, respectively.

High field spin functions $|m\rangle$ ($m = 0, \pm 1$) are related to those at zero field by

$$|\pm 1\rangle = \frac{1}{\sqrt{2}}(|x\rangle \pm i|y\rangle)$$

$$|0\rangle = |z\rangle \qquad (H \parallel z) \tag{3}$$

Thus, for $H \parallel z$, populating and depopulating rates P_m and k_m at high field are given by,[6]

$$P_0 = P_z$$
$$P_{\pm 1} = \tfrac{1}{2}(P_x + P_y)$$
$$k_0 = k_z$$
$$k_{\pm 1} = \tfrac{1}{2}(k_x + k_y) \tag{4}$$

where k contains both radiative and non-radiative parts.

Similar relations are obtained for $H \parallel x$ and y by cyclic permutation of subscripts. Introducing orientation-dependent spin–lattice relaxation rates $w_1^{(i)}$, $w_2^{(i)}$ and $w_3^{(i)}$ ($i = x, y, z$) and Boltzmann factors $\exp(E_m - E_{m'})/kT \approx 1 + \delta(m - m')$, where $\delta \equiv (E_m - E_{m'})/kT$, the kinetic parameters are given in the level scheme of Figure 1. The steady-state signal amplitudes S for each of the

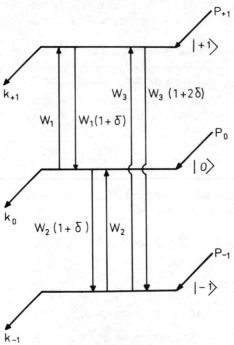

Figure 1   Kinetic scheme for triplet magnetic sublevels $|0\rangle$ and $|\pm 1\rangle$ at high field. For definition of symbols, see text

six $\Delta m = \pm 1$ transitions are obtained by solving the rate equations for the populations N_0 and $N_{\pm 1}$[7] for $H \parallel x$, y and z with the conditions

$$w_1^{(i)} = w_2^{(i)} = w_3^{(i)} = w^{(i)} \tag{5a}$$

$$\delta \ll 1 \tag{5b}$$

Although (5a) may be a rather serious approximation, it does not affect the qualitative conclusions.

Assuming $g \approx 2$ and expressing in cm^{-1} units,

$$D, E \ll \frac{g\beta}{hc} H_0 \ll \frac{kT}{hc}$$

in the case at hand. Under these conditions, the ratio of signal amplitudes S_i^+ and S_i^- at field positions H_i^+ and H_i^- can conveniently be expressed as a *polarization ratio*

$$R_i \equiv \frac{S_i^+ - S_i^-}{S_i^+ + S_i^-}$$

Realizing that

$$\begin{aligned} S_i^{\pm} &\sim \pm(N_0 - N_{\pm 1})(i = x, y) \\ S_z^{\pm} &\sim \pm(N_{\pm 1} - N_0) \end{aligned} \tag{6}$$

and assuming $w^{(i)} > k_0, k_1$, we find

$$R_i = \frac{k_1 P_0 - k_0 P_1 - P_1 k_0 \delta}{w^{(i)} \delta(P_0 + 2P_1) + P_1 k_0 \delta} \tag{7}$$

In the absence of relaxation ($w^{(i)} = 0$) R_i can be shown to go to infinity, expressing the fact that S_i^+ is emissive (<0), if S_i^- is absorptive (>0), and *vice versa*. On the other hand, $R_i = 0$ for $w^{(i)} \to \infty$ and no spin polarization is observed. Note that *no polarization* can be observed, when the product $w^{(i)}\delta$ is much larger than $k_1 P_0 - k_0 P_1$.

For fast relaxation R_i has the property

$$R_x + R_y + R_z = 0 \tag{8}$$

if there is no anisotropy in the relaxation rate: ($w^{(i)} = w$) as is evident from (7).

As observed by Sixl and Schwoerer,[7] the pumping cycle may be selective in P, in k or in both.

Table 1 reviews the possible spectral signs for a k-selective pumping cycle, expressed by inequalities $P_i > P_j$, P_k and $k_i \gg k_j$, k_k, in the absence of relaxation. With $H \parallel$ axis i, steady state populations of the spin levels $|\pm 1\rangle$ and $|0\rangle$ are given by[4]

$$\frac{P_j + P_k}{k_j + k_k} \quad \text{and} \quad \frac{P_i}{k_i}$$

Some lines appear as absorptive (A), others as emissive (E), leading to $R = \pm\infty$, where the sign of R is determined by (7).

Low-field transitions ($H < H_0$) have opposite signs with respect to their high-field counterparts. Furthermore, it turns out that x- and y-transitions have different signs if the populating mechanism involves at least one of the spin functions $|x\rangle$ and $|y\rangle$; if they have the same sign $|z\rangle$ must be involved. At sufficiently low temperature (<4 K) and in weakly interacting lattices,[8] $\Delta m = \pm 1$ spectra of the predicted type are observed.[7,9]

In most solvents at ~ 77 K the spin system is only partially polarized due to fast relaxation and one observes no emission lines, but an asymmetric intensity distribution, as in Figures 2–4. The component out of a pair of lines (e.g. H_x^+

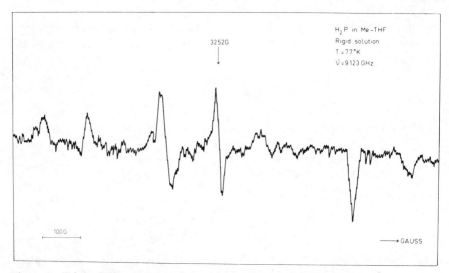

Figure 2   Triplet ESR spectrum of porphin free base in MTHF (4×10^{-5} M) at 77 K indicating the presence of spin polarization. Microwave power 0·05 mW, modulation amplitude 16 gauss, modulation frequency 100 KHz. Spectrum recorded with continuous illumination

and H_x^-) with *lowest* intensity will appear as emissive if relaxation is sufficiently slowed down. Biological molecules with short-lived triplet states are likely candidates to show spin polarization at relatively high temperature. We obtain the following results:

Porphin free base (MTHF; 77 K):

$$R_x = +0.30\,(\pm 0.05),\ R_y = -0.50\,(\pm 0.06),\ R_z \approx 0 \text{ (Figure 2)}$$

Chlorophyll-b (MTHF; 94 K):

$$R_x = +0.26\,(\pm 0.03),\ R_y = -0.59\,(\pm 0.06).\ R_z \approx 0 \text{ (Figure 3)}$$

From these results we arrive at the conclusion that in both cases transitions at H_x^- and H_y^+ would be emissive and those at H_x^+ and H_y^- would be absorptive in the absence of relaxation. Applying Table 1, we find that the spin level $|y\rangle$

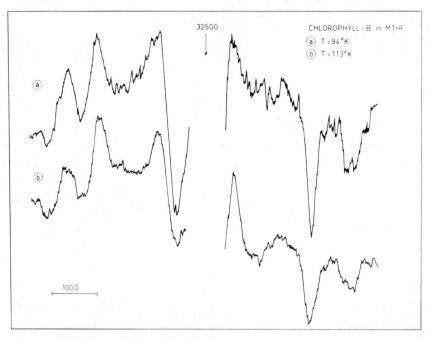

3250G ↓

CHLOROPHYLL-B in MTHF
ⓐ T = 94°K
ⓑ T = 113°K

ⓐ

ⓑ

100G

Figure 3 Triplet ESR spectra of chlorophyll b in MTHF (2×10^{-4} M) at two different temperatures recorded with continuous illumination; the central radical signal is omitted. Microwave power is: 0·5 mW (a) and 5 mW (b), microwave frequency 9099 MHz, modulation amplitude 16 gauss, modulation frequency 100 KHz. In this region the shape of the observed spectrum is independent of microwave power

Table 1 Sign of ESR transitions for a k-selective pumping cycle; E = emissive, A = absorptive signal. It is assumed that P_x, P_y, P_z, k_x, k_y and k_z all have a non-zero value, and that relaxation is absent

Populating rates	Calculated sign of ESR triplet signals at indicated field position						Remarks
	H_z^-	H_x^-	H_y^-	H_y^+	H_x^+	H_z^+	
$\boxed{P_x} > P_y > P_z$	A	A	E	A	E	E	$\boxed{k_x} \gg k_y, \; k_z$
$P_x < \boxed{P_y} > P_z$	A	E	A	E	A	E	$\boxed{k_y} \gg k_x, \; k_z$
$P_x < P_y < \boxed{P_z}$	E	E	E	A	A	A	$\boxed{k_z} \gg k_x, \; k_y$
$\boxed{P_x}, \; \boxed{P_y} > P_z$	A	A/E	E/A	A/E	E/A	E	$k_z \ll \boxed{k_x} \cdot \boxed{k_y}$
$P_x < \boxed{P_y}, \; \boxed{P_z}$	E/A	E	E/A	A/E	A	A/E	$k_x \ll \boxed{k_y} \cdot \boxed{k_z}$
$\boxed{P_x} > P_y < \boxed{P_z}$	E/A	E/A	E	A	A/E	A/E	$k_y \ll \boxed{k_x} \cdot \boxed{k_z}$

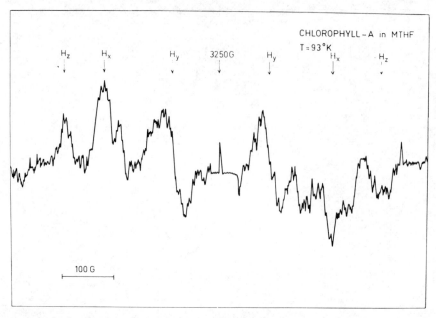

Figure 4 Triplet ESR spectrum of chlorophyll *a* in MTHF (2×10^{-4} M) recorded with continuous illumination. Background subtracted by using a Hewlett–Packard 5480 signal analyser. Arrows indicate positions of triplet ESR lines used in calculating *D*- and *E*-values. Microwave power: 1·5 mW, microwave frequency 9099 MHz, modulation amplitude 20 gauss, modulation frequency 100 KHz

decays faster than $|x\rangle$ and/or $|z\rangle$, whereas the same level is involved in the populating mechanism.

The spin–lattice relaxation rates $w_1^{(i)}$, $w_2^{(i)}$, $w_3^{(i)}$ depend on the direction of H with respect to the molecular frame as is expected for a spin moving in an anisotropic molecular field.

The observed polarization strongly depends on temperature as is shown in Figure 3, probably by a change in solvent viscosity. Assuming that molecular symmetry implies a difference between P_x or P_y and P_z, the result is obtained that

$$P_z < P_x, P_y$$

$$k_z < k_x, k_y$$

This agrees with the observation[9] of the EAEA sequence for porphin free base in an *n*-octane lattice, where relaxation is slow.

The spectrum of chlorophyll *b* (Figure 3) suggests that the populating and depopulating pathway is very similar to that of porphin free base, and not to that of porphyrins with heavier metal ions.[10] As is shown in Figure 4 no steady state spin polarization was observed for chlorophyll *a*. This may be attributed to experimental conditions different from those for chlorophyll *b*. Zero-field parameters for the three compounds of interest in MTHF and ethanol glasses at 77 K and 95 K are collected in Table 2. Typical spectra are shown in Figures 2–4 (solvent MTHF). For ethanol similar spectra are obtained. There is good agreement with published data.[11–13] Note that the zero-field parameters for

Table 2   Zero-field parameters of free base porphin and chlorophyll a and b. D-values were calculated from the separation of the outer peaks; E-values from the average of the separation between x- and y-positions in the low and high field portions of the spectrum. All measurements were carried out at 77 K, except for chlorophyll a, where $T = 95$ K. The variations in D and E in this temperature range are within the measuring error

Compound	Solvent	$D \times 10^4 \, cm^{-1}$	$E \times 10^4 \, cm^{-1}$
Porphin	Ethanol	437 ± 7	66 ± 5
	MTHF	436 ± 6	65 ± 3
Chlorophyll b	Ethanol	287 ± 5	36 ± 4
	MTHF	289 ± 4	49 ± 3
Chlorophyll a	MTHF	281 ± 6	39 ± 3

porphin are insensitive to a change in solvent in contrast to chlorophyll b where E changes with $\sim 30\%$ upon change of solvent (Table 2). D appears to be much more insensitive for differences in environment. Chlorophyll a has a lower E value than chlorophyll b, in agreement with previous conclusions from $\Delta m = 2$ spectra.[12] Temperature effects on D and E are within measuring error in the range 77–110 K. These findings can be rationalized by realizing that among other factors it is the distance between the two lowest *electronic* triplet states in chlorophyll which governs the expectation value of E.[14,15] This distance is affected by the interaction of the magnesium ion or the side groups in chlorophyll with solvent ligands, in contrast with porphin free base.

3 TRANSIENT SPIN-POLARIZATION

With standard ESR detection and continuous illumination, quantitative information on the kinetic scheme cannot be obtained in a simple manner. These data could provide valuable checks on the well-developed theory of the electronic structure of compounds based on the porphyrin skeleton.[16] Following Lhoste[17] and Levanon[18] we have employed a chopped light source and phase-sensitive detection with reference to the chopping frequency. It appears to be useful to study both the time dependence of each of the $\Delta m = \pm 1$ transitions, as well as the shape of the spectrum after a certain time has elapsed since the exciting light has been switched on or off, using box-car integrator.[19] Treatment of our results goes along the same lines as in previous studies on slower systems by Schwoerer and Sixl[7] and Clarke.[20]

Solving the differential equations leads to a sum of exponentials for the time-dependence of the various signal amplitudes. If the relaxation rates are of comparable magnitude or larger than the decay rates, signals S_i^+ and S_i^- have unequal amplitude.[7] For $H \parallel x, y$ and z this is what we observe. After switching *off* the exciting light, the S_y^+ signal *changes sign* from A to E and then decays to zero. The same phenomenon, but less pronounced, is found for S_z^+. This also indicates $k_y > k_x, k_z$.

Figure 5 represents the observed time-dependence for the signal amplitudes S_x^+, S_y^+ and S_z^+. The time-resolution is probably limited by the 0·3 ms RC-time

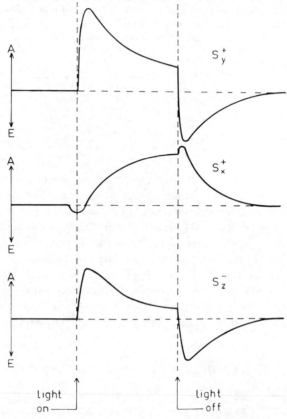

Figure 5   Observed time-dependence of S_x^+, S_y^+, S_z^r (see text) of chlorophyll *b* in MTHF. Response time spectrometer 0·3 ms

of the spectrometer. When switching on the exciting light, signal S_i^+ initially grows proportional to $P_i - \frac{1}{2}(P_j + P_k)$. Thus

$$S(H) = \lim_{\Delta t \to 0} \int_0^{\Delta t} S(t) \, dt$$

represents a spectrum *in statu nascendi* with normalized signal amplitudes reflecting the *difference* in the populating rates P_0 and $P_{\pm 1}$.

For chlorophyll *b* such a gated spectrum is shown in Figure 6. Note the large amplitude of the $S_y^\pm$ signal confirming the previous conclusion that $P_y > P_x, P_z$.

The opposite behaviour of a transition with continuous or pulsed excitation can now be understood. In the former case, the S_y^+ transition for instance, would be

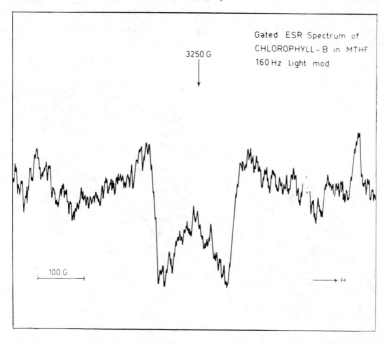

Figure 6   Gated ESR spectrum of chlorophyll b in MTHF. Response time ESR spectrometer 0·3 ms; gatewidth: 1 ms; delay with respect to edge of lightpulse: 1 μs. Response time gated amplifier: 10 s. Chopping frequency: 160 Hz

emissive in the absence of relaxation, even when the *onset* of the same transition is *absorptive*. This is due to the fact that, although P_y is larger than $\frac{1}{2}(P_x + P_z)$, P_y/k_y can be smaller than $(P_x + P_z)/(k_x + k_z)$.

Since the pumping-cycle is k-selective with fast relaxation, the kinetics of the signal-amplitude $S_i(t)$ after switching off the exciting light, are determined[7] by $\frac{1}{3}(k_x + k_y + k_z)$, found to be $320 \pm 30 \, \mathrm{s}^{-1}$. This result can be combined with the fact that S_y^+ and S_x^- have emissive character during continuous excitation, resulting in

$$k_y > 520 \, \mathrm{s}^{-1}; \qquad k_y > k_x, k_z$$

For spin-polarization to be observable the spin–lattice relaxation time, defined[7] as $T_1 \equiv (3w)^{-1}$, should be in the region

$$(k_0 - k_1)^{-1}\delta < T_1 < (k_x + k_y + k_z)^{-1}, \text{ i.e. } 0·005 \, \mathrm{ms} < T_1 < 1 \, \mathrm{ms}$$

Using phase-sensitive detection, with reference to the chopping frequency, generally leads to distorted ESR spectra, because transients of different shape are integrated for $H \parallel x, y$ or z. Some transitions may be even absent from the spectrum. This is observed in Figure 7, where the S_x^+ and S_x^- transitions do not

324

appear. As a check on the model, we tried to simulate the transients with an equivalent electronic network[21]. Figure 8 represents a superposition of the actually observed and the simulated transient for the S_y^- transition. It is clearly seen that it starts as emissive and changes to absorptive during illumination before returning to zero after switching off the light.

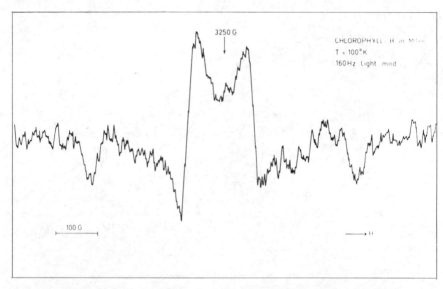

Figure 7 Chlorophyll b triplet ESR spectrum, using 160 Hz chopped excitation and phase sensitive detection at chopping frequency. Response time ESR spectrometer $\sim$180 μs; PSD response time: 10 s Phase-angle: 150°. Single scan

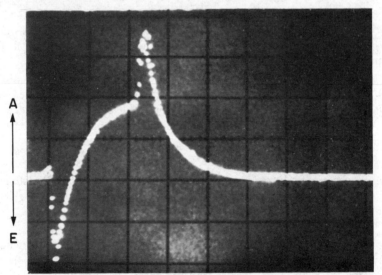

Figure 8 Superposition of simulated and observed transient S_y^- of chlorophyll b in MTHF at chopping frequency 15 Hz, ESR response time 0·3 ms. Accumulation of $\sim$1000 transients; 1 division = 5 ms

4 SUMMARY

The results can be summarized as follows:

(a) Steady state spin polarization at and above 77 K as observed in chlorophyll b is solvent-dependent, probably through interaction of the solvent with the fifth coordination bond of the Mg^{2+} ion, or with the sidegroups. This explanation is in agreement with the absence of solvent dependence for porphin free base.

(b) Kinetic behaviour of chlorophyll b is very similar to that of porphin free base and different from porphyrins with heavier metal ions. Spin polarization arises from a fast-decaying $|y\rangle$ level, which also has the largest populating rate.

(c) Spin–lattice relaxation rates are larger than the decay-rates for any orientation of the magnetic field, but smaller than $10^6 \, s^{-1}$.

(d) Zero-field magnetic parameters are obtained for porphin free base, chlorophyll a and b, the latter having equal D, but significantly different E-values.

5 EXPERIMENTAL

ESR spectra were obtained with a standard Varian E-6 spectrometer, equipped for continuous and chopped excitation. Compounds were pure, as checked by TLC and visible spectra.[22] Samples were free from moisture and oxygen. 2-Methyl-tetrahydrofuran (MTHF) was dried by distillation from a sodium mirror. All preparations were carried out in grease-free glassware.

6 ACKNOWLEDGEMENT

We want to thank Dr J. H. van der Waals for permission to refer to unpublished results obtained in his group. One of us (T. J. S.) is indebted to Drs J. H. van der Waals and M. Gouterman for inspiring discussions on the electronic structure of porphyrins. Mr P. A. de Jager has provided us with valuable technical assistance.

REFERENCES

1. A. N. Terenin, *Izv. Akad. Nauk. SSSR, Ser. Biol.*, no. 3, 369 (1947).
2. A. A. Krasnovskii, *Izv. Akad. Nauk. SSSR, Ser. Biol.*, no. 3, 377 (1947).
3. W. G. van Dorp, T. J. Schaafsma, M. Soma and J. H. van der Waals, *Chem. Phys. Letters*, **21**, 221 (1973).
4. A. T. Gradyushko, A. N. Sevchenko, K. N. Solovyov and M. P. Tsvirko, *Photochem. Photobiol.*, **11**, 387 (1968).
5. C. A. Parker, *Photoluminescence of Solutions*, p. 373 ff. Elsevier Publ. Co., Amsterdam, 1968.

6. J. H. van der Waals and M. S. de Groot, in *The Triplet State*, p. 125 (ed. A. B. Zahlan), Cambridge Univ. Press, London, 1967.

7. H. Sixl and M. Schwoerer, *Z. Naturforsch.*, **25a**, 1383 (1970); **24a**, 952 (1969); M. Schwoerer, in *Proc. XVII Congress Ampère*, p. 143 (ed. V. Hovi), North Holland Publ. Co., Amsterdam.

8. E. V. Shpolskii and L. A. Klimova, *Optics & Spectroscopy* **7**, 499 (1959); E. V. Shpolskii, *Soviet Phys. Usp.* **3**, 372 (1960); **5**, 522 (1962); **6**, 411 (1963).

9. M. Soma, unpublished results.

10. I. Y. Chan, W. G. van Dorp, T. J. Schaafsma and J. H. van der Waals, *Mol. Phys.*, **22**, 741 (1971); **22**, 753 (1971).

11. J. M. Lhoste, *C. R. Acad. Sci. Paris* (*Série D*), **266**, 1059 (1968).

12. G. T. Rikhireva, L. A. Sibel'dina, Z. P. Gribova, B. S. Marinov, L. P. Kayushin and A. A. Krasnovskii, *Dokl. Nauk. USSR* (*Biophysical Section*), **181**, 103 (1968).

13. G. T. Rikhireva, Z. P. Gribova, L. P. Kayushin, A. V. Umrikhina and A. A. Krasnovskii, *Dokl. Akad. Nauk. USSR*, **159**, 196 (1964).

14. J. H. van der Waals, A. M. D. Berghuis and M. S. de Groot, *Mol. Phys.*, **13**, 301 (1967); **21**, 497 (1971),

15. J. M. Lhoste, C. Hélène and M. Ptak, *The triplet State*, p. 487 (ed. A. B. Zahlan), Cambridge Univ. Press, London, 1967.

16. M. Gouterman, *Excited States of Matter*, Vol. 2, pp. 1–174 (ed. C. W. Shoppee), Grad. Studies Texas Tech. Univ., 1973 and references cited therein.

17. J. M. Lhoste, *Stud. Biophys. Berlin*, **12**, 135 (1968).

18. H. Levanon and S. I. Weissman, *J. Am. Chem. Soc.*, **93**, 4309 (1971); *Israel J. Chem.*, **10**, 1 (1972); *Chem. Phys. Lett.*, **24**, 96 (1974).

19. M. Plato and K. Möbius, *Messtechnik*, **8**, 224 (1972).

20. R. H. Clarke, *Chem. Phys. Letters*, **6**, 413 (1970).

21. P. A. de Jager, J. F. Kleibeuker and T. J. Schaafsma; to be published.

22. *The Chlorophylls*, (eds. L. P. Vernon and G. R. Seely), Academic Press, New York; U. Eisner and R. P. Linstead, *J. Chem. Soc.*, 3749 (1955); G. P. Gurinovich, A. N. Sevchenko and K. N. Solovyov, *Spectroscopy of Chlorophyll and Related Compounds*, Science and Engineering Publishing House, Minsk, 1968 (Engl. transl. by Natl. Techn. Inform. Service, Springfield, Va. 22151, U.S.A.).

C.8. Magnetically induced circular emission spectra of some metalloporphyrins

A. J. McCaffery, R. Gale, R. A. Shatwell and K. Sichel

School of Molecular Sciences, University of Sussex, Brighton BN1 9QJ, U.K.

In a series of recent papers,[1-3] we have described the measurement of magnetically induced circular emission (MCE) of a number of molecular systems. In this contribution, we report the results of a study on a number of metalloporphyrins in order to investigate the role of the metal ion in modifying the properties of the excited states of the porphyrin system.

The experimental arrangement has been described in a recent publication.[4] In the case of the metalloporphyrins discussed herein, the samples were dissolved in rigid matrices of polymethyl methacrylate which were cut and polished for optical examination and mounted in the bore of a superconducting solenoid in contact with liquid helium. Measurement of the sample temperature with a carbon resistance thermometer indicated a temperature of around 30 K. The layout of the equipment is shown in Figure 1. Exciting light from an argon ion

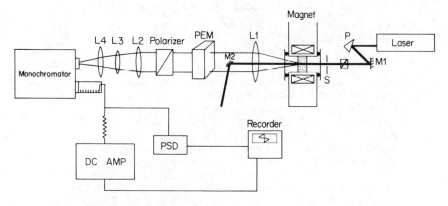

Figure 1   Experimental arrangements for measuring magnetically induced circular emission

laser or from a xenon arc source passes through a small monochromator and is incident on the sample mounted in the cryostat. The emitted radiation is collimated and is polarization analysed by a photoelastic modulator set for $\lambda/4$ retardation, followed by a calcite polarizer. The radiation is wavelength analysed by a Spex 1406 double grating monochromator followed by an EMI 9558 QA photomultiplier which has provision for cooling. The d.c. photocurrent is amplified and recorded directly to give the emission spectrum. The a.c. voltage generated in the photomultiplier anode load in phase with the modulation frequency is proportional to the circular polarization of the emission. It is amplified by a phase sensitive detector and displayed simultaneously to the emission spectrum, on a double-pen chart recorder.

The theory of MCE spectra may be expected to be very similar to that developed for the absorption analogue, magnetic circular dichroism[5] (MCD). In treating the emission from porphyrin molecules we shall be concerned with the so-called C term which arises from an initial state degeneracy, as illustrated in Figure 2. The magnitude of the predicted MCE is inversely proportional to absolute temperature and the magnitude of the circular polarization for the transition state a to j can be shown to be given by the product

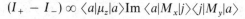

$$(I_+ - I_-) \propto \langle a|\mu_z|a\rangle \text{Im} \langle a|M_x|j\rangle\langle j|M_y|a\rangle$$

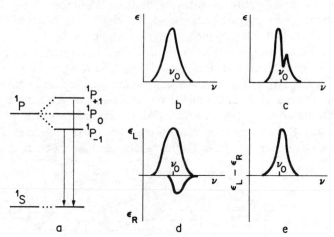

Figure 2   Illustration of circular polarization of emission from a degenerate state. Transition from excited state ^1P to ^1S gives rise to the emission spectrum (a) which, in the presence of a magnetic field, may show a Zeeman splitting (b) provided the line width is narrow compared to the Zeeman energy. The emission from the Zeeman components of ^1P is circularly polarized and the intensities of left and right circular components (I_- and I_+) are shown in (c) for a thermal distribution between the split levels. (d) Illustrates the circular intensity difference ($I_+ - I_-$) measured in the MCE experiment. Note that in the general case where the Zeeman splitting is small compared to the line width, the minor circular component is not observed

where μ is the magnetic dipole moment operator, M the electric dipole moment operator, and a and j are the initial and final states, respectively.

As in MCD theory, it is convenient to divide the circular polarization intensity by the emission intensity $(I_+ + I_-)$, the area under the corresponding emission band. $(I_+ + I_-)$ is proportional to the product of the two electric dipole matrix elements above and thus the quantity

$$\frac{I_+ - I_-}{I_+ + I_-}$$

gives the initial state magnetic moment directly. The MCE parameters may be obtained from the experimental line shapes by the method of moments.

Figure 3 shows the emission spectrum and MCE of palladium octaethyl-

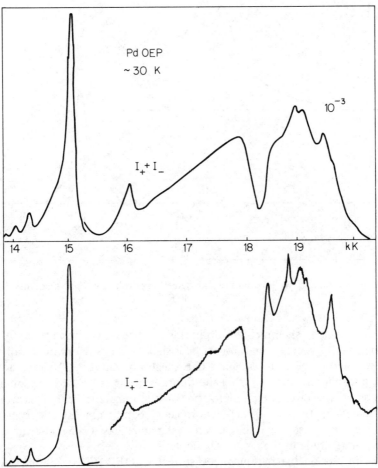

Figure 3 Emission (upper curve) and MCE spectra of palladium octaethyl porphyrin (Pd OEP) in polymethylmethacrylate at 30 K. Intensities are in arbitrary units

porphyrin (Pd OEP) excited by the 5145 Å line of Ar^+. Carbon resistor measurements indicate a temperature of 30 K for low laser powers and this may be increased to 65 K by increasing the laser power. We find the MCE to be a linear function of $1/T$ in this range, indicating that the interpretation in terms of a major contribution from the emitting state degeneracy to the MCE is justified. The Figure shows the very weak fluorescence and the strong phosphorescence characteristic of this d^8 metalloporphyrin.[6] The theory of the electronic

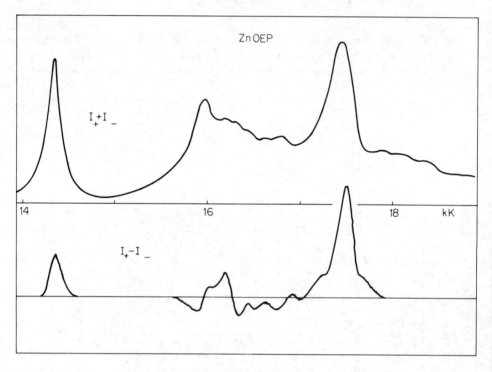

Figure 4   Emission and MCE spectra of zinc octaethyl porphyrin in polymethylmethacrylate at ~40 K

structure of open shell metalloporphyrins has been given recently by Ake and Gouterman,[7] whereas that for closed shell systems is more established.[8] Figure 4 shows the emission and MCE spectra of ZnOEP at around 30–40 K. Both phosphorescence from ^3E_u and fluorescence from ^1E_u to the ground $^1A_{1g}$ state show strong circular polarization in the magnetic field. This would be anticipated for the 18,500 cm^{-1} fluorescence band from MCD[9] experiments. However, it is also clear that the lowest triplet has a magnetic moment of similar magnitude to that of the Q_0 singlet state.

The results on the paramagnetic system CuOEP (Figure 5) show a more complex pattern in emission and MCE. The emission spectrum has been interpreted as arising from an equilibrium between emitting spin doublet and quartet

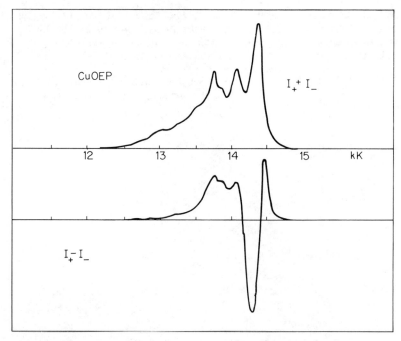

Figure 5   Emission and MCE spectra of copper octaethyl porphyrin in polymethylmethacrylate at ~ 40 K

states, the latter increasing its population as the temperature is lowered. Both states arise from the coupling of the metal ground state doublet with the π-electron triplet, differing in energy by an exchange integral. The MCE appears to confirm this picture, since as temperature is reduced the central doublet MCE diminishes leaving a broad background structure arising from the quartet state.

The theory of the d^8 porphyrins, PdOEP (Figure 3) is less well established. These systems are diamagnetic, but have open d-shells and hence spin–orbit coupling may play an important role in determining luminescence properties. It is known that in the series of porphyrins NiOEP, Pd OEP, Pt OEP, fluorescence quantum yields drop and phosphorescence quantum yields increase until for Pt OEP $\phi_p \sim 1\cdot0$. A detailed analysis of the MCE results for d^8 and d^9 systems is given in reference 10 together with a deeper discussion of the theory of MCE.

REFERENCES

1. R. A. Shatwell and A. J. McCaffery, *J.C.S. Chem. Comm.*, 546 (1973).
2. A. J. McCaffery, P. Brint, R. Gale and R. A. Shatwell, *Chem. Phys. Letters*, **22**, 600 (1973).
3. R. Clark, S. R. Jeyes, A. J. McCaffery and R. A. Shatwell, *Chem. Phys. Letters*, **25**, 74 (1974).
4. R. A. Shatwell and A. J. McCaffery, J. *Phys. E*, to be published (1974).

5. (a) A. D. Buckingham and P. J. Stephens, *Ann. Rev. Phys. Chem.*, **17**, 399 (1966).
   (b) P. N. Schatz and A. J. McCaffery, *Quart. Rev. Chem. Soc.*, **23**, 552 (1969).
6. D. Eastwood and M. Gouterman, *J. Mol. Spectroscopy*, **30**, 437 (1968).
7. R. L. Ake and M. Gouterman, *Theor. Chem. Acta.*, **15**, 20 (1969).
8. M. Gouterman, *J. Mol. Spectroscopy*, **6**, 138 (1961).
9. R. Gale, A. J. McCaffery and M. D. Rowe, *J. Chem. Soc.*, (*Dalton*), 596 (1972).
10. R. A. Shatwell, A. J. McCaffery, R. Gale and K. Sichel. *J. Amer. Chem. Soc.* to be published.

C.9. A simple stochastic model for the investigation of the fate of excitation energy in the photosynthetic system

Susan Gray and Patrick Williams

Biophysics Laboratory, Chelsea College, Manresa Road, London, S.W.3, U.K.

SUMMARY

The de-excitation pathways for energy absorbed by the photosynthetic pigments are described in terms of a simple stochastic model. A set of equations is derived which relates the probabilities of the various de-excitation mechanisms both at the level of the photosynthetic unit and of the photosynthetic pigment systems. Using these equations and experimental data available in the literature, probability values are calculated for the individual de-excitation pathways for the photosystems for a range of wavelengths. The value and reliability of such data is dicussed.

1 INTRODUCTION

The choice of a suitable model for investigating the fate of excitation energy in photosynthesis is restricted by the limited amount of detailed experimental data relating to the structural organization and operation of the photosynthetic apparatus available in the literature. Whilst it is important that the model reflects the basic operation of the photosynthetic system, it is equally important that it should not be more complex than is justified by the quantity and quality of the available data.

In this paper, a simple stochastic model is used to investigate the fate of the excitation energy. The model is based only on the simplest of organizational assumptions and contains no assumptions of detailed structure. Used in combination with experimental data on the wavelength dependencies of photosynthetic oxygen evolution and fluorescence taken from the literature, it provides a series of probability values for the individual de-excitation pathways in photosynthesis.

The paper contains a description of the model and an account of its theoretical basis. The method of calculating the de-excitation probabilities from the available experimental data using the model is detailed and the resulting values listed. The potential applications for such data are also briefly discussed.

2 THE MODEL AND ITS THEORETICAL BASIS

The photosynthetic pigments are organized in two pigment systems, one of which is associated with photosystem I (PS I) and the other with photosystem II (PS II). The pigments associated with a given photosystem are believed to be arranged in aggregates termed photosynthetic units. Each photosynthetic unit is thought to consist of 300–400 'light-harvesting' pigments which pass on their excitation energy to a specialized chlorophyll molecule—the 'trap' or 'reaction centre'—where it is utilized in photochemistry.[1] The photosynthetic units are envisaged as being spatially separate, but sufficiently close for energy transfer to take place between them. Energy transfer, both between units within the same photosystem (intrasystem transfer),[2–4] and between units in different photosystems (intersystem transfer, usually termed spill-over),[5–6] have been postulated.

The possible fates for excitation energy absorbed within a given photosynthetic unit are thus: internal conversion, fluorescence, photochemistry, intrasystem transfer and intersystem transfer. Let the probabilities of each of these fates be denoted by p_h, p_f, p_c, p_i and p_s respectively. In order to distinguish between light absorbed in photosynthetic units of PS I and PS II, the relevant quantities will, when necessary, be subscripted by $_I$ or $_{II}$. Let the corresponding fates of excitation energy within a given photosystem, as opposed to a photosynthetic unit, be denoted by P_h, P_f, P_c and P_s. There is, of course, no corresponding quantity P_i as intrasystem energy transfer does not constitute a loss of

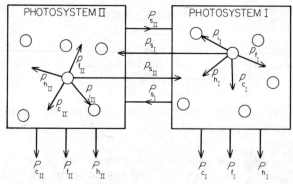

Figure 1 A stochastic model for the fate of excitation energy in photosynthesis. The photosynthetic system is assumed to be organized into two photosystems each containing separate functional units—photosynthetic units. The various possible de-excitation pathways for the individual units and photosystems are as indicated

excitation energy to the system. These quantities are summarized, in the form of a simple stochastic model, in Figure 1.

If the excitation energy is considered to be delocalized throughout the whole photosynthetic unit, intrinsic lifetimes characteristic of such units can be defined for each of the de-excitation processes. Let these intrinsic lifetimes (i.e. the lifetimes of the excited unit if the given process is the only de-excitation process), be denoted by τ_h, τ_f, τ_c, τ_i and τ_s. If the actual lifetimes is τ:

$$\frac{1}{\tau} = \frac{1}{\tau_h} + \frac{1}{\tau_f} + \frac{1}{\tau_c} + \frac{1}{\tau_i} + \frac{1}{\tau_s} \tag{1}$$

The probabilities of the individual de-excitation processes are given by equations of the form:

$$p_f = \frac{\text{number of fluorescent transitions from excited state/unit time}}{\text{total number of transitions from excited state/unit time}} \tag{2}$$

or in terms of lifetimes:

$$p_f = \frac{\tau}{\tau_f} \tag{3}$$

As de-excitation must take place via one or other of the pathways, the sum of the individual probabilities must equal one:

$$p_h + p_f + p_c + p_i + p_s = 1 \tag{4}$$

Expressed in terms of lifetimes:

$$\tau \left(\frac{1}{\tau_f} + \frac{1}{\tau_c} + \frac{1}{\tau_h} + \frac{1}{\tau_i} + \frac{1}{\tau_s} \right) = 1 \tag{5}$$

The values of τ_h and τ_f will be independent of structural considerations but those of τ_c, τ_i and τ_o, corresponding to processes involving the interaction of units with each other, will be structure-dependent.

The probabilities of internal conversion, fluorescence, etc. for a photosystem can be defined in terms of the corresponding probabilities for the individual photosynthetic units. Thus the probability of trapping de-excitation energy and utilizing it for photochemistry is given by:

$$P_c = (p_c + p_i p_c + p_i p_i p_c + \cdots) = \frac{p_c}{(1 - p_i)} \tag{6}$$

that is P_c equals the probability of trapping and utilizing the energy in the unit in which the energy is originally absorbed plus the sum of the products of the probabilities of its transfer and trapping in the units to which it is transferred.

In terms of intrinsic lifetimes, from equations (3) and (6)

$$P_c = \frac{\tau}{\tau_c (1 - p_i)} \tag{7}$$

P_h, P_f and P_s can be defined by similar equations. Again the sum of the probabilities of the various de-excitation processes must equal one, hence:

$$P_h + P_f + P_c + P_s = 1 \tag{8}$$

or:

$$\frac{\tau}{(1 - p_i)}\left(\frac{1}{\tau_h} + \frac{1}{\tau_f} + \frac{1}{\tau_c} + \frac{1}{\tau_s}\right) = 1 \tag{9}$$

On substituting τ/τ_i for p_i and rearranging, equation (9) reduces—as it should—to equation (5).

3 CALCULATION OF PROBABILITY VALUES

The values of P_h, P_f, P_c and P_s cannot be calculated directly from the equations set out in the previous section, as the values of many of the parameters in the equations are unknown. However, examination of the available experimental data indicates that enough information is present in the literature to calculate values first for P_f and P_c and thence to estimate value for P_h and P_s.

The wavelength dependence of the quantum yields of fluorescence,[7] ϕ_f, and of photochemistry,[8] ϕ_c, of the green alga *Chlorella Pyrenoidosa* have been investigated in detail. If these data are combined with available data for the distribution of excitation energy between the two photosystems,[7,9] values for P_f and P_c can easily be calculated.

Let the fraction of the total incident radiation, for a given wavelength, absorbed in PS II equal α and the fraction absorbed in PS I equal $(1 - \alpha)$. As practically all chlorophyll *a* fluorescence arises from PS II:[7]

$$P_{f_{II}} \simeq \phi_f/\alpha \quad \text{and} \quad P_{f_I} \simeq 0 \tag{10}$$

The quantum yield of photochemistry is given by:

$$\phi_c = \frac{\text{number of molecules of } O_2 \text{ evolved}}{\text{number of quanta absorbed}} \tag{13}$$

In order to evolve one molecule of O_2, four quanta have to be trapped in each of the photosystems. The relative absorption wavelengths of PS II and PS I pigments are such that transfer of excitation energy can take place from PS II to PS I but not in the reverse direction.[5] Spillover from PS II will not affect the value of ϕ_c as long as the turnover of PS I and not PS II is rate-limiting (this is in fact the case for most of the spectral range below 685 nm).[6] The value of $P_{c_{II}}$ can thus be taken to be:

$$P_{c_{II}} = \frac{4\phi_c}{\alpha} \tag{11}$$

As PS I is usually rate-limiting, excitation energy transferred from PS II to PS I will tend to increase ϕ_c. The value of P_{c_I} is thus given by:

$$P_{c_I} = \frac{4\phi_c}{(1 - \alpha) + \alpha P_{s_{II}}} \tag{12}$$

Given that P_{s_I} is equal to zero and P_{f_I} is effectively zero, the values of P_{h_I} can be calculated from the values of P_{c_I} using equation (8). The values of $P_{h_{II}}$ and $P_{s_{II}}$ are much less easily accessible. However, if the values for one can be deduced, those for the other can again be calculated using equation (8).

Values for $P_{h_{II}}$ can be estimated by the following procedure. $P_{h_{II}}$ and $P_{f_{II}}$ are defined by equations analogous to equation (7). The ratio of these probabilities is, therefore, given by:

$$\frac{P_{h_{II}}}{P_{f_{II}}} = \frac{\tau_{f_{II}}}{\tau_{h_{II}}} \tag{13}$$

Under conditions of zero photochemistry (inhibited photosynthesis) P_c equals zero. Experiments performed by Bonaventura and Myers[10] suggest that spillover of excitation energy takes place in response to an imbalance in the distribution of excitation energy between the two photosystems. Their experiments also indicate that spillover is at a minimum at the onset of illumination (i.e. in the absence of previous photochemistry). $P_{s_{II}}$ would, therefore, be expected to

Table 1 The probabilities for the competing processes for de-excitation of photosystem II. The values for $P_{f_{II}}$, $P_{c_{II}}$, $P_{h_{II}}$ and $P_{s_{II}}$ were calculated from fluorescence and oxygen evolution data taken from the literature.[7,8] The relative quantum yield of fluorescence values in Reference 7 was converted into absolute quantum yields by assuming the average quantum yield in the range 540–650 nm to equal 0·03.[1] The corresponding value for poisoned algae was taken to equal three times that for fresh algae.[7] The values for α were also taken from Reference 7

λ(nm)	α	$P_{f_{II}}$	$P_{c_{II}}$	$P_{h_{II}}$	$P_{s_{II}}$
540	0·55	0·055	0·60	0·25	0·09
560	0·55	0·055	0·63	0·25	0·06
580	0·58	0·053	0·62	0·24	0·09
600	0·59	0·052	0·61	0·24	0·10
620	0·57	0·053	0·63	0·24	0·08
640	0·59	0·051	0·60	0·23	0·12
645	0·60	0·051	0·58	0·23	0·14
650	0·63	0·048	0·55	0·22	0·18
660	0·57	0·050	0·60	0·23	0·12
680	0·53	0·051	0·68	0·23	0·04

be effectively zero for poisoned algae in which the initial light distribution between the two systems is equal (i.e. $\alpha = 0.5$). Under these conditions:

$$P_{h_{II}} + P_{f_{II}} = 1 \qquad (14)$$

Combining equations (10), (13) and (14)

$$\frac{\tau_{f_{II}}}{\tau_{h_{II}}} = \frac{(1 - \phi_f^*/0.5)}{(\phi_f^*/0.5)} \qquad (15)$$

where ϕ_f^* is the quantum yield of poisoned algae at a wavelength for which $\alpha = 0.5$.

This value of $\tau_{f_{II}}/\tau_{h_{II}}$ can be substituted into equation (13). On substitution for $P_{f_{II}}$, using equation (10), this yields a general expression for $P_{h_{II}}$:

$$P_{h_{II}} = \frac{\phi_f(1 - \phi_f^*/0.5)}{\alpha(\phi_f^*/0.5)} \qquad (16)$$

The values of P_h, P_c, P_f and P_s for the two photosystems of *Chlorella* in the wavelength range 540–680 nm, calculated on the basis of the above equations, are listed in Tables 1 and 2. These values correspond to the fates of excitation energy absorbed within a given photosystem. It is sometimes easier to appreciate the significance of these values in terms of the total absorption rather than absorption in the individual photosystems. The values in Tables 1 and 2 expressed as fractions of total number of absorbed quanta are, therefore, listed in Table 3.

The accuracy of the values listed in the tables is, of course, limited by the validity of the assumptions in the present analysis, the experimental errors and assumptions used in the calculation of the physiological data and the heterogeneity of the material from which the original data was collected. The values, however, probably represent a good approximation to the true values.

Table 2 The probabilities for the competing processes for de-excitation of photosystem I. The values given in the Table were calculated from the equations given in the text. Literature data as for Table 1

λ(nm)	$1 - \alpha$	P_{c_I}	P_{h_I}
540	0.45	0.66	0.34
560	0.45	0.72	0.28
580	0.42	0.76	0.24
600	0.41	0.77	0.23
620	0.43	0.76	0.24
640	0.41	0.74	0.26
645	0.40	0.72	0.28
650	0.37	0.72	0.28
660	0.43	0.69	0.31
680	0.47	0.73	0.26

Table 3 The fate of excitation energy absorbed in the photosynthetic system. Data from Tables 1 and 2 expressed as the fractions of the total number of incident quanta. The fractional input of excitation energy for PS II is equal to the fractional absorption of PS II. The fractional input for PS I is the fractional absorption of PS I plus the fraction of incident quanta that spills over from PS II to PS I

λ(nm)	Photosystem	Fractional input	Photochemistry	Fluorescence	Internal conversion	Spillover
540	II	0·55	0·33	0·03	0·14	0·05
	I	0·50	0·33	—	0·17	—
560	II	0·55	0·35	0·03	0·14	0·03
	I	0·48	0·34	—	0·14	—
580	II	0·58	0·36	0·03	0·14	0·05
	I	0·47	0·36	—	0·11	—
600	II	0·59	0·36	0·03	0·14	0·06
	I	0·47	0·36	—	0·11	—
620	II	0·57	0·36	0·03	0·14	0·04
	I	0·48	0·36	—	0·12	—
640	II	0·59	0·35	0·03	0·14	0·07
	I	0·48	0·35	—	0·13	—
645	II	0·60	0·35	0·03	0·14	0·08
	I	0·48	0·35	—	0·13	—
650	II	0·63	0·35	0·03	0·14	0·11
	I	0·48	0·34	—	0·14	—
660	II	0·57	0·34	0·03	0·13	0·07
	I	0·50	0·34	—	0·16	—
680	II	0·53	0·36	0·03	0·12	0·02
	I	0·49	0·36	—	0·13	—

The values, apart from their own intrinsic interest, have considerable practical use. Firstly, they present the fragmented data available in the literature in a coherent pattern. Secondly, as they are derived from a model that involves only a minimum number of assumptions they can be used as a basis for testing more complex models. The simple stochastic model presented in this paper can readily be extended by introducing a more complex organizational and structural framework. These models in turn can be used to test more detailed hypotheses on the distribution of energy in the photosynthetic system or to predict values for structural parameters that can be checked by direct measurement.

ACKNOWLEDGEMENT

The support of the Royal Society is gratefully acknowledged.

REFERENCES

1. E. Rabinowitch and Govindjee, *Photosynthesis*, John Wiley New York, 1969.
2. A. Joliot and P. Joliot, *C. R. Acad. Sci. Paris.*, **258**, 4622 (1964).
3. P. Joliot, A. Joliot and B. Kok, *Biochim. Biophys. Acta.*, **153**, 635 (1968).
4. P. Joliot, P. Bennoun and A. Joliot, *Biochim. Biophys. Acta.*, **305**, 317 (1973).
5. J. Myers and J. R. Graham, *Plant Physiol.*, **38**, 105 (1963).
6. R. K. Clayton, *J. Theoret. Biol.*, **5**, 497 (1963).
7. W. P. Williams, N. R. Murty and E. Rabinowitch, *Photochem. Photobiol.*, **9**, 455 (1969).
8. R. Emerson and C. M. Lewis, *Amer. J. Bot.* **30**, 165 (1943).
9. W. P. Williams, *Biochim. Biophys. Acta* **153**, 484 (1968).
10. C. Bonaventura and J. Myers, *Biochim. Biophys. Acta*, **189**, 366 (1969).

C.10. Nanosecond irradiation studies of ubiquinone and derivatives

E. Amouyal,† R. Bensasson,† C. Salet‡ and E. J. Land§

†Université de Paris VI, ER 98, Laboratoire de Chimie Physique,
91405 Orsay, France
‡Institute de Pathologie Cellulaire, Hôpital de Bicêtre,
94270 Le Kremlin-Bicêtre, France
§Paterson Laboratories, Christie Hospital and Holt Radium Institute,
Manchester M20 9BX, U.K.

ABSTRACT

Ubiquinone, an isoprenoid quinone present in mitochondria and in bacterial chromatophores, has been studied in solution by laser flash photolysis and pulse radiolysis. The triplet excited states of ubiquinone and of derivatives in which the various ring substituents of ubiquinone-30 are progressively altered, have been characterized. The physico-chemical properties determined include triplet absorption spectrum, extinction coefficient, lifetime, energy level and quantum efficiency of formation. It is deduced that the low triplet energy and quantum efficiency of formation of triplet ubiquinone-30 is caused by the presence of the two adjacent methoxy substituents.

The low quantum efficiency of triplet formation found here for ubiquinone is nonetheless higher than the quantum yield for irreversible photomodification of ubiquinone *in vivo* in *Escherichia coli* B found by Werbin *et al.* (1973). The triplet state could therefore be involved in such photomodifications, albeit inefficiently, due to the action of cellular quenchers. Such quenchers could be porphyrins or carotenes which have even lower triplet energy levels than ubiquinone.

The low quantum efficiency and triplet level also make very inefficient any change of the redox state of ubiquinone, in the normal functioning of the bacterial chromatophore, via ubiquinone excitation and subsequent H or e⁻ abstraction from the surrounding medium. Thus the only source for ubisemiquinone formation is indirect or direct electron transfer from excited chlorophyll.

C.11. The determination of localized energy states in β-carotene by A.C. conduction studies

T. J. Lewis and R. Pethig

School of Electronic Engineering Science, University College of North Wales, Bangor, Gwynedd, U.K.

SUMMARY

Electrical conductivity measurements, through the frequency range 10 Hz to 33 GHz and through the temperature range 290 K to 363 K, are reported for the all-*trans*, the oxygen complex and the *cis-trans* glass form of β-carotene. Over a significant frequency range the conductivity varies with angular frequency ω as $A\omega^s$, with A and s being temperature dependent. This behaviour is analysed in terms of a simple activated charge-carrier-hopping process, from which a physical understanding of the observed experimental results can be obtained. A new feature of this theoretical model, as compared with earlier theories in the literature, is that an understandable physical basis is established for the frequency exponent s having a value greater than unity. It is concluded, that a detailed analysis of such a.c. conductivity measurements as reported here can provide a valuable insight into both localized and delocalized electronic transport processes in biologically important materials in particular, and in organic molecular solids in general.

1 INTRODUCTION

It is well known[1,2] that a large number of poorly conducting materials exhibit an electrical conductivity frequency dependence of the form

$$\sigma(\omega) = \sigma_0^1 + \sigma^1(\omega) \tag{1}$$

where the steady state (DC) contribution is given by σ_0^1 and the frequency dependent part often has the empirically determined form

$$\sigma^1(\omega) = A\omega^s \tag{2}$$

with the exponent s close to unity.

342

Pollack and Geballe[3] were the first to propose an interpretation of this phenomenon in terms of charge carrier hopping between localized sites distributed randomly in space and energy. Previous measurements have been reported[4] for the all-*trans* and *cis-trans* glass forms of β-carotene where it was shown that the a.c. conductivity followed the form given by equation (2) over the frequency range 10 Hz to 10 GHz. A more extensive account and interpretation of the a.c. conduction properties of β-carotene is given here, together with a theoretical analysis relating the value of the exponent s in equation (2) to the energy distribution profile of the localized electronic states considered to be involved in the dominant charge transport mechanism.

2 EXPERIMENTAL

The conductivities of the various samples were evaluated over the frequency range 10 Hz to 100 kHz using the General Radio 1621 Bridge in the three-terminal configuration. Where sensitivity permitted, measurements were also obtained at 1 MHz using the Wayne Kerr B201 Bridge, and through the range 1 MHz to 300 MHz using the Marconi Q-meter assembly. Microwave measurements at the nominal frequencies of 10 GHz and 33 GHz were also made, using a resonating cavity technique as previously outlined.[5]

For the frequency range 10 Hz to 300 MHz measurements were made in the temperature range 290 K to 363 K on discs of the order 1 mm thick (compressed at around 1.4×10^7 kg/m^2) of both the all-*trans* and *cis-trans* glass form of β-carotene (obtained from Koch-Light Labs Ltd.) in a vacuum of 1.33×10^{-2} N/m^2 and in a dry oxygen atmosphere of pressure around 2.7×10^4 N/m^2. For the microwave measurements, results were obtained for smaller disc fragments in vacuum at 9.95 GHz and in a dry oxygen rich atmosphere at normal atmospheric pressure at 8.6, 11.46 and 32.8 GHz. In the one evacuable cavity operating at 9.95 GHz, the differences observed for oxygen-only and oxygen-rich atmospheres were found to be negligible.

The glass form was produced by allowing an all-*trans* melt to cool slowly under vacuum to produce deep red shiny glass fragments which were then ground into a powder and compressed into discs. Evidence for the transition from the all-*trans* to the *cis-trans* form was obtained by optical absorption measurements showing a decrease in the 400–500 nm absorption peak and the appearance of the 340 nm (*cis*-peak) absorption as previously described by Rosenberg.[6] Apart from the microwave measurements, which were electrodeless, polished copper pressure electrodes were used.

3 RESULTS

Some of the results obtained are shown in Figure 1. For the all-*trans* form under vacuum the results are seen to follow the law described by equation (2)

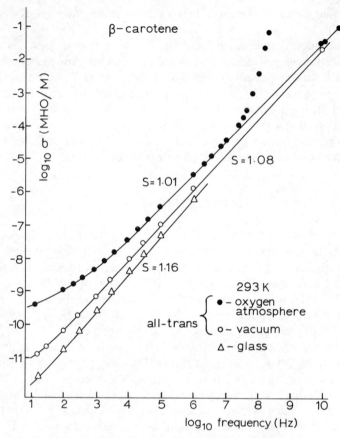

Figure 1 The logarithmic variation of the a.c. conductivity with frequency for the all-*trans* and glass form of β-carotene. The slope values *s*, obtained for the linear portions of the plots, are indicated

over the frequency range from about 500 Hz to 10 GHz. The presence of oxygen resulted in a conductivity increase, particularly at the lowest frequencies of measurement, a result probably attributable to the well-known electron-accepting charge-transfer action of oxygen on β-carotene.[7,8] The departure from the ω^s law above about 20 MHz was considered to result from electrode effects for two reasons; firstly, the results in the range 20 MHz to 200 MHz varied with electrode area and pressure; and secondly, the electrodeless measurements obtained at frequencies around 10 GHz and 33 GHz were seen to fall accurately on the extended plot passing through the results obtained below 10 MHz. For the *cis-trans* glass specimens the effect of oxygen was small over the measured frequency range 10 Hz to 1 MHz, the conductivity increase being no greater than around 5%.

The slope values *s* from Figure 1 (corresponding to the frequency exponent of equation (2)) were found to vary with temperature as shown in Figure 2. The merging of the *s*-values for the all-*trans* vacuum and oxygen atmosphere results

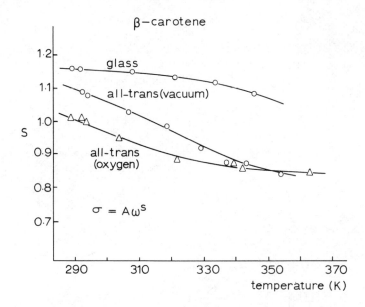

Figure 2   The variation of the frequency exponent *s* with temperature

at 350 K is understandable in terms of oxygen desorption, as observed by
Rosenberg for his d.c. semiconduction studies on β-catotene.[7]

It was found that the measured conductivity $\sigma(\omega)$, determined at constant
frequency values, was a process with constant activation energy according to
the law

$$\sigma(\omega) = \sigma_0 \exp \frac{-\Delta E}{2kT} \tag{3}$$

and ΔE decreased with increasing frequency, as shown in Figure 3. The activation energies for the all-*trans* material in oxygen were obtained over a limited
temperature range to avoid the effects associated with oxygen desorption at the
higher temperatures. It can be seen that the activation energy for the all-*trans*
material in vacuum tended to a value of the order 1·5 eV at the lowest frequencies
of measurement, a value in close agreement with that of 1·52 eV[7] and 1·45 eV[8]
obtained previously for d.c. conduction in all-*trans* β-carotene.

The activation energy for the all-*trans* material in oxygen tended to a value
of 1·2 eV at the lowest frequency of measurement, which again appears to
compare favourably with Rosenberg's d.c. value[7] of 1·29 eV. Rosenberg[7]
assumed that his activation energies of 1·52 eV and 1·29 eV for all-*trans* β-
carotene and the oxygen complex, respectively, referred to a surface conduction
process, whereas Chapman *et al.*,[8] using guarded single crystals, concluded
that they were observing a bulk effect. The low frequency activation energy of the
order 0·6 eV for the glass form is not in good agreement with the d.c. value of
2·49 eV obtained previously by Rosenberg.[6]

β-carotene

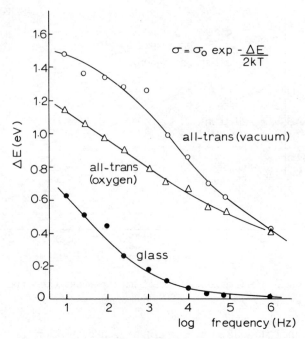

$$\sigma = \sigma_0 \exp -\frac{\Delta E}{2kT}$$

Figure 3   The variation of the activation energy ΔE with frequency

4 DISCUSSION

The hopping mechanism proposed by Pollak and Geballe[3] can essentially be described as phonon-assisted tunnelling of charge carriers between spatially localized states. The theory[3] for this mechanism predicts the relationship given by equation (2) with an s-value around 0·8, and with s being almost invariant with temperature and A only weakly temperature-dependent. This behaviour has been found to occur for impurity conduction in n-type silicon[3] and for electronic conduction in As_2Se_3.[9] On the other hand, many materials have been found to exhibit the relationship given by equation (2), where A increases significantly with temperature and the value for s is of the order of unity and approaches the value 0·8 at high temperatures. This behaviour is exhibited by amorphous selenium[10] and scandium oxide.[11] The measurements reported here for β-carotene also appear to fall into this latter category. To account for such a frequency-dependent conductivity, Pike[11] derived a model involving the classical hopping of charge carriers between localized states over potential barriers with a random height distribution. The maximum value for s was found to be 1·0.

It has recently been emphasized that models based on tunnelling or classical hopping mechanisms are not the only ones which could give a conductivity roughly proportional to $\omega^{1 \cdot 0}$. This type of conductivity is characteristic of any relaxation process with a wide distribution of the relaxation time τ, with τ an exponential function of a random variable ξ

$$\tau = \tau_0 \exp(\xi)$$

Because of this exponential dependence of τ on ξ, then even a distribution of ξ uniform over only a relatively narrow range will cause the approximate $\omega^{1 \cdot 0}$ behaviour to extend over a large frequency range. If ξ can vary in the region between ξ_1 and ξ_2, then the conductivity will vary approximately linearly with frequency in the frequency range

$$\exp(-\xi_2) < \omega\tau_0 < \exp(-\xi_1)$$

The random variable ξ corresponds to the hopping distance for the tunnelling model, and to the barrier energy height (E/kT) for the classical hopping model. A combination of both these forms for ξ could be involved for electron hopping within a lattice defect where lattice relaxation, and hence the Franck–Condon principle, may be an important consideration.[12]

It is possible that the distinction between strictly localized charge transport and long-range transport may come from the detailed analysis of the temperature variation and low-frequency behaviour of the observed conductivity. For localized transport such as dipolar rotation, the rotational motion of polar molecules or a localized electronic or ionic hopping mechanism, the conductivity must become vanishingly small at frequencies lower than the reciprocal of the longest relaxation time. From Pollak's analysis,[1] such a process could result in an abrupt transition from an $\omega^{1 \cdot 0}$ to an $\omega^{2 \cdot 0}$ conductivity dependence as the frequency is reduced to low values, as has been found[13] for $(V_2O_5)_2$ $(P_2P_5)_2$ BaO, for example. For a delocalized conduction mechanism, however, the conductivity should approach a constant d.c. value (ω^0 dependence) at low frequencies corresponding to a limiting multiple tunnelling or classical hopping mechanism for the charge carriers. This behaviour has been found[14] to occur for As_2Te_3. It should be noted, however, that if the steady state d.c. and a.c. conductivities result from completely different conduction mechanisms, then as the d.c. conductivity component becomes sufficiently dominant, the conductivity will approach an ω^0 dependency as the frequency is lowered. Such a situation could mask any low-frequency ω^2 effects due to a localized conduction mechanism. It is also possible that the transition from a multiple hopping mechanism at low frequencies to essentially a near single hopping mechanism at high frequencies results in the observed conductivity $\sigma(\omega)$ not following the simple relationship given by equation (1). Precise measurements in the transition region between the ω^0 and $\omega^{1 \cdot 0}$ conductivity dependencies could obviously be important.

Interfacial electron tunnelling mechanisms[15, 16] can also result in a near $\omega^{1 \cdot 0}$ conductivity dependence, but the effect would be temperature independent. The exponential (E/kT) dependence found here for all-*trans* β-carotene would indicate that the localized energy states involved in the conduction process are far from the Fermi energy (see Rockstad (1971)[14]), possibly near to the conduction or valence band edge. Since oxygen is known to introduce electron accepting states into all-*trans* β-carotene, and since oxygen was observed to modify the value of s, then this would indicate that states near to the valence band edge are involved. A hopping mechanism involving states near the Fermi energy results in $\sigma^1(\omega) \propto$ temperature,[1] which appears to be the case for the glass form at 1 MHz.

Chapman *et al.*[8] in Figure 8 of their paper give the current–voltage characteristic obtained for their β-carotene crystals, showing that the current changes from an ohmic dependence to a square-law space-charge-limited dependence at a voltage of the order 100 volts. The ohmic free charge carrier concentration n can be derived from the expression[17]

$$V = 8qa^2n/9\varepsilon_r\varepsilon_0$$

Taking a value[18] for the crystal thickness a of 3×10^{-4} m, and an estimated value for ε_r of 3, then the derived carrier concentration is of the order 2×10^{17} m^{-3}. Using the 20 °C resistivity value[8] of 9×10^{14} ohm m, the derived charge carrier mobility has a value of $3\cdot5 \times 10^{-14}$ m^2/Vs. Such a low mobility value is typical of an activated hopping process between localized energy states. In this way, an analysis of the electronic conduction in β-carotene in terms of an activated hopping process should be of some relevance. However, instead of adopting one of the theories outlined above, we propose a simpler model which unlike those earlier theories,[1,3,11] allows a meaningful interpretation of s-values greater than 1·0.

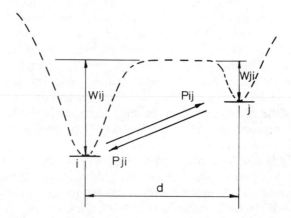

Figure 4   The potential energy diagram of the localized two-state element. P_{ij} and P_{ji} denote the charge carrier transition rates between the states i and j. The major contribution to the a.c. conductivity arises from such 'active' elements when $W_{ij} = W_{ji} = W$ in zero electric field

5 THEORY

The model involves an ensemble of elements, the characteristic features of which are shown in Figure 4. Each element when populated (i.e. active) contains a charge carrier q which can exist in either of two localized states i and j with a spatial separation d. Transitions of the carrier between the states require activation energies W_{ij} and W_{ji} respectively. By considering appropriate distributions of d and of the energies W_{ij}, W_{ji} charge carrier 'hopping' between, for example, localized states within or at the ends of single β-carotene molecules, or associated with defects within or at the boundaries of crystallites, can be represented.

The transition rate between states i and j at a temperature T is given by

$$P_{ij} = v \exp[-(f(d) + W_{ij}/kT)] = v \exp(-u_{ij}) \qquad (4)$$

where v is the product of a characteristic phonon frequency and the degree of carrier–phonon coupling, and the function $f(d)$ is determined by the wave-function overlap between the i and j states. It is convenient to use a generalized parameter u_{ij} as indicated above. The hopping of a carrier between i and j constitutes an elementary current and under an applied alternating field, which modulates the transition rates p_{ij}, p_{ji} about their equilibrium value, it is possible to determine an a.c. conductivity which is a function of the frequency ω. Details of the procedure have been given by Lewis.[19]

In practice the ensemble will be characterized by a distribution in W_{ij}, W_{ji} and in d, and the a.c. conductivity of the ensemble $\sigma(\omega)$ will result from the sum of the elementary contributions.[19] It can be shown, using an argument similar to that of Gevers,[20] that it is a reasonable approximation to assume that the major contribution comes from symmetrical elements for which $W_{ij} = W_{ji} = W$, or $u_{ij} = u_{ji} = u$. Then for $\omega \ll v(\sim 10^{11}\text{–}10^{13}\ \text{s}^{-1})$

$$\sigma^1(\omega) = \frac{\pi q^2 \bar{d}^2}{4kT} \omega n(u_m) \qquad (5)$$

where $\bar{d}$ is the mean spacing, $n(u)\,du$ is the number density of elements in the interval du and

$$u_m = \ln 2v - \ln \omega \qquad (6)$$

The important deduction is that the distribution function $n(u_m)$ is obtainable directly from measurement of $\sigma^1(\omega)$ over a range of frequencies. According to the results for β-carotene in Figure 1, the conductivity at $T = 293$ K obeys the law $\sigma^1(\omega) = A\omega^s$ over a wide range of ω, and thus using equations (5) and (6) we deduce that for this range

$$n(u) = \frac{A4kT}{\pi q^2 \bar{d}^2}(2v)^{s-1} \exp -(s-1)u \qquad (7)$$

Since $s > 1$ in Figure 1, the distribution in u at constant temperature T decreases with increasing u. At higher temperatures (Figure 2) it is seen that s decreases and can become less than unity so that $(s - 1)$ goes through zero to negative values. Unlike other interpretations of the ω^s dependence, there is no restriction on the range of s in the present treatment, and values greater or less than unity are acceptable, corresponding to a distribution decreasing or increasing with u, respectively. Furthermore, experiments at constant frequency but increasing temperature allow the distribution at increasingly large values of energy W to be investigated (assuming $f(d)$ does not change with T), as also do experiments at constant temperature but decreasing frequency. According to Figure 1 and for the case $f(d) \ll W/kT$, then from equation (6) with $v = 10^{12}$ s^{-1}, the range of energies W over which equation (7) applies at $T = 293$ K is 0·12 eV to 0·66 eV. For example, the 293 K measurement at 1 MHz 'spot-lights' those transitions occurring over potential barrier heights of 0·39 eV.

If s is not constant (for example the low-frequency tail of the characteristic for the all-*trans* material in oxygen, Figure 1), then defining a slope $s(\ln \omega)$ by

$$s(\ln \omega) = d(\ln \sigma^1(\omega)/d(\ln \omega)$$

gives from equation (5)

$$d[\ln n(u)]/du = 1 - s(\ln \omega)$$

i.e.

$$n(u) = C \exp\left[\int_{-\infty}^{\ln \omega} (s - 1)\,d(\ln \omega)\right]$$

where C is a constant. Thus in the case of all-*trans* β-carotene in oxygen (Figure 1) there will be a departure from the law given by equation (10) at frequencies less than 200 Hz (i.e. energies W greater than about 0·62 eV) at 293 K.

It is now necessary to consider the fact that the conductivity of the β-carotenes at constant ω follow equation (3), the activation energy ΔE decreasing with increasing frequency (Figure 3). Since when ω is constant u is also constant (see equation (6)), an activated process can arise from the factor $n(u_m)$ in equation (5) only if s in equation (7) is of the form $L/kT + M$, where L and M are appropriate constants. In fact it is possible to fit an expression of this form to the results of Figure 2 with reasonable accuracy and at constant ω, the resultant activation energy ΔE would be

$$2Lu = 2L \ln(2v/\omega)$$

This decreases with increasing frequency but the decrease is much less than that found experimentally, especially for the glass form (Figure 3). It is clear that the major contribution to ΔE most probably comes from the factor A and that it contains a Boltzmann factor of the form $\exp(-\Delta E^1/kT)$, which determines the probability that a given element is active. As the frequency of measurement is increased, the range of u investigated for a given temperature range must shift to lower values and it seems therefore that ΔE^1 correspondingly decreases. The exact form of the relationship between the change in ΔE^1 and the shift in the range u is complicated and work is in progress on this problem. However, our preliminary studies suggest that a solution of this problem will reveal the energy distribution profile for the total number of localized energy states present in the energy scheme of the various forms of β-carotene.

6 CONCLUSIONS

It has been found experimentally for the various forms of β-carotene here, that over a wide frequency range the a.c. conductivity varies with frequency as $A\omega^s$. The essential features of the frequency and temperature variation of the observed conductivities can be explained using a simple activated hopping model in which the value of s is not restricted.

The concept of a conduction process for organic molecular solids involving the activated hopping of charge carriers is not new.[21] However, we believe that an analysis of a.c. conduction, as outlined here, is a powerful technique for the investigation of the electronic properties of such biologically relevant materials as β-carotene.

In the theory used above we have used the approximation of a symmetrical double-well element. This is obviously a restrictive feature for many solids where asymmetrical double wells might be more representative of the true situation. In the more general treatment[19] it is found that the major contribution will come from elements in which

$$P_{ij} = P_{ji} = \omega$$

However, it is not unlikely that in many organic and biological systems the symmetrical double-well element will effectively represent the physical situation. For example, for a molecular solid such as β-carotene, the electron of a charge-transfer exciton may be able to hop between identical excited energy states in the neighbourhood, and within the range of the coulombic attraction, of the positive 'hole'. An electron hopping within or between the ends of a β-carotene molecule could also represent the symmetrical double-well situation. Electron transport between and within closely related cytochrome structures in the photosynthetic and respiratory electron transport chains could also involve near symmetrical well elements.

ACKNOWLEDGEMENTS

Some of the measurements were made with the assistance of Mr A. N. M. N. U. Chaudhury. Messrs K. Beardsell, I. Wynne and W. Taylor assisted in the design and construction of the various microwave cavities and conductivity cells.

352

REFERENCES

1. M. Pollak, *Phil. Mag.*, **23**, 519 (1971).
2. A. K. Jonscher, *J. Non-cryst. Solids*, **8–10**, 293 (1972).
3. M. Pollak and T. H. Geballe, *Phys. Rev.*, **122**, 1745 (1961).
4. R. Pethig, *3rd Int. Symp. Chemistry of the Organic Sold State*, Glasgow, 1972.
5. D. D. Eley and R. Pethig, *Disc. Faraday Soc.*, **51**, 164 (1971).
6. B. Rosenberg, *J. Chem. Phys.*, **31**, 238 (1959).
7. B. Rosenberg, *J. Chem. Phys.*, **34**, 812 (1961).
8. D. Chapman, R. J. Cherry and A. Morrison, *Proc. Roy. Soc.*, **A301**, 173 (1967).
9. E. B. Ivkin and B. T. Kolomiets, *J. Non-cryst. Solids*, **3**, 41 (1970).
10. L. I. Lakatos and M. Abkowitz, *Phys. Rev.*, **B3**, 1791 (1971).
11. G. E. Pike, *Phys. Rev.*, **B6**, 1571 (1972).
12. M. Pollak and G. E. Pike, *Phys. Rev. Letters.*, **28**, 1449 (1972).
13. H. Namikawa, *J. Ceram. Assoc. Japan*, **74**, 34 (1966); **77**, 46 (1969).
14. H. K. Rockstad, *Solid State Comms.*, **7**, 1507 (1969); **9**, 2233 (1971).
15. C. Cherki, R. Coelho and R. Nannoni, *Phys. Stat. Sol (a)*, **2**, 785 (1970).
16. H. J. Wintle, *J. Appl. Phys.*, **44**, 2514 (1973).
17. D. R. Lamb, *Electrical Conduction Mechanisms in Thin Insulating Films*, p. 25, Methuen 1967.
18. R. J. Cherry, private communication.
19. T. J. Lewis, *Electrical Properties of Organic Solids*, Scientific Papers of Wroclaw Technical University, No. 7, 145–161 (1974).
20. M. Gevers, *Philips Research Rpts.*, **1**, 298 (1945/6).
21. D. D. Eley, *J. Polymer Sci. C.*, **17**, 73 (1967).

C.12. Steady state electrical conduction in β-carotene and Rose-Bengal under high electric field conditions

D. K. Das Gupta and M. K. Barbarez

School of Electronic Engineering Science, University College of North Wales, Bangor, Gwynedd, U.K.

SUMMARY

Experiments were performed to investigate the conduction mechanisms in β-carotene, Rose-Bengal and a mixture of the two over the field range 4×10^5 to 3×10^6 Vm^{-1} and at temperatures ranging from 200 to 295 K. The conduction mechanism for β-carotene appears to satisfy a relationship of the form $J \propto V^2$ suggesting space-charge-limited conduction, and β-carotene–Rose-Bengal mixture behaves in a similar manner. Rose-Bengal, however, seems to follow the relationship $J \propto V^{1.25}$. The activation energy for β-carotene was found to be 0·44 eV at high field and 0·61 eV at low field.

1 INTRODUCTION

The problem of phototransients and electronic conduction in organic polymers has some relevance to problems of energy transfer in chloroplasts, and possibly visual excitation. Photoconductivity of carotenoid crystals has been studied in chlorophyll systems by Calvin and coworkers. Photocurrent can also be sensitized under optical illumination by Rose-Bengal in the presence of oxygen in colourless nerve fibres.

We have as yet, unfortunately, no comprehensive knowledge of electronic conduction mechanisms in organic materials. The quantum-mechanical approach determines the width of energy bands by calculating the overlap integral between the π-electron molecular wave function in adjacent molecules. This approach, however, gives rather narrow conduction bands and low mobility values for the carriers.

It has been suggested[1,2,3,4] from the evaluation of the band gap and the energy of the triplet state of the isolated molecule, measured from the ground state,

353

that the molecular triplet states contributes to conduction. This assumption, however, is subject to criticism, and Fox[1] has proposed a trapping model for electronic conduction in organic semiconductors. In such studies considerable importance has always been given to the preparation and purity of the specimen, although little is known of the relative merits of the different techniques.

In this work samples were in the form of pellets of compressed crystalline powder. Workers such as Rosenberg[2,3] have used this technique, whereas Chapman et al.[5] used single crystals of β-carotene. The electrical measurements included a determination of activation energies for the three materials investigated, and also calculations of mobility and free carrier density for the β-carotene assuming a value for the relative permittivity ε_r.

2 EXPERIMENTAL

The β-carotene used was that supplied by Koch-Light Laboratories as pure all-*trans* β-carotene (1,600,000 I.V 3g, m.p. 183°C, >95%). The crystalline powder was kept in a refrigerator at 0 °C to minimize decomposition, the crystal containers having been filled with nitrogen and sealed.

The powder was compressed under 1.55×10^8 kg m^{-2} to form disks of 13 mm diameter and 1·0 mm thickness. Gold contact electrodes of area 0·95 cm^2 were evaporated onto the samples. Since it was not necessary for the electrodes to transmit light they could be made quite thick to ensure good contact. Under vacuum the samples were maintained at a temperature of 0 °C for a day under an applied field of 1×10^6 Vm^{-1} as a conditioning process, which has been shown to produce good repeatability of results. Measurements were made at a chamber pressure of 10^{-6} torr, the sample current being measured with a Keithley 602 electrometer.

Similar dark conductivity measurements were conducted with samples of Rose-Bengal (tetra-chloro-tetra-iodo-fluorescein) and samples prepared from a mixture of β-carotene and 25% by weight Rose-Bengal. In the case of pure Rose-Bengal sample breakdown occurred at fields much greater than 10^6 Vm^{-1}, which was somewhat lower than for β-carotene.

3 RESULTS

The d.c. conductivity of all the samples was very sensitive to temperature variation. Typical values at room temperature under an applied field of 10^6 Vm^{-1} were $3 \times 10^{-11} \Omega^{-1}$m^{-1} for β-carotene and $3 \times 10^{-10} \Omega^{-1}$m^{-1} for Rose-Bengal.

From the plots (Figures 3, 4 and 5) of $\log_{10} J$ vs. $1/T$ the values of the activation energy E were determined and are listed in Table 1, assuming that the conductivity obeys the general relationship,

$$\sigma = \sigma_0 \exp\frac{-E}{2kT}$$

Table 1   Field-assisted activation energy

Sample	Field (Vm^{-1})	Activation energy E (eV)
β-Carotene	3×10^6	0·44 eV
	8×10^5	0·61 eV
Rose-Bengal (R.B.)	8×10^5	0·42 eV
β-Carotene + 25 % R.B.	3×10^6	0·37 eV
	8×10^5	0·52 eV

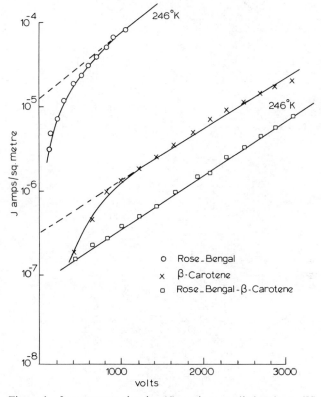

Figure 1   Log current density (J) against applied voltage (V)

where J is the current density, k is the Boltzmann constant and T is the temperature.

Above 10^6 Vm^{-1} the conduction of the β-carotene appears to be space-charge-limited, i.e.

$$J = \tfrac{9}{8}\varepsilon_0\varepsilon_r\mu\frac{V^2}{d^3}$$

356

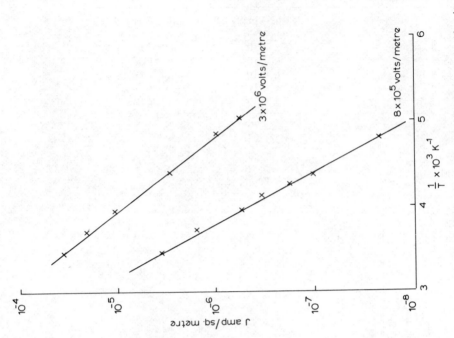

Figure 3 β-Carotene. Arrhenius plot of log current density (J) against reciprocal temperature (1/T)

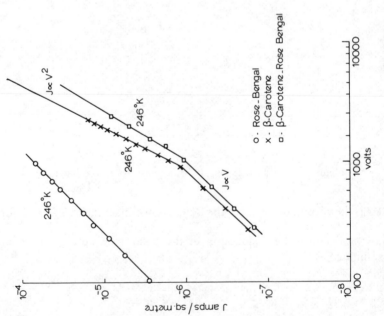

Figure 2 Log current density (J) against log applied voltage (V)

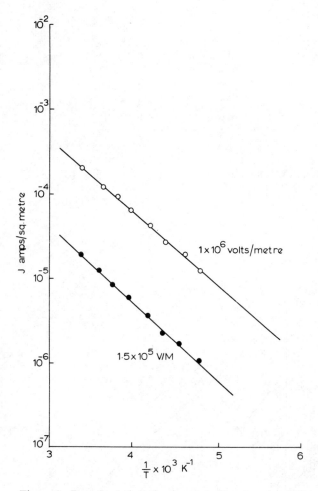

Figure 4   Rose-Bengal. Arrhenius plot of log J against $1/T$

where μ is the mobility, V is the applied voltage and d is the inter-electrode spacing.

From the plot of $\log_{10} J$ vs. V (Figures 1 and 2) in which the high-field linear region is extrapolated to zero voltage the mobility μ is calculated, assuming a value 3·0 for the relative permittivity ε_r. We have that

$$\log_{10} J_0 = \log_{10}\left(\frac{9}{8}\frac{\varepsilon_0\varepsilon_r}{d^3}\right) + \log_{10}\mu + 2\log_{10} V$$

from which the mobility of the β-carotene is found to be $\mu = 6\cdot6 \times 10^{-8}\,\mathrm{m^2/Vs}$. Also from Figures 1 and 2 we have that at the break point where $V = 1000\,\mathrm{V}$

$$6 = \frac{Jd}{V} = ne\mu$$

358

thus

$$\tfrac{9}{8}\varepsilon_0\varepsilon_r\mu\frac{V}{d^2} = ne\mu$$

and

$$n = \tfrac{9}{8}\varepsilon_0\varepsilon_r\frac{V}{ed^2}$$

For β-carotene this analysis yields a figure of $n = 5 \times 10^{17}\, m^{-3}$.

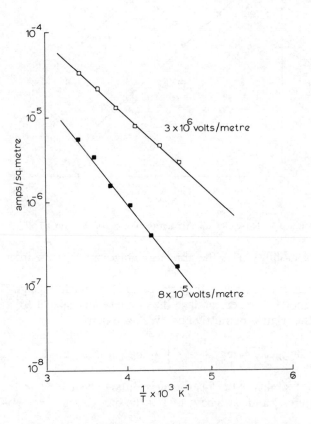

Figure 5 β-Carotene + 25% by weight Rose-Bengal.
Arrhenius plot of log J against $1/T$

4 DISCUSSION

The linearity of the log J–log V characteristics suggests space-charge-limited conduction, i.e. $J \propto V^a$ where a for β-carotene was found to be 2·1 and 1·6 for high and low fields, respectively. This is in agreement with Chapman et al.,[5] but Rosenberg[2,3] found a to be 1·5, so that there seems to be agreement with Chapman et al.[5] at high fields and with Rosenberg[2,3] at low fields.

The measured activation energy of β-carotene was 0·44 eV at high fields and 0·61 eV at the lower range. These figures are in disagreement with the results of all other workers who have performed this measurement. Rosenberg[2] obtained values of 3·1 eV and 1·45 eV for β-carotene powder sandwiched between conductive glass and heated slowly through. The 1·45 eV value, according to Rosenberg,[3] is due to much increased surface currents brought about by the presence of adsorbed oxygen. Chapman et al.[5] using single crystals of all-*trans* β-carotene found the activation energy to be 1·32 eV which appears to agree with Rosenberg's[3] figure for the surface conduction activation energy. The fact that Chapman et al. used a guard ring arrangement, however, makes it very unlikely that they were measuring the same thing. The activation energy determined by the authors suggests the possibility of a new trapping level situated at approximately 0·6 eV below the conduction band edge. The low-field figure has been taken here because the 0·41 eV will correspond with the field-lowered activation energy, whereas the 0·6 eV is likely to give a much closer approximation to the actual trap depth.

The conductivity of the Rose-Bengal samples was found to be at least an order of magnitude greater than that for β-carotene. The low-field activation energy was correspondingly smaller, being 0·42 eV. The activation energy for the β-carotene: 25% Rose-Bengal mixture was 0·37 eV for the high-field case and 0·52 eV for low fields. Rose-Bengal, which is a typical organic dye bound together with β-carotene in abundance in nature, was used here as an impurity, the effect of which was to increase the conductivity of the β-carotene greatly. The type of conduction mechanism, however, seems to have remained unchanged.

5 CONCLUSION

The results suggest a conduction mechanism following the law $J \propto E^a$, where a for β-carotene was found to be approximately 2, indicating space-charge-limited current.

The activation energy was found to be about 0·6 eV for the β-carotene, which can be accounted for only by the suggestion of the existence of a trapping level at this depth.

The modification of the conductivity of β-carotene by Rose-Bengal is perhaps the most interesting result of this work. It is hoped that this might provide a basis for greater understanding of mechanisms such as photosynthesis found in nature.

ACKNOWLEDGEMENTS

The authors would like to thank Mr E. Hall for carrying out the experimental work.

REFERENCES

1. D. Fox, in *Electrical Conductivity in Organic Solids*, p. 239 (eds. H. Kallman and M. Silver), Interscience, New York, 1961.
2. B. Rosenberg, *J. Chem. Phys.*, **31**, 238 (1959).
3. B. Rosenberg, *J. Chem. Phys.*, **34**, 812 (1961).
4. D. C. Northrop and O. Simpson, *Proc. Roy. Soc.*, **A234**, 124 (1956).
5. D. Chapman, R. J. Cherry and A. Morrison, *Proc. Roy. Soc.*, **A301**, 173 (1967).

Session D
Excited states of proteins and amino-acids

D. What we have learnt about proteins from the study of their photoexcited states

Gregorio Weber

Roger Adams Laboratory, University of Illinois, Urbana, Illinois 61801, U.S.A.

1 INTRODUCTION

The investigations of the structure of protein crystals by X-ray diffraction carried out in the last ten years[1] have demonstrated the existence of specific protein structures in which the *average* positions of most amino-acid residues are clearly defined. In such average positions the surroundings of any given amino-acid are found to be quite heterogeneous, and the constancy and heterogeneity of environment confer properties to the excited states of the amino-acids or of the protein prosthetic groups, which can hardly be simulated by inclusion of the chromophores in a pure solvent, a solvent mixture or even in an ordered crystal structure.† These features, of constancy and heterogeneity of environment, should permit, in the long run, an analysis of the spectroscopic properties of the chromophores far more detailed and penetrating than that emerging today from the study of solutions, in which the environment has, to a smaller or larger extent, a statistical character, and the absorption and emission properties result often from the superposition of several significant configurations. It is likely that in the future the theoretical chemist will be able to draw upon this uniqueness of environment to analyse more fully the properties of particular excited states or even general properties of excited states. Up to the present, however, the work carried out on the optical spectroscopy of proteins has been mostly phenomenological, and more often than not it has been directed to reveal the presence of chromophores, the character of their environment and the changes upon addition of sundry reagents that interact in many diverse fashions with the proteins. It is therefore appropriate at this time to ask what this rather large collection of observations has taught us about the general properties of proteins that could not have been obtained by the

---

† As a characteristic example one can mention the spectrum of the flavin prosthetic group in D-amino-acid oxidase, which has been extensively studied.[2-5]

application of other methods. On analysing the available experimental material it becomes clear that the unique contribution from observations of the properties of the excited states refer to those processes in the protein that occur in the nanosecond or subnanosecond time scale. Fluorescence observations offer only marginal possibilities when it comes to space resolution, certainly incomparably less than proton magnetic resonance, but this insufficiency arises, at least in part, from the property that is the origin of their greatest usefulness: the existence of a very brief, but accurately measurable, time interval between the absorption and emission of light. During the fluorescence lifetime, which for the intrinsic protein chromophores is $\sim$2–6 ns, molecular motions of considerable magnitude can take place. A simple application of Einstein's diffusion relation, giving the mean quadratic displacement as a function of time, indicates that in 5 ns a structure with the diffusion coefficient of a small molecule (10^{-5} cm^2 s^{-1}) will move on average a distance of 20 Å if placed in a medium of the viscosity of water. The motions of the structural elements surrounding the protein fluorophores are evidently not free to move, and if contact with the fluorophore is at all possible, it is determined by an energy of activation E, giving the contact rate the typical form $A \exp(-E/RT)$. If A is taken to be 10^{13} s^{-1}, then a contact rate equal or greater than the typical radiation rate (2×10^8 s^{-1}) implies an energy of activation which can reach 7 kcal. This is not a negligible quantity, and proteins would require to be held uniformly together by remarkably strong bonds for their structures to be considered invariant during an interval of a few nanoseconds. We shall have occasion to see that this is not the case, and that in consequence anybody trying to obtain hard structural information by the unqualified use of fluorescence experiments *is indeed on the wrong track*. In fact a careful examination of the literature results that purport to extract structural information from fluorescence data shows that the conclusions have to be qualified by the possibility of time-dependent effects in every instance, and that in those cases in which the actual dynamics has been worked out this has turned out to be more interesting and important than the originally sought structural counterpart.

2 RATE OF THERMALIZATION OF EXCITED STATES IN PROTEINS

Amongst the simplest and most direct examples of dynamic properties revealed by fluorescence stands the observation that the tryptophan emission in proteins has no more dependence upon the excitation wavelength[6] than that of tryptophan and its simple derivatives in solution.[7] In a protein with tryptophan emission at 340 nm, excitation at 250 nm results in dissipation of an energy close to 30 kcal, and the independence of the tryptophan emission from the exciting wavelength indicates that the thermalization of the excited state is virtually complete[8] in a time short compared to the fluorescence lifetime, and therefore in the subnanosecond time range. This result, which indicates that the excess vibrational energy is just as rapidly dissipated in a protein molecule as in a solvent, would be indeed expected from phenomena as varied as the thermal conductivities of liquids and solids, the velocity of sound in liquids and the vibrational frequencies of the atoms that form the protein.† Nevertheless it has been hypothesized at various times, but particularly very recently,[9,10] that vibrational energy quanta could be stored by proteins for periods sufficiently long to permit their useful utilization in metabolism. However unlikely this hypothesis, one would be hard put to find *direct* experimental evidence against it, apart from the quite decisive one of the rapid thermalization of the photoexcited states of proteins.

3 RELAXATION PHENOMENA OF THE FLUOROPHORE SURROUNDINGS

Another aspect of protein investigations where the implications of the results of the fluorescence observations were previously not fully realized, but where present theory and practice emphasizes them, relates to the determination of the character of the environment of the fluorophores by the emission spectrum. In the original work of Oochika[11] and of Lippert[12] in which the polarity of the environment was related to the decrease in energy of the excited state, and therefore to the displacement of the emission spectrum to the red, it was implicitly assumed that the reorientation of the solvent molecules around the excited fluorophore occurred in a time short compared to the fluorescence lifetime, and therefore that the emission state practically represented an equilibrium state of the lowest excited singlet. Consequently, the equations derived by these authors to describe these phenomena do not contain either the rate of solvent reorientation or the rate of emission, but only quantities denoting stationary properties like the difference in dipole moment between the ground and excited states and the dielectric constant and refractive index of the environment. It was pointed out by Bakshiev[13] that such equations cannot hold true unless the solvent relaxation rate (k_s) is much faster than the rate of radiation (k_r). In practice, in the case of many fluid solvents, the solvent relaxation rate can be slowed down by simply decreasing the temperature to the point that it becomes comparable to the rate of radiation or even longer than this. The emission spectrum shows a maximal shift to the red when $k_s \gg k_r$, moves progressively to the blue in the region where k_s and k_r become comparable and reaches a short-wave limit when $k_s \ll k_r$. These effects may be shown just as well by an increase in k_r, through changes in viscosity and temperature, as by the shortening of the fluorescence lifetime brought about by addition of a quencher molecule. In the latter case an estimate of the rate of solvent relaxation

---

† Quite recently Laubereau et al.[39] have been able to measure directly the decay of vibrational excitations in liquids, and in complete agreement with expectations they found decay times of a few picoseconds.

in fluid solvents is possible, and Weber and Lakowicz[14] have shown by the use of oxygen quenching that the relaxation of ethanol molecules around fluorophores is accomplished within 30–60 ps.

Particularly interesting have been the attempts at application of these ideas to the case of proteins. Following the original work of Oochika and of Lippert, and the observation that indole belongs to the category of molecules where a change in dipole moment in the fluorescent state is demonstrable, Weber, Teale and others[15–17] emphasized the existence of several environments for the tryptophans in protein molecules, and concluded that these environmental differences explain the wide range of emission maxima observed in proteins, going from 315 nm in azurin[17] to 345 nm in pepsin.[6] Even in a single protein the existence of more than one tryptophan environment may be shown by fluorescence techniques. These observations appear to furnish a good example of the detection of structural differences by the observation of excited state properties, but difficulties appear in trying to characterize more closely the structural differences in the environment. One cannot be sure to what extent the dynamic effects pointed out by Bakshiev come into play in any given case, permitting or restricting the environmental relaxation. The structures surrounding the tryptophan in the native protein can differ both in the polarity and in the rate of reorientation following the change in dipole moment of the tryptophan, but the two factors appear difficult to separate. Lakowicz and Weber,[18] in a study of oxygen quenching of the protein fluorescence, regularly observed a blue shift of the spectrum upon quenching. Before pronouncing this effect as being due to a less complete relaxation of the surroundings of the fluorophore, the question of the heterogeneous character of the emission must be considered. If the fluorophores emitting at longer wavelengths are those of longer lifetime they will be preferentially quenched. The existence of such an effect is not in doubt because in Human serum albumin, in which tyrosine contributes to the total fluorescent emission almost as much as the single tryptophan,[16] the displacement of the spectrum to the blue is the most marked of all the proteins examined, and tyrosine emission is both shorter lived and placed at shorter wavelengths than tryptophan emission. In proteins without appreciable tyrosine emission a similar relation appears to hold among the tryptophan residues, so that the observations cannot be used to deduce unequivocally the existence of relaxation in the fluorophore environment during the excited state. I have discussed these effects at some length because we encounter here the most common, the most difficult and the most important problem arising in the spectroscopy of biological systems: how to detect, characterize and give due weight to the members of a heterogeneous population. Unless this problem is solved no reliable correlation of the spectroscopic effects to their physical causes appears possible. Yet no other problem is more consistently ignored or assumed to be unimportant. In fact the most common practice continues to be that of attributing whatever spectroscopic changes are recorded—and these are often *very* small—to changes in the most prevalent species present.

4 DETECTION OF RELAXATION IN REAL TIME

The measurements of the fluorescence spectrum under stationary conditions are capable of revealing the existence of relaxation phenomena and even of permitting the calculation of the rate of solvent relaxation, but they are less powerful and informative than the method of direct time-resolved spectroscopy introduced by Ware and coworkers[19] and recently applied to observations of protein–fluorophore complexes by Brand and Gohlke.[20] By the use of this method the progressive changes in the fluorescence spectrum during the lifetime of the excitation may be followed, and as practised at present the emission spectrum followed some 1·5 ns after excitation represents the useful limit of the method. Brand and Gohlke have observed a conspicuous time-dependent relaxation of the fluorescence of toluidino naphthalene sulphonate (TNS) bound to serum albumin. Although the method of time-resolved spectroscopy, as practised today, permits the recognition of the protein relaxation, it does not seem suitable for its quantitative characterization over subnanosecond intervals of time. For this purpose, we have employed† successfully the methods of phase fluorometry. If the fluorescence lifetime is measured across the fluorescence band it is seen to be shorter at the blue edge and longer at the red edge, when compared with the broad-band value for the whole spectrum. The difference in lifetime between the blue and red edges is obtained directly by differential phase fluorometry. The fluorescence is excited with light modulated at mega-cycle frequency and the phase difference Δ between the currents from two phototubes observing simultaneously the red and blue edges of the fluorescence through suitable optical filters is measured. A minimum relaxation time of some 50 ps can be detected with our present arrangement. An analysis of the relaxation phenomenon as involving transitions between two levels, Franck–Condon and relaxed, shows that $\tan \Delta = \omega/(\lambda + h)$. ω is the circular frequency of excitation, λ the rate of emission and h the rate of *antirelaxation*, that is the rate of transitions from the relaxed to the Franck–Condon level. Thus Δ measures directly the rate at which the structure surrounding the fluorophore returns to the conditions present in the ground state. The sensitivity and precision of the measurements is sufficient for a study of the temperature dependence of the antirelaxation rate of the ANS–albumin system (Figure 1). From the plot of $\log h$ against $1/T$ an energy of activation of antirelaxation of 3·4 kcal is obtained. These observations have value beyond the particular case, since Lakowicz and Weber[21] have shown that the rate of quenching of ANS adsorbed on serum albumin is $3 \times 10^8 \ M^{-1} s^{-1}$, five times smaller than the rate of quenching by oxygen of the protein's own tryptophan residues[19] ($1\cdot5 \times 10^9 \ M^{-1} s^{-1}$). Since ANS is better protected by the protein structure from the diffusion of oxygen to its immediate vicinity, the energy of interaction of ANS with its surroundings must be greater than that of the tryptophan residues

† D. Kolb, G. W. Mitchell and G. Weber, to be published.

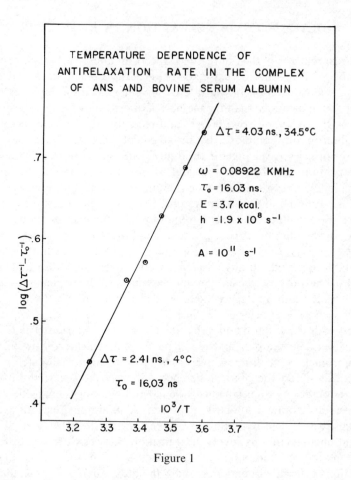

Figure 1

with its own surroundings. Consequently we have every reason to expect relaxation of the surroundings of tryptophan during the fluorescence lifetime, just as in the case of ANS. It is of interest to recognize that the presence of heterogeneity can easily lead to a false interpretation of the results of the experiments whether these are carried out by static or time-dependent methods. In the latter case however, the existence of apparent relaxation as a result of heterogeneity of emissions and lifetimes can be revealed, because the supposed time-dependent effects will persist even when the rate of radiation is made very fast in comparison with the rate of relaxation by lowering the temperature of the sample. The experiments that I have described are still preliminary but we can expect that in the next few years the improvement in the experimental techniques and in the methods of analysis will make it possible to detect and characterize in real time the processes of relaxation of the protein structure that follow the creation of excited states of tryptophan or prosthetic fluorophores.

5 THE FREEDOM OF MOTION OF THE TRYPTOPHAN RESIDUES REVEALED BY FLUORESCENCE POLARIZATION

The fluorescence lifetime of the tryptophan residues is too short for the polarization of the radiation to be affected appreciably by the tumbling of the whole protein molecule. According to Perrin's depolarization law[22] if the fluorescence lifetime is 4 ns and the rotational relaxation time of the protein is 100 ns only marginal depolarization by the Brownian rotations will be observed. Substantial depolarization can only be expected if the tryptophan residues themselves have freedom of rotation and this is seen when the protein is denatured by guanidine or urea[23] or by changes in pH.[24] The results obtained by fluorescence polarization give for these cases rotational relaxation times of the order of 1 ns corresponding therefore to the expected free rotational motion of the tryptophan residues. Similar quantitative conclusions have been reached recently by the use of ^{13}C NMR spectroscopy,[25] but the actual computations are not without ambiguities requiring in some cases the use of the values derived from fluorescence polarization for their resolution. The low polarization of the fluorescence observed in native proteins on excitation in the region of 270–280 nm is due partially to electronic energy transfer among residues (tyrosine–tryptophan as well as tryptophan–tryptophan) with a contribution usually difficult to evaluate, from rotational motion of the tryptophans. Within the uncertainty introduced by the presence of energy transfer it appears possible to estimate the separate motions of individual tryptophan residues. Thus, Teichberg and Shinitzky[26] have attempted to isolate the fluoresence polarization spectrum of tryptophan 108 in lysozyme from the differences in fluorescence intensity and polarization of native and iodine-oxidized lysozyme. In the latter protein the fluorescence of residue 108 disappears following its conversion to an oxindole derivative. These authors arrive in this way to a rotational relaxation of 4 ns for tryptophan 108 in the native protein, which increases to 7 ns when tetra-acetyl glucosamine is bound to the enzyme.

6 QUENCHING OF PROTEIN FLUORESCENCE BY OXYGEN: NANOSECOND FLUCTUATIONS OF PROTEIN STRUCTURE

I will deal in greater detail, because they are a recent development, with observations that indicate indirectly, but I believe unequivocally, the existence of fluctuations in protein structure taking place in the nanosecond range. The observations of X-ray diffraction, and in fact all ground-state spectroscopy, reveal only the average conditions at the time of the absorption or scattering, in accordance with the Franck–Condon principle. If there are local distortions of the structure, and an energy of activation E is required to produce them, the fraction of structures differing from the normal will be $\exp(-E/RT)$ and an energy E of only 2 kcal will reduce their number to 0·03 of the total, making their observation impossible, or at least uncertain by most methods. On the other hand if the distorted state appreciably modifies the fluorescence emitted by the structure, the contribution of such distorted structures will be expected, in the case mentioned, to be higher than 90% of the total, because there will be 90% chances that the distortion will happen within the fluorescence lifetime of a few nanoseconds. Although in principle observations of the ground state can give similar information by recording only the fluctuations of the average, as done recently by Magde and coworkers[27] for the fluorescence, and such methods are promising, for the moment, and perhaps for all time, they are not comparable in scope or precision to the direct measurement of excited state properties. The method that we have used to detect structural fluctuations in proteins relies upon the quenching of protein fluorescence by oxygen. A decrease in emission intensity to $\frac{1}{2}$ to $\frac{1}{6}$ of the unquenched value can be obtained by equilibrium oxygen pressures of 100 atmospheres, which result in an oxygen concentration in water of 0·13 M. Using this range of oxygen pressures Lakowicz[18] examined the behaviour of some dozen proteins and the results have been sufficiently uniform to permit general conclusions of interest. The X-ray analysis of protein crystals[28–30] shows the proteins to be compact masses hardly penetrated by solvent. In some cases the tryptophan residues occupy positions in the interior of the protein, seemingly away from the perturbing effect of other solutes. The emission maximum of these proteins is at shorter wavelengths (325–335 nm) than in those proteins like lysozyme (maximum at 341 nm) in which the X-ray structures show tryptophans in direct contact with the solvent. In all the cases studied by Lakowicz, which included examples of both types of proteins as well as intermediate cases, the quenching constants K ranged from 4·1 to 28·6 M^{-1}. The unquenched fluoresecnce lifetimes, τ_0 ranged from 1·5 ns to 6·2 ns and the ratio K/τ_0 from 2×10^9 $M^{-1} s^{-1}$ to $5·2 \times 10^9$ $M^{-1} s^{-1}$. Thus the rate of quenching varied by a factor of less than three and was only $\frac{1}{2}$ to $\frac{1}{5}$ of the diffusion-controlled rate of 1×10^{10} $M^{-1} s^{-1}$ observed for the quenching of the fluorescence of tryptophan and other low molecular weight fluorophores in water.[21] Quenching of the fluorescence of copolymers of tryptophan with glutamic acid or lysine by oxygen gives results identical to the free tryptophan case, showing that the high charge density of these polymers has no effect upon oxygen contact and quenching. These results stand in contrast to the observations of Lehrer[31] on the quenching of the fluorescence of the same polymers by iodide: The glutamic acid–tryptophan copolymer is quenched some 50 times less effectively than the oppositely charged lysine–tryptophan copolymer, in which case the apparent quenching constant (2×10^{10} $M^{-1} s^{-1}$) exceeds the diffusion controlled value. In contrast, iodide quenching of the proteins[31,18] is 10–30 times less efficient than oxygen quenching. In attempting to explain the remarkable efficiency of oxygen quenching we are again faced with the usual problem of separating static and time-dependent—or dynamic—effects. The relative fluorescence efficiency, F/F_0 can be related to the molar concentration of quencher, [Q] by a modified Stern–Volmer equation.[32,33]

$$\frac{F_0}{F} = (1 + K_e[Q])(1 + k^*\tau_0[Q]) \tag{1}$$

K_e is the equilibrium constant of a supposed 'dark complex', that is one pre-dating the excitation, between fluorophore and quencher, k^* is the rate of quenching encounters in the excited state and τ_0 the unquenched fluorescence lifetime. The first factor in equation (1) represents static quenching, independent of the processes starting at excitation and therefore of the fluorescence lifetime, while the second describes the time-dependent process of quencher-excited fluorophore encounters. Evidently the Stern–Volmer linear relation between F_0/F and [Q] is followed when either of the two processes is the only operative one, and departure from linearity requires the presence of both processes, but does not indicate their relative importance. This can be established unequi-vocally by independent measurements of the fluorescence lifetime upon quench-ing. If the 'dark complexes' do not contribute appreciably to the emission, which is by all accounts the common case,[33]

$$\frac{\tau_0}{\tau} = 1 + k^*\tau_0[Q] \tag{2}$$

where τ is the fluorescence lifetime in the presence of the quencher. Thus, once again the determination of the part played by the time-dependent processes is indispensable to refer the fluorescence observations to their physical cause with certainty. In the quenching of simple fluorophores by oxygen a parallel study of F_0/F and τ_0/τ shows that the quenching in water solutions is all dynamic, and that in organic solvents a very small static component may sometimes be observed.† In proteins the interpretation of the results is further complicated by the heterogeneity of the emission, since most proteins studied had more than one tryptophan per molecule and the individual tryptophan residues may be expected to differ both in their fluorescence lifetimes and quenching rate constants. In these cases the relation between the observed lifetime, $\bar{\tau}$ and the relative yield $\bar{q} = (F/F_0)$, is given by[18]

$$\bar{\tau} = \frac{1}{\bar{q}}\sum_i f_i q_i^2 \tau_{0i} \tag{3}$$

where τ_{0i} and q_i are respectively the unquenched lifetime and the quenched yield of the ith residue, contributing a fraction f_i to the unquenched fluorescence intensity. A homogeneous population quenched exclusively by the dynamic process follows the relation $\tau = \tau_0\bar{q}$, or $F_0/F = \tau_0/\tau$. For a heterogeneous population equation (3) predicts[18] that identity of quenching rates, k_i^* and heterogeneity of fluorescence lifetimes τ_{0i} should give rise to $\tau_0/\tau > F_0/F$, while homogeneity of unquenched lifetimes and divergence in quenching rates results in the opposite effect: $\tau_0/\tau < F_0/F$. Supposing that the heterogeneities of the quenching rates and lifetimes of the tryptophans within any one protein are at most as large as the difference in the average quantities among the different proteins, that is that in neither case they exceed a ratio of 3:1, we can expect only minor deviations in either direction from the relation $F/F_0 = \tau/\tau_0$, which

† In dodecane solutions of perylene quenched by oxygen, for example[21] $k^*\tau_0/K_e = 112/2\cdot6$.

applies to homogeneous dynamic quenching. Table 1 shows that this is precisely the case with the proteins studied. We find in it representatives of the three possible classes. It appears that, in agreement with expectations, heterogeneity of quenching constants is somewhat more important than heterogeneity of lifetimes, although neither are large enough to produce deviations from the Stern–Volmer plot, as is observed, for example in the quenching of pepsin fluorescence by iodide.[31] Rapid exchange of energy among the tryptophan residues could certainly falsify the results by making them appear more accessible to oxygen than they really are, but consideration of the cases of azurin and human serum albumin that have only one tryptophan, and the further observation that only minor differences are observed on excitation at the red edge of the absorption (305 nm) where transfer of electronic energy is known to fail,[34,35] demonstrates that this factor cannot have other than marginal significance. The uniform and very effective quenching of the fluorescence of proteins by oxygen could be explained by supposing that these residues become exposed to the solvent environment with sufficient frequency, but the observation that iodide is less effective a quencher of protein fluorescence by over an order of magnitude, although both quenchers have comparable efficiencies in the quenching of indole fluorescence,[18] shows that this explanation is insufficient. Oxygen must be able to penetrate the protein structure, helped by its liposolubility, smaller size and absence of electric charge, and to diffuse within it, so as to reach the quenchable residues during the 2–6 nanoseconds of the fluorescence lifetime. An estimation of the likelihood of this process, based on very general energetic considerations, would go as follows: Considering the protein structure as a homogeneous medium, the rate of quenching of interior tryptophan residues of 3×10^9 M^{-1} s^{-1} observed in the 'blue emitters' aldolase and azurin would be expected to result if the energy of activation for the diffusion of oxygen is only 0·8 kcal larger than that for diffusion in water (3·1 kcal). Although this figure of 0·8 kcal carries some uncertainty arising from the assumption that the equilibrium concentration of oxygen in the protein

Table 1   From data of Lakowicz and Weber[18]

Protein	Wavelength of maximum emission (nm)	$K(\bar{\tau})$ (M^{-1})	$K(\bar{q})$ (M^{-1})
Chymotrypsin	332	5·3	4.1
Carbonic anhydrase	341	14	11·4
Aldolase	328	6·1	6·1
Trypsin	335	10·2	10·0
Carboxypeptidase	330	4·7	6·4
Azurin	315	7·0	10·0
Bovine serum albumin	342	9·0	15·2
Human serum albumin	342	6·5	14·7

$K(\tau)$: Stern–Volmer constant from the plot of τ_0/τ against $[O_2]$.
$K(\bar{q})$: Stern–Volmer constant from the plot of F_0/F against $[O_2]$.

interior is that of the solution as a whole, this uncertainty could at worst rise the value by an additional 0.75 kcal, giving for the energy of activation of oxygen diffusion inside the protein a value of 4–4.5 kcal. Such a figure is far from surprising: only the internal hydrogen bonds linking portions of the backbone would be larger, while the usual dipole interactions and dispersion forces among the non-polar residues would certainly be smaller, probably around 3 kcal. If we assume that the rate of energy exchange among neighbouring amino-acid residues is of the order of 10^{13} s^{-1}, the rate of breaking of residue interactions necessary for the diffusion would be $\sim 10^{10}$ s^{-1} explaining quite well the observed phenomena. These observations deserve to be discussed in detail because they put a definite limit to the use of static models in explaining the properties of protein molecules. They demonstrate that structural fluctuations must be taken into consideration, and that in certain cases like the present one, they are appreciable, even in the nanosecond time-range. It does not seem that at present such rapid structural changes can be demonstrated by other methods, although structural fluctuations extending into the microsecond time range have to be postulated to explain the results of isotopic hydrogen exchange in proteins.[36]

7 ROTATIONAL DIFFUSION AND ENERGY TRANSFER IN PROTEIN–FLUOROPHORE CONJUGATES

In this survey of results I have left out of consideration two important areas: the use of protein fluorophore conjugates or adsorbates to study the rotational diffusion properties, and the use of energy transfer as an absolute method for the determination of distances between chromophores. These two aspects are too extensive to be reviewed here in any detail, and no doubt they will receive much attention in this conference. I will nevertheless point out that they both need to be brought into the main scope of this lecture: as shown by the oxygen quenching experiments, local fluctuations of structure in the nanosecond time-scale are widespread in proteins and thus, for the purpose of interpreting the data of fluorescence polarization, proteins cannot be accurately represented by a rigid hydrodynamic ellipsoid, the precise determination of which awaits only the improvement of present techniques. It appears necessary to consider the motions of the fluorophore and its surroundings as well as those of the protein as a whole. Such restricted flexible models, to which Wahl[37] first called attention, appear to require special refinements of theory and observation. Not very different remarks can be applied to the study of energy transfer, since precise, invariant orientations of the chromophores with respect to each other during the fluorescence lifetime cannot be accepted as a suitable hypothesis. The situation here is quite complicated,[38] but I believe that its study will finally give one more example of the basic philosophy that I have tried to illustrate in this lecture: the full exploitation of the properties of the excited states in biology always requires detailed consideration of time-dependent processes and the more interesting results from such studies are those that reveal obscure or unexpected dynamic properties of the biological systems.

REFERENCES

1. R. E. Dickerson, *Ann. Rev. Biochem.*, **41**, 815 (1972).
2. V. Massey and H. Ganther, *Biochemistry*, **4**, 1161 (1965).
3. T. Shiga, M. Layani and P. Douzou, in *Flavins and Flavoproteins*, p. 140 (ed. K. Yagi), University of Tokyo-Park Press, 1968.
4. K. Yagi, O. Takayuki, M. Naoi and A. Kotaki, in *Flavins and Flavoproteins*, p. 237 (ed. K. Yagi), University of Tokyo-Park Press, 1968.
5. C. Veeger, in *Flavins and Flavoproteins*, p. 252 (ed. K. Yagi), University of Tokyo-Park Press, 1968.
6. F. J. W. Teale, *Biochem. J.*, **76**, 381 (1960).
7. G. Weber and F. J. W. Teale, *Biochem. J.*, **65**, 476 (1957).
8. N. A. Borisevich, 'Intramolecular redistribution of vibrational energy in an excited state,' in *Elementary Photo-processes in Molecules*, p. 39 (ed. B. S. Neporent), Eng. Trans. Consultants Bureau, New York, 1968.
9. C. W. F. McClare, *J. Theoret. Biol.*, **5**, 569 (1972).
10. J. L. Sohet and S. A. Reible, *Ann. N. Y. Acad. Sci.*, **227**, 641 (1974).
11. Y. Oochika, *J. Phys. Soc. Japan*, **9**, 594 (1954).
12. E. Lippert, *Z. Electrochem.*, **61**, 962 (1957).
13. G. N. Bakshiev, *Opt. Spectrosc.* (Eng. Trans.), **16**, 446 (1964).
14. G. Weber and J. R. Lakowicz, *Chem. Phys. Letters*, **22**, 419 (1973).
15. G. Weber, 'Excited states of proteins', in *Light and Life*, p. 82 (eds. W. D. McElroy and B. Glass), Johns Hopkins Press, Baltimore, 1961.
16. F. J. W. Teale, *Biochem. J.*, **76**, 381 (1960).
17. A. Finazzi-Agro, G. Rotilio, L. Avigliano, P. Guerrieri, V. Boffi and B. Mondovi, *Biochemistry*, **9**, 2009 (1970).
18. J. R. Lakowicz and G. Weber, *Biochemistry*, **12**, 4171 (1973).
19. W. R. Ware, P. Chow and S. K. Lee, *Chem. Phys. Letters*, **2**, 356 (1968).
20. L. Brand and J. R. Gohlke, *J. Biol. Chem.*, **246**, 2317 (1973).
21. J. R. Lakowicz and G. Weber, *Biochemistry*, **12**, 4161 (1973).
22. G. Weber, *Adv. Prot. Chem.*, **8**, 415 (1953); R. F. Steiner and H. Edelhoch, *Chem. Rev.*, 457 (1962).
23. G. Weber, *Biochem. J.*, **75**, 345 (1960).
24. S. R. Anderson and G. Weber, *Arch. Biochem. and Biophys.*, **116**, 207 (1966).
25. V. Glushko, P. J. Lawson and F. R. N. Gurd, *J. Biol. Chem.*, **247**, 3176 (1972).
26. V. E. Teichberg and M. Shinitzky, *J. Mol. Biol.*, **74**, 519 (1973).
27. D. Magde, E. Ellson and W. W. Webb, *Phys. Rev. Letters*, **29**, 705 (1972).
28. H. Muirhead, J. M. Cox, L. Mazzarella and M. F. Perutz, *J. Mol. Biol.*, **28**, 117 (1967).
29. F. M. Richards and B. Lee, *J. Mol. Biol.*, **55**, 379 (1971).
30. D. C. Phillips, *Proc. Ntl. Acad. Sci. US*, **57**, 484 (1967).
31. S. Lehrer, *Biochemistry*, **10**, 3254 (1971).
32. A. Weller, *Discussions Faraday Soc.*, **27**, 28 (1959).
33. W. M. Vaughan and G. Weber, *Biochemistry*, **9**, 464 (1970).
34. G. Weber, *Biochem. J.*, **75**, 335 (1960).
35. G. Weber and M. Shinitzky, *Proc. Ntl. Acad. Sci. U.S.*, **65**, 823 (1970).
36. A. Hvidt and S. O. Nielsen, *Adv. Prot. Chem.*, **21**, 288 (1966).
37. P. Wahl, Doctoral Thesis, U. of Strasbourg (1963). See also P. Wahl and G. Weber, *J. Mol. Biol.*, **30**, 371 (1967).
38. J. Eisinger and R. A. Dale, *J. Mol. Biol.*, **84**, 643 (1974).
39. A. Laubereau, D. von der Linde and W. Kaiser, *Phys. Rev. Letters*, **28**, 1162 (1972).

D.1. Fluorescence quantum yield of the aromatic amino-acids as a function of excitation wavelength

I. Tatischeff and R. Klein

Institut du Radium-Laboratoire Curie†
11, rue Pierre et Marie Curie—75231 Paris–Cedex 05 France

SUMMARY

Taking the fluorescence quantum yield of quinine sulphate in 0·1 N sulphuric acid as independent of the excitation wavelength over the 200–400 nm range, it was found that the fluorescence quantum efficiencies of tryptophan, tyrosine and phenylalanine in neutral aqueous solution depend on excitation wavelength in a similar manner to that of benzene. For each aromatic amino-acid, the yield is constant in the first absorption band, falls abruptly, then remains constant in the second absorption band. This was checked by using sodium salicylate as a quantum counter instead of quinine sulphate.

The absolute quantum yields measured, at room temperature, in the first absorption band, by the optically dilute method with quinine sulphate as a standard, were 0·13, 0·15 and 0·02 for tryptophan, tyrosine and phenylalanine, respectively.

1 INTRODUCTION

The statement that the fluorescence quantum efficiency is independent of the excitation wavelength (for non-ionizing radiation) is known as Vavilov's law. It is used to evidence efficient intramolecular radiationless transitions from the higher excited electronic states. As a consequence is the general failure to observe fluorescence from higher excited singlet states, which led to Kasha's rule.[1] The first exception to Vavilov's law was observed by Braun, Kato and Lipsky,[2] who found that the fluorescence quantum yield did depend on the excitation wavelength in the case of benzene and its derivatives. As confirmed by further studies,[1,3] the deviation from Vavilov's law seems to be an inherent property of the benzene molecular structure.

† Laboratoire Associé au C.N.R.S. n° 198.

Concerning the three aromatic amino-acids, tryptophan, tyrosine and phenyl-alanine in water, it is generally acknowledged, following Teale and Weber,[4] that the quantum yields are independent of the excitation wavelength over the 320–200 nm range.[5-8] However, Shore and Pardee,[9] using a rather rough method, found a variation with wavelength, which was ascribed to systematic errors.[4] More recently, Dodonova et al.[10,11] working with thin layers of the aromatic amino-acids excited by vacuum ultraviolet at room temperature or at 90 K, claimed that the yield of fluorescence falls appreciably towards short wavelengths. The same had been observed by Augenstein et al.[12] for tryptophan powder at 110 K.

The aim of this paper is to reinvestigate the subject of the fluorescence quantum yield of the aromatic amino-acids in aqueous solutions using a method somewhat different from that used by Teale and Weber,[4] whose reliability has been strongly questioned recently,[13,8] at least with regard to the absolute quantum yields obtained.

2 EXPERIMENTAL

The fluorescence quantum yield at a given excitation wavelength λ_0, for a substance in solution, is defined as the ratio of the number of emitted photons to the number of quanta absorbed:

$$Q(\lambda_0) = \frac{I_f(\lambda_0)}{I_a(\lambda_0)} \qquad (1)$$

For each aromatic amino-acid, we were interested in measuring:

(1) the relative yield as a function of excitation wavelength:

$$Q_r(\lambda_0) = \frac{Q(\lambda_0)}{Q_1} \qquad (2)$$

where Q_1 represents the yield in the first absorption band;

(2) the absolute yield Q_1, with quinine sulphate as a standard, in order to compare with already published values.

2.1 Expression of fluorescence yield $Q(\lambda_0)$

(a) *Calculation of* $I_a(\lambda_0)$

If $I_{ex}(\lambda_0)$ is the number of photons with excitation wavelength λ_0 that reach the sample in the cell C, and if $B(\lambda_0)$ is the fraction absorbed in the first two mm ($l = 0.2$ cm), 'viewed' by monochromator M_2 (Figure 1), then

$$I_a(\lambda_0) = I_{ex}(\lambda_0)B(\lambda_0) \qquad (3)$$

and

$$B(\lambda_0) = \frac{\alpha}{\alpha + \beta}(1 - e^{-2 \cdot 3(\alpha + \beta)}) \qquad (4)$$

with

$$\alpha = \varepsilon(\lambda_0)cl, \quad \beta = k(\lambda_0)l$$

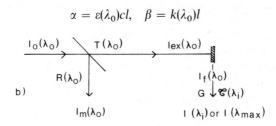

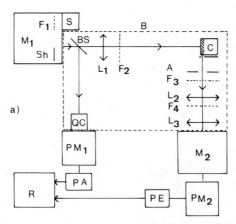

Figure 1 Experimental arrangement. (a) S:XBO 150 W xenon arc or original Hanau D–200F deuterium light sources; M_1: excitation vacuum ultraviolet grating monochromator (McPherson 218); M_2: analysing grating monochromator (Bausch and Lomb high intensity); C:10 × 10 mm Helema quartz cell, 'viewed' at 90° from excitation beam, in the first two mm; L_1: non-achromatic fluorite lens focusing on cell C; L_2, L_3: quartz lenses imaging the cell C on the analysing monochromator; F_1, F_2, F_3, F_4: filters to test for stray light, second-order light or to eliminate excitation light from fluorescence measurements; BS: fluorite beam splitter at 45° from excitation beam; QC: sodium salicylate quantum counter; Sh:shutter; B:light-tight box; PM_1:E.M.I. 9524S photomultiplier; PM_2:E.M.I. 6256S photomultiplier; PA: Tekelec picoammeter; PE:slow 'pulse' electronics with integrator; R:potentiometric double-beam Speedomax Meci Recorder. (b) symbols used (see text)

ε is the molar extinction coefficient of the substance of molar concentration c. The values of ε are obtained by means of an absorption spectrophotometer, either a Beckman DK2A or a Cary 17, and are compared with published values.[14] The term kl accounts for solvent absorption and is effective for water below 190 nm.

One of the problems encountered is to measure the excitation intensity $I_{ex}(\lambda_0)$. Light from monochromator M_1, $I_0(\lambda_0)$, reaches the fluorite beam splitter BS (Figure 1). A fraction $R(\lambda_0)$ is reflected towards the light intensity monitor, which consists of a sodium salicylate quantum counter QC associated with a photomultiplier PM_1 measuring $I_m(\lambda_0)$. A fraction $T(\lambda_0)$ is transmitted through the beam splitter, fluorite lens and cell window, and provides the excitation $I_{ex}(\lambda_0)$ on the sample in the cell C:

$$I_{ex}(\lambda_0) = \frac{T(\lambda_0)}{R(\lambda_0)} I_m(\lambda_0) = f(\lambda_0) I_m(\lambda_0) \tag{5}$$

The experimental determination of $f(\lambda_0)$, which represents a correction factor between the measured intensity $I_m(\lambda_0)$ and $I_{ex}(\lambda_0)$ will be given later. From (3) and (5),

$$I_a(\lambda_0) = f(\lambda_0) B(\lambda_0) I_m(\lambda_0) \tag{6}$$

(b) *Calculation of* $I_f(\lambda_0)$

The measured number of emitted photons at wavelength λ_i is:

$$I(\lambda_i) = I_a(\lambda_0) Q(\lambda_0) \Delta F(\lambda_i) G \mathscr{T}(\lambda_i) \tag{7}$$

with

$$\Delta F(\lambda_i) = F(\lambda_i) \Delta\lambda \quad \text{and} \quad \int F(\lambda_i) \, d\lambda = 1$$

G is a geometry factor, $\mathscr{T}(\lambda_i)$ is the transmission characteristic of the analysing spectrometer, $F(\lambda_i)$ is the fluorescence wavelength distribution and $\Delta\lambda$ is the bandwidth used for observation.

Two kinds of spectra were measured:

(i) *Fluorescence excitation spectra.* The analysing monochromator M_2 is set at the maximum, λ_{max}, of the fluorescence emission spectrum and the excitation wavelength λ_0 is varied. The measured excitation spectrum is:

$$A(\lambda_0) = \frac{I(\lambda_{max})}{I_m(\lambda_0)}$$

The true excitation spectrum is:

$$A_c(\lambda_0) = \frac{I(\lambda_{max})}{I_{ex}(\lambda_0)} = \frac{A(\lambda_0)}{f(\lambda_0)} \tag{8}$$

From (5), (6), (7):

$$A_c(\lambda_0) = B(\lambda_0)Q(\lambda_0)\,\Delta F(\lambda_{max})G\mathcal{T}(\lambda_{max})$$

$$A_c(\lambda_0) = K B(\lambda_0)Q(\lambda_0)$$

(ii) *Fluorescence emission spectra.* The excitation monochromator M_1 is set at the chosen excitation wavelength λ_0 and the wavelength λ_i of the analysing monochromator is varied. The total number of emitted photons is:

$$I_f(\lambda_0) = \frac{1}{G}\sum_i \frac{I(\lambda_i)}{\mathcal{T}(\lambda_i)} = I_a(\lambda_0)Q(\lambda_0) \tag{9}$$

From (1), (6) and (9),

$$Q(\lambda_0) = \frac{(1/G)\sum_i I(\lambda_i)/\mathcal{T}(\lambda_i)}{f(\lambda_0)B(\lambda_0)I_m(\lambda_0)} \tag{10}$$

2.2 Measurement of relative yield $Q_r(\lambda_0)$

The variation of relative quantum yield $Q_r(\lambda_0)$ versus wavelength is obtained by comparing the corrected excitation spectrum with the fractional absorption spectrum. To obtain true excitation spectra, instead of using an optically dense quantum counter to measure the light intensity $I_{ex}(\lambda_0)$, we preferred to work with optically dilute solutions. The excitation fluorescence spectrum for quinine sulphate 10^{-6} M in 0·1 N sulphuric acid was measured and the method of Argauer and White[15] was used to deduce the correction factor $f(\lambda_0)$. This method is based on the assumption that fluorescent quantum yield is independent of wavelength. This independence was disputed in the case of quinine sulphate;[13] however, the recent careful investigations of Gill[16] and Fletcher[17] revealed no deviations from a constant value in the range 200–400 nm. Therefore, we assumed a constant yield for quinine sulphate in 0·1 N sulphuric acid, measured its excitation fluorescence spectrum and compared it with its fractional absorption spectrum in order to obtain the relative values of the correction factor $f(\lambda_0)/f_1$. From (8),

$$f(\lambda_0) = \frac{A'(\lambda_0)}{K'Q'B'(\lambda_0)}$$

$$\frac{f(\lambda_0)}{f_1} = \frac{A'(\lambda_0)}{B'(\lambda_0)} \times \frac{B'_1}{A'_1} \tag{11}$$

Then, we did the same comparison with each aromatic amino-acid in order to decide whether there was a variation of fluorescence quantum yield with wavelength or not. For the aromatic amino-acid investigated:

$$Q(\lambda_0) = \frac{A(\lambda_0)}{K f(\lambda_0)B(\lambda_0)}$$

and:

$$Q_r(\lambda_0) = \frac{Q(\lambda_0)}{Q_1} = \frac{A(\lambda_0)}{f(\lambda_0)B(\lambda_0)} \times \frac{f_1 B_1}{A_1} \propto \frac{A_c(\lambda_0)}{B(\lambda_0)} \qquad (12)$$

The variation of yield between two absorption bands was also corroborated using fluorescence spectra. Two excitation wavelengths λ_1 and λ_2 were chosen near the first two absorption maxima of the investigated amino-acid. Four fluorescence spectra were recorded at these wavelengths and at such concentrations as to work with optically dilute conditions ($\varepsilon cl \lesssim 10^{-2}$): two for quinine sulphate and two for the aromatic amino-acid. From (9) and (6), for quinine sulphate:

$$I_f'(\lambda_1) = \frac{1}{G}\sum_i \frac{I_1'(\lambda_i)}{\mathscr{T}'(\lambda_i)} = f(\lambda_1)I_m(\lambda_1)B'(\lambda_1)Q'$$

$$I_f'(\lambda_2) = \frac{1}{G}\sum_i \frac{I_2'(\lambda_i)}{\mathscr{T}'(\lambda_i)} = f(\lambda_2)I_m(\lambda_2)B'(\lambda_2)Q'$$

The ratio $f(\lambda_1)/f(\lambda_2)$ is obtained from the area ratio of measured fluorescence spectra. For the investigated aromatic amino-acid, the same ratio leads to $[f(\lambda_1)/f(\lambda_2)] \cdot [Q(\lambda_1)/Q(\lambda_2)]$, whence the possibility to evidence, in a somewhat different way, the variation of yield observed by means of excitation fluorescence spectra. Note that the ratios are for measured fluorescence spectra, as the transmission characteristic $\mathscr{T}(\lambda_i)$ is independent of λ_0 for a given fluorescent substance. Note also that, in all these experiments, the monitor function is not to obtain $I_{ex}(\lambda_0)$, but to control eventual variations of intensity between the quinine sulphate and amino-acid experiments.

2.3 Measurement of absolute yield Q_1

Finally, we were interested in absolute fluorescence quantum yield at a given excitation wavelength. Therefore we chose the relative quantum yield method with quinine sulphate as a standard, ascribing to it the most commonly assumed value[13] given by Melhuish[18] for excitation at $\lambda_0 = 365$ nm in $1 \cdot 0$ N sulphuric acid, that is $0 \cdot 55$. Since we observed a yield 7% higher in the $1 \cdot 0$ N solution than in the $0 \cdot 1$ N solution, in agreement with Eisenbrand and Dawson, but in disagreement with Turner,[13] the $0 \cdot 55$ value was corrected to $0 \cdot 51$, to allow for utilization of $0 \cdot 1$ N sulphuric acid. In order to obtain the absolute quantum yield, the measured fluorescence spectrum had to be corrected by the transmission factor $\mathscr{T}(\lambda_i)$. This factor was obtained by using a MgO screen instead of the cell C, at $35°$ to the exciting beam. The scattered excitation light was successively measured without the analysing device, using rhodamine B or sodium salicylate as quantum counters, then with the analysing monochromator. The method is similar to those developed by Melhuish[19] and Parker[20] and used by Chen.[21]

At a given wavelength λ_0, the corrected emission spectra for the standard and for the unknown compound are, respectively:

$$\sum \left(\frac{dQ}{d\bar{v}}\right)' = \sum_i \frac{I'(\lambda_i) \cdot \lambda_i^2}{G\mathcal{T}'(\lambda_i)} = I_m(\lambda_0)f(\lambda_0)B'(\lambda_0)Q'$$

$$\sum \left(\frac{dQ}{d\bar{v}}\right) = \sum_i \frac{I(\lambda_i) \cdot \lambda_i^2}{G\mathcal{T}(\lambda_i)} = I_m(\lambda_0)f(\lambda_0)B(\lambda_0)Q(\lambda_0)$$

in units of relative quanta per unit frequency interval. The quantum yield of the unknown substance is given by:

$$Q(\lambda_0) = \frac{\sum (dQ/d\bar{v})}{\sum (dQ/d\bar{v})'} \times \frac{B'(\lambda_0)}{B(\lambda_0)} \times Q' \tag{13}$$

3 RESULTS

3.1 Experimental conditions

The materials were purchased from: Touzart and Matignon (quinine sulphate); Merck (sulphuric acid); Cyclochemicals (L-tryptophan); Schuchardt (L-tyrosine); Calbiochemicals (L-phenylalanine). All these compounds were used without further purification. Measurements were performed at room temperature ($\simeq 22\,°\text{C}$), in unbuffered neutral aqueous solutions, prepared with triple distilled water and not deoxygenated before use. The optically dilute measurements rest on Beer's law:

$$I_{ex}(\lambda_0)B(\lambda_0) = I_{ex}(\lambda_0)(1 - e^{-2\cdot3\varepsilon(\lambda_0)cl})$$

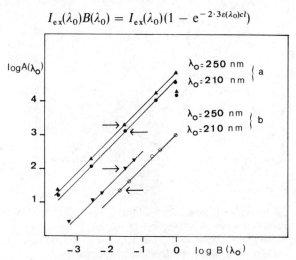

Figure 2 Test of proportionality of measured fluorescence intensity $A(\lambda_0)$ to fractional absorption $B(\lambda_0)$. (a) Quinine sulphate, (b) phenylalanine (arrows correspond to the concentrations used in fluorescence excitation measurements)

382

In our experimental arrangement, the proportionality of $A(\lambda_0)$ to $B(\lambda_0)$ was checked for quinine sulphate and phenylalanine, over a great range of concentrations, at wavelengths $\lambda_1 = 250$ nm and $\lambda_2 = 210$ nm (Figure 2).

3.2 Variation of relative fluorescence quantum yield $Q_r(\lambda_0)$

Figures 3 to 5 show relative yield variations versus excitation wavelength, respectively, for tryptophan, tyrosine and phenylalanine. For each aromatic amino-acid, $Q_r(\lambda_0)$ is constant in the first absorption band, falls abruptly, and then remains constant in the second absorption band.

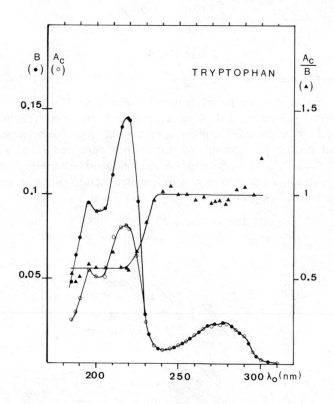

Figure 3   Relative fluorescence yield of Tryptophan (10^{-5} M) in aqueous solution as a function of excitation wavelength. $B(\lambda_0)$: fractional absorption spectrum; $A_c(\lambda_0)$: corrected fluorescence excitation spectrum in arbitrary units.

$$\frac{A_c}{B}(\lambda_0) \propto Q_r(\lambda_0): \text{relative fluorescence yield}$$

(Bandwidths: excitation $\Delta\lambda_0 = 2 \cdot 6$ nm; emission $\Delta\lambda = 9 \cdot 6$ nm)

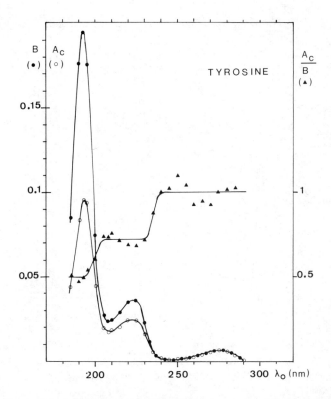

Figure 4 Relative fluorescence yield of Tyrosine (10^{-5} M) in aqueous solution as a function of excitation wavelength. (B, A_c and A_c/B as in Figure 3, $\Delta\lambda_0 = 2.6$ nm, $\Delta\lambda = 4.8$ nm)

(a) Tryptophan (Figure 3)

Results for tryptophan were first obtained using the xenon (XBO 150 W) excitation light source, referred to quinine sulphate 10^{-6} M in 0.1 N sulphuric acid. The main relative yield variation, as described, was already observed, but the $Q_r(\lambda_0)$ was not so 'smooth' as the one presented on Figure 3, obtained with the deuterium light source and referred to sodium salicylate as quantum counter. This is an opportunity to point out that one of the major difficulties encountered in these experiments is the problem of stray light. The xenon light source yields quite low emission below 250 nm, and at a given λ_0 stray light mainly from higher wavelengths may disturb the measurements. This was taken into account in two ways: one, by measuring stray light at 150 nm, and the other by means of a filter transmitting above 250 nm (MTO J280). Both methods gave similar results with regard to corrected excitation spectra $A_c(\lambda_0)$.

Another difficulty lay in the lack of accuracy of correction factors $f(\lambda_0)$ about 270 nm, where quinine sulphate absorption is very low. When using the

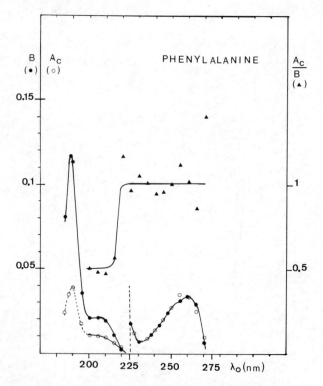

Figure 5 Relative fluorescence yield of Phenylalanine in aqueous solution as a function of excitation wavelength. The concentration was 5×10^{-4} M for $\lambda_0 > 225$ nm and 5×10^{-6} M for $\lambda_0 < 225$ nm (B, A_c and A_c/B as in Figure 3, $\Delta\lambda_0 = 5\cdot2$ nm, $\Delta\lambda = 4\cdot8$ nm)

deuterium light source, stray light was much less troublesome and it was then possible to use solid sodium salicylate, sprayed on a glass plate from a methanol solution,[22] as a quantum counter, in order to obtain more precise correction factors and check the optical dilute measurements with quinine sulphate. In this arrangement, a mirror reflected the fluorescence from sodium salicylate on to the analysing monochromator M_2, set at the maximum emission wavelength. This comparison was quite satisfactory as it only 'smoothed' the $Q_r(\lambda_0)$ variation, without modifying the drop from first to second absorption band.

(b) *Tyrosine (Figure 4)*

The results presented for tyrosine were also obtained using the deuterium light source. Under 200 nm, the correction factors could then be calculated by using sodium salicylate instead of quinine sulphate and the $B(\lambda_0)$ values were calculated using ε-values from.[14] There is good evidence for a drop of fluorescence yield from second to third absorption band, as the third band is much more clearly dissociated from the second one than in tryptophan (Figure 3).

(c) *Phenylalanine* (*Figure* 5)

For phenylalanine, measurements were performed with the xenon light source. Correction factors $f(\lambda_0)$ were calculated with quinine sulphate excited with the same bandwidth ($\Delta\lambda_0 = 5\cdot2$ nm), the bandwidth for emission being then 9·6 nm. It was checked by means of quinine sulphate that using different bandwidths for absorption and excitation fluoresence measurements was not a source of errors. Although the third absorption band is well resolved from the second one, the yield variation between the two bands is not given, as the corrected excitation spectrum is obtained with extrapolated values of $f(\lambda_0)$, quinine sulphate being inadequate below 200 nm.

In all these measurements, no attempts were made to correct for refractive index variations with wavelength. Such variations were shown to be relatively small in benzene.[2]

These results, obtained by fluorescence excitation measurements, were corroborated by fluorescence emission measurements at two different wavelengths and comparison with fluorescence spectra of quinine sulphate at the same wavelengths. Table 1 gives the conditions and results of these experiments for each aromatic amino-acid.

Table 1 Relative fluorescence quantum yield variations obtained by fluorescence emission measurements (xenon XBO 150 W light source)

Amino-acid	Relative quantum yield $\dfrac{Q(\lambda_2)}{Q(\lambda_1)}$	Quantum yield Q_1
Phenylalanine ($c = 5 \times 10^{-6}$ M and 5×10^{-4} M)	$\dfrac{Q(210)}{Q(250)} = 0\cdot61 \pm 0\cdot14$	$Q(250) = 0\cdot02$
Tryptophan ($c = 10^{-5}$ M)	$\dfrac{Q(219)}{Q(250)} = 0\cdot5$	$Q(250) = 0\cdot13$
	$\dfrac{Q(219)}{Q(279)} = 0\cdot7$	
Tyrosine ($c = 10^{-5}$ M)	$\dfrac{Q(219)}{Q(279)} = 0\cdot6$	$Q(279) = 0\cdot15$

3.3 Absolute fluorescence quantum yield Q_1 in the first absorption band

On Figure 6 are reported the corrected fluorescence emission spectra of the three aromatic amino-acids and of quinine sulphate, in units of relative quanta per unit frequency interval versus wavenumber (μm^{-1}), excited at $\lambda_0 = 250$ nm and reduced to constant absorption ($\varepsilon c l = 0\cdot01$).

In Table 1 are also given the absolute yield Q_1, obtained as described in the experimental part of this paper. The obtained values: 0·13, 0·15 and 0·02 for

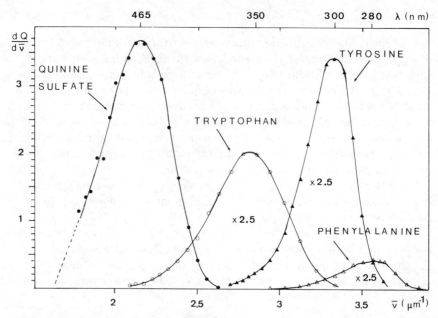

Figure 6 Corrected fluorescence spectra excited at $\lambda_0 = 250$ nm and reduced to constant absorption ($\varepsilon cl = 0.01$)

tryptophan, tyrosine and phenylalanine are much lower than those of Teale and Weber[4] (0.20, 0.21 and 0.04). However, they are in good agreement with those of Chen,[23] also obtained by comparison with quinine and actually considered as the most reliable (0.13, 0.14 and 0.024).

Inasmuch as the observed drop in relative yield from first to second absorption band is about 40%, the absolute fluorescence yields in the second bands should be about 0.08 for tryptophan, 0.09 for tyrosine and 0.01 for phenylalanine.

4 CONCLUSION

Although many experimental difficulties were encountered in measuring true excitation light and in calculating correction factors, the present results give good evidence for a drop of fluorescence quantum yield between the first and second absorption bands, amounting to about 40% for the three aromatic amino-acids in neutral non-deoxygenated aqueous solutions at room temperature. This deviation from Vavilov's law is to be compared with the one observed for benzene and some derivatives, and ascribed to an inherent property of the benzene molecular structure.[1-3] More experiments are indeed needed to give a somewhat elaborated explanation of these observations, which actually only indicate that, for the three aromatic amino-acids, intramolecular radiationless transitions from higher excited electronic states are not so efficient as was predicted from Vavilov's law.

ACKNOWLEDGEMENTS

The authors wish to acknowledge Professor M. Duquesne for stimulating discussions and Mrs J. Duquesne for her help to improve the quality of the manuscript.

REFERENCES

1. J. B. Birks, *Photophysics of Aromatic Molecules*, pp. 142–177, Wiley–Interscience, London and New York, 1970.
2. C. L. Braun, S. Kato and S. Lipsky, *J. Chem. Phys.*, **39**, 1645 (1963).
3. C. Fuchs, Thesis Strasbourg, 1972.
4. F. W. J. Teale and G. Weber, *Biochem. J.*, **65**, 476 (1957).
5. A. D. McLaren and D. Shugar, *Photochemistry of Proteins and Nucleic Acids*, p. 51, Pergamon Press, Oxford, 1964.
6. S. V. Konev, *Fluorescence and Phosphorescence of Proteins and Nucleic Acids*, pp. 22, 54, 57, Plenum Press, New York, 1967.
7. Yu. A. Vladimirov, *Photochemistry and Luminescence of Proteins*, p. 27, I.P.S.T., Jerusalem, 1969.
8. I. Weinryb and R. F. Steiner, *Excited States of Proteins and Nucleic Acids*, pp. 288–293, Plenum Press, New York, 1971; J. W. Longworth, *ibid*, p. 325.
9. V. G. Shore and A. B. Pardee, *Arch. Biochem. Biophys.*, **60**, 100 (1956).
10. N. Ya. Dodonova, I. P. Vinogradov and L. V. Kravchenko, *Vest. Leningr. Univ. Ser. Fiz. Khim.*, **16**, 76 (1968).
11. I. P. Vinogradov and N. Ya Dodonova, *Biophysics*, **16**, 355 (1971).
12. L. Augenstein, E. Yeargers, J. Carter and De Vaughn Nelson, *Rad. Res. Suppl.*, **7**, 128 (1967).
13. J. N. Demas and G. A. Crosby, *J. Phys. Chem.*, **75**, 991 (1971).
14. *Handbook of Biochemistry*, The Chemical Rubber Co., 1970; D. B. Wetlaufer, *Advanc. Protein Chem.*, **17**, 320 (1962).
15. R. J. Argauer and C. E. White, *Anal. Chem.*, **36**, 368 (1964); S. Udenfried, *Fluorescence Assay in Biology and Medicine*, Vol. II, p. 190, Academic Press, New York and London, 1969.
16. J. E. Gill, *Photochem. Photobiol.*, **9**, 313 (1969).
17. A. N. Fletcher, *Photochem. Photobiol.*, **9**, 439 (1969).
18. W. H. Melhuish, *J. Phys. Chem.*, **65**, 229 (1961).
19. W. H. Melhuish, *J. Opt. Soc. Am.*, **52**, 1256 (1962).
20. C. A. Parker, *Anal. Chem.*, **34**, 502 (1962).
21. R. F. Chen, *Anal. Biochem.*, **20**, 339 (1967).
22. J. A. R. Samson, *Techniques of Vacuum Ultraviolet Spectroscopy*, J. Wiley and Sons, Inc., New York and London, 1967.
23. R. F. Chen, *Anal. Letters*, **1**, 35 (1967).

D.2. Effect of temperature on the fluorescence of proteins

G. Laustriat and D. Gerard

Laboratoire de Biophysique de la Faculté de Pharmacie,
Université Louis Pasteur, Strasbourg, France

SUMMARY

The thermal dependence of the fluorescence quantum yield ϕ and lifetime τ of several proteins was investigated and characterized by the temperature coefficient at 30 °C (relative variation of ϕ or τ per degree).

Tryptophan residue emissions of various spectral ranges were studied: (i) short-wavelength fluorescence ($\lambda_{max} \sim 330$ nm) from buried residues, as displayed by native *E. Coli* alkaline phosphatase and β-lactoglobulin; (ii) long-wavelength fluorescence ($\lambda_{max} \sim 350$ nm) from exposed residues, as observed on the same proteins denatured; (iii) intermediate-range fluorescence ($332 < \lambda_{max} < 345$ nm) typical of lysozyme, trypsin and *E. Coli* ribosomal protein S_7. Tyrosine residue emission was studied on insulin and *E. Coli* ribosomal protein S_8.

It is shown the thermal behaviour of the protein fluorescence can be understood by analysing the temperature effect on model systems: indole in water or in dioxane, for exposed or buried tryptophan residues respectively; phenol in water, for exposed tyrosine residues. In particular, quenching processes, the influence of which has been studied on model systems, are shown to be an important cause of the thermal effect on protein emission.

1 INTRODUCTION

Numerous works have been devoted to the temperature effect on the fluorescence of proteins,[1,2] most often with a view to demonstrating conformation changes, since the passage of tryptophan residues from a hydrophobic to an aqueous environment induces a pronounced red shift of the emission spectrum. On the other hand, the temperature dependence of the fluorescence quantum

388

yield of aromatic amino-acids and derivatives, in various solvents, has also received considerable attention,[3–7] notably as a method of determining the rate parameters of the non-radiative deactivation processes in these molecules. A marked influence of the solvent on the magnitude of the temperature effect on the fluorescence yield of indole derivatives has been reported.[5,6] However with the exception of the initial work of Gally and Edelman,[3] no attempt has been made to correlate the two sets of data and to interpret the results obtained with proteins in the light of the temperature effect on monomers.

For this purpose, we have considered the influence of temperature on the fluorescence yield, and in some cases lifetime, of several solutions which can *a priori* be thought to be model systems for the tryptophan and tyrosine residues, in aqueous and non-aqueous environments, respectively. The magnitude of the temperature effect has been evaluated in each case by means of a temperature coefficient, and compared with the effect observed on emissions of tryptophan and tyrosine residues in proteins.

2 METHODS

2.1 Choice of model solutions

(a) *Solvents*

In a simplified picture of protein conformation, emitting residues may be considered as being located either in an aqueous environment ('exposed' residues) or in an intraglobular medium ('buried' residues). As a representative solvent of this last undefined medium, we selected dioxane since solutions of tryptophan in this solvent showed similar fluorescence properties (λ_{max} spectrum width, lifetime) as an enzyme in which the accessibility of emitting tryptophan residues to quenching ions had been found negligible (*E. Coli* alkaline phosphatase[8]).

(b) *Solutes*

Model molecules for the residues may be of various kinds, according to the influence of the polypeptidic chain and charged groups on the excited residues in the protein. In this work we studied:

(i) indole and phenol, which may be thought as good models if the aromatic residues suffer no quenching;

(ii) *N*-acetyl-Trp (or Tyr)-amide, compounds possessing two amide bonds able to interact with the excited ring, as would do peptide bonds of the polypeptidic chain; and

(iii) the amino-acids, most often studied in previous work, which may be interesting models, since they are submitted to the quenching effect of charged groups.

2.2 The temperature coefficient

The temperature effect on fluorescence quantum yield and lifetime may generally be interpreted by means of a simple kinetic scheme, involving three intramolecular processes: emission (rate constant: k_f), temperature-independent (k_{nr}^0) and -dependent ($k_{nr}^T = A \exp(-E/kT)$) non-radiative deactivation (k = Boltzmann constant, T = absolute temperature).

In order to compare the magnitude of thermal effects in various systems, we used the temperature coefficient C, that is the relative variation of the quantum yield ϕ or lifetime τ per unit temperature interval.

In the absence of external quenching, the expression of this coefficient (C_0) can be obtained from those of ϕ_0 and τ_0:

$$\tau_0 = [k_f + k_{nr}^0 + A \exp(-E/kT)]^{-1} \tag{1}$$

$$\phi_0 = k_f \tau_0 \tag{2}$$

leading to:

$$C_0 = \phi_0^{-1} \frac{d\phi_0}{dT} = \tau_0^{-1} \frac{d\tau_0}{dT} = -\left(\frac{AE}{kk_f}\right) \frac{\phi_0}{T^2} \exp(-E/kT) \tag{3}$$

The value of C_0 is seen to be temperature dependent, and it is therefore necessary to fix the experimental conditions of its determination. This could be achieved by drawing the tangent to the yield (or lifetime) versus temperature curve at a given temperature. Since such plots are often linear or display small curvatures, we choose in this work to estimate C_0 from the variations $\Delta\phi_0$ and $\Delta\tau_0$ observed for a temperature increment of 15° around 30 °C.†

In presence of a diffusion controlled quenching process, the lifetime of excited states is given by:

$$\tau_q = \{k_f + k_{nr}^0 + A \exp(-E/kT) + k_q[Q]\}^{-1} \tag{4}$$

where the additional term $k_q[Q]$ is the probability per unit of time of deactivation by a quencher at molar concentration $[Q]$. The rate constant k_q is temperature-dependent and may be written as:[9]

$$k_q = k_d \exp(-E_d/kT) \tag{5}$$

where k_d is a constant and E_d the activation energy for the molecular diffusion (E_d is assumed to be identical for the two interacting molecules). Differentiating (4) with respect to temperature and rearranging leads to the temperature coefficient C_q:

$$C_q = \tau_q^{-1} \frac{d\tau_q}{dT} = \phi_q^{-1} \frac{d\phi_q}{dT} = C_0 \frac{1 + K[Q](E_d/C_0 kT^2)}{1 + K[Q]} \tag{6}$$

---

† The temperature effect was in fact explored in a much larger range (typically 5–50 °C). Detailed results of this study will be published elsewhere.

where K is the Stern–Volmer constant ($K = k_q\tau_0$). In this equation C_q and C_0 are *absolute values* and will hereafter be considered as such.

Like C_0, C_q is temperature dependent and must therefore be determined in the same conditions (around 30 °C in the present case) if both coefficients are to be compared.

Inspection of (6) shows that C_q may be greater or smaller than C_0 according to the respective values of E_d and C_0: at 30 °C ($kT^2 \approx 8$ eV dg), $C_q > C_0$ for $E_d/C_0 > 8$ eV dg and vice versa.

Values of E_d in the case of the model systems used here have not been experimentally determined to our knowledge. However, analysing the quenching of indole by histidine in water, Kirby and Steiner[6] deduced from their results a value of 3·3 kcal/mole (i.e. $\sim 0·14$ eV). Values of E_d in dioxane must be smaller, since Van der Waals forces in organic solvents are weaker than in water. For proteins, any *a priori* estimation of E_d would be hazardous since constraints affecting the thermal motion of residues are difficult to appraise.

3 MATERIALS AND TECHNIQUES

3.1 Materials

Model solutions were prepared in air equilibrated bidistilled water or spectroquality dioxane (Merck, Uvasol). L-tryptophan, L-tyrosine and L-histidine (Sigma, Σ grade), N-acetyl-L-tryptophanamide (Cyclo Chem.), N-acetyl-L-tyrosinamide (Cyclo Chem. and Sigma, Σ grade), phenol (Merck) were used without further purification. Indole (Prolabo) was twice recrystallized in bidistilled water.

Native protein solutions were prepared with pure commercially available samples, used without purification: *E. Coli* alkaline phosphatase (Worthington and Sigma) in pH 8, 0·05 M Tris buffer; β-lactoglobulin (Sigma) in pH 6·8, 0·02 M Tris buffer; lysozyme (Worthington) in pH 5·5 and 7·5 phosphate buffer; trypsin (Worthington) and insulin (Sigma) in pH 3·4, citric acid–phosphate buffer; *E. Coli* ribosomal proteins S_7 and S_8 (gift of Dr Lemieux) in the reconstitution buffer: 0·35 M KCl, 0·02 M Mg^{++}, 0·02 M Tris, pH 7·4.

Denaturation of proteins was achieved either by 6 M guanidine HCl for β-lactoglobulin, or by 6 M urea at pH 3·5 for alkaline phosphatase (studied at pH 8).

3.2 Techniques

Absorption spectra were recorded with a Cary 15 spectrophotometer.

Fluorescence spectra were obtained with an apparatus built in the laboratory and previously described[8] or with an absolute spectrofluorimeter (Fica 55). Indole derivatives and tryptophan residues were excited at 280 nm (bandwidth: 5 nm) or 295 nm (bandwidth: 2·5 nm), phenol derivatives and tyrosine residues at 275 nm (bandwidth: 2·5 nm). Relative fluorescence quantum yields were determined from the area under corrected emission spectra of solutions of equal optical density at the excitation wavelength; absolute yields were obtained using the reported value of 0·14 for L-tryptophan[10] and L-tyrosine[11] at 25 °C.

Lifetimes were measured by the single photoelectron technique with an apparatus designed in the laboratory and previously described.[8]

Quartz vessels containing solutions were placed in a metallic sample holder which could be heated or cooled by fluid circulation. Temperature of solutions, measured by a thermocouple, was maintained constant within 0·5 °C in the range 0–70 °C.

4 RESULTS

4.1 Model systems

The temperature coefficients obtained on model systems are presented in Table 1, which also includes some results derived from published data.

(a) *Indole derivatives*

Values of C for indole in dioxane and for tryptophan in water are in good agreement with those which can be deduced from the results of Kirby[6] and Gally[3] respectively. However, for indole in water, we found a larger temperature dependence of ϕ than reported by Walker et al.[5] It is to be noted that the temperature effect is considerably less important in dioxane than in aqueous solutions, as already mentioned by some authors.[5,6]

(b) *Phenol derivatives*

The coefficient given in Table 1 for phenol and tyrosine in water ($C = 0.85 \times 10^{-2}$) is somewhat larger than those derived from the data of Gally[3] and Turoverov[7] ($C = 0.70 \times 10^{-2}$ and 0.50×10^{-2} respectively for tyrosine; $C = 0.70 \times 10^{-2}$ for phenol[7]). Solutions in dioxane had not yet been studied. Two main features emerge from these results: (i) the small influence of the solvent polarity on the temperature effect, in the case of unquenched compounds (phenol and tyrosine), (ii) the increased value of C, in water, in the presence of quenching (external by histidine or intramolecular by amide bonds).

4.2 Proteins

The present study was restricted to proteins for which the emission is due to only one type of residue (tryptophan or tyrosine). In each case, results were obtained in a temperature range where no conformational change could be detected (by spectral shift and variation of the spectrum width notably); they are therefore assumed to reflect intrinsic properties of a given conformational state of the protein in solution.

Table 1 Temperature dependence of fluorescence quantum yield of model systems

Systems	$\phi_{25\,°C}$	$\tau_{25\,°C}$ (ns)	Temperature[a] coefficient $\times 10^2$
(a)			
Indole/dioxane	0.42^6	5.4	0.45 ± 0.1
Tryptophan/dioxane	0.29	4.5	0.8 ± 0.1
N-Acetyl tryptophan amide/dioxane	0.30	4.6	0.7 ± 0.1
(b)			
Indole/water	0.28^6	4.5	3.0 ± 0.2
Indole + histidine (0.25 M)/water (6)	0.06	—	2.1 ± 0.2
Tryptophan/water	0.14^{10}	3.0	2.5 ± 0.2
N-Acetyl tryptophan amide/water	0.14	3.0	2.2 ± 0.2
(c)			
Phenol/dioxane	0.19	5.2	0.55 ± 0.1
N-acetyl tyrosine amide/dioxane[b]	0.19	—	0.6 ± 0.1
(d)			
Phenol/water	0.14	3.5	0.85 ± 0.1
Phenol + histidine (0.25 M)/water	0.03	—	2.0 ± 0.2
Tyrosine/water	0.14^{11}	3.0	−85 ± 0.1
N-acetyl tyrosine amide/water	0.065	1.5	1.3 ± −1

[a] Absolute values.
[b] The tyrosine/dioxane system has not been studied, due to the insolubility of the amino-acid in dioxane.

Table 2 Fluorescence characteristics and temperature coefficient of some proteins

Protein	λ_{exc} (nm)	λ_{max} (nm)	$\phi_{25\,°C}$	$\tau_{25\,°C}$ (ns)	Temperature[b] coefficient $\times 10^2$	Results obtained from
(a)						
E. Coli alkaline phosphatase (pH 8)	280	325	0.13[a]	4.5	0.55 ± 0.05	ϕ, τ
β-Lactoglobulin A B (pH 6·8)	296	330	0.056[17]	2	2.0 ± 0.2	ϕ, τ
(b)						
E. Coli alkaline phosphatase (6 M urea)	280	350	0.11[a]	3.5	2.0 ± 0.2	ϕ, τ
β-Lactoglobulin (6 M guanidine HCl)	296	352	0.116[17]	—	1.9 ± 0.2	ϕ
(c)						
Lysozyme (pH 5·5, 7·5)	280	342	0.045	2.4	2.2 ± 0.2	ϕ, τ
Trypsin (pH 3·4)	295	335	0.09[17]	2.4	1.9 ± 0.2	ϕ
E. Coli ribosomal protein S7 (pH 7·4)	295	342	0.16	4.5	1.65 ± 0.02	ϕ
(d)						
Insulin (pH3·4)	275	302	—	1.5	0.85 ± 0.1	ϕ
E. Coli ribosomal protein S8 (pH 7·4)	275	302	0.045	3	1.45 ± 0.1	ϕ

[a] ϕ of tryptophan residues.
[b] Absolute values.

(a) *Temperature dependence of tryptophan emission*

This part of our work was performed on class B proteins in the emission of which the tyrosine fluorescence was absent or rendered negligible by excitation at 295 nm. In these conditions, three groups of proteins may be distinguished according to the microenvironment of their emitting tryptophyl residues.

(i) *Proteins whose emission originates from buried residues only* (Table 2a). The fluorescence of such proteins is characterized by a spectrum of short wavelength ($\lambda_{max} \sim 330$ nm) and small half-width (48–49 nm, Burstein criterion[12]), and by a very weak sensibility to external quenchers such as potassium iodide.

Two proteins were found to satisfy this set of conditions: *E. Coli* alkaline phosphatase, previously studied in this laboratory,[8,13] and β-lactoglobulin studied by Burstein *et al.*[12]

Table 2a shows that although possessing the same type of fluorophores, the two proteins behave quite differently: for alkaline phosphatase the temperature coefficient is about the same as that of the indole–dioxane system, but it is much greater in the case of lactoglobulin.

(ii) *Proteins whose emission originates from exposed residues only* (Table 2b). No proteins in the native state have a fluorescence spectrum as red-shifted as that of indole derivatives in water (350–355 nm).

The temperature effect on exposed residues was therefore examined on the two preceding proteins after complete denaturation. Both proteins then emit at 350–352 nm and exhibit the same temperature coefficient ($C = 1.9$–2.0×10^{-2}), which nevertheless is inferior to that found for the indole–water system ($C = 3.0 \times 10^{-2}$).

(iii) *Proteins of 'intermediate' emission* (Table 2c). The fluorescence of tryptophan may appear in an intermediate spectral range ($332 < \lambda_{max} < 345$ nm) if a protein contains either buried and exposed residues which both emit, or tryptophan residues located in an environment of intermediate polarity and/or viscosity. We examined three proteins exhibiting such an intermediate emission: lysozyme (pH 5.5 and 7.5), α-trypsin and *E. Coli* ribosomal protein S_7. The first two had already been studied: for lysozyme, our results are in very good agreement with those of Gally,[3] the data of whom lead to the same coefficient ($C = 2.2 \times 10^{-2}$); for trypsin which undergo a conformational change above 45 °C, we observed the same temperature dependence ($C = 1.9 \times 10^{-2}$) as Arrio,[14] although Turoverov[15] reports a smaller effect (corresponding to $C = 1.50 \times 10^{-2}$).

(b) *Temperature dependence of tyrosine emission* (Table 2d)

Two class A proteins (containing no tryptophan residue) have been investigated: insulin and *E. Coli* ribosomal protein S_8. We found for the former the same thermal effect as Turoverov;[7] the latter had not been studied previously.

5 DISCUSSION

The temperature coefficients found for the four groups of proteins will be discussed in connection with results obtained on the corresponding model systems.

5.1 Protein emission from buried tryptophan residues

Results obtained on *model compounds* (Table 1a) show that in dioxane the temperature coefficient is greater in the presence of quenching (tryptophan, tryptophanamide) than in the absence of quenching (indole). This is consistent with (6), since $C_q > C_0$ implies $E_d/C_0 > 8$ eV dg, that is, taking for C_0 the value found for indole ($C_0 = 0.45 \times 10^{-2}$):

$$E_d > 0.035 \quad \text{or} \quad E_d \gtrsim \tfrac{3}{2}kT$$

which condition is certainly satisfied.

Considering now results obtained on *proteins* (Table 2a) it is seen that alkaline phosphatase exhibits about the same coefficient (0.55×10^{-2}) as indole and derivatives in dioxane, from which it may be inferred that no important dynamic quenching of the tryptophan residues occurs in this protein. This conclusion is corroborated by the relatively high value of the fluorescence lifetime of this protein (4.5 ns), which is close to that of tryptophanamide in dioxane (4.6 ns). The low quantum yield of alkaline phosphatase ($\phi = 0.13$) as compared with this model system ($\phi = 0.30$) must reflect the occurrence of a static quenching, which had already been deduced previously from different experiments.[8]

Another protein, the goat lactalbumin studied by Sommers et al.,[16] behaves like alkaline phosphatase since it presents a short-wavelength emission spectrum ($\lambda_{max} = 330$ nm) and an analysis of the authors' data leads to a temperature coefficient of 0.55×10^{-2}.

β-lactoglobulin however, has a much higher coefficient ($C = 2.0 \times 10^{-2}$), indicating according to (6) the occurrence of an efficient intraglobular dynamic quenching. This is in agreement with the low quantum yield ($\phi = 0.056$) reported by Kronmann[17] and the low lifetime (2 ns) we found.

5.2 Protein emission from exposed tryptophan residues

In aqueous solutions, the influence of quenching on the temperature coefficient of *model compounds* (Table 1b) is opposite to that observed in dioxane: the highest coefficient (3.0×10^{-2}) is exhibited by the unquenched compound (indole). This fact is again in agreement with (6) since, taking for C_0 the value of the coefficient for indole (3.0×10^{-2}) and considering the reported order of magnitude of E_d in water (3 kcal/mole, i.e. 0.13 eV), the ratio E_d/C_0 is now inferior to 8 eV dg, which is the condition for $C_0 > C_q$. It is furthermore interesting to note, in the particular case of the indole + histidine (0.25 M)– water system, that introducing in (6) the values of $K[Q]$ and C_q which can be deduced from the data of Kirby and Steiner (4·6 and 2·2 × 10^{-2} respectively) gives $E_d = 0.14$ eV, in excellent agreement with the value found by these authors by a different method.

As for tryptophan residues exposed to water *in denatured proteins* (Table 2b), it may be seen that the temperature coefficients of alkaline phosphatase and lactoglobulin—which were very different for the native proteins—now have very close values (1·9 × 10^{-2}–2·0 × 10^{-2}). This first indicates that the particular quenching process evidenced for lactoglobulin in the native state disappears upon denaturation. On the other hand, the common value of C is smaller than that of indole in water, meaning that in both denatured proteins exposed tryptophan residues are subject to some dynamic quenching from neighbour residues. For alkaline phosphatase, this conclusion is consistent with the value of τ (3·5 ns) which is lower than the fluorescence lifetime of the indole–water system (4·5 ns).

5.3 Protein (tryptophan) emission of intermediate spectral range

The temperature effect on *proteins* of intermediate emission (Table 2c) is more difficult to analyse since they may contain the two types (buried and exposed) of tryptophan residues which, as we saw, may themselves exhibit temperature coefficients of the same order of magnitude in the presence of quenching. This situation is met with lysozyme and trypsin, for which the low value of τ (2·4 ns) indicates that a dynamic quenching indeed occurs.

The case of ribosomal protein S_7 is however interesting to discuss further. On one hand, in effect, the high value of the lifetime (4·5 ns) rules out the possibility of an important dynamic quenching. On the other hand, the pronounced red shift of the fluorescence spectrum ($\lambda_{max} = 342$ nm) implies a predominance of exposed residues, which in the absence of quenching should display a much greater coefficient (3·0 × 10^{-2}) than the observed one (1·65 × 10^{-2}). A possible explanation of this discrepancy could be the presence in ribosomal protein S_7 of the particular kind of tryptophan residues recently proposed by Burstein.[12] Such residues (type II), 'immobilized at the protein surface and in limited contact with water', according to this author, would present an emission culminating at 342 nm with a quantum yield of 0·21 and a lifetime of 4·4 ns. The two first parameters being indeed found in the ribosomal protein, such residues could be responsible for its emission.

5.4 Protein emission of tyrosine residues

Solutions in water are likely to represent the correct *model systems* for class A proteins, since emission from buried tyrosine residues has never been ascertained.

Table 1d shows that phenol and tyrosine in aqueous solution have the same temperature coefficient (0·85 × 10^{-2}), as expected from (3) since the amino-acid presents the same quantum yield as indole and is therefore unquenched. In the presence of quenching (tyrosinamide and phenol + histidine), the temperature effect is enhanced ($C_q > C_0$), as predicted from (6) which leads to $E_d/C_0 >$ 8 eV dg if one takes $E_d \approx 0.13$ eV and $C_0 = 0.85 \times 10^{-2}$. It is furthermore to be noted that values of C_q and $K[Q]$ found for the phenol-histidine (0.25 M) system at 30 °C (2.0 $\times$ 10^{-2} and 5.7 respectively) lead, when substituted in (6), to $E_d = 0.17$ eV, which value is of the same order of magnitude as that previously deduced from the study of the indole–histidine system.

As for the two proteins studied, results obtained with insulin ($C = 0.85 \times 10^{-2}$) if compared to the coefficient of phenol in water solution, would indicate an absence of quenching, which is in contradiction with the fact that its lifetime (1.5 ns) is smaller than that of tyrosine and phenol in water (3 ns). However, if compared with the value of C_0 in dioxane solution ($C_0 = 0.55 \times 10^{-2}$), then the protein coefficient implies the existence of a quenching process, in agreement with the lifetime value. We are then led to conclude that in insulin the emitting tyrosine residues—or at least some of them—are buried, as has been proposed by Menendez et al.[19]

The temperature coefficient of ribosomal protein S_8 is more difficult to interpret. It has been previously reported that the three tyrosine residues of this protein are exposed. If this is so, the coefficient value of 1.45 $\times$ 10^{-2}, compared to $C_0 = 0.85 \times 10^{-2}$, would indicate the occurrence of a dynamic quenching process, but the high lifetime of this protein (3 ns) is in conflict with this proposition. Further study is currently in progress, to try to explain this surprising result, which is perhaps related to particular properties of tyrosine in this ribosomal protein.

6 CONCLUSION

This study shows that the effect of temperature on the fluorescence quantum yield and lifetime of proteins depends on three main factors: the kind of the emitting residues (tryptophan, tyrosine or both), the hydrophilic or hydrophobic nature of their microenvironment and the occurrence of dynamic and static quenching processes.

The relative importance of these factors—and notably the magnitude of quenching processes—may be established by comparing the thermal dependence of protein emissions with that of indole and phenol in water or dioxane. These solutions indeed appear to be good models for tryptophan and tyrosine residues in aqueous or non-polar environment respectively.

The temperature coefficient, which can be easily determined, is a convenient means to carry out such comparison and thus offers a new and interesting tool for the investigation of protein conformation by fluorescence methods.

ACKNOWLEDGEMENTS

The authors wish to thank Professor H. Lami for helpful discussions and Mrs B. Lux for her assistance during this work, which was supported by the C.N.A.M.T.S. (Caisse Nationale de l'Assurance Maladie des Travailleurs Salariés).

REFERENCES

1. S. V. Konev, *Fluorescence and Phosphatase of Proteins and Nucleic Acids*, Plenum Press, New York, 1967.
2. J. W. Longworth, in *Excited States of Proteins and Nucleic Acids* (eds. R. F. Steiner and I. Weinryb, MacMillan Press, London and Basingstoke, 1971.
3. J. A. Gally and G. M. Edelman, *Biochim. Biophys. Acta.*, **60**, 499 (1962).
4. E. Leroy, H. Lami and G. Laustriat, *Photochem. Photobiol.*, **13**, 411 (1971).
5. M. S. Walker, T. W. Bednar, R. Lumry and F. Humphries, *Photochem. Photobiol.*, **14**, 147 (1971).
6. E. P. Kirby and R. F. Steiner, *J. Phys. Chem.*, **74**, 4480 (1970).
7. K. K. Turoverov and B. V. Shchelchkov, *Biofizika*, **15**, 800 (1970).
8. D. Gerard, G. Laustriat and H. Lami, *Biochim. Biophys. Acta.*, **263**, 482 (1972).
9. S. Glasstone, K. J. Laidler and H. Eyring, *The Theory of Rate Processes*, McGraw-Hill Book Company, New York and London, 1941.
10. J. Eisinger and G. Navon, *J. Chem. Phys.*, **50**, 2069 (1969).
11. R. F. Chen, *Anal. Letters*, **1**, 35 (1967).
12. E. A. Burstein, N. S. Vedenkina and M. N. Ivkova, *Photochem. Photobiol.*, **18**, 263 (1973).
13. G. Laustriat and D. Gerard, *Ann. Phys. Biol. et Med.*, **4**, 177 (1972).
14. B. Arrio, M. Hill and C. Parquet, *Biochimie*, **55**, 283 (1973).
15. K. K. Turoverov and B. V. Shchelchkov, *Biofizika*, **15**, 965 (1970).
16. P. B. Sommers, M. J. Kronman and K. Brew, *Biochem. Biophys. Res. Comm.*, **52**, 98 (1973).
17. M. J. Kronman and L. G. Holmes, *Photochem. Photobiol.*, **14**, 113 (1971).
18. G. Lemieux, Thesis, University of Strasbourg (1973).
19. C. J. Menendez, T. T. Herskovits and M. Laskowski, Jr., *Biochemistry*, **8**, 5044 (1969).

D.3. Luminescence of lysozyme-lactalbumin family of proteins †

V. L. Koenig

Division of Biochemistry, University of Texas Medical Branch, Galveston, Texas 77550, U.S.A.

J. W. Longworth

Biology Division, Oak Ridge National Laboratory, Oak Ridge, Tennessee 37830, U.S.A.

ABSTRACT

Hen and human lysozyme and cow α-lactalbumin are believed to fold into a homologous conformation, the lysozyme fold, even though at certain positions different amino-acids occur. Our interest lies in the tryptophan residues, of which the four in cow lactalbumin are in common with four out of five in human lysozyme and four out of six in the hen lysozyme. Other workers have used chemical modification to show that in hen lysozyme only two tryptophan residues, Trp_{62} and Trp_{108}, fluoresce; the human enzyme lacks Trp_{62}, and the fluorescence is consistent with emission from the single residue Trp_{108}. The fluorescence decay of hen lysozyme has two components, the human is less fluorescent and has only a single, short-lived component. The hen and human lysozymes are atypical in possessing a phosphorescence with a short decay time; the hen phosphorescence is fine structured, the human diffuse. Denaturation with guanidinium hydrochloride restores the typical tryptophan phosphorescence decay and structure. A model is proposed which accounts for the variations in fluorescence and phosphorescence and the lack of any anisotropy depolarization. The luminescence of cow lactalbumin is consistent with this interpretation of lysozyme luminescence.

The fluorescence yields of these three native and denatured proteins were studied between 100 and 350 K. The influence of temperature could be accounted for by considering a temperature-independent fluorescence and radiationless

† Research sponsored by a USPHS Special Fellowship and by the U.S. Atomic Energy Commission under contract with the Union Carbide Corporation.

conversion and a temperature-dependent radiationless process. Fluorescence lifetimes were measured at 298 and 77 K, and the individual rates were evaluated. A linear relation was found between the pre-exponential and exponential terms in the Arrhenius factor. A compensation law applied to the activated internal conversion, with an isokinetic temperature of 356 K.

Phosphorescence rise times were investigated, and shortening was detected in native proteins, which was absent in denatured proteins. It is possible that the shortened phosphorescence rise times arise from a triplet photoionization in native proteins.

D.4. Luminescence of Bence Jones proteins and light chains of immunoglobulins †

J. W. Longworth

Biology Division, Oak Ridge National Laboratory, Oak Ridge, TN 37830, U.S.A.

Carla L. McLaughlin and Alan Solomon

Memorial Research Center and Hospital, University of Tennessee, Knoxville, TN 37920, U.S.A.

ABSTRACT

The fluorescence spectra of Bence Jones proteins have proved to be extremely diverse in their amplitude, spectral distribution and lifetime. No two proteins had the same spectral properties in comparative studies on over 50 light chains. The polypeptide chain is believed to fold into an homologous conformation, the immunoglobulin fold, even though no two proteins have identical amino-acids at all positions. The structure consists of two approximately equally sized domains, with analogous structural features. For most of the kappa-type chains, there are only two tryptophan residues in each structural domain and they are located at homologous conformational positions. The N-terminal domain is subject to variation in amino-acids, the C-terminal domain is constant to all light chains of a given type. Hence the diversity of fluorescence is a result of perturbation of a single tryptophan residue in the variant domain.

The phosphorescence is equally diverse, in particular the decay is often composed of two components. One of the components is short lived, the other lies in the typical range of values for protein tryptophan residues. Certain chains decay exclusively with the short lifetime value. The existence of a short-lifetime tryptophan phosphorescence decay is believed to result from a continuous

† Research sponsored by the U.S. Public Health Service, National Science Foundation, American Cancer Society and the U.S. Atomic Energy Commission under contract with the Union Carbide Corporation.

disulphide linkage. The molecular structures of light chains show that there is such a potential contiguity in both structural domains. Hence we attribute the diversity to the variation in the perturbation by disulphide of the variant domain tryptophan.

We have found that the apparent natural lifetime of fluorescence from tyrosine and tryptophan of light chains falls into a series of groups, where there is an eightfold variation. This appears to result from a variable internal conversion, perhaps associable with specific structural features.

The fluorescence spectra have been studied over a wide range of temperature (100–350 K). The changes in yield and the dependence of the fluorescence maximum upon temperature were also diverse. The phosphorescence yield decreased between 85 and 170 K, and above 150 K shifted to the red as a result of solvent relaxation. The protein environment creates an activated radiationless conversion.

D.5. Fluorescence properties of tryptophyldiketopiperazine solutions. A study by pulse fluorometry

P. Gauduchon, B. Donzel and Ph. Wahl

Centre de Biophysique Moleculaire—45 045 Orleans CEDEX—France

SUMMARY

It is shown in this paper that, contrary to the behaviour of their parent compounds, 3-methylindole and acetyltryptophanamide, the fluorescence decays of cycloglycyltryptophan and cycloalanyltryptophan in dimethylsulphoxide (DMSO) and in aqueous solutions are not single exponentials, and their shape varies across the emission spectra.

These decays can be considered as the sums of two exponential functions, the time constants of which are independent of the emission wavelength. For the DMSO solutions, the longest time constant has a value which is close to the decay time of 3-methylindole and acetyltryptophanamide. Its relative intensity increases with the emission wavelength.

These results may be explained if we assume that the quenching by the peptide ring depends on the conformation of the diketopiperazine molecule. A quantitative interpretation is proposed, in which quenching occurs only in the folded conformer and not in the unfolded conformers. One has to assume that the equilibrium constant in the ground state, which is determined by NMR experiments, is different from the equilibrium constant in the excited state. It is then possible to determine the rate constant of folding and unfolding, and the equilibrium constant in the excited state. One can also reconstitute the emission spectrum of the two conformers which are shifted by 10 nm in DMSO.

On the other hand, our results suggest that, in the quenching conformer, the quenching group can closely approach the indole ring. The mechanism of quenching by the peptide group must then be due to a mesomeric effect rather than to an inductive effect.

1 INTRODUCTION

In order to understand the influence of primary, secondary and tertiary structures on the fluorescence of proteins, several authors have been led to study simple model compounds.[1,16] It has been shown for example that the quantum yield of aqueous solutions of oligopeptides containing aromatic residues is lower than the quantum yield of the corresponding amino-acid solutions.[2,3] This quenching by the peptide bond has been attributed to an inductive effect.[3] More recently, Cowgill has suggested that the quenching involves direct interaction of the peptide carbonyl with the aromatic ring and requires strict spatial conditions.[4] Tournon and El Bayoumi[5] have found that the fluorescence of the phenylalkylcarboxylic acids is quenched and that their phosphorescence is enhanced when they are compared to toluene. This is explained by the presence of an intramolecular charge transfer complex occurring between the carboxylic group and the phenyl ring.

The diketopiperazines are model compounds especially interesting for studies of protein conformations. They are simple molecules which have a restricted rotational freedom. It has been shown by NMR[6-8] that the preferential conformation in solution of diketopiperazines containing an aromatic residue is a folded one in which the aromatic ring faces the peptide ring. The two open conformers in which the aromatic side chains are away from the dipeptide ring, are unfavoured at room temperature.

Edelhoch et al.[9] have found that the fluorescence of these compounds is strongly quenched in water solution; in contrast, the quantum yield of diketo-piperazines in dimethylsulphoxide (DMSO) is very similar to the quantum yield of the parent-compound acetyltryptophanamide. To explain this result, the authors assume that the equilibrium between the folded and unfolded form is different in the ground and excited states.

We thought that it would be interesting to investigate the aromatic diketo-piperazines further by using pulse fluorometry methods. In this paper, we give a brief report concerning studies performed with cycloglycyltryptophan and cycloanalyltryptophan. Further details on these studies will be published elsewhere.[10]

Abbreviations used

AcTrypNH$_2$:   acetyltryptophanamide
CycloGlyTryp: diketopiperazine of glycine and tryptophan
CycloAlaTryp: diketopiperazine of alanine and tryptophan
DMSO: dimethylsulphoxide

2 MATERIAL AND METHODS

The experimental details concerning the compounds used and the methods employed are given elsewhere,[10] and we shall confine ourselves to information about the pulse fluorometry experiments. In these experiments, the fluorescence decay is determined by the single photoelectron counting method. The wavelength of the exciting pulse selected by the monochromator is 280 nm with a 10 nm bandwidth. A second monochromator selects the fluorescence wavelength with a bandwidth of 5·4 nm.

The time response function of the apparatus is obtained by using a reference compound (here, p-terphenyl or 2,5-diphenyloxazole) having a well-determined decay time. The technique has been explained elsewhere.[11] It is used for eliminating the systematic error inherent in the usual method of assuming the response function to be identical to the instrumental response to the exciting flash.

The decays are analysed as sums of exponential functions. The time constants and the amplitude are determined by a computer programme based on the modulating function method.[12] The convolution is then recalculated and the fit of the computed and experimental decays are checked by examining the weighed average residue.

3 EXPERIMENTAL RESULTS

We first examined the cyclopeptides in DMSO solutions at a concentration of 10^{-4} M and a temperature of 18 °C. For comparison, we measured the fluorescence decays of the two parent-compounds, 3-methylindole and AcTrypNH$_2$. Their decays are single exponential functions. The decay times are independent of the emission wavelength and equal to 7·5 ns and 7·3 ns, respectively.

In contrast, the decays of the diketopiperazines cannot be fitted with a single exponential function. Furthermore, the shape of these decays varies along the emission spectrum (Figure 1). These decays can be considered as the sums of two exponential functions, which may be expressed by the following relation:

$$I(\lambda) = C_\alpha(\lambda)\,\mathrm{e}^{-t/\tau_\alpha} + C_\beta(\lambda)\,\mathrm{e}^{-t/\tau_\beta} \tag{1}$$

where the decay times τ_α and τ_β are independent of the emission wavelength, while the amplitudes $C_\alpha(\lambda)$ and $C_\beta(\lambda)$ are functions of this wavelength.

The longer decay time τ_β is close to the decay time of the parent compound ($\tau_\beta = 7$ ns for CycloGlyTryp and 7·3 ns for CycloAlaTryp), while the other decay time τ_α is considerably shorter ($\tau_\alpha = 1·9$ ns for CycloGlyTryp and 2·7 ns for CycloAlaTryp).

These results mean that there are two fluorescent species, each characterized by its fluorescent spectrum and by its decay. They must correspond to different conformers of the diketopiperazine molecules.

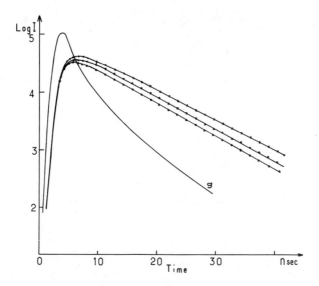

Figure 1   Fluorescence decay of AcTrypNH$_2$ in DMSO
×, emission wavelength 350 nm, and of CycloGlyTryp in
DMSO at 2 emission wavelengths: 320 nm ● and 360 nm ■.
Excitation is at 280 nm. Temperature is 18 °C

If one assumes that there is no conformation change during the excited state, and that the two kinds of conformers have single exponential decays of time constant τ_α and τ_β, the spectra of these conformers are given by:

$$F_\alpha(\lambda) = \frac{F(\lambda)C_\alpha(\lambda)\tau_\alpha}{C_\alpha(\lambda)\tau_\alpha + C_\beta(\lambda)\tau_\beta}$$

$$F_\beta(\lambda) = \frac{F(\lambda)C_\beta(\lambda)\tau_\beta}{C_\alpha(\lambda)\tau_\alpha + C_\beta(\lambda)\tau_\beta}$$

where $F(\lambda)$ is the observed composite fluorescence spectrum. An example of the separation of the two spectra is given in Figure 2.

The ratio of concentrations of the conformers is then given by the equilibrium constant K_1 which can be experimentally determined by the following relation:

$$K_1 = \frac{S_\alpha}{S_\beta} \cdot \frac{\tau_\beta}{\tau_\alpha}$$

S_α and S_β are the areas of the spectra $F_\alpha(\lambda)$ and $F_\beta(\lambda)$. The numerical values of K_1 are 0·19 for CycloGlyTryp and 0·26 for CycloAlaTryp. Let us see how we can correlate these results with those obtained by NMR.

It seems natural to attribute the time constant τ_β to the unfolded conformer and τ_α to the folded conformer. For this last conformer, the peptide bounds forming the diketopiperazine ring is close to the indole ring. However, the value

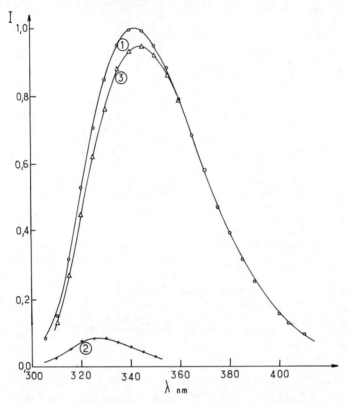

Figure 2 Fluorescence spectrum of CycloGlyTryp in DMSO at 18 °C (curve 1); decomposition into two components corresponding to the two decay times τ_α (curve 2) and τ_β (curve 3)

of the equilibrium constant K_2 determined by NMR is much higher than K_1 (namely $K_2 = 1\cdot9$ for CycloGlyTryp and $2\cdot3$ for CycloAlaTryp).

We then assume that the equilibrium constant is different in the ground state (case of the NMR experiments) and in the excited state (case of the fluorescence experiments). Furthermore, during the lifetime of the excited state, there is a possibility of unfolding and folding with rate constants k and k'.

Then we apply a kinetic scheme similar to the scheme used in the case of exciplex formation.[13] As a result, the decay of each conformer is now a sum of two exponentials, the time constant of which are the experimental values τ_α and τ_β. These time constants are functions of the intrinsic time constants τ_1 and τ_2 of the isolated conformers and of k and k'. If the intrinsic time constant τ_2 of the unfolded form is taken to be equal to the time constant of AcTrypNH$_2$, one possesses enough experimental data to calculate the other three parameters τ_1, k and k', and the equilibrium constant in the excited state $K_e = k'/k$.

In this way we obtained for CycloGlyTryp: $\tau_1 = 4\cdot2$ ns, $k = 2\cdot7 \times 10^8\,\mathrm{s}^{-1}$, $k' = 2\cdot3 \times 10^7\,\mathrm{s}^{-1}$, $K_e = 0\cdot085$.

The spectra $F_1(\lambda)$ and $F_2(\lambda)$ of the two conformers can also be determined by an appropriate linear combination of $F_\alpha(\lambda)$ and $F_\beta(\lambda)$.[10] It is found that they are shifted by 10 nm.

We have also studied the emission of CycloGlyTryp in water at 6 °C. Here again, the decay of 3-methylindole and AcTrypNH$_2$ are single exponentials. Their time constants, 12·9 ns and 3·8 ns respectively, are independent of the emission wavelength.

The decays of CycloGlyTryp are also sums of two exponentials and can be represented by the formula (1) with $\tau_\alpha = 0\cdot3$ ns and $\tau_\beta = 3$ ns. $F_\alpha(\lambda)$ and $F_\beta(\lambda)$ have been computed and the constant K_1 is found to be equal to 14·5.

There is no determination of the ground state equilibrium constant by NMR for the aqueous solution of CycloGlyTryp, since the solubility of the product is very low in that solvent. However, one can apply the kinetics scheme already used for the DMSO solution. It seems more appropriate to identify τ_2 with the decay time of 3-methylindole than to the decay time of AcTrypNH$_2$. The low decay time of this compound indicates an important quenching by the amide and the carbonyl groups. A quantitative discussion allows one then to conclude that τ_1 and k' are practically independent of K_2. Their values are respectively 0·3 ns and $2\cdot6 \times 10^8$ s^{-1}.

One can conclude that two causes contribute to give a stronger quenching in water solution than in DMSO. First, τ_1 is about 14 times smaller in water than in DMSO, which indicates that the peptide bond is a stronger quencher in water than in DMSO. Secondly, the rate constant of folding is 10 times greater in water than in DMSO. It seems reasonable to assume that the rate constant of unfolding is of the same order of magnitude in the two solvents. Then the equilibrium constant in the excited state must be considerably higher in water than in DMSO.

4 DISCUSSION

The important experimental finding of our work is that the fluorescence decay of tryptophyl cyclopeptides are not single exponential functions, and that they depend on the emission wavelength. These results inevitably entail that the quantum yield and the emission spectrum of indole fluorescence depend on the molecular conformation.

It is interesting to try correlating these emission properties with the different conformations of the molecule. We have proposed an interpretation in which the unfolded conformers have an intrinsic decay time equal to the 3 methylindole decay time; while the folded conformer is quenched by interaction with the peptide ring. In order to correlate the pulse fluorometry experiments with the NMR experiments, one has to assume that the ground state equilibrium is perturbed by changes of conformation during the lifetime of the excited state.

In this interpretation, the apparent equilibrium constant K_1, determined by pulse fluorometry, must become closer to the ground state constant K_2

as the temperature decreases. Measurements are now in progress along this line.

In any case, our results strongly suggest that efficient quenching requires that the peptide bond should be in close contact with the indole ring.

Let us consider again the decay of $AcTrypNH_2$ in aqueous solution. We have seen that this decay is a single exponential and that its time constant is independent of the emission wavelength. However, its value is more than three times lower than the decay time of 3-methylindole.

This may be explained if the rate of conformation change is much greater than the rate of fluorescence emission or, more precisely, if the lifetime of the conformers is smaller than the resolution time of the pulse fluorometry method used. This must also be the case of most of the small linear peptides containing tryptophan.[14,16]

On the contrary, for rigid peptides and for proteins one can expect to find, in some cases, multiexponential fluorescence decays as a result of the quenching by peptide bonds. This might be the case in human serum albumin where the fluorescence of the single tryptophan has a biexponential decay.[17]

REFERENCES

1. F. W. J. Teale and G. Weber, *Biochem. J.*, **65**, 476 (1957).
2. A. White, *Biochem. J.*, **71**, 217 (1959).
3. R. W. Cowgill, *Arch. Biochem. Biophys.*, **100**, 36 (1963).
4. R. W. Cowgill, *Biochim. Biophys. Acta*, **200**, 18 (1970).
5. J. Tournon and M. A. El Bayoumi, *J. Chem. Phys.*, **56**, 5128 (1972).
6. K. D. Kopple and D. M. Marr, *J. Amer. Chem. Soc.*, **89**, 6193 (1967).
7. K. D. Kopple and M. Ohnishi, *J. Amer. Chem. Soc.*, **91**, 962 (1969).
8. B. Donzel, Thesis, Zürich (1971).
9. H. Edelhoch, R. S. Bernstein and M. Wilcheck, *J. Biol. Chem.*, **243**, 5985 (1968).
10. B. Donzel, P. Gauduchon and Ph. Wahl, *J. Amer. Chem. Soc.*, **96**, 801 (1974).
11. Ph. Wahl, J. C. Auchet and B. Donzel, *Rev. Sci. Instrum.*, **45**, 28 (1974).
12. B. Valeur and J. Noirez, *J. Chim. Phys.*, **70**, 500 (1973).
13. J. B. Birks, *Photophysics of Aromatic Molecules*, Wiley–Interscience, London and New York, 1970.
14. W. B. De Lauder and Ph. Wahl, *Biochemistry*, **9**, 2750 (1970).
15. R. A. Bradley and F. W. J. Teale, *J. Mol. Biol.*, **44**, 71 (1969).
16. I. Weinryb and R. F. Steiner, in *Excited States of Proteins and Nucleic Acids*, p. 277 (eds. R. F. Steiner and I. Weinryb), Macmillan, 1972.
17. Ph. Wahl and J. C. Auchet, *Biochim. Biophys. Acta*, **285**, 99 (1972).

D.6. Studies of the conformation and self-association of cyclic decapeptides by triplet–triplet energy transfer†

C. F. Beyer, L. C. Craig and W. A. Gibbons

Rockefeller University, New York, N.Y. 10021, U.S.A.

J. W. Longworth

Biology Division, Oak Ridge National Laboratory, Oak Ridge, Tenn. 37830, U.S.A.

SUMMARY

We have studied intra- and inter-molecular triplet energy transfer from ionized tyrosine residues to tryptophan residues in the cyclic antibiotic decapeptides, the tyrocidines. Chromophore separations deduced from transfer efficiencies have been used to help define peptide secondary and quaternary structure. Our major results and conclusions are the following. (1) No triplet transfer in monomeric tyrocidine B (in which tyrosinate and tryptophan are separated by three intervening residues in the primary structure) indicates that the peptide does not have a secondary structure which brings the donor and acceptor chromophores into close relation. A flat, antiparallel *beta*-structure for the peptide backbone, which has been proposed from nuclear magnetic resonance experiments, is consistent with the observed transfer results. (2) Extensive triplet transfer in monomeric tyrocidine C (tyrosinate and tryptophan separated by two residues in the sequence) requires that the separation between the donor and acceptor be close to 8 Å. This distance can be accommodated to the *beta*-structure model for the tyrocidine backbone conformation by rotating the chromophores about their C_α–C_β bonds to bring them into closest proximity. Thus the orientations of the tyrosinate and tryptophan side chains relative to the rigid peptide backbone are established. (3) Selective triplet transfer to the red tryptophan phosphorescence component of tyrocidine B under conditions

† Research supported in part by Grant AMO2493 from the U.S. National Institutes of Health. Oak Ridge National Laboratory is operated by Union Carbide Corporation under contract with the U.S. Atomic Energy Commission.

411

where the peptide is partially aggregated is interpreted as intermolecular transfer between chromophores of peptide molecules within the aggregate. This suggests that the separation between aromatic rings of different peptide molecules in an aggregate can be quite small, and lends support to the proposed micelle-like structure for the peptide aggregate.

1 INTRODUCTION

The efficiency of intramolecular energy transfer between intrinsic chromophores of peptides and proteins is potentially of use in obtaining conformational information because of the dependence of transfer on chromophore separation and orientation. Singlet transfer between tyrosine and tryptophan residues has been most frequently used for this purpose.[1] Another kind of energy transfer, which has received less attention, is triplet–triplet transfer from ionized tyrosine to tryptophan. Such transfer has been observed in model peptide systems[2,3,4,5,6a] and in two proteins.[6b,7] Transfer from extrinsic chromophores to tryptophan at the triplet level is also known.[8,9]

Triplet transfer occurs by an electron exchange mechanism and thus has a very strong distance dependence. Chromophore separations determined by this method are complementary to those obtained by other transfer mechanisms, and in some situations triplet transfer may be more informative than singlet transfer. We have studied intramolecular triplet energy transfer from ionized tyrosine to tryptophan in the tyrocidine series of cyclic decapeptide antibiotics. Our goal was to estimate distances between chromophores and use the distances to draw specific conclusions about peptide conformation.

The tyrocidines are well-suited for conformational study by energy transfer methods. The primary structures of the A, B and C peptides are known;[10,11,12] the differences among them are specific substitutions of aromatic amino-acids. Their sequences are $cyclo(-Val_1-Orn_2-Leu_3-D-Phe_4-Pro_5-X_6-D-Y_7-Asn_8-Gln_9-Tyr_{10}-)$ where X and Y are phenylalanine in tyrocidine A, X is tryptophan and Y is phenylalanine in tyrocidine B, and X and Y are tryptophan in tyrocidine C. Chromophore separations derived from transfer efficiencies have direct conformational relevance in this system for two reasons. First, because of their covalent cyclic structure, the available conformations of the tyrocidine peptide backbone are limited, and one or a few conformations may be strongly favoured over all others. This is in contrast to linear peptides of the same size class which very likely have a large number of allowed, rapidly interconverting, conformations. Second, much information about the secondary structures of the tyrocidines has recently become available from proton magnetic resonance studies.[13,14] This was of particular importance for the interpretation of our triplet transfer results.

The tyrocidines are of additional interest because of their marked tendency to self-associate in aqueous solution. Indeed, the basis of their antibiotic activity is thought to be the disruption of cell membrane structure by a detergent-like solubilization of non-polar membrane components. Tyrocidine self-association has been studied by several physical techniques, notably ultracentrifugation[15] and thin-film dialysis,[16] and a partial understanding of the process has emerged.[17] The luminescence of the tyrocidines is strongly influenced by self-association.[18] The dependence of chromophore environment on aggregation, and the occurrence of intermolecular triplet energy transfer within the aggregate, are presented in this paper.

2 EXPERIMENTAL PROCEDURES

The tyrocidines were purified by countercurrent distribution followed by adsorption chromatography on Sephadex G-25.[12] L-tyrosyl-L-tryptophan was purchased from Cyclo Chemical Company.

All emission spectra were obtained on an instrument previously described,[19] according to general procedures given by Longworth.[20] A dewar with an un-silvered tail section was used to hold the samples, in 3-mm internal diameter quartz tubes, at liquid nitrogen temperature. Sample concentrations were about 10^{-4} M which corresponded to absorptions at 280 nm at room temperature in one-cm cuvettes of 0·4 to 2. All basic solutions were used within an hour of preparation. Excitation and emission bandwidths were never greater than 5 nm and were usually smaller. Phosphorescence lifetimes were determined as previously described.[21]

3 RESULTS

Triplet–triplet energy transfer is a process in which a donor is converted from its lowest triplet state to its ground singlet state with the simultaneous excitation of an acceptor from its ground singlet to its first triplet state. Transfer may be either intra- or inter-molecular. The mechanism of transfer is electron exchange, which requires some overlap of the electron densities of donor and acceptor. The distance between donor and acceptor must therefore be short. Triplet transfer is reviewed extensively in References 22, 23 and 24.

Transfer of donor triplet energy will reduce the rates of other deactivational processes operative in the absence of transfer. One of these is donor phosphorescence, and transfer will therefore result in a quenching of donor phosphorescence. At the same time, transfer to an acceptor will lead to a sensitization of acceptor phosphorescence. Quenching of the donor and sensitization of the acceptor are most easily demonstrated if the singlet excitation energy of the donor is less than that of the acceptor. In this situation, the donor may be selectively excited to its triplet state by singlet absorption and intersystem

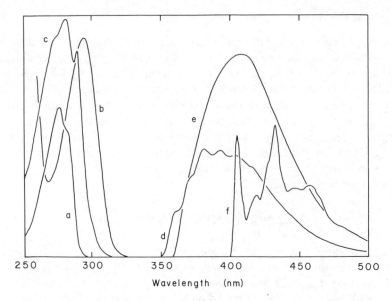

Figure 1 Room temperature absorption and low-temperature phosphorescence emission spectra of tyrosine, tyrosinate and tryptophan. a: Absorption of glycyl-L-leucyl-L-tyrosine; d: emission of N-acetyl-L-tyrosinamide. b: absorption of glycyl-L-leucyl-L-tyrosinate; e: emission of N-acetyl-L-tyrosinate amide. c: Absorption of glycyl-L-tryptophan; f: emission of N-acetyl-L-tryptophanamide. N-acetyl-L-tyrosinamide was dissolved in dimethyl-sulphoxide-ethylene glycol–water (4:5:1); all other compounds were dissolved in methanol–water (1:1). Basic solutions were 0·1 N in KOH. The relative heights of the various spectra are not significant. Phosphorescence (77 K) was excited at 280 nm

crossing, and acceptor phosphorescence will only be observed if triplet transfer occurs.

In peptides and proteins, the intrinsic chromophores which satisfy these conditions are ionized tyrosine (triplet donor) and tryptophan (triplet acceptor). The room-temperature absorption spectra and 77 K phosphorescence emission spectra of model compounds containing tyrosine, ionized tyrosine and tryptophan are shown in Figure 1. Tyrosine absorption is at higher energies than tryptophan, so experimentally it is not a good candidate for the triplet donor. Ionization of the phenolic hydroxyl of tyrosine to produce the tyrosinate anion causes a large red shift of the absorption spectrum. Tyrosinate has strong absorption at 300 nm, where tryptophan absorption is weak. The phosphorescence emission of tyrosine is characterized by an onset wavelength of 350 nm, peak emission at about 390 nm and a decay time of 2·9 s. Tyrosinate has an onset wavelength of 360 nm, a maximum at about 410 nm and a decay time of 1.5 s. Tryptophan phosphorescence has several sharp vibrational bands, especially the $0 - 0$ transition at the blue edge of the spectrum, which permits its identification even in the presence of simultaneous tyrosinate emission.

Tryptophan phosphorescence has an onset wavelength of 401 nm and a decay time of about 6 s, depending on chromophore environment.

Tyrocidine A is the ideal reference peptide for obtaining the luminescence spectra of tyrosine and tyrosinate residues, because it contains a tyrosine residue, but no tryptophan, in a chemical environment which must be very similar to that of the tyrosine residue in tyrocidines B and C. The upper panel of Figure 2 presents the total luminescence spectra of tyrocidine A dissolved in methanol. The fine structure of the phosphorescence spectrum excited at 280 nm is similar to the spectrum of N-acetyl-L-tyrosinamide (Figure 1). The phosphorescence onset wavelength and lifetime are characteristic of tyrosine. Excitation at 300 nm, where tyrosine does not absorb, produces essentially no emission.

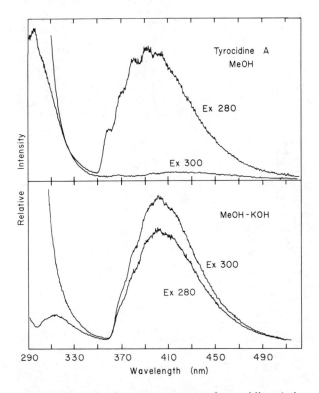

Figure 2 Total luminescence spectra of tyrocidine A in neutral and basic methanol at 77 K. *Upper panel*: tyrocidine A dissolved in pure methanol, excited at 280 and 300 nm. *Lower panel*: tyrocidine A dissolved in methanol containing 0·1 N KOH, excited at 280 and 300 nm. In each panel, the relative intensities of the two spectra are correct

The addition of KOH to the methanolic solution of tyrocidine A causes ionization of the tyrosine phenolic hydroxyl, and a major change in the emission properties of the peptide (Figure 2, lower panel). The onset wavelength, lifetime

416

and fine structure of the phosphorescence emission are now characteristic of tyrosinate. The significant absorption of tyrosinate at 300 nm is reflected in the strong emission for excitation at this wavelength.

The total luminescence spectra of L-tyrosinate-L-tryptophan, tyrocidine B, and tyrocidine C, all in basic methanol solutions, are presented in Figure 3. L-tyrosinate-L-tryptophan was examined as a control because the proximity of the donor tyrosinate and acceptor tryptophan was known to result in complete energy transfer.[2,3] There is no tyrosinate emission from L-tyrosinate-L-tryptophan (Figure 3, upper panel; cf. Figure 2), indicating that tyrosinate

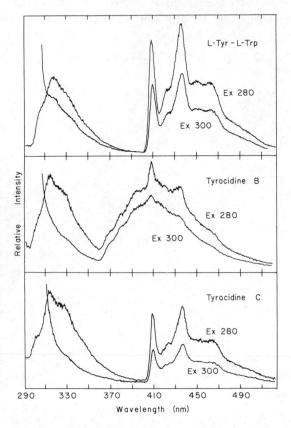

Figure 3   Total luminescence spectra of L-tyrosinate-L tryptophan, tyrocidine B, and tyrocidine C in basic methanol at 77 K. *Upper panel*: L-tyrosinate-L-tryptophan dissolved in methanol containing 0·1 N KOH, excited at 280 and 300 nm. *Middle panel*: tyrocidine B dissolved in methanol containing 0·1 N KOH, excited at 280 and 300 nm. *Lower panel*: tyrocidine C dissolved in methanol containing 0·1 N KOH, excited at 280 and 300 nm. The relative intensities of the two spectra in each panel are correct.

phosphorescence is completely quenched. At the same time, there is a sensitization of tryptophan phosphorescence, since the intensity of the tryptophan emission excited at 300 nm is greater than can be accounted for by the fractional absorption of tryptophan at this wavelength. These observations indicate that triplet energy transfer occurs with high efficiency from ionized tyrosine to tryptophan in this dipeptide.

Tyrocidine B on the other hand shows no triplet transfer (Figure 3, middle panel). There is a strong emission from tyrosinate, and the intensity of the tryptophan emission is no greater than in neutral methanol. The lifetime of the phosphorescence emission observed at 390 nm is 1·5 s, characteristic of tyrosinate, whereas at 435 nm the emission contains two components, a short-lived one from tyrosinate, and a long-lived one from tryptophan. When the sample is excited at 300 nm, the sharp peaks of the tryptophan emission are very weak on a strong tyrosinate background, consistent with the fractional absorptions of these residues at the excitation wavelength and the absence of triplet transfer.

The spectra of tyrocidine C (Figure 3, lower panel) are quite distinct from those of tyrocidine B, and resemble closely the spectra of the dipeptide L-tyrosinate-L-tryptophan. There is very little or no tyrosinate emission from tyrocidine C, and the intensity of tryptophan emission for 300 nm excitation is approximately half of its value for 280 nm excitation, indicating a sensitization of tryptophan emission. Thus there is essentially complete triplet energy transfer from ionized tyrosine to tryptophan in tyrocidine C. Since the tyrocidines are known to be monomeric in methanol solution at low concentrations, this transfer must be intramolecular.

The tyrocidines possess a dual tryptophan phosphorescence emission under certain conditions.[18] The observed peptide spectrum can be resolved into two overlapping tryptophan emissions having characteristically different wavelength positions, excitation spectra and phosphorescence lifetimes. The relative intensities of the two emissions can be manipulated by changing various factors such as solvent composition or peptide concentration. An example of this is presented in Figure 4 which shows the 0 − 0 region of the tryptophan phosphorescence of tyrocidine C in methanol:water mixtures. In pure methanol (curve a), a single tryptophan emission is observed, and its wavelength distribution is similar to that of model compounds such as N-acetyl-L-tryptophanamide under the same conditions. As the solvent composition is changed to include more water (curves b to e), a second tryptophan emission 8 nm to the red of the first emission progressively appears. Tyrocidines B and C behave identically in such a solvent titration. In a previous paper,[18] we presented evidence that the red component is emitted by tryptophan residues of aggregated peptide molecules and the blue component by residues of monomeric peptide molecules. We were therefore able to ask the question whether triplet transfer in tyrocidine B can occur selectively from ionized tyrosine to one of the tryptophan populations.

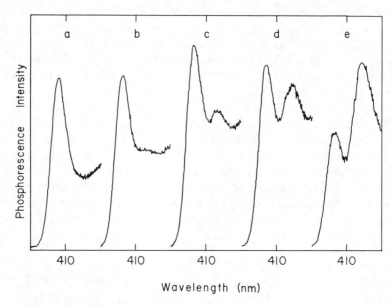

Figure 4 Blue edge of the tryptophan phosphorescence emission of tyrocidine C in methanol–water mixtures, at 77 K. Solvent compositions: a, pure
methanol; in b through e, the methanol–water volume ratios were: b, 8:2;
c, 7:3; d, 6:4; e, 5:5. Only the relative intensities of the blue and red components in each solvent are significant

We chose methanol:water (1:1) as the solvent for these experiments because
in it the intensities of the resolved emission components are about equal.
Figure 5 presents the luminescence spectra of tyrocidine B in neutral and basic
methanol:water (1:1). The dual tryptophan phosphorescence emission is
readily apparent when the sample is excited at 280 nm (Figure 5, upper panel).
In agreement with our previous results,[18] excitation at higher wavelengths
(300 nm) under neutral conditions selectively excites the red emission component, because of the slight red shift of the excitation spectrum of the red component relative to that of the blue component. When the solution is made basic
by the addition of KOH, the luminescence spectra change as shown in the
lower panel of Figure 5. There is considerable emission from tyrosinate residues,
but there is also a selective sensitization of the phosphorescence of the red
tryptophan component. Intensities of the various emissions after resolution of
the spectra are given in Table 1. Selective sensitization of the red component is
seen most clearly from the spectrum excited at 300 nm, a wavelength where the
fractional absorption of tyrosinate relative to tryptophan is large. The blue
tryptophan component is barely discernable from the total emission envelope
whereas the red component is nearly as intense as for 280 nm excitation. After
allowing for the greater absorption of the red tryptophan population at the
higher excitation wavelength, the values in Table 1 make it clear that the strong

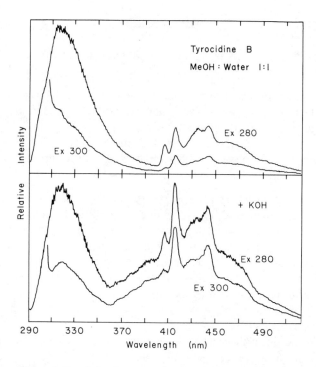

Figure 5 Total luminescence spectra of tyrocidine B in neutral and basic methanol–water (1:1), at 77 K. *Upper panel*: tyrocidine B dissolved in methanol–water (1:1) excited at 280 and 300 nm. *Lower panel*: tyrocidine B dissolved in methanol–water (1:1) containing 0·1 N KOH, excited at 280 and 300 nm. The relative intensities of the two spectra in each panel are correct

intensity of the red emission must be due to triplet energy transfer from tyrosinate.

Table 1 Phosphorescence intensities of resolved red and blue tryptophan components of tyrocidine B in methanol:water (1:1) under neutral and basic conditions. The intensities are in arbitrary units

Excitation wavelength (nm)	Phosphorescence intensity			
	neutral		basic	
	blue	red	blue	red
280	10·0	11·5	9·5	24·0
300	2·0	5·5	3·5	19·0

4 DISCUSSION

Our results may be summarized in the following statements:

(a) Triplet–triplet energy transfer at low temperature from ionized tyrosine to tryptophan does not occur in monomeric tyrocidine B dissolved in methanol.

(b) Triplet–triplet energy transfer does occur with a high efficiency in monomeric tyrocidine C dissolved in methanol.

(c) Triplet–triplet energy transfer occurs selectively to the red tryptophan phosphorescence component in tyrocidine B dissolved in methanol:water (1:1).

Recent proton magnetic resonance experiments[13,14] have led to a model of the backbone conformation of tyrocidine A in dimethylsulphoxide, which serves as a useful starting point for the correlation of transfer efficiencies with molecular structure. The covalently cyclic peptide backbone forms an anti-parallel *beta*-structure with four strong intramolecular hydrogen bonds. A schematic illustration of the proposed structure is shown in Figure 6. The backbone conformation of tyrocidine A is very similar to that of tyrocidine S (also called gramicidin S–A), which was previously determined by magnetic resonance experiments.[25]

The conformation of the backbone ring and the presence of a D-amino-acid at position 7 in the sequence result in a clustering of most of the hydrophobic

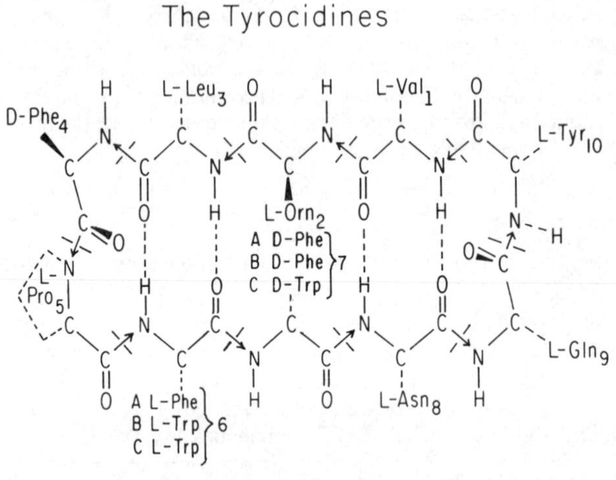

The Tyrocidines

Figure 6 Schematic illustration of the peptide backbone structure of the tyrocidines derived from proton magnetic resonance.[13,14] All C_α–C_β bonds are directed below the plane of the page except those of Orn_2 and $D-Phe_4$. The substitutions of aromatic residues in the A, B and C peptides are shown at positions 6 and 7. Hydrogen bonds between the two strands of the antiparallel *beta*-structure are indicated by dashed lines. The regularity of the *beta*-structure is idealized in this drawing

side chains on one side of the plane formed by the backbone ring. On this side are found the side chains of Val_1, Leu_3, Pro_5, Phe_6, and D-Phe_7, as well as those of Asn_8, Gln_9, and Tyr_{10}. The side chains of only two residues, Orn_2, containing the only charged group in the peptide, and D-Phe_4, are on the other side.

The backbone conformation derived from nuclear magnetic resonance suggests how the C_α–C_β bond of each residue is oriented relative to the backbone, but it does not define side-chain rotation about the C_α–C_β bond. This may be determined for certain side chains by intramolecular energy-transfer studies such as those reported here. Such studies therefore complement other techniques useful in conformational analysis, and provide additional restrictions which must be met by any proposed conformation.

Trp_6 and Tyr_{10} in tyrocidine B are separated in the primary structure by three intervening residues. This separation is too great to allow triplet transfer unless the peptide has a secondary structure which brings the two chromophores into close relation. The absence of intramolecular transfer indicates that tyrocidine B does not possess such a secondary structure, and is consistent with a flat, *beta*-structure for the peptide backbone. The distance between the chromophores in the *beta*-structure model is 12 to 18 Å, depending on the precise rotation of each about its C_α–C_β bond. Even at the minimum distance, the separation is too great for orbital overlap and triplet energy transfer to occur.

There are two residues between D-Trp_7 and Tyr_{10} in the amino-acid sequence of tyrocidine C. This is close enough to permit triplet transfer if side-chain rotations are favourable. According to the *beta*-structure model of tyrocidine C, rotations of the D-Trp_7 and Tyr_{10} chromophores produce separations of 8 to 15 Å. The experimental observation of extensive intramolecular triplet energy transfer from ionized tyrosine to tryptophan requires that the side chains be more or less fixed relative to each other in the positions which minimize their separation. Minimum separation is achieved when both side chains are folded in over the hydrophobic side of the peptide backbone ring. This orientation makes chemical sense because it provides the least exposure of the aromatic chromophores to polar solvent molecules. It is also reasonable to expect that the bulky aromatic rings will have essentially fixed positions relative to each other and to the peptide backbone because rotations about the C_α–C_β bonds, especially of D-Trp_7, are highly hindered. The tight clustering of non-polar side chains on one side of the peptide ring limits the rotational freedom of the chromophores. There are only a few available positions for D-Trp_7, and the high transfer efficiency suggests that it occupies the position which places it closest to Tyr_{10}.

Before considering the triplet transfer results for tyrocidine B in methanol:water (1:1), a brief description of tyrocidine self-association must be given. The tyrocidines form micelle-like aggregates which are stabilized by hydrophobic interactions. This conclusion has been derived from a series of studies by several techniques including chemical modification of amino-acid side chains,[26] thin-film dialysis[16] and ultracentrifugation.[15] Under certain conditions, at least, the initial stages of aggregation can be described as a mono-mer–n-mer–nj-mer equilibrium, where n is about 9 and j is about 3.[15,17] An apparent critical micelle concentration can be detected by use of a fluorescent probe,[27] and the fluorescence characteristics of probe molecules bound to aggregated tyrocidine B suggest that the probe is sequestered in a hydrophobic environment.[27] A micelle-like structure for the tyrocidine aggregate is con-sistent with the secondary structure of individual peptide molecules discussed above, because of the distinct difference in hydrophobicity of the two sides of the peptide ring. The aggregate is pictured as consisting of peptide molecules ar-ranged such that their hydrophobic sides face inward in contact with neigh-bouring molecules, resulting in the exclusion of polar solvent from the interior, while the polar sides of the peptides face outward towards the solvent.

In our previous work on tyrocidine phosphorescence heterogeneity,[18] we found that the intensity of the red tryptophan emission component increased under conditions which were known to promote peptide aggregation. From this information and comparisons with model compounds in various solvents, we proposed that the blue component was emitted by monomeric tyrocidine molecules whose tryptophan residues were largely exposed to the polar solvent, whereas the red component was emitted by tryptophan residues buried in the hydrophobic interior of peptide aggregates.

The triplet transfer results for tyrocidine B in methanol:water (1:1) may now be understood in terms of peptide self-association. We know for the case of tyrocidine B in methanol (where only the blue emission component is present) that there is very little or no triplet transfer in monomeric peptide molecules. We also find no evidence of transfer to the blue tryptophan population of tyrocidine B in methanol:water (1:1), as expected if the blue emission is indeed from peptide monomers. Transfer does occur from tyrosinate to the red trypto-phan population. This is interpreted as intermolecular transfer from a tryosinate of one molecule to a tryptophan of a neighbouring molecule within the peptide aggregate.

This interpretation is consistent with our previous conclusion[18] that the red emission came from aggregated peptide molecules, and suggests that the distance between tyrosinate and tryptophan residues of different molecules in the aggregate can be quite small (of the order of 8 Å). This is perfectly compatible with the view that the hydrophobic sides of individual tyrocidine molecules are in close contact in the aggregate to form the interior of a micelle-like structure.

In conclusion, we have shown how triplet–triplet energy transfer studies can yield conformational information about peptides. The technique is most useful when combined with others, especially nuclear magnetic resonance, which yield complementary information on different aspects of the conformational problem. In our study, the covalent cyclic structure of the tyrocidines was a great advantage in that it drastically limited the possible conformations of the peptide backbone. We are currently investigating other types of intramolecular energy transfer between the intrinsic chromophores of the tyrocidines. These include singlet–singlet transfer from tyrosine to tryptophan and from tryptophan to ionized tyrosine. Singlet transfer occurs by a different mechanism than triplet transfer and has different distance and orientation dependencies. Thus additional relationships between the chromophores will be revealed. The results of all the transfer studies, combined with information from other techniques, should lead to a detailed understanding of the structure and self-associated properties of the tyrocidines.

ACKNOWLEDGEMENT

We thank Dr Herman Wyssbrod for discussions of the proposed model of tyrocidine A derived from proton magnetic resonance experiments.

REFERENCES

1. I. Z. Steinberg, *Ann. Rev. Biochem.*, **40**, 83 (1971).
2. C. Hélène, M. Ptak and R. Santus, *J. Chim. Phys.*, **65**, 160 (1968).
3. R. F. Steiner and R. Kolinski, *Biochemistry*, **7**, 1014 (1968).
4. D. Mantik, J. E. Maling and M. Weissbluth, *Biophys. J.*, **9**, A160 (1969).
5. R. Santus, M. Bazin, M. Aubailly and R. Guermonprez, *Photochem. Photobiol.*, **15**, 61 (1972).
6. J. W. Longworth, in *Excited States of Proteins and Nucleic Acids*, (a) pp. 352–353, (b) pp. 410–411 (eds. R. F. Steiner and I. Weinryb), Plenum Press, New York, 1971.
7. C. A. Ghiron, J. W. Longworth and N. Ramachandran, *Proc. Nat. Acad. Sci. USA*, in press (1974).
8. W. C. Galley and L. Stryer, *Proc. Nat. Acad. Sci. USA*, **60**, 108 (1968).
9. R. M. Purkey and W. C. Galley, *Biochemistry*, **9**, 3569 (1970).
10. A. Paladini and L. C. Craig, *J. Amer. Chem. Soc.*, **76**, 688 (1954).
11. T. P. King and L. C. Craig, *J. Amer. Chem. Soc.*, **77**, 6627 (1955).
12. M. A. Ruttenberg, T. P. King and L. C. Craig, *Biochemistry*, **4**, 11 (1965).
13. H. R. Wyssbrod, M. Fein, J. Dadok, R. F. Sprecher, C. F. Beyer, L. C. Craig, P. Ziegler and W. A. Gibbons, Abstracts of 1973 American Chemical Society National Meeting, Chicago, *Abstr. Biol.*, 184 (1973).
14. H. R. Wyssbrod and W. A. Gibbons, manuscript in preparation.
15. R. C. Williams, Jr., D. A. Yphantis and L. C. Craig, *Biochemistry*, **11**, 70 (1972).
16. M. Burachik, L. C. Craig and J. Chang, *Biochemistry*, **9**, 3293 (1970).
17. S. L. Laiken, Ph.D. Thesis, The Rockefeller University, New York (1970).
18. C. F. Beyer, J. W. Longworth, W. A. Gibbons and L. C. Craig, *J. Biol. Chem.*, in press (1974).
19. J. W. Longworth and R. O. Rahn, *Biochim. Biophys. Acta*, **147**, 526 (1967).
20. J. W. Longworth, *Photochem. Photobiol.*, **8**, 589 (1968).
21. C. Hélène and J. W. Longworth, *J. Chem. Phys.*, **57**, 399 (1972).
22. J. B. Birks, *Photophysics of Aromatic Molecules*, p. 537, Wiley–Interscience, London, 1970.
23. R. G. Bennett and R. E. Kellogg, *Progress in Reaction Kinetics*, Vol. 4, p. 215 (ed. G. Porter), Pergamon Press, New York, 1967.
24. A. A. Lamola, in *Energy Transfer and Organic Photochemistry*, p. 17 (eds. A. A. Lamola and N. J. Turro), Wiley–Interscience, New York, 1969.

25. A. Stern, W. A. Gibbons and L. C. Craig, *Proc. Nat. Acad. Sci. USA*, **61**, 734 (1968).
26. M. A. Ruttenberg, T. P. King and L. C. Craig, *Biochemistry*, **5**, 2857 (1966).
27. C. F. Beyer, L. C. Craig and W. A. Gibbons, *Biochemistry*, **11**, 4920 (1972).

D.7. Laser study of triplet porphyrin quenching by oxygen in porphyrin–globins

B. Alpert

Institut de Biologie Physico-chimique, 13, rue Pierre et Marie Curie, 75005 Paris, France

L. Lindqvist

Laboratoire de Photophysique Moléculaire du C.N.R.S., Université de Paris-Sud, 91405 Orsay, France

1 INTRODUCTION

Molecules in electronically excited states have an important application in biology as probes of dynamic processes, and fluorescence properties have been particularly studied. We wish to demonstrate the possibilities of using the triplet state as a means of probing the diffusion of oxygen in hemoproteins.

The fixation of oxygen on hemoproteins is known to vary over a wide range.[1] The extent of fixation is determined on the one hand by the complexing power of the heme iron and on the other hand by the rate of diffusion of oxygen into (and out of) the pocket of the globin in which the heme is imbedded. Since data on diffusion of oxygen in proteins are very scarce it is difficult to evaluate the relative importance of these two factors.

We have made an attempt to determine these dynamics by studying the quenching of the triplet state of the heme (Fe^{2+}-protoporphyrin IX complex) group by oxygen. Such triplet quenching studies are most conveniently made by laser or flash photolysis techniques, that is, by pulse excitation and measurement of the decay of the transient triplet state absorption. However, it has not been possible to observe the heme triplet even with these fast techniques,[2] presumably because of the rapid transition to the ground state. Removal of the heme iron is expected to increase this lifetime to the range normally found for porphyrin triplets (in the order of 10^{-4} to 1 s).[3–6]

We have therefore undertaken a laser study of porphyrin–globins (obtained from hemoglobin A and horse myoglobin). The results show that the triplet state of the porphyrin is detectable by its light absorption and that it can indeed be used to determine the rate of diffusion of oxygen into the globin pocket.

425

2 EXPERIMENTAL

2.1 Laser photolysis apparatus

The laser photolysis equipment has been described briefly.[7,8] A Q-switched Nd^{3+}-doped glass laser with second-harmonic generation was used as pulse excitation light source. The 529 nm emission was isolated from the infrared by a 4 mm thick Schott BG 38 glass filter. The pulsewidth at half-maximum was 30 ns. The light was directed at normal incidence onto one of the shorter sides of a 10×5 (or $\times 1$ or $\times 10$) mm spectrophotometry cell with polished windows containing the solution to be studied. Transient light absorption changes were measured along the short axis of the cell using a xenon flash lamp (approximately 1 ms duration) as measuring light source. The light intensity of the xenon flash was constant within 5% during the maximal time interval of measurement, $50 \mu s$. The wavelength band of measurement (bandwidth 2 nm) was selected by means of a monochromator. The transmitted light was measured as a function of time using a photomultiplier tube connected to one channel of a double-beam oscilloscope. The detector time resolution was approximately 10 ns. The laser pulse energy was measured by deviating a small fraction of the second-harmonic beam onto a vacuum photodiode. The diode signal was integrated and recorded on the second channel of the oscilloscope.

2.2 Preparation of solutions

Human hemoglobin (HbA) and horse myoglobin (Mb) were prepared by the usual methods.[9,10] The globins of these proteins were obtained by precipitation in acid acetone at $-20\,°C$ and they were slowly regenerated by severalfold dialysis.[11] The globins were associated to the porphyrin[12,13] at pH 7 (0.1 M potassium phosphate). The excess of porphyrin and denatured proteins was eliminated by standard chromatography on DEAE cellulose. The globin–porphyrin complexes last 2.4 days between 1 and 5 °C. The preparation obtained by this method shows one homogeneous component on starch gel electrophoresis.

Crystalline protoporphyrin IX (Calbiochem) was dissolved in spectral grade p-dioxane (Carlo Erba). The solutions were degassed by 4 cycles of pumping and saturation by argon. The porphyrin concentration was 5×10^{-5} M in all the runs.

3 RESULTS

3.1 Triplet assignments and spectra

Laser pulse excitation at 529 nm of the porphyrin–globins, or the free porphyrin in aerated solutions, produced transient changes in the absorption spectra.

Figure 1 is a typical oscillogram obtained for porphyrin–globin (Hb), showing the transient light transmission decrease at 465 nm on laser excitation. The transmission returned after a few microseconds to its original value. This complete return was observed at all wavelengths where the ground state absorbs, and one may thus conclude that no permanent photooxidation or other product was formed by one laser pulse, within the detection limit. The decay kinetics were identical at all wavelengths studied (see Kinetics) strongly indicating that only one transient species is formed. In deaerated solutions of the porphyrin–globins the optical density (OD) change produced by laser excitation lasted without decaying at least over a period of 50 μs. Only the transient OD of free porphyrin in dioxane presented a very slight decrease over 20 μs.

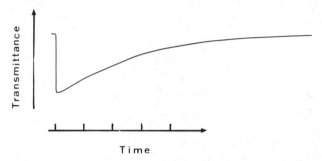

Figure 1 Oscilloscope trace: kinetics of triplet quenching by oxygen in porphyrin–globin (from hemoglobin)

Transient absorptions in porphyrins due to flash irradiation have been reported long ago[3,5,6] and are attributed to triplet state population. To check this assignment in the present case, cis-piperylene was added to a solution of protoporphyrin IX in deaerated solution and the decay of the transient measured. The decay of the transient absorption was found to be strongly accelerated in the presence of piperylene which is known as a very efficient triplet quencher. One may thus conclude that the transient species is indeed the triplet state of the porphyrin.

The assignment of the transient species obtained from the porphyrin–globins was made by comparing their absorption spectra to that of the triplet state of the free porphyrin. The transient spectra of the porphyrin–globins were measured at high laser energy, in conditions of light saturation. Study of the transient absorption change as a function of laser pulse energy showed that the change reached a limit above a certain energy value. This means that above this energy all the porphyrin molecules were converted to the transient species, that is, the triplet concentration was equal to the overall porphyrin concentration. The OD variation at high energies then directly gives a measure of the difference in extinction coefficient of the transient species and that of the ground-state porphyrin–globin. Figure 2 (a and b) shows spectra of the transient species obtained in this manner. In the case of dioxane, it was not possible to saturate the solutions by light in the same way because of the appearance of a disturbing

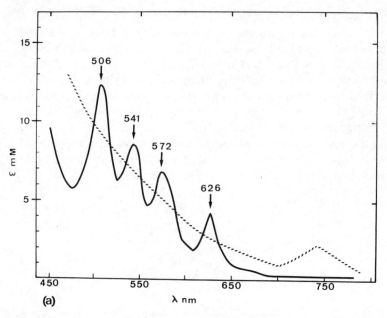

(a)

shock-wave at high energies in this solvent, and the spectral variations were
instead measured at a low laser energy. Normalizing to a mean laser energy
value gave the spectrum shown in Figure 2(c) (points ×). The dotted spectrum
in this figure was obtained by extrapolation of the normalized spectrum to
higher triplet population until the ground-state absorption structure disappeared (factor 4·3). Superposition of the transient spectra of the porphyrin–

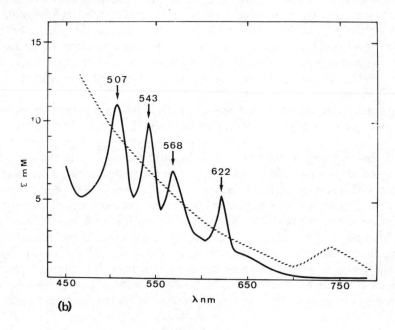

(b)

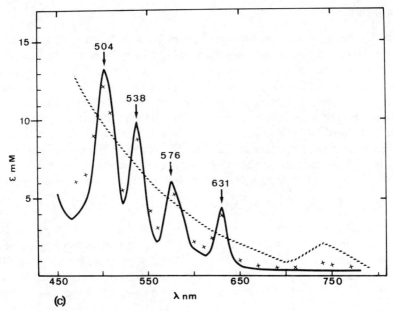

(c)

globins on the triplet porphyrin spectrum shows that all the spectra are identical within the error of measurements (Figure 2d). This result shows that the triplet state is populated also in the porphyrin–globins.

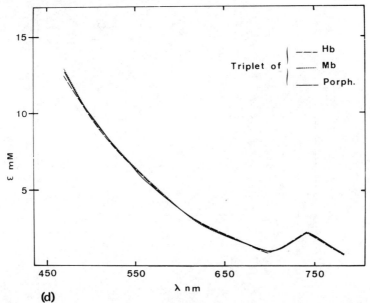

(d)

Figure 2 Spectra of ground state (full lines) and transient species (dotted lines) of porphyrin–globin complexes of (a) horse heart myoglobin (Mb) and (b) human hemoglobin A (HbA) and of (c) free protoporphyrin in dioxan (Porph). (d) Comparison of the triplet spectra

430

3.2 Kinetics

The kinetics of decay of the triplet porphyrins were measured in aerated solutions. Figure 3 shows semilogarithmic plots of the transient OD change at a number of wavelengths as a function of time for the different compounds. The linearity of the curves shows that the quenching is first-order; the parallelism within each system indicates strongly that only one transient component is involved. Oxygen is known as a very efficient quencher of triplet states; singlet oxygen is formed (at least in part) in the quenching process if the triplet energy is higher than that of the $^1\Sigma_g^+$ or $^1\Delta_g$ state of oxygen. Phosphorescence studies show that the triplet energy is above $10,000 \text{ cm}^{-1}$ for porphyrins, that is, higher than the $^1\Delta_g$ level of oxygen (8000 cm^{-1}). The following quenching process

$$^3P + {}^3O_2 \rightarrow P + {}^1\Delta_g O_2 \text{ (or } {}^3O_2) \quad \text{(rate constant } k_q)$$

is thus expected to occur.

The measured first-order rate constants give the k_q values on division by the oxygen concentration. Using known values[14] of oxygen solubilities ($4\cdot30 \times 10^{-4}$ M in water at $2 \,^\circ C$, $1\cdot50 \times 10^{-3}$ M in dioxane at $23 \,^\circ C$), the quenching rate constants given in Table 1 were obtained. Quenching by oxygen is a dynamic

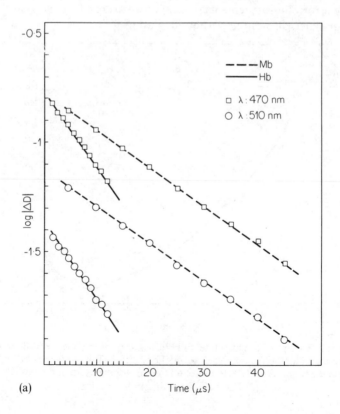

(a)

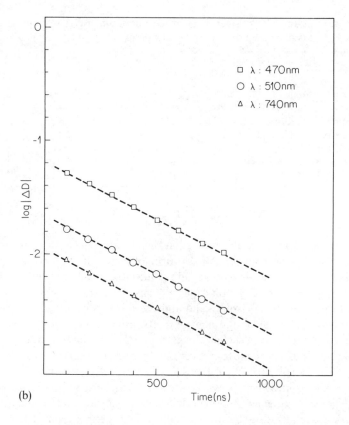

Figure 3 Decay of the porphyrin triplet, induced by oxygen diffusion. (a) Porphyrin–globin complexes, (b) free porphyrin

process requiring collisions between the reaction partners. The quenching rate constant therefore is dependent on the rate of solute diffusion in the solvent, that is, on the viscosity η and the temperature T. As an approximation one may set

$$k_q = \text{const.} \frac{T}{\eta}$$

The viscosity η of water at 2 °C is 1·67 cP; that of dioxane at 23 °C is 1·20 cP. To compare the k_q values in water at 2 °C and in dioxane at 23 °C the product $k_q\eta/T$ is also included in the Table. It is seen that the normalized rate constants thus obtained are an order of magnitude lower for the porphyrin–globins than for the free porphyrin. There is also a significant difference between the k_q values of the two porphyrin–globins, indicating differences in the oxygen diffusion rate into the globin pocket.

Table 1   Rate constants (k_q) of quenching of the porphyrin triplet state by molecular oxygen, obtained from the triplet decay rate ($k_q[O_2]$)

System	$k_q[O_2] \times 10^{-4}$ (s^{-1})	$k_q \times 10^{-7}$ (M^{-1} s^{-1})	$k_q \dfrac{\eta}{T} \times 10^{-5}$ (M^{-1} s^{-1} cP K^{-1})
Protoporphyrin IX in dioxane at 23 °C	240	160	65
Porphyrin–globin (Hb) in aqueous solution (pH 7) at 2 °C	7·8	18	11
Porphyrin–globin (Mb) in aqueous solution (pH 7) at 2 °C	4·0	9·3	5·6

4 DISCUSSION

The porphyrin triplet spectra reported in the literature have the same general features as that of the transient absorption observed in the present study: broad absorption in the visible extending into the near infrared. However, radicals or radical ions of porphyrins[15,16] have spectra which are not very different from those of the triplet state and the absorption structure therefore does not allow a definitive assignment of the transient absorption to the triplet state. Since the porphyrin in the globins is close to a histidine residue,[17] there is indeed the possibility of photoinduced hydrogen or electron transfer from histidine to the porphyrin; such a photoreduction of porphyrins by suitable substrates is a well-known process. However, it has been shown that histidine is quite inefficient as quencher of several triplet dye sensitizers.[18] If the transient absorption were indeed due to radicals one would not expect complete recovery of the starting material, and in particular not when the radicals are quenched by oxygen. These arguments, in addition to those presented in the Results section lead us definitely to discard the possibility of radical formation in the irradiated porphyrin–globins.

The results show that the decay of the porphyrin triplet state absorption is due to the quenching action of oxygen; the corresponding quenching rate constant is directly related to the rate of approach of oxygen close to the porphyrin. The observed variations of this time period for the compounds studied show that the probability of collision between an oxygen and a porphyrin molecule depends on the structure surrounding the porphyrin. It is usually assumed that porphyrin–globins have the same molecular weight[12,17] and structure as the corresponding deoxygenated proteins.[19] Thus the kinetics of quenching of the triplet state is expected to reflect the oxygen penetration into the heme pocket in the functional deoxy structure.[20,21]

Gibson[22] and Hartridge and Roughton[23] have shown that the rate constants for O_2-binding to hemoglobin and myoglobin are smaller by a factor of 10 than those we report for oxygen diffusion through the globin to the porphyrin site, thus the size and shape of the protein pocket[24] cannot be thought of as governing the kinetics of ligand-binding to the given proteins. It is not possible to discuss this effect in more detail in the absence of knowledge of the actual tertiary and quaternary structures of the porphyrin–globins. However, even in the possible case of a heme pocket undergoing structural modifications upon replacement of heme by porphyrin, it does not appear very probable that these modifications would be important enough to strongly alter the order of magnitude of the rate constants for oxygen diffusion to the pocket site in the protein.

The results reported here show that a drastic slowing of the oxygen diffusion rate may occur if the chromophore is surrounded by an efficient natural protection. Even though a quantitative interpretation of the values of the diffusion constants still appears difficult, the experimental results clearly show that the various specific oxygenation properties of different hemoproteins cannot be attributed to any steric hindrance of oxygen from penetrating into the heme pocket.

Similar studies of the dynamics of approach of the oxygen molecule close to the porphyrin site should lead to a better understanding of the roles of protein and porphyrin respectively, in the process of oxygen binding to respiratory hemoproteins.

REFERENCES

1. J. Barcroft, *The Respiratory Function of the Blood*, Cambridge University Press, 1914.
2. B. Alpert, R. Banerjee and L. Lindqvist, *Biochem. Biophys. Res. Comm.*, **46**, 913 (1972).
3. R. Livingston, *J. Amer. Chem. Soc.*, **77**, 2179 (1955).
4. J. B. Callis, J. M. Knowles and M. Gouterman, *J. Phys. Chem.*, **77**, 154 (1973).
5. H. Linschitz and K. Sarkanen, *J. Amer. Chem. Soc.*, **80**, 4826 (1958).
6. S. Claesson, L. Lindqvist and B. Holmström, *Nature*, **183**, 661 (1959).
7. H. Lutz, E. Bréhéret and L. Lindqvist, *J. Phys. Chem.*, **77**, 1758 (1973).
8. B. Alpert, L. Lindqvist and R. Banerjee, *Dynamic Aspects of Conformation Changes in Biological Macromolecules*, p. 171, D. Reidel Publ. Co., Boston, 1973.
9. D. L. Drabkin, *J. Biol. Chem.*, **164**, 103 (1946).
10. P. George and D. H. Irvine, *Biochem. J.*, **52**, 511 (1952).
11. A. Rossi-Fanelli, E. Antonini and A. Caputo, *Biochim. Biophys. Acta*, **30**, 608 (1958).
12. A. Rossi-Fanelli, E. Antonini and A. Caputo, *Biochim. Biophys. Acta*, **35**, 93 (1959).
13. E. Breslow, R. Koehler and A. W. Girotti, *J. Biol. Chem.*, **242**, 4149 (1967).
14. *Landolt-Börnstein, Zahlenwerte und Funktionen*, Springer-Verlag, Berlin, 1962.
15. A. K. Chibisov, *Photochem. Photobiol.*, **10**, 331 (1969).
16. H. Seki, S. Arai, T. Shida and M. Imamura, *J. Amer. Chem. Soc.*, **95**, 3404 (1973).
17. M. R. Mauk and A. W. Girotti, *Biochem.*, **12**, 3187 (1973).
18. I. Kraljic and L. Lindqvist, *Photochem. Photobiol.*, **20**, 351 (1974).
19. R. W. Noble, G. L. Rossi and R. Berni, *J. Mol. Biol.*, **70**, 689 (1972).
20. M. F. Perutz, H. Muirhead, J. M. Cox, L. C. G. Goaman, F. S. Mathews, E. L. McGandy and L. E. Webb, *Nature*, **219**, 29 (1968).
21. J. C. Kendrew, H. C. Watson, B. E. Strandberg, R. E. Dickerson, D. C. Phillips and V. C. Shore, *Nature*, **190**, 666 (1961).
22. Q. H. Gibson, *Progr. Biophys.*, **9**, 29 (1959).
23. H. Hartridge and F. J. W. Roughton, *Proc. Roy. Soc.*, **A107**, 654 (1925).
24. M. F. Perutz, *Nature*, **228**, 726 (1970).

D.8. The study of spin-states of heme proteins by near infrared magnetic circular dichroism

P. J. Stephens, J. C. Sutherland† and J. C. Cheng

Department of Chemistry, University of Southern California,
Los Angeles, California 90007, U.S.A.

W. A. Eaton

Laboratory of Chemical Physics, National Institute of Arthritis,
Metabolism and Digestive Diseases, National Institutes of Health,
Bethesda, Maryland 20014, U.S.A.

SUMMARY

Near infrared (600–2000 nm) absorption and magnetic circular dichroism (MCD) are reported for high-, low- and intermediate-spin derivatives of methemoglobin. The MCD spectra are shown to be highly characteristic of the spin state. High- and low-spin bands of intermediate-spin hydroxymethemoglobin are distinguished by their MCD. The origin of the MCD is illustrated by model calculations.

1 INTRODUCTION

Spin-state equilibria have been extensively discussed in heme protein work[1] and have been related to biological function in some cases. We have studied the near infrared (IR) magnetic circular dichroism (MCD) of a number of methemoglobin derivatives of high, low and intermediate spin.[2] The MCD is found to be very sensitive to the spin state and provides a new method of (1) discriminating and assigning near IR absorption bands of high- and low-spin species, (2) monitoring spin-state populations and (3) examining conformational and other perturbations on the heme geometrical and electronic structure. In this paper

† Current address: Department of Physiology, California College of Medicine, University of California, Irvine, California 92664, U.S.A.

we report the near IR MCD of two high-spin, two low-spin and one inter-mediate-spin methemoglobin derivatives and distinguish bands of high- and low-spin species in the latter. We then present model theoretical calculations which exhibit the origins of the MCD. Some ramifications of our work are also discussed.

2 EXPERIMENTAL

MCD was measured with a near IR CD/MCD instrument described else-where.[3] The basic components include a 500 W tungsten–halogen lamp, a 0·75 m grating monochromator, a Glan–Taylor calcite polarizer, an infrasil quartz photoelastic modulator and a liquid-nitrogen-cooled InAs detector. Magnetic fields up to 50 kilogauss are provided by a superconducting magnet. For measurements in the 600–1200 nm region the instrument sensitivity has been enhanced by addition of a dry-ice-cooled S1 photomultiplier tube. All experiments reported here were performed on room-temperature (300 K) solutions in 1 cm infrasil cells in fields of $\sim$ 40 kilogauss. Absorption spectra were measured simultaneously with MCD spectra on a Cary 14 spectrometer.

Absorption and MCD spectra are shown in Figures 1–5. We report absor-bance (A) and differential absorbance ($\Delta A = A_L - A_R$, where L and R denote left and right circularly polarized light). ΔA is normalized to a field of $+$ 10 kilogauss = 1 tesla, positive fields being parallel to the light propagation vector. A and ΔA relate to ε and $\Delta \varepsilon$ via

$$A = \varepsilon c l; \quad \Delta A = (\Delta \varepsilon) c l \qquad (1)$$

The significant quantity in MCD is the *ratio* $\Delta A/A$ ($= \Delta \varepsilon/\varepsilon$) and this is indepen-dent of concentration and pathlength. This factor also determines the instru-mental signal-to-noise (S/N) ratio. Instrumental ΔA sensitivity is typically $\sim 10^{-5}$ and optimum A is $\sim$ 1. Hence, the S/N ratio is optimally $\sim 10^5 \Delta A$.

Horse hemoglobin was used excepting for cyanomethemoglobin, where human hemoglobin was employed. Solutions of oxyhemoglobin were prepared by the method of Perutz,[4] oxidized with $K_3Fe(CN)_6$ and passed through a Sephadex G25 column to yield methemoglobin (HbH_2O). Other derivatives (HbX; X = F, CN, N_3, OH) were prepared from HbH_2O as follows: HbF and HbN_3 were obtained by addition of buffered NaF and NaN_3 solutions respectively in excess; solid KCN was added to form HbCN; HbOH was generated by addition of glycinate buffer to a pH of 10. In the case of HbCN, the methemoglobin was prepared initially in D_2O to allow measurements as far as 2000 nm.

3 DISCUSSION

The near-IR absorption spectra of the high- and low-spin methemoglobin derivatives examined here are a function of spin state. The high-spin aquo and fluoro derivatives (HbX; X = H_2O, F) exhibit a broad band at $\sim$ 10,000 cm^{-1} (1·0 μm^{-1}) and $\sim$ 12,000 cm^{-1} (1·2 μm^{-1}) respectively (Figures 1 and 2). The low-spin cyano and azido derivatives (HbX; X = CN, N_3) have structured bands at much lower energy, shifting to higher energy from X = CN to X = N_3 (Figures 3 and 4).

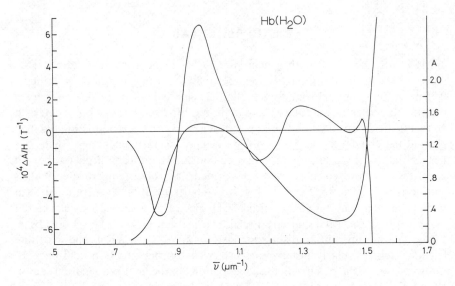

Figure 1 Absorption (below) and MCD (above) of 2·0 mM HbH$_2$O in H$_2$O

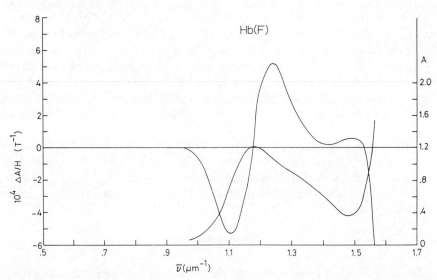

Figure 2 Absorption (below) and MCD (above) of 1·0 mM HbF in H$_2$O

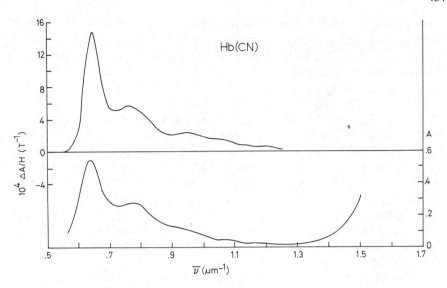

Figure 3 Absorption (below) and MCD (above) of 1·3 mM HbCN in D_2O

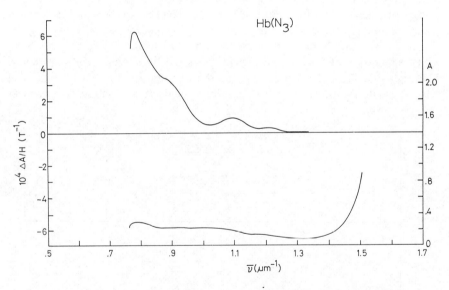

Figure 4 Absorption (below) and MCD (above) of 1·0 mM HbN_3 in H_2O

The near-IR MCD further distinguishes high- and low-spin derivatives. In the low-spin compounds the MCD replicates the absorption spectrum in form very closely ($\Delta A/A$ is almost independent of wavelength). On the other hand, the high-spin compounds exhibit MCD with sign oscillating through the absorption band ($\Delta A/A$ varying rapidly with wavelength). The low-spin MCD is intrinsically somewhat larger than the high-spin MCD. All else being equal this leads, *inter alia*, to higher S/N ratios in the low-spin compounds.

The intermediate-spin hydroxy derivative HbOH shows more complex near-IR absorption and MCD spectra than either pure high- or low-spin derivatives (Figure 5). Examination of the MCD leads immediately to the conclusion that the $10,000 \, cm^{-1}$ ($1.0 \, \mu m^{-1}$) band is a low-spin transition and the 12,000 and $14,000 \, cm^{-1}$ (1.2 and $1.4 \, \mu m^{-1}$) bands originate in a high-spin ground state. The $10,000 \, cm^{-1}$ ($1.0 \, \mu m^{-1}$) band shows a larger MCD than the more intense higher-energy bands and a low-spin dispersion form. The 12,000 and 14,000 cm^{-1} (1.2 and $1.4 \, \mu m^{-1}$) bands give MCD very similar in magnitude and dispersion to HbH_2O and HbF; differences in the high-energy region can be attributed to changes in overlap with the rail of the visible MCD.

The interpretation of the MCD spectra requires spectroscopic assignments of the bands and detailed theoretical calculations. These will be presented elsewhere.[5] Here we attempt to describe diagrammatically the essential origin of the MCD phenomena.

A molecular-orbital diagram for the heme moiety is shown in Figure 6. C_{4v} is a first approximation to the heme symmetry and will be assumed here for simplicity. The D_{4h} labels are also given because porphyrin spectra are frequently discussed within this group. The ground and low excited states of high- and low-spin Fe(III) hemes are also shown in Figure 6. The low-spin ground state has a configuration $\ldots b_2^2 e^3$ and is ^2E. The low lying configuration $\ldots b_2^1 e^4$ gives rise to a $^2B_2$ state. Spin–orbit coupling splits ^2E into two Kramers doublets with a separation of $\zeta \sim 400 \, cm^{-1}$, where ζ is the Fe $3d$ spin–orbit parameter. The high-spin ground state belongs to the configuration $\ldots b_2^1 e^2 a_1^1 b_1^1$ and is $^6A_1$. These descriptions differ from reality in neglecting rhombic distortion in

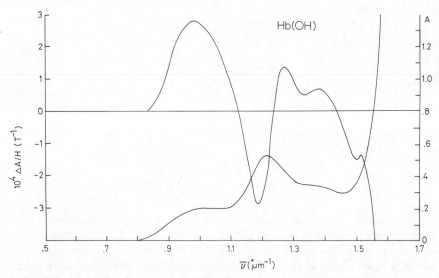

Figure 5 Absorption (below) and MCD (above) of $1.0 \, mM$ HbOH in H_2O

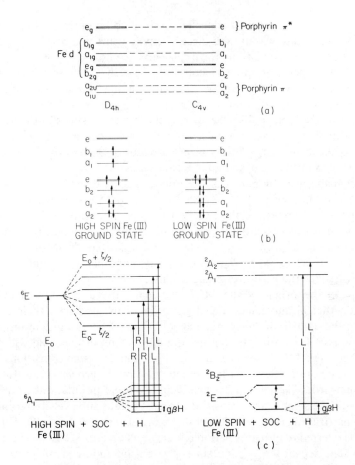

Figure 6(a) Heme molecular orbitals in D_{4h} and C_{4v} symmetries. (b) Electronic configurations of the high- and low-spin Fe(III) ground states. (c) Ground and excited electronic states of high- and low-spin compounds, including spin–orbit coupling (SOC) and axial magnetic fields (H). L and R denote left and right circular polarizations

the low-spin compounds and zero-field splitting in the high-spin derivatives. From EPR[6,7] it is known that a rhombic distortion is present in HbCN and HbN₃, intermixing the ^2E and $^2B_2$ levels. However, the actual wavefunction of the lowest Kramers doublet in these compounds is still close to the tetragonal function and in this simplified discussion the latter is adequate. Likewise, the high-spin $^6A_1$ ground state suffers a zero-field splitting of a few cm^{-1} due to second-order spin–orbit coupling,[8] but for our purposes here we can neglect this.

We assume that all near-IR bands discussed here are porphyrin-to-Fe charge transfer. In the low-spin case the lowest excitations are $a_1 \rightarrow e$ and $a_2 \rightarrow e$, leading to $^2A_1$ and $^2A_2$ excited states respectively, both of which are accessible from the ground state via allowed electric dipole transitions. In high spin, the lowest charge transfer transitions are a_1, $a_2 \rightarrow b_2$, e. These give rise to a variety of states; to first order in spin–orbit coupling the only states to which electric dipole transitions are allowed are two ^6E states arising from $a_1 \rightarrow e$ and $a_2 \rightarrow e$. These can interact by configuration interaction (CI) just as the lowest porphyrin $\pi \rightarrow \pi^*$ transitions,[9] but unlike the corresponding upper states in the low-spin configuration. The ^6E states are also split by first-order spin–orbit coupling. It is easy to show that this splitting is independent of the nature of the ^6E state and is as shown in Figure 6, spanning a range ζ in all.

We assign the near-IR bands of the low-spin derivatives to $^2E \rightarrow {}^2A_1, {}^2A_2$ transitions and those of the high-spin compounds to $^6A_1 \rightarrow {}^6E$, where the high-spin ^6E state is a mixture of the two ^6E excited configurations. We note that these assignments are essentially those assumed for some time,[1] that they are in-plane polarized in accord with single-crystal polarized absorption spectra[10] and that they are allowed in D_{4h} symmetry also and hence are unlikely to be highly sensitive to the Fe out-of-plane distance.

To explain the origin of the MCD we make the unphysical simplifying assumption that all molecules are aligned identically and with their C_4 axes parallel to the light beam and magnetic field. The effects of random molecular orientation are of quantitative, but not qualitative, importance and will not be pursued here. The MCD observed in both high- and low-spin transitions is due predominantly to the effects of the magnetic field on the *ground* states: these are split by first-order Zeeman effects producing unequal populations in different Zeeman sublevels. Since the transitions from different sublevels have differing circular polarization, the population differences give rise to a net CD. This type of MCD is referred to either as paramagnetic or as C-term MCD.[11] The magnitude and sign depend on ground state g-values (including sign); ground and excited state symmetries and, hence, circularly polarized selection rules; and kT, with an inverse temperature dependence. The dispersion is the same as that of the absorption band.

The Zeeman effects in the high- and low-spin ground states are shown in Figure 6. Within our models $g \sim 2$ and 4 for high- and low-spin states respectively. Examination of the selection rules for the low-spin $^2E \rightarrow {}^2A_1, {}^2A_2$ transitions shows that the transitions from the lowest Zeeman component in positive magnetic fields are entirely left circularly polarized, irrespective of the upper state. The MCD C terms are then positive in sign, as observed. For the high-spin system, on the other hand, it is found that the selection rules from the ground state Zeeman components vary with ^6E spin–orbit component. This is shown in more detail in Figure 6. The MCD C terms are thus negative in sign for the three lower spin–orbit components, and of opposite sign for the three higher components. Consequently, the MCD is expected to change in sign through the band, being negative on the low-energy side; this is as observed.

The structure observed in the low-spin spectra is attributed to vibrational effects and the presence of two electronic transitions. That in the high-spin spectra is attributed to vibrational effects (and possibly other electronic transitions not considered here).

The discussion presented here is oversimplified and inadequate in detail. However, the description of the MCD as C terms in porphyrin-to-Fe charge-transfer transitions is fully substantiated by a proper theoretical discussion.[5]

4 CONCLUSION

MCD provides a new method for studying a number of problems in the heme protein field. First, the spin-state origin of near-IR porphyrin-to-Fe charge transfer bands can be assigned. We have demonstrated this here for HbOH. In fact, this assignment has already been arrived at in this case[12] by studying the change in relative intensities of the bands from HbOH to the myoglobin analogue, whose spin-state equilibrium is further to the high-spin side. It could also have been reached by measuring the temperature variation of the relative intensities.[13] However, the MCD technique has the advantage of involving only one experiment.

Secondly, given assignment of bands to specific spin states, MCD provides a new monitor of perturbing effects on spin-state equilibria. Further, when the perturbation also changes the geometrical and/or electronic structure of a given spin state, MCD should assist in the understanding of the changes taking place in the absorption spectrum. This appears to be more simply carried out in the near-IR region than in the visible–UV spectral region owing to the greater complexity and overlap of high- and low-spin bands in the latter. The effects of various perturbations on the absorption spectrum of Fe(III) hemes in hemoglobins have recently been much used to elucidate the cooperativity mechanism in this protein.[14]

Lastly, our spectra provide further data which can be used as a reference for judging the closeness of modelling achieved in synthetic model compounds for heme proteins. These can be expected to be reported with increasing frequency in the near future.

We emphasize that the MCD experiments are carried out on room temperature solutions and do not require special sample treatment or low temperatures—in contrast to EPR and Mossbauer techniques. Further, the phenomenon is spectroscopic and responds to individual species, even in a solution mixture, differing from magnetic susceptibility which is a bulk property.

REFERENCES

1. See, for example, D. W. Smith and R. J. P. Williams, *Structure and Bonding*, 7, 1 (1970); T. Iizuka and T. Yonetani, *Adv. in Biophys.*, 1, 157 (1970).
2. A preliminary report of part of this work has appeared: J. C. Cheng, G. A. Osborne, P. J. Stephens and W. A. Eaton, *Nature*, 241, 193 (1973).
3. G. A. Osborne, J. C. Cheng and P. J. Stephens, *Rev. Sci. Inst.*, 44, 10 (1973).

442

4. M. F. Perutz, *J. Crystal Growth*, **2**, 54 (1968).
5. J. C. Sutherland, P. J. Stephens and W. A. Eaton, to be published.
6. R. G. Shulman, S. H. Glarum and M. Karplus, *J. Mol. Biol.*, **57**, 93 (1971).
7. J. F. Gibson and D. J. E. Ingram, *Nature*, **180**, 29 (1957); J. S. Griffith, *Nature*, **180**, 30 (1957); G. Harris, *Theoret. Chim. Acta*, **5**, 379 (1966).
8. G. C. Brackett, P. L. Richards and W. S. Caughey, *J. Chem. Phys.*, **54**, 4383 (1971).
9. M. Gouterman, *J. Chem. Phys.*, **30**, 1139 (1959).
10. M. W. Makinen and W. A. Eaton, to be published.
11. A. D. Buckingham and P. J. Stephens, *Ann. Rev. Phys. Chem.*, **17**, 399 (1966); P. J. Stephens, *J. Chem. Phys.*, **52**, 3489 (1970).
12. P. O. 'D. Offenhartz, *J. Chem. Phys.*, **42**, 3566 (1965).
13. P. George, J. Beetlestone and J. S. Griffith, *Rev. Mod. Phys.*, **36**, 441 (1964). We have confirmed the assignment in this manner by measurements at 77 K.
14. M. F. Perutz, *Nature*, **237**, 495 (1972); M. F. Perutz, P. D. Pulsinelli and H. M. Ranney, *Nature New Biology*, **237**, 259 (1972).

D.9. *Measurement of the lifetime of protein fluorescence*†

J. W. Longworth and S. S. Stevens

Biology Division, Oak Ridge National Laboratory, Oak Ridge, TN 37830

ABSTRACT

A nanosecond light flash, with a 1 kHz repetition, was used to excite fluorescence from protein solutions. Single fluorescence photon events were detected with a photomultiplier. A delayed coincidence was made to analyse the time intervals between the lamp flash and single photoelectron events. The nanoflash was formed in hydrogen at 0·5 atmosphere and produced the necessary quartz UV continuum to derive the monochromatic excitation wavelengths. Zirconium electrodes were used and permitted up to 10^9 discharges without maintenance. The lamp was enclosed within a transmission line structure beneath a coaxial capacitor and mounted upon a coaxial ceramic hydrogen–thyratron switch-tube. The system response to single photoelectron events was dominated by the photomultiplier at early times and the lamp beyond 25.ns. Timing discrimination was achieved by forming a constant fraction pulse at the input of a differential discriminator functioning in the low-level timing mode. This procedure eliminated serious ground loops associated with the thyratron switching.

The single electron response was bimodal in time; at 12 ns a later output distribution can be readily detected. Late pulses which appear at this time are clearly not preceded because a delayed coincidence is adopted. A FWHM of 1·4 ns is routinely achieved, late pulse frequencies of $1·7 \times 10^{-2}$ and $2·9 \times 10^{-4}$ at 12 and 24 ns with FWHMs of 3 and 6 ns. The lamp discharge also possesses a long-lived process, 0·1 % with a 42 ns relaxation time.

We have adopted an extremely simple but laborious approach to analyse the data because of the notorious difficulties of interpretation of multiexponential data. A background of $c.\ 10^{-4}$ is estimated beyond 150 ns after the flash, and subtracted. A 42 ns slope is then fit and subtracted from the data after the late

† Research sponsored by the U.S. Atomic Energy Commission under contract with the Union Carbide Corporation.

pulse. The fluorescence decay then comprises two segments, prior to and after the late pulse. A single decay component leads to parallel line fits. Two exponentials can be resolved when 10^6 events are collected in the peak channels.

Protein decays have been determined for many examples at both 300 K and 77 K, and range between 1 and 20 ns. Examples have been found where there are two tryptophan decays and others where there is a sum of tyrosine and tryptophan. The relatively low excitation resolution, 7 nm FWHM, precludes selective excitation of tryptophan in the presence of tyrosine.

D.10. Photoionization and recombination delayed fluorescence of aromatic amines in rigid glasses. Application to tryptophan-containing proteins

M. Bazin, M. Aubailly and R. Santus

*Laboratoire de Biophysique, Museum National d'Histoire Naturelle,
61 rue Buffon, 75005 Paris, France*

A. Mousnier,† R. Lesclaux and M. Ewald‡

*Départment de Chimie Physique A, Université de Bordeaux I,
351 Cours de la Liberation—33405 Talence, France
Equipe de Recherche Associée au C.N.R.S. n° 167*

SUMMARY

The purpose of this paper is to show how the study of recombination delayed fluorescence in rigid matrices at 77 K can contribute to the understanding of the mechanisms of photoionization and electron trapping in proteins.

In a first approach the information available from the observation of recombination delayed fluorescence is described for model systems, e.g. N, N, N', N'-tetramethylparaphenylenediamine (TMPD), diphenylamine, carbazole, indole, tryptophan in polar and in non-polar solvents. In these cases two-photon UV photoionization and recombination delayed fluorescence are observed. The intermediate state is identified to be the lowest triplet state of the molecule. The nature of the electron traps in the polar and non-polar matrices are discussed on the basis of the decay and the build-up of kinetics of the delayed fluorescence. The luminescence, optically stimulated by the excitation of trapped electrons, is also studied. The effect of the oxygen contents of the aerated samples will be discussed.

In a second approach, tryptophan recombination delayed fluorescence is observed directly in the proteins, e.g. pepsin, trypsin, bovine carbonic anhydrase, horse liver alcohol deshydrogenase (HLAD), at 77 K in water–ethylene glycol

---

† This work is taken in part from the *Thèse de troisième cycle*, Bordeaux (1973) of A. Mousnier, who died accidentally in December 1973.

‡ To whom correspondence should be addressed.

glasses. This new way to observe directly the two-photon photoionization and the electron trapping into the proteins is described. The interest of this method for the understanding of the photochemistry of proteins is discussed.

1 INTRODUCTION

The mechanism of low-energy photoionization of solutes in glassy matrices is now well known to be a two-step absorption process involving the first triplet state of the solutes as the intermediate step (for more details see the recent reviews[1,2]).

Of interest to us in this work are the events that follow photoionization which may be described schematically as follows:

(a) The photoejected electrons are trapped in the glass as detected by the characteristic spectra of solvated electrons.[4,5,10]

(b) The trapped electrons recombine with the radical cations giving back the neutral molecule in a luminescent first excited singlet or triplet state, depending on the energy requirements between the ionic species and the solvent and on the process of detrapping.[6] The recombination is most often geminate, involving the photoejected electron and its parent radical cation.[2]

The detrapping mechanism is still not well elucidated, being most likely quite complex, involving thermal activation, solvent trap relaxation and destruction, tunnel effect, optical or electric field stimulation . . . etc.

Combining the efforts of our two laboratories, we tried to assemble a large number of data on the nature of trapping in matrices of widely different nature, non-polar and polar, by the same method of investigation in order to get a unified view of the problem. Furthermore, our aim was to show that in the case of simple aromatic solutes and of proteins containing tryptophan residues, the observation of recombination delayed fluorescence (written hereafter as DF) using a phosphoroscope to eliminate the ordinary fast fluorescence (10^{-9}– 10^{-7} s) is a powerful method to apprehend the primary step of photoionization and the nature of the electron traps involved in the recombination delayed luminescence processes (isothermal spontaneous recombination, delayed fluorescence (DF) or phosphorescence (DP), optically stimulated recombination fluorescence (OSF) or phosphorescence (OSP)).

It should be recalled that, according to the model proposed by Albrecht,[2] a recombination time constant may be associated with a trapping distance to the potential well of the parent radical cation. Thus polyexponential decays are related to different electron populations with their relative proportions and decay-time constants (or trapping distances). Our experimental set-up permits us to make observations prior to the elapse of the ordinary phosphorescence as opposed to most of the recombination delayed luminescence work published so far in this field. [1,2,6,15] Thus only the apparatus response time limits us in the study of the recombination DF components shorter than a millisecond (the resolution time of our phosphoroscope) and longer than the ordinary fluorescence decay (10^{-9} to 10^{-7} s).

It should be noted that the DF observed in our study is distinct from the one also produced by a two-photon process, seen in glassy systems, due to triplet–triplet annihilation (also called P-type DF) which is important only when high concentration of solutes are used ($c > 10^{-3}$ M).[7] Moreover, it has been experimentally observed that the exponential build-up of the triplet–triplet annihilation DF (τ_{DF}^{R}, where superscript R = rise) is faster than that of the phosphorescence (τ_{P}^{R}) and that the difference $\tau_{P}^{R} - \tau_{DF}^{R}$ clearly increases with the concentration of the solute molecules.[7,16] This phenomenon, presently under investigation, is not observed here.

The work presented here summarizes our results obtained on:

(a) Seven aromatic solutes—$N,N,N'N'$ tetramethyl-paraphenylene-diamine (TMPD), diphenylamine (DPA), carbazole, indole, tryptophan (TRP), tyrosine, phenylalanine.

(b) Several proteins—bovine carbonic anhydrase, horse liver alcohol deshydrogenase, pepsin, trypsin.

(c) Eight solvents—3-methylpentane(3MP), methylcyclohexane, diethyl ether, EPA (diethyl ether, isopentane, ethanol) (5v–5v–2v), absolute ethanol, 95% ethanol, EGW (ethylene glycol, water (1·2v–1v), aqueous NaOH 10 M, aqueous $CaCl_4$ 4 M.

For conciseness and clarity only the system DPA–3MP (10^{-3} M), TRP–EGW (7×10^{-4} M) and the bovine carbonic anhydrase BCA in EGW (10^{-4} M, equivalent to 7×10^{-4} M in TRP residues) will be described.

Diphenylamine (DPA) is an Eastman Kodak Lab. product which had been submitted to a twenty-fold zone-refined purification in helium atmosphere. L-tryptophan is a Mann Research Lab. product and was used without purification. 3-Methylpentane (3MP) obtained from Fluka (*purum*) was purified by two successive passages on activated silica and kept in the dark over molecular sieves (Merck, 3A). The BCA enzyme was obtained salf-free, lyophilized from Koch-Light. BCA apoenzyme and Cd–BCA protein were prepared according to the method of Linskog and Malström.[14] Activity assays carried out on BCA samples showed that freezing and thawing a sample does not produce any permanent denaturation. Ethylene glycol was a fresh Merck 'pro analysis' product. Deionized and twice-distilled water was used, with 0·22 M KCl and either phosphate buffers or potassium or sodium hydroxide or HCl for pH adjustments. All protein solutions were further diluted with 1·2 volume of ethylene glycol.

Interfering luminescence from the solvents and from the suprasil quartz tubes (2 to 3 mm internal diameter) was negligible. The reproductibility of kinetic data was on the average about 10% or better. Our experimental set-ups are classical spectrophosphorimeters (Aminco Bowman, Hitachi–Perkin Elmer MPF 3 and a home-made apparatus).[7–9,11–13]

2 RESULTS ON THE MODEL SYSTEMS

Figures 1(a), 1(b), 4(a) and 4(b) show the absorption spectra, the ordinary luminescences (fluorescence and phosphorescence) and the two-photon dependent DF respectively, observed in the DPA–3MP and in the TRP–EGW systems. The electron trapping in the glasses are observed either by absorption or by optical stimulation (OSF and OSP). Radical cations of the DPA have been observed in diethyl ether in the presence of an electron scavenger, CCl_4.[12] No attempt has been made to obtain the TRP^+ absorption spectrum. A quench-

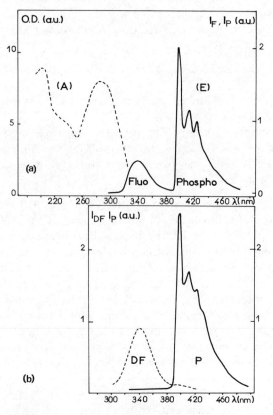

Figure 1 (a) Absorption (A) at room temperature and ordinary uncorrected emissions (E) (fluorescence and phosphorescence) at 77 K of the DPA $(10^{-3}$ M) in 3MP. Luminescence monochromatic excitation at 313 nm. (b) Corresponding emissions of the delayed fluorescence (DF) (at high sensitivity) and of the phosphorescence (P) observed with a phosphoroscope at 77 K. UV monochromatic excitation at 313 nm. Two MTO (France) H325a filters in front of the emission monochromator for the DF observation

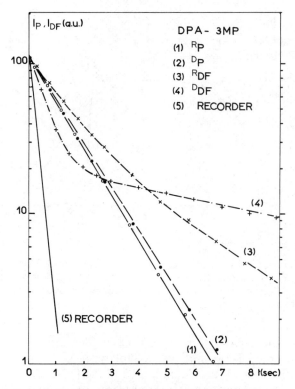

Figure 2 Diphenylamine (DPA), 10^{-3} M in 3MP, 77 K, $\lambda_{ex} = 310$ nm. Rise (R) and decay (D) of the luminescence observed through a phosphoroscope; phosphorescence (P) and recombination delayed fluorescence (DF)

ing of the DF, OSF and OSP emission is observed in the presence of electron scavengers (CCl_4 in 3MP, $CHCl_3$ in EGW). The kinetics of the DF are classically described in a semi-logarithmic plot by comparison with those of the phosphorescence (Figures 2 and 5).

A typical linear recording of the DF is presented in Figure 3 with the DPA–3MP system. It is possible to distinguish between the build-up of the DF intensity under UV excitation followed by a stationary state and the decay of the DF intensity when the UV light is cut-off. During the decay it is possible to observe an OSF spike when the sample is irradiated by a visible or infrared stimulation light. (In some cases[13] only without the phosphoroscope.)

All these observations confirm that the stationary DF intensity obtained under UV irradiation is due, in our experimental conditions, to a two-photon recombination DF. The nature of the intermediate state (the slowest step in the kinetic succession of events due either to the triplet population growth risetime τ_P^R or to the trapping lifetimes, if they are long compared to τ_P^R) is seen from the semi-logarithmic plot of the DF and P build-up (Figures 2 and 5). In the TRP–

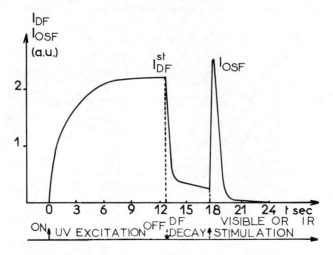

Figure 3 Linear plot of the rise and decay of the DF intensity. I_{DF}^{st} is the intensity of the DF obtained in the stationary state. I_{OSF} is the intensity of the optically stimulated fluorescence (see text)

EGW system the triplet intermediate is easily identified since τ_{DF}^{R} and τ_{P}^{R} are equal. The difference between τ_{P}^{D} (where superscript D = decay) and τ_{P}^{R} depends on the intensity of the UV excitation.[13,15] In the DPA–3MP system it can be inferred from the relative importance of long trapping lifetimes, compared to the short trapping lifetimes observed in the TRP–EGW system, that the trapping durations are the slowest steps for almost all the DF build-up (in the DPA–diethyl ether system, 10^{-3} M, however, the DF build-up is exponential and $\tau_{DF}^{R} = \tau_{P}^{R}$).

The results of the DF build-up kinetics cannot be discussed without a parallel description of the DF decays. In fact the DF decay is faster in EGW than in 3MP. This supports, as already mentioned, the interpretation of the DF build-up compared to that of the phosphorescence. It is important to remark that the DF decay in the two systems may be described by a superposition of several exponential decays. At different and distinct times of observation after the UV irradiation cut-off, there corresponds an exponential decay of the DF with a rate constant characteristic of the solvent. A given rate constant can be associated to a particular population of electrons with a typical trapping distance from the parent cation (in the Albrecht model[2]).

It is clear from our results on the DF decays that:

(a) The proportion of the electron populations with long decay rate constants is much greater in 3MP than in EGW. This important result can be extended to all non-polar solvents and all hydro-alcoholic solvents used in our work.

(b) The importance of the electron populations with short decay rate constants is definitely not negligible compared to the longer ones. In the case of

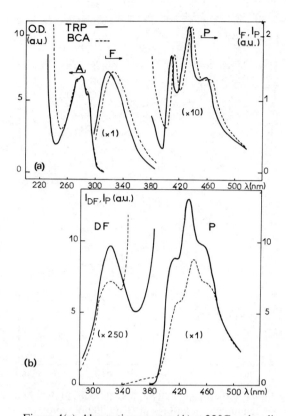

Figure 4(a) Absorption spectra (A) at 25 °C and ordinary uncorrected emissions (fluorescence (F) ($\times$ 1) and phosphorescence (P) ($\times$ 10)) at 77 K in EG-W (1·2–1) glass at neutral pH of the tryptophan molecule (7×10^{-4} M) (—) and of the BCA protein (10^{-4} M in BCA equivalent to 7×10^{-4} M in tryptophan residues) (- - -). Luminescence excitation at 270 nm. All the curves (A, F, P) are normalized to have the same maximum value for the bands studied. (b) Corresponding emission of the delayed fluorescence (DF) ($\times$ 250) and of the phosphorescence (P) ($\times$ 1) observed with a phosphoroscope at 77 K. UV excitation with an Osram HBO 200 W ultra-high-pressure mercury lamp with an MTO (France) 'H235a' optical filter

EGW it is the predominant phenomenon. Short decay constants in the range of a millisecond to a second have been measured in the TRP (5×10^{-3} M)–$CaCl_2$(4 M)–H_2O glassy system.[11] Here we are limited by the recorder time response as shown on the figures.

In conclusion, the electrons with the shorter decay time constants play an important role in the total stationary DF emission. Shorter range distances

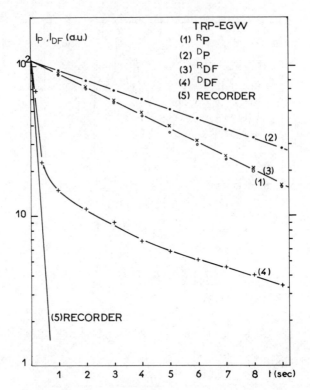

Figure 5 Tryptophan (TRP), 7×10^{-4} M, EGW, 77 K, UV excitation with an Osram HBO 200 W lamp and an MTO (France) 'H325a' filter. Rise (R) and decay (D) of the luminescence observed with a phosphoroscope, phosphorescence (P) and recombination delayed fluorescence (DF)

for the electron traps may be suggested in EGW compared to 3MP but they are not absent in 3MP.[12]

Moreover our results suggest, for at least the system (DPA–3MP), that:

(1) The DF decay depends upon the concentration of the solution in air before freezing. The decay time constant observed between 2 and 10 s, after the UV irradiation cut-off is slower in the absence of air (freeze–thawing pumping technique) than in aerated matrices. This result is also observed in the TMPD–3MP (10^{-3} M) system. These data are consistent with a model where the oxygen molecules from the air-containing samples are the principal scavengers of the trapped electrons.

(2) The DF decay depends drastically upon the viscosity of the solvent at 77 K as shown for the 3MP–isopentane glass. The long decay time constant component of the DF, observed between 2 and 10 s, disappears in the same way as the proportion of isopentane in 3MP is increased, which corresponds to a

huge decrease in the glass viscosity.[17] This suggests that solvent relaxation plays an important role in the trapping ability of the glass as its viscosity decreases.

3 RESULTS ON PROTEIN SYSTEMS

The study of proteins like BCA (see Figures 4a, 4b and 6) and trypsin prepared in neutral aqueous solutions at room temperature with a concentration of TRP equivalent to 7×10^{-4} M and then cooled at 77 K, was made in the mixed solvent EGW, as already described (see also Reference 13). The results are compared with those obtained from the TRP–EGW (7×10^{-4} M) model system. It is observed that the ordinary emissions (fluorescence and phosphorescence) and the delayed luminescences (two-photon recombination DF, isothermal delayed luminescences—after the UV irradiation cut-off—optically stimulated luminescences) of the proteins are due essentially to the TRP residues. These emissions

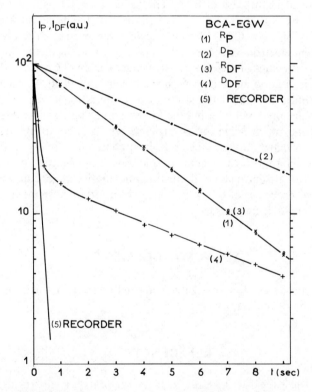

Figure 6 Bovine carbonic anhydrase (BCA), 10^{-4} M (equivalent to 7×10^{-4} M in tryptophan residues), EGW, 77 K, UV excitation with an Osram HBO 200 W lamp and an MTO (France) 'H325a' filter. Rise (R) and decay (D) of the luminescence observed with a phosphoroscope, phosphorescence (P) and recombination delayed fluorescence (DF)

show the same qualitative characteristics as those of the TRP alone in EGW. Nevertheless, the insertion of TRP residues in some proteins and the presence of other aromatic amino-acids (e.g. tyrosine) may modify in certain cases the quantitative characteristics of these emissions (for more details see Reference 13).

The risetimes of the DF and the P emissions in the BCA–EGW system, compared to those of the TRP alone in EGW, and the pH effect on the DF yield suggest that an energy transfer occurs from tyrosine to TRP, both in the singlet state and in the triplet state. This point is discussed in detail elsewhere.[13]

It is known[8,9] that electron scavengers quench the DF of tryptophan. In the case of trypsin, the quenching of the DF by the $CHCl_3$ molecules is 4 times less efficient than for the DF of the TRP alone, but the DF yield is the same without scavengers. The total TRP concentration is the same in the two cases, in the protein and in the model TRP solution (7×10^{-4} M). This shows that the photoejected electrons which are involved in the recombination DF, are probably trapped in the protein and not in the EGW solvent. An interesting result is obtained in the case of BCA. The yield of the DF is four times lower for the Cd^{++}–BCA than in Zn^{++}–BCA (native). It has been verified that the substitution of Zn^{++} by Cd^{++} does not change the protein structure.[18] Cd^{++} is a very efficient electron scavenger[11,13] and the value of the quenching constant suggests that the radius of the effective scavenger sphere of Cd^{++} is of the order of 30 Å. This is in good agreement with the dimensions of BCA protein.[19]

These examples show that the study of recombination DF, produced via a two-step primary process including the lowest triplet state as intermediate, is a sensitive method of investigation of proteins which is complementary to the study of the ordinary fluorescence and phosphorescence (one-step processes). This new method applied to proteins is able to give information on the protein structure. It may also give some interesting information regarding the primary photochemical processes of other biologically important systems like protein–DNA or protein–dye complexes.

ACKNOWLEDGEMENT

We thank Professor Joussot-Dubien for his interest in this work and for helpful criticism of the manuscript.

REFERENCES

1. J. Joussot-Dubien and R. Lesclaux, *Israël J. Photochem.*, **8**, 181 (1970); *Pure and Applied Chem.*, **34**, 265 (1973); R. Lesclaux and J. Joussot-Dubien, *Organic Molecular Photophysics*, Vol. 1, p. 457 (ed. J. B. Birks), Wiley–Interscience, London and New York, 1973.
2. A. C. Albrecht, *Accounts Chem. Res.*, **3**, 238 (1970).
3. A. Bernas, M. Gauthier, D. Grand and G. Parlant, *Chem. Phys. Letters*, **17**, 439 (1972); A. Bernas, M. Gauthier and D. Grand, *J. Phys. Chem.*, **76**, 2236 (1972).

4. T. Shida, S. Iwata and T. Watanabe, *J. Phys. Chem.*, **76**, 3683 (1972).
5. H. Hase, T. Higashimura and M. Ogasawara, *Chem. Phys. Letters*, **16**, 214 (1972).
6. F. Kieffer and M. Magat, *Actions Chimiques et Biologiques des Radiations*, **14**, 135 (ed. M. Haissinsky), Masson, Paris, 1970.
7. M. Ewald, Thèse d'Etat, Université de Paris-Sud (1972); M. Ewald, D. Muller and G. Durocher, *Spectrochimica Acta*, **29A**, 1051 (1973).
8. M. Aubailly, M. Bazin and R. Santus, *Chem. Phys. Letters*, **14**, 422 (1972).
9. M. Aubailly, M. Bazin and R. Santus, *Chem. Phys.*, **2**, 203 (1973).
10. J. Moan, *Acta Chem. Scandinavica*, **26**, 897 (1972).
11. M. Aubailly, M. Bazin and R. Santus, *Chem. Phys. Letters*, **12**, 70 (1971).
12. A. Mousnier, Thèse de troisème cycle, Université de Bordeaux I (1973); M. Ewald, R. Lesclaux and A. Mousnier, *VII Int. Conf. Photochem.*, Jerusalem (1973).
13. M. Bazin, M. Aubailly and R. Santus, *Photochem. Photobiol.*, accepted for publication (1974).
14. S. Linskog and B. G. Malström, *J. Biol. Chem.*, **237**, 1129 (1962).
15. J. Moan and H. B. Steen, *J. Phys. Chem.*, **75**, 2887 and 2893 (1971).
16. M. Ewald, D. Muller and G. Durocher, *Izvestia Akad. Nauk. SSSR*, **37**, 470 (1973); *Luminescence of Crystals, Molecules and Solutions*, p. 98 (ed. F. Williams), Plenum Publishing Corporation, 1973; M. Ewald, D. Muller and G. Durocher, *J. Luminescence*, **5**, 69 (1972).
17. J. R. Lombardi, J. Raymonda and A. C. Albrecht, *J. Chem. Phys.*, **40**, 1148 (1964).
18. R. W. Henskens, G. D. Watt and J. M. Sturtevant, *Biochemistry*, **8**, 1874 (1969).
19. S. Lindskog, L. E. Henderson, K. K. Kannan, A. Liljas, P. O. Nyman and B. Strandberg, in *The Enzymes*, Vol. 5, (ed. P. D. Boyer), Academic Press, New York, p. 587 (1972).

D.11. The interaction of 7,8-anilinonaphthalenesulphonate with glycoprotein hormones and its application to subunit interactions†

K. C. Ingham*, S. M. Aloj‡ and H. Edelhoch

National Institute of Arthritis, Metabolism, and Digestive Diseases, National Institutes of Health, Bethesda, Maryland 20014, U.S.A.

SUMMARY

Human chorionic gonadotropin (hCG), human luteinizing hormone (hLH), and bovine thyrotropin (bTSH) all self-associate in the presence of the fluorescence probe, 1,8-anilinonaphthalenesulphonate (ANS). The interaction is accompanied by enhanced ANS fluorescence. The concentration of ANS required for half-maximal fluorescence is about 1 μM for hLH, 20 μM for hCG, and 250 μM for bTSH. By contrast, the α and β subunits of these hormones have little or no effect on ANS fluorescence. Consequently, changes in ANS fluorescence can be used to measure the rates of dissociation and recombination of the subunits. A comparison of the kinetics of these reactions in the three hormones is of interest because of the extensive homology among the primary structures of both subunits. All three hormones dissociate readily in dilute acid or in concentrated urea solutions, but are relatively stable in alkali. The subunits of all three hormones can be recombined with reaction half-times on the order of 1–3 hours. The dependence of the recombination rates on subunit concentration is different for the three hormones. A kinetic model for the recombination reaction is proposed.

1 INTRODUCTION

The pituitary hormones, thyrotropin (TSH), luteinizing hormone (LH), follicle stimulating hormone (FSH) and the placental hormone, chorionic

† This work was supported in part by the Italy–USA Scientific Cooperation Program.

* Present address: Blood Research Laboratory, American National Red Cross, Bethesda, Md. 20014, U.S.A.

‡ Present address: Centro do Endocrinologia Ed Oncologia Sperimentale del C.N.R.: Istituto di Patologia Generale, University of Naples, Naples, Italy.

gonadotropin (CG) are glycoproteins each consisting of two subunits, α and β, held together by non-covalent bonds.[1-9] The primary structures of the α subunits are very similar.[9-13] The β subunits also have extensive homology,[9,13-17] but are sufficiently different to confer target organ specificity on the hormone. This is confirmed by hybridization studies in which the α subunit from one hormone is combined with the β subunit of a different hormone to yield a hybrid with biological activity characteristic of the hormone from which the β subunit was derived.[18-23]

Very little is known about the nature of the subunit interactions in these hormones. The individual subunits possess little, if any, biological activity.[1-9,24] This is compatible with the known changes in tertiary structure which accompany the dissociation and recombination reactions, as evidenced by the significant changes in circular dichroic spectra[25-29] and ultraviolet absorption spectra,[30-32] as well as decreased accessibility of tyrosine residues towards reaction with tetranitromethane[33] or iodinating reagents.[34-35] Apparently, the hormonally active conformation is stabilized by the interaction between the subunits when they combine.

A detailed study of the kinetics of the dissociation and recombination reactions in these hormones (and their hybrids) would be useful for understanding the nature of the subunit interactions. A few of the important questions which might be answered are: (1) What are the relative stabilities of the various hormones towards dissociation? (2) How does a minor change in the primary structure of one subunit affect its interaction with the other subunit? (3) Do the carbohydrate moieties have any effect on the subunit interactions? In addition, the results of kinetic studies would obviously aid in the refinement of preparative procedures used to produce purified subunits.

In order to measure accurately the dissociation and recombination rates, one must have a convenient method of distinguishing between the native (intact) hormone and its dissociated subunits. Many of the potentially useful methods, such as gel filtration and electrophoresis, sedimentation, light scattering, etc., are either too slow, too difficult to quantitate, or require excessive amounts of rather precious material. The observation[36] that human CG (hCG) and human LH (hLH) enhance ANS fluorescence while their subunits do not, led to the development of a fast, sensitive method which requires relatively small amounts of material.[37,38] The details of this method, which was later extended to bovine TSH (bTSH),[32] are discussed in the following sections.

2 INTERACTION OF ANS WITH GLYCOPROTEIN HORMONES

ANS (1,8-anilinonaphthalene sulphonate) is an organic dye molecule which is almost non-fluorescent in aqueous solutions but fluoresces strongly when dissolved in organic solvents or when bound to hydrophobic sites on a variety of proteins. Several excellent reviews of the fluorescent properties of naphthalene sulphonate compounds and their use as probes in protein structure analysis have appeared in recent years.[39–41] The exact mechanism by which binding to a protein enhances ANS fluorescence is still not clear. The low fluorescence yield of ANS in aqueous solution has been attributed to a rapid reorientation of the polar water molecules surrounding the dye, in response to the increase in its permanent dipole moment following excitation.[42] The binding of ANS to a hydrophobic site on a protein probably excludes water molecules from the immediate environment of the dye, thus preventing the solvent relaxation process responsible for the quenching.

The titration of dilute solutions of hLH, hCG and bTSH is illustrated in Figures 1 and 2. The ANS concentration required for half-maximal fluorescence increases from less than 1 μM with hLH to about 10 μM with hCG and to greater than 250 μM with bTSH. The sigmoid shape of the bTSH curve (Figure 2) can also be observed with hCG at higher ionic strength (e.g. 0·1 M KCl). With hLH, the fluorescence titration curve is essentially hyperbolic and is not affected by

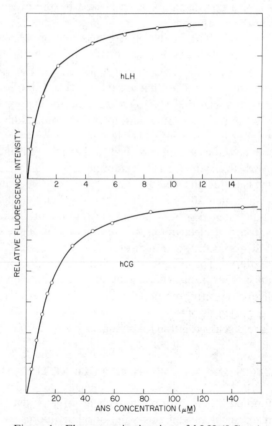

Figure 1   Fluorometric titration of hLH (0·5) μM) and hCG (1·5 μM) with ANS in 0·02 M potassium phosphate buffer, pH 7·4, at 25 °C. Excitation and emission at 390 and 480 nm, respectively

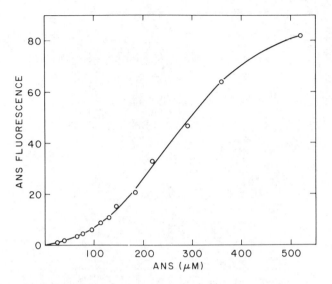

Figure 2   Fluorometric titration of bTSH (2·0 μM) at 25 °C in
0·01 M potassium phosphate buffer, ph 7·1. Excitation and
emission at 420 and 480 nm, respectively

addition of 0·1 M KCl. However, there are large time effects in the ANS fluorescence signals with hLH, especially at low ANS and hLH concentration where as much as 1 hour is required for the fluorescence to reach a plateau. These observations prompted a further investigation of the effect of ANS on the molecular behaviour of these hormones.

Gel filtration experiments on Sephadex G-100 columns at 4 °C indicate that all three hormones self-associate in the presence of ANS. This is illustrated in Figure 3 by the decrease in the elution volume which occurs when the columns are equilibrated with saturating levels of ANS. Sedimentation equilibrium measurements at 25 °C indicate that both hormones exist primarily as dimers in the presence of saturating levels of ANS (0·1 mM for hLH and 1·0 mM for hCG). The similar elution volumes of hLH and bTSH (Figure 3) in the presence of excess ANS suggest that the latter also associates to form a dimer.

3 ANS AS A PROBE FOR SUBUNIT INTERACTIONS IN GLYCOPROTEIN HORMONES

In spite of the rather complicated nature of the ANS interaction, it is still possible to use ANS fluorescence to study subunit dissociation and recombination rates. The enhancement of ANS fluorescence which accompanies the addition of bTSH to a concentrated (0·5 mM) solution of the dye is illustrated in Figure 4. Under these conditions the fluorescence intensity is seen to be directly proportional to the hormone concentration. The lower curve in Figure 1 shows

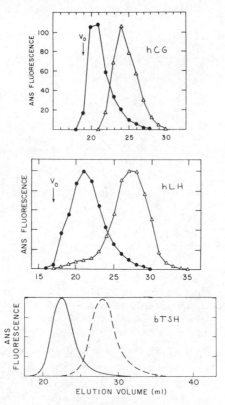

Figure 3 Gel filtration of hCG, hLH and bTSH on sephadex G-100 in the absence (right) and presence (left) of saturating concentrations of ANS at 4 °C in 0·01 M potassium phosphate buffer, pH ~7. Three different columns, approximately 1 × 50 cm, were used. When the columns were equilibrated with buffer alone, fractions (~1/ml) were analysed for ANS fluorescence by the aliquot method (see text). When the columns were equilibrated with ANS (100 μM for hCG and hLH, 500 μM for bTSH), fluorescence was measured directly in the fraction tubes. Approximately 1 mg of hormone was applied to each column

that the subunits have an order of magnitude smaller effect. Similar results are observed for hCG[37] and hLH[38] except that in these cases, the subunits have little, if any, effect on ANS fluorescence. It is this greater enhancement of ANS fluorescence by the native hormones, as opposed to their subunits, which provides the basis for the measurement of dissociation and recombination rates.

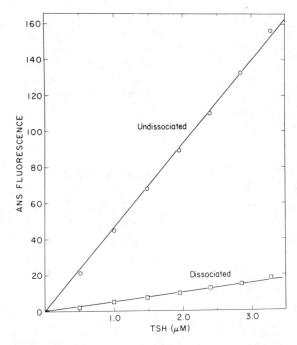

Figure 4 Enhancement of ANS fluorescence by bTSH
(○) and its subunits (□) at 25 °C in 0·01 M potassium
phosphate buffer, pH 7·1. The subunits were prepared by
incubation of bTSH at pH 2·5 and 37 °C for 1 hour. The
ANS concentration was 500 μM. Excitation and emission
at 420 and 480 nm, respectively

For measuring dissociation rates, the 'aliquot method' is used. Small aliquots
(25–50 μl) are periodically removed from the reaction solution and diluted into
buffered solutions of ANS (pH ∼ 7) which are kept at 0 °C. The appropriate
ANS concentration in these solutions is 500 μM for bTSH, 250 μM for hCG and
40 μM for hLH. When the reaction is complete, the solutions are then warmed
to room temperature and the ANS fluorescence measured. The dissociating
agent (acid or urea) is sufficiently dilute in these solutions to prevent further
dissociation at 0 °C. This was confirmed by comparison of fluorescence measurements made immediately after diluting the aliquots with measurements made
on solutions which were stored overnight at 0 °C.
Recombination rates can be measured by the aliquot method just described
or by a 'direct method' in which ANS is included in the reaction mixture. In the
latter case, the reaction takes place in a thermostated cylindrical quartz micro-
cell which is kept in the fluorimeter. The ANS fluorescence can then be measured as frequently as desired. It was shown that ANS had no effect on the
recombination rate at neutral pH and 37 °C.[32,37,38] The direct method, when
appropriate, is superior to the aliquot method because less handling is required,

462

smaller amounts of hormone are needed, and more data can be obtained. However, this method cannot be used for measuring dissociation rates since ANS retards the rate of reaction.[38]

For hCG, purified subunits were used to study the recombination reaction. With hLH and bTSH, purified subunits were unavailable. In these cases, native hormone was dissociated into subunits by incubating in acid at 37 °C. The solutions were then neutralized and the reverse reaction followed by the same procedure as with hCG.

4 RATES OF DISSOCIATION

All three hormones (hCG, hLH and bTSH) can be easily dissociated into their subunits either in dilute acid or in concentrated urea solutions at neutral pH. The reactions are first order and the rates strongly dependent on the

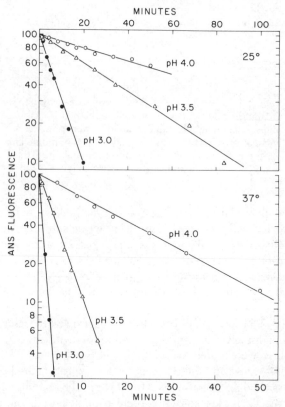

Figure 5   First-order plots illustrating the effects of acid pH and temperature on the rate of dissociation of bTSH into its subunits. Note the different time scales for the upper and lower panels. The buffer was 0·01 M phosphate, 0·01 M acetate and 0·01 M glycylglycine

Table 1 First-order dissociation rate constants $(k, 10^{-3}\,min^{-1})$ for hCG, hLH and bTSH: The effect of pH, urea concentration and temperature[a]

pH	bTSH		hCG		hLH	
	25 °C	37 °C	25 °C	37 °C	25 °C	37 °C
4·0	12	42	—	—	—	8
3·5	26	220	—	—	—	—
3·0	120	950	~1	30	~2	28
2·0	—	—	6	220	14	230
Urea concentration (pH 8)						
4 M	—	—	—	—	10	30
6 M	—	—	2	20	35	130
8 M	11	37	21	125	90	570
10 M	24	250	200	—	—	—

[a] The buffer for the acid dissociation experiments was potassium phosphate (0·01 M), sodium acetate (0·01 M) and glycylglycine (0·01 M). For the urea solutions, the buffer was 0·01 M phosphate.

temperature. The qualitatively similar behaviour of these hormones with respect to subunit dissociation is not surprising in view of the extensive homology among their primary structures. However, the exact dependence of the rates on pH or urea concentration reveals differences in the three hormones which must ultimately be related to the relatively minor variations in primary structure.

A typical example of the first-order dissociation process is given in Figure 5, which illustrates the dissociation of bTSH as a function of pH at 25 °C and 37 °C. The first-order rate constants for the three hormones in acid at several pH values and in urea at pH 8·0 are summarized in Table 1. The order of stability of the three hormones towards dissociation in urea is bTSH > hCG > hLH. By contrast, bTSH is the least stable towards dissociation in acid. For example, at pH 3·0 and 37 °C, the dissociation rate constant for bTSH is ~ 30 times greater than for hCG and hLH. All three hormones are remarkably stable in alkali; incubation at 25 °C and pH 12 gave no evidence for dissociation after 1 hour. Increasing the temperature to 37 °C had very little effect on hCG and bTSH, but caused hLH to dissociate readily with a half-life of 4·3 minutes.

5 RATES OF RECOMBINATION

It was known from measurements of biological potency that the subunits of hCG,[8] hLH,[3] bTSH[1] and other homologous hormones[21,22] could be recombined under appropriate conditions. These experiments usually involved incubation of the subunits at neutral pH and 37 °C for 12–40 hours after which time part or most of the biological potency of the native hormone was recovered.

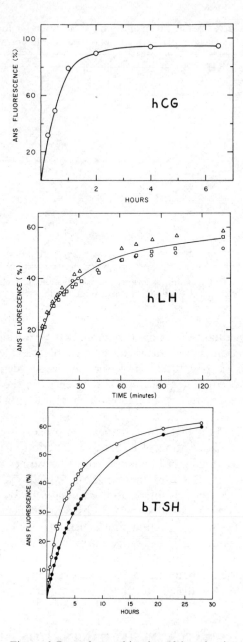

Figure 6 Rate of recombination of the subunits of hCG, hLH and bTSH at neutral pH and 37 °C in 0·01 M potassium phosphate buffer. The concentration of both hCG subunits was 74 μM. The subunit concentrations were 10 (□), 20 (△) and 40 (○) μM for hLH, and 8 (●) and 32 (○) μM for bTSH

However, very little was known about the rate of the reaction. The time dependence of the recombination of the subunits of hLH, hCG and bTSH at neutral pH and 37 °C is illustrated in Figure 6. In the case of hCG, there is complete recovery of ANS fluorescence as well as biological activity.[37] With hLH and bTSH, only about 60–70% of the ANS fluorescence and potency is usually recovered.[32,38] The rates of recombination of all three hormones are strongly dependent on temperature. No recombination is observed at 0 °C but the rate increases steadily between 25°C and 45°C. From an Arrhenius plot of the initial rate, an activation energy of $\sim$ 22 kcal/mole was estimated for bTSH.[32]

The rates of recombination of the three hormones differ in their dependence on subunit concentration. With hLH, the rate of recombination at pH 7 and 37 °C was essentially the same over a subunit concentration range of 10–40 μM (Figure 3). With bTSH, a four-fold increase in the concentration of both subunits caused about a two-fold decrease in the half-time of the reaction. A four-fold decrease in half-time would be expected if the reaction were second order. The recombination of hCG is also intermediate between first and second order. Increasing the concentration from 30 to 90 μM caused a 2·2-fold decrease in half-time.[43]

A plausible mechanism, capable of reconciling the various effects of subunit concentration on the rate of recombination, is shown below:

$$\alpha + \beta \xrightarrow{k_1} \alpha\beta$$

$$\alpha\beta \xrightarrow{k_2} H$$

$$H + ANS \rightarrow \text{enhanced ANS fluorescence}$$

The first step is a bimolecular combination to form a complex between the subunits. The rate of this step will be proportional to the concentration of each subunit. The second step is a first-order (concentration-independent) conformational rearrangement in which the subunits refold to form the native structure (H). The third step is a rapid binding of ANS by the hormone. Depending on the subunit concentration and the relative magnitudes of k_1 and k_2, the overall rate could vary from second order (at very low subunit concentration and/or $k_2 \gg k_1$) to first order (very high concentration and/or $k_2 \ll k_1$). Presumably the latter situation is applicable to hLH whereas hCG and bTSH fall into an intermediate case. The differential equations corresponding to this kinetic scheme are not soluble in exact form. However, current efforts are being made to evaluate quantitatively the magnitudes of k_1 and k_2 by numerical analysis.

6 DISCUSSION

The fluorescence probe, ANS, has made it possible to measure the rates of dissociation and recombination of hLH, hCG and bTSH under a variety of conditions. Hopefully, the method can be extended to other hormones of this type from various animal sources. The extensive homology which exists among the primary structures of these hormones provides us with an interesting model for studying subunit interactions. The data presented here indicate that relatively minor variations in primary structure can measurably influence the stability of these hormones towards dissociation into subunits. As purified subunits from other species become more available, it should be possible to prepare hybrids whose dissociation rates can be compared to those of the native hormones.

The failure to obtain complete recovery of ANS fluorescence is a problem for which there is no obvious explanation. Incomplete recovery of potency has also been encountered by others. One possibility is that the structure of the recombinant is not identical to the native hormone and has a lower affinity for ANS. However, in the case of hLH it was found that the recombinant was indistinguishable from the native hormone in terms of its affinity for ANS. Furthermore, the recombinant could be redissociated in acid (pH 3·0, 37 °C) at a rate identical to that of the native hormone. The alternative that some of the subunits are irreversibly damaged, although possibly correct for hLH,[38] seems unlikely for bTSH since the extent of recombination did not seem to vary with the pH or incubation time of the dissociation conditions. A third possibility, that the hormone exists in equilibrium with its subunits at neutral pH and 37 °C, is excluded by the observations that the extent of recombination was independent of hormone concentration. An adequate explanation of the lack of complete recombination must await further investigation. Hopefully, conditions will be found which lead to complete recovery since this would simplify attempts to evaluate quantitatively the kinetics of recombination.

REFERENCES

1. J. G. Pierce, T. Liao, S. M. Howard, B. Shome and J. S. Cornell, *Rec. Progr. Horm. Res.*, **27**, 165 (1971).
2. J. G. Pierce, T. Liao and R. B. Carlsen, *Hormonal Proteins and Peptides*, Vol. 1, p. 17 (ed. C. H. Li), Academic Press, New York, 1973.
3. D. N. Ward, L. E. Reichert, W. Liu, H. S. Nahm, J. H. Hsia, W. M. Lamkin and N. S. Jones, *Rec. Progr. Horm. Res.*, **29**, 533 (1973).
4. H. Papkoff, *Hormonal Proteins and Peptides*, Vol. 1, p. 59 (ed. C. H. Li), Academic Press, New York, 1973.
5. B. B. Saxena and P. Rathnam, *J. Biol. Chem.*, **246**, 3549 (1971).
6. O. P. Bahl, *Hormonal Proteins and Peptides*, Vol. 1, p. 171 (ed. C. H. Li), Academic Press, New York, 1973.
7. F. J. Morgan and R. E. Canfield, *Endocrinology*, **88**, 1045 (1971).
8. R. E. Canfield, F. J. Morgan, S. Kammerman, J. J. Bell and G. M. Agosto, *Rec. Progr. Horm. Res.*, **27**, 121 (1971).
9. T. Liao and J. G. Pierce, *J. Biol. Chem.*, **246**, 850 (1971).
10. R. Bellisario, R. B. Carlsen and O. P. Bahl, *J. Biol. Chem.*, **248**, 6809 (1973).
11. M. R. Sairam, H. Papkoff and C. H. Li, *Biochem. Biophys. Res. Comm.*, **48**, 530 (1972).
12. M. R. Sairam, H. Papkoff and C. H. Li, *Arch. Biochem. Biophys.*, **153**, 554 (1972).
13. M. R. Sairam, H. Papkoff and C. H. Li, *Arch. Biochem. Biophys.*, **153**, 572 (1972).

14. F. J. Morgan, S. Birken and R. E. Canfield, *FEBS Letters*, **31**, 101 (1973).
15. B. Shome and F. Parlow, *J. Clin. Endocrinol. Metab.*, **36**, 618 (1973).
16. R. Carlsen, O. P. Bahl and N. Swaninathan, *J. Biol. Chem.*, **248**, 6810 (1973).
17. J. Closset, G. Hennen and R. M. Lequin, *FEBS Letters*, **29**, 97 (1973).
18. J. G. Pierce, O. P. Bahl, J. S. Cornell and N. Swaninathan, *J. Biol. Chem.*, **246**, 2321 (1971).
19. L. E. Reichert, M. A. Rasco, D. N. Ward, G. D. Niswender and A. R. Midgley, *J. Biol. Chem.*, **244**, 5110 (1969).
20. J. L. Vaitukaitus, G. T. Ross and L. E. Reichert, *Endocrinology*, **92**, 411 (1973).
21. H. Papkoff and M. Ekblad, *Biochem. Biophys. Res. Comm.*, **40**, 614 (1970).
22. P. Rathnam and B. B. Saxena, *J. Biol. Chem.*, **246**, 7087 (1971).
23. L. E. Reichert, A. R. Midgley, G. D. Niswender and D. N. Ward, *Endocrinology*, **87**, 534 (1970).
24. K. J. Catt, M. L. Dufau and T. Tsuruhara, *J. Clin. Endoc. Metb.*, **36**, 73 (1973).
25. T. A. Bewley, M. R. Sairam and C. H. Li, *Biochem.*, **11**, 932 (1972).
26. J. C. Pernolet and J. Garnier, *FEBS Letters*, **18**, 189 (1971).
27. M. Ekblad, T. A. Bewley and H. Papkoff, *Biochem. Biophys. Acta*, **221**, 142 (1970).
28. P. Rathnam and B. B. Saxena, *Protein and Polypeptide Hormones*, Vol. 2, p. 320 (ed. M. Margoulies and F. C. Greenwood), Excerpta Medica, 1972.
29. W. E. Merz, U. Hilgenfeldt, P. Brockerhoff and R. Brossmer, *Eur. J. Biochem.*, **35**, 297 (1973).
30. K. F. Mori, *Biochem. Biophys. Acta*, **257**, 523 (1972).
31. W. H. Bishop and R. S. Ryan, *Biochem.*, **12**, 3076 (1973).
32. K. C. Ingham, S. M. Aloj and H. Edelhoch, *Arch. Biochem. Biophys.*, **163**, 589 (1974).
33. K. W. Cheng and J. G. Pierce, *J. Biol. Chem.*, **247**, 7163 (1972).
34. K. P. Yang and D. N. Ward, *Endocrinology*, **91**, 317 (1972).
35. R. E. Canfield, F. J. Morgan, S. Kammerman and G. R. Ross, *Protein and Polypeptide Hormones*, Vol. 2, p. 341 (ed. M. Margoulies and F. C. Greenwood), Excerpta Medica, 1972.
36. S. M. Aloj, K. C. Ingham and H. Edelhoch, *Arch. Biochem. Biophys.*, **155**, 478 (1973).
37. S. M. Aloj, H. Edelhoch, K. C. Ingham, F. J. Morgan, R. E. Canfield and G. T. Ross, *Arch. Biochem. Biophys.*, **159**, 497 (1973).
38. K. C. Ingham, S. M. Aloj and H. Edelhoch, *Arch. Biochem. Biophys.*, **159**, 596 (1973).
39. L. Brand and J. R. Gohlke, *Ann. Rev. Biochem.*, **41**, 843 (1972).
40. G. K. Radda, *Recent Advances in Bioenergetics*, **4**, 81 (1971).
41. G. M. Edelman and W. O. McClure, *Accts. Chem. Res.*, **1**, 65 (1968).
42. S. K. Chakrabarti and W. R. Ware, *J. Chem. Phys.*, **55**, 5494 (1971).
43. K. C. Ingham, H. A. Saroff and H. Edelhoch, *Biochem.*, in press (1975).

D.12. Luminescence of model indole-disulphide compounds†

J. W. Longworth

Biology Division, Oak Ridge National Laboratory, Oak Ridge, Tennessee 37830, U.S.A.

Claude Hélène

Centre de Biophysoqie Moléculaire, F-45045-Orléans-Cedex, France

SUMMARY

Cowgill found that the fluorescence of *bis*-(indole-3-methylene)disulphide was significantly quenched by the presence of the vicinal disulphide linkage, whereas a monocyclic heterodetic peptide, Cys Trp Cys (compound I) was perturbed to a much lesser extent. We have investigated the fluorescence and phosphorescence of these same compounds at 77 K. At 77 K the fluorescence is subject to a large quenching for indole compound I but not for the cyclic tripeptide. The phosphorescence of compound I is typical in its fluorescence-to-phosphorescence yield ratio, fine structure and phosphorescence decay. The phosphorescence of *bis*-(indole-3-methylene)disulphide is significantly perturbed. The yield is markedly reduced, there is only diffuse structure and the peak wavelength is shifted. The decay is shortened and non-exponential in form. A similar behaviour is known for the linear heterodetic tetrapeptide, L-cystyl-*bis*-L-tyrosine.

Red-shifted, short-lived phosphorescence with no fine structure are known from tyrosine and tryptophan residues in certain proteins. Denaturation of these proteins restores the typical tyrosine and tryptophan phosphorescence characters. The tyrosine residues of RNase and tryptophan residues of Bence–Jones proteins and lysozomes are known to be placed in contiguity with disulphide links in the molecular structure of the protein.

† Research sponsored by the U.S. Atomic Energy Commission under contract with the Union Carbide Corporation.

The nature of the interaction is not established, but it may arise from a charge-transfer interaction.

1 INTRODUCTION

Cowgill[1] noted in 1964 that the fluorescence of oxytocin, a small monocyclic heterodetic polypeptide hormone, had an atypically small yield at room temperature, compared to simple peptides. Since the sole tyrosine residue (Tyr$_2$) is vicinal to the intramolecular disulphide link (Cys$_1$–Cys$_6$), the low tyrosine yield was attributed to the presence of the neighbouring disulphide link. Proton[2] and deuteron[3] nuclear magnetic resonance spectroscopy has shown that in aqueous solvents oxytocin is a flexible molecule, and there is no precise knowledge of the exact relationship between the phenoxyl ring and the disulphide link.

The fluorescence observations were subsequently extended by Cowgill[4] to a model compound, bis-(p-methoxybenzyl)disulphide, and a disulphide-linked linear heterodetic tetrapeptide, L-cystyl-bis-L-tyrosine (Tyr Cys Cys Tyr). Both compounds have low fluorescence yields, but this was significantly increased upon reduction of the disulphide link. Next, diethyldisulphide was found to act as an efficient fluorescence quencher for tyrosine and phenols.[5] At low temperatures, the fluorescence of oxytocin[6] and the heterodetic tetrapeptide[7a,8] were significantly enhanced and reached values found for other tyrosine peptides at 70 K. The two heterodetic peptides were quite similar in phosphorescence,[7a] though they differed from tyrosine by lacking any noticeable diffuse structure and by having a greater Stokes' shift. It was suggested that the vicinal disulphide link had created an enhanced internal quenching mechanism at the singlet level. The perturbation of the triplet state was reflected in the decay behaviour; Churchich and Wampler[8] reported a short-lived decay through oscillography, and Longworth[7b] who used a 100 ms responding X–Y recorder as a detector, found a long-lived component. The nature of the interactions between the disulphide and tyrosyl singlet and triplet levels is not well established, but additional features are to be found in photochemical studies, which have shown a tyrosyl-sensitized disulphide photolysis.[9,10,11] Flash photolytic spectroscopy has recently demonstrated that a photoionization occurs from the triplet state ultimately to form a phenoxyl radical and a negative disulphide ion radical, either directly or through the intermediary of an ejected solvated electron.[12]

Less extensive investigations are available upon the indole-disulphide interaction, but an obvious parallel exists between tyrosine and tryptophan. Cowgill[6] noted that an indole model compound, bis-(indole-3-methylene)-disulphide was poorly fluorescent at 298 K, whereas L-methionyl-L-tryptophan had a typical yield value. Diethyldisulphide was also shown to be an effective quencher of indole fluorescence in fluid media. Rather than studies upon simple compounds, the only relevant flash photolytic investigations are upon hen lysozyme. The molecular structure of this protein[13] is known, and it showed that several tryptophyl residues within the protein have a disulphide link in their immediate environment. Grossweiner and Usui[14] used flash photolytic spectroscopy to demonstrate the formation of an indole radical and the concomitant formation of the electron adduct of a disulphide link after a microsecond flash. It remains to be settled whether or not a triplet state is an intermediary in this photoionization process.[15]

Atypically short phosphorescence decays have been found from three tyrosyl-fluorescent proteins, ribonuclease A, insulin and pancreatic trypsin inhibitor (all from cattle) when native, but not when denatured.[7b] The molecular structures[16] of all three proteins are known, and several of the tyrosyl residues have disulphide links in their immediate environments. Other proteins, which lack any disulphide links, such as the subtilisins Carlsberg and BPN' and *Staphylococcus aureus* endonuclease, have a tyrosyl phosphorescence that has fine structure and a typical decay time, one comparable to the decay times found for simple tyrosine compounds. Similarly, with tryptophyl residues in proteins, examples are known, hen lysozyme and most Bence–Jones proteins, where a short-lived phosphorescence decay occurs.[17,18,19,20]

The particular spatial requirements of the indole-disulphide interaction were found by Cowgill. A monocyclic heterodetic tripeptide, cyclo-dithio-L-cysteinyl-L-tryptophyl-L-cystine does not have a significantly quenched fluorescence at 298 K, whereas the linear heterodetic pentapeptide L-cystyl-L-tryptophyl-L-cystine does. The molecular structure of the cyclic heterodetic tripeptide prevents a direct encounter between the indole ring and the disulphide link, as the rigid structure provides a 0·7 nm separation.

In this paper we describe studies at low temperature of these model peptides of tyrosine and tryptophan and relate them to an atypical behaviour of both residues as observed in certain proteins.

2 MATERIALS AND METHODS

L-Cystyl-L-tyrosine was obtained from Cyclo Chemical Corp. and used as supplied. Dr R. W. Cowgill kindly supplied us with *bis*-(indole-3-methylene)-disulphide, *bis*-(*p*-methoxybenzyl)disulphide and cyclo-dithio-L-cysteinyl-L-troptophyl-L-cystine. Ethylene glycol was of chromatoquality obtained from Matheson–Coleman–Bell, dimethyl sulphoxide was spectroquality, from the same supplier. Ethanol was U.S. Pharmacopaeia quality. Dimethyl sulphoxide in equivolume mixture with ethylene glycol forms a glass at 77 K. Samples were studied in quartz cells at 1 mg/ml concentration.

Phosphorescence lifetimes were determined with a multichannel analyser operating in a stable averaging mode after several sweeps of the memory, following the procedure described by Hélène and Longworth.[21] The multiexponential decays were fitted with a non-linear least-squares algorithm due to Marquardt, described by Bevington.[22] Initial parameters were derived from manual graph plots and prepared for an IBM 360-75/91 by a Hewlett–Packard HP45, kindly purchased by Lynne and Gill Longworth.

3 RESULTS

The low-temperature luminescence spectra of L-cystyl-*bis*-L-tyrosine were previously reported by Longworth[7a] and by Churchich and Wampler[8] in ethylene glycol:water glasses. No significant differences were found in ethanol glass. The phosphorescence has a 30-fold reduced quantum yield and is red-shifted compared to the spectrum of simple tyrosine peptides and derivatives. A similar spectrum and location are found for oxytocin, insulin, and ribonuclease A. We show in Figure 1 that the phosphorescence of the model tyrosine compound *bis*(p-methoxybenzyl)disulphide, which behaves like cystyl-*bis*-tyrosine, is red-shifted and has an atypically small yield.

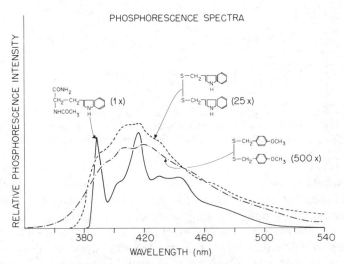

Figure 1 Phosphorescence spectra of model tyrosine and indole-disulphide compounds. Solutions were 1 mg/ml in ethanol at 77 K and were excited at 280 nm with a 4 nm FWHM spectral resolution

An analogous red shift and a markedly reduced fine structuring are found in the phosphorescence of *bis*-(indole-3-methylene)disulphide when compared to N-acetyl-L-tryptophan amide. The spectra are given in Figure 1. Unlike the tyrosine compounds, there is a dramatic reduction in yield and the fluorescence of *bis*-(indole-3-methylene)disulphide is not shifted, but the yield is markedly reduced in value at 77 K. The fluorescence at 77 K and the fluorescence-to-phosphorescence ratio of the indole-disulphide model can be compared with the spectra obtained from N-acetyl-L-tryptophan amide and cyclo-L-tryptophyl-L-tryptophan in Figure 2. The phosphorescence decay of L-cystyl-*bis*-L-tyrosine and *bis*-(p-methoxybenzyl)disulphide obtained in ethanol glass at 77 K are compared in Figure 3. Detailed parameters of the decay are collected in Table 1. The non-exponential decay is well fitted with a two-exponential decay curve. A Chi-square probability of 1·0–0·98 is found, with the data equally distributed about the fit curve and a reduced Chi-square of less than 0·50.

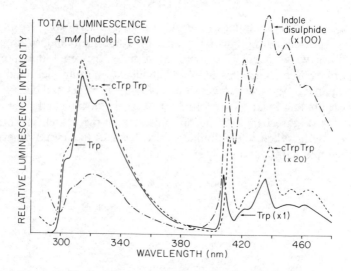

Figure 2   Total luminescence spectra of indole compounds. Solutions were 4 mM in indole residues in ethanol at 77 K and were excited at 280 nm with 3 nm FWHM spectral resolution

The phosphorescence decay of *bis*-(indole-3-methylene)disulphide measured in ethanol is compared in Figure 4 with the decay obtained from cyclo-dithio-L-Cys-L-Trp-L-Cys in dimethyl sulphoxide:ethylene glycol glass. The total luminescence and phosphorescence (Figure 5) of Cys Trp Cys was fine structured and typical of indole derivatives, though the phosphorescence decay consisted of two components, one of which was short-lived, the other long, and in the typical range (Table 1).

Table 1   Phosphorescence decay times[a]

Compound	Short-lived		Long-lived	
	fraction	decay time(s)	fraction	decay time(s)
(1) N-Ac–Tyr–NH$_2$	—	—	—	2·83 ± 0·01
(2) (p-MeOB$_3$–3–)$_2$	0·67 ± 0·02	0·176 ± 0·005	0·33 ± 0·03	1·15 ± 0·01
(3) Tyr Cys Cys Tyr	0·32 ± 0·01	0·160 ± 0·008	0·68 ± 0·02	1·10 ± 0·01
(4) N-Ac–Trp–NH$_2$	—	—	—	6·77 ± 0·01
(5) (Ind–3–CH$_2$–S–)$_2$	0·84 ± 0·04	0·201 ± 0·012	0·16 ± 0·01	1·31 ± 0·04
(6) C$_s$–Cys Trp Cys	0·32 ± 0·004	2·23 ± 0·08	0·68 ± 0·01	6·57 ± 0·01

[a] Mean values ± standard deviations.

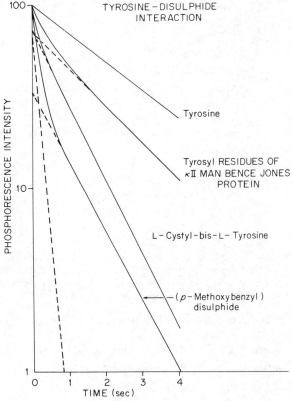

Figure 3 Phosphorescence decay of tyrosine compounds. Solutions were 1 mg/ml in ethanol at 77 K and were excited at 280 nm and observed at 400 nm with 4 nm FWHM spectral resolution

4 DISCUSSION

The primary influence of a vicinal disulphide on the phosphorescence of indole compounds is the development of non-exponential decay. There is a fast component accompanied by a much longer value, which may not be significantly less than that observed for indole of tryptophan. The indole phosphorescence is characterized by a well-developed fine structure, and this becomes diffuse in certain disulphide-indole compounds. The exact spatial relationship of the indole ring and the disulphide link are crucial in some manner for the removal of fine structuring. Thus, Cys Trp Cys and hen lysozome are typically fine structured, yet decay with two components, while *bis*-(indole-3-methylene)disulphide and human lysozome have only diffuse structure and also decay by two processes. In the absence of any detailed conformational information upon the model compounds, it is not yet apparent what particular orienta-

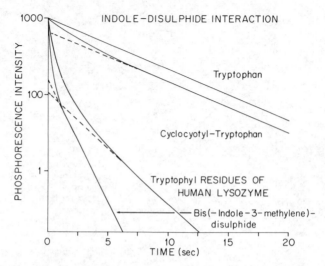

Figure 4 Phosphorescence decay of indole compounds. Solutions were 1 mg/ml in ethanol at 77 K and were excited at 280 nm, and observed at 440 nm, with 4 nm FWHM spectral resolution

tion and/or distance is necessary for the onset of dual-component decay and for the elimination of fine structure.

A clue to the nature of the interaction between indole triplet and a disulphide link is offered in the studies of von Schutz et al.[19] at 4·2 K. Those workers found

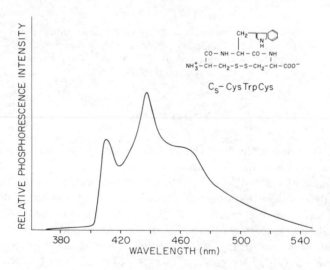

Figure 5 Phosphorescence spectrum of cyclo-dithio-L-cystyl-L-tryptophyl-L-cysteine. The solution was a mixture of dimethyl sulphoxide:ethylene glycol:water, 1:2:2, at 77 K, 1 mg/ml, which was excited at 280 nm with 2 nm FWHM spectral resolution

that hen lysozyme phosphorescence decayed at 6·3 s at 4·2 K, and this was a single exponential value. Grossweiner and Usui[14] have emphasized the ability of a triplet of indole to eject an electron, particularly into a vicinal disulphide link. It has yet to be shown whether this photoionization requires only a single photon. The absence of the luminescence decay perturbation at 4·2 K suggests there is a barrier to the formation of the necessary interaction. The precise atomic positions found for the majority of the tryptophyl and disulphide links in hen lysozyme at 298 K[23] suggests that there is little flexibility in the positions of the interior atoms in this protein. There remains the possibility of a charge-transfer interaction, and this may also account for the reduction of the phosphorescence decay magnitude.

Without exception, the phosphorescence decays of kappa Bence–Jones proteins have been found to contain two decay components. Many individual proteins have decay behaviours in their native conformations that are analogous to the decays of these simple model compounds. The lysozome fold conformational family also exhibits this phenomenon. Certain proteins in both conformational families also have a tryptophyl spectrum that lacks fine structure. These features are simulated by the simple indole-disulphide model compounds, and it is only necessary to suggest that features of this nature arise from a contiguity in the polypeptide fold of indole and disulphide links. The remarks apply equally well to tyrosyl phosphorescence from proteins, including lysozomes and Bence–Jones proteins.

REFERENCES

1. R. W. Cowgill, *Arch. Biochem. Biophys.*, **104**, 84–92 (1964).
2. A. I. Richard-Brewster and V. J. Hruby, *Proc. Nat. Acad. Sci. U.S.A.*, **70**, 3806–3809 (1973).
3. J. A. Glasel, V. J. Hruby, J. F. McKeivy and A. F. Spatsley, *J. Mol. Biol.*, **79**, 555–575 (1973).
4. R. W. Cowgill, *Biochim. Biophys. Acta*, **140**, 37–44 (1967).
5. R. W. Cowgill, *Biochim. Biophys. Acta*, **204**, 556–559 (1970).
6. J. W. Longworth, *Photochem. Photobiol.*, **7**, 587–596 (1968).
7. J. W. Longworth, *Excited States of Proteins and Nucleic Acids* (ees. R. F. Steiner and I. Weinryb), Plenum Press, New York, 1971. (a) pp. 362–364, (b) p. 390.
8. J. E. Churchich and J. Wampler, *Biochim. Biophys. Acta*, **243**, 304–311 (1971).
9. R. Shapira and G. Stein, *Science*, **162**, 1489–1490 (1968).
10. K. Dose, *Photochem. Photobiol.*, **8**, 331–335 (1968).
11. Sh. Arian, M. Benjamini, J. Feitelson and G. Stein, *Photochem. Photobiol.*, **12**, 481–487 (1970).
12. J. Feitelson and E. Hayon, *Photochem. Photobiol.*, **17**, 265–274 (1973).
13. T. Imoto, L. N. Johnson, A. C. T. North, D. C. Phillips and J. A. Rupley, *The Enzymes*, 3rd edn., Vol. 7, pp. 682, 717 (ed. P. Boyer), Academic Press, New York, 1972.
14. L. I. Grossweiner and Y. Usui, *Photochem. Photobiol.*, **13**, 971–982 (1971).
15. R. Santus and L. I. Grossweiner, *Photochem. Photobiol.*, **15**, 101–105 (1972).
16. Appendix: Stereo Plates, *Cold Spring Harbor Symp. Quant. Biol.*, **36**, 595–632 (1972).
17. J. W. Longworth, *Biopolymers*, **4**, 1131–1168 (1966).
18. J. E. Churchich, *Biochim. Biophys. Acta*, **92**, 194–197 (1964).

19. J. V. von Schultz, J. Zuchlich and A. H. Makai, *J. Amer. Chem. Soc.*, **96**, 714–718 (1974).
20. J. W. Longworth, C. L. McLaughlin and A. Solomon, paper D4, this volume.
21. C. Hélène and J. W. Longworth, *J. Chem. Phys.*, **37**, 399–408 (1972).
22. P. R. Bevington, *Data Reduction and Error Analysis for the Physical Sciences*, 2nd edn., pp. 204–246, McGraw-Hill, New York, 1969.
23. R. Diamond, *J. Mol. Biol.*, **82**, 371–391 (1974).

D.14. Formation of hydrated electrons from higher excited states of tryptophan in aqueous solution

H. B. Steen and M. Kongshaug†

Biophysics Department, Norsk Hydro's Institute for Cancer Research, The Norwegian Radium Hospital, Montebello, Oslo 3, Norway

SUMMARY

The UV-induced oxidation of Fe^{2+} in acid, aerated solutions of tryptophan at room temperature has been measured as a function of the wavelength of the exciting light. The effect of N_2O showed that the oxidation was induced by e_{aq}^- originating from excited tryptophan. The quantum yield of e_{aq}^- formation from tryptophan was found to have a constant value $\Phi(e_{aq}^-) = 0.015 \pm 0.005$ down to 250 nm. Below this wavelength it increased rapidly to $\Phi(e_{aq}^-) = 0.2$ at 210 nm. It is concluded that formation of e_{aq}^- from the S_2 state of tryptophan competes efficiently with $S_2 - S_1$ internal conversion. $\Phi(e_{aq}^-)$ appears to increase rapidly with the vibrational excitation energy of the S_2 state, whereas this is not the case for the e_{aq}^- formation from S_1. The e_{aq}^- formation from higher excited levels commences at $\geqslant 2.8$ eV below the gas phase ionization potential, demonstrating that e_{aq}^- is not formed via a free electron state. The mechanism of this photoionization is discussed and some possible implications for the photochemistry and radiolysis of related compounds are noted.

1 INTRODUCTION

It has been reported that excitation with UV light of various indole derivatives in aqueous solution may lead to ejection of electrons.[1,2] It is assumed that this electron ejection occurs from a complex formed between the S_1 state of indole and the surrounding water. The nature of this complex is not known. Flash photolysis studies of tryptophan and other indole derivatives and proteins indicate that the electron ejection is a major primary process in the photochemistry of these compounds.[3-5]

† Fellow of the Norwegian Cancer Society.

We report here measurements of the quantum yield of formation of e_{aq}^- from tryptophan as a function of the wavelength of the exciting light. The electron yield has been determined quantitatively by measurement of the oxidation of Fe^{2+} in acid, aerated solutions of tryptophan. It is found that the quantum yield is constant above 250 nm, whereas below this wavelength it increases rapidly with decreasing wavelength. Hence, e_{aq}^- formation also takes place from higher excited levels in competition with internal conversion, and with a much larger quantum yield and probably by a somewhat different mechanism than the formation of e_{aq}^- from the S_1–H_2O complex.

2 PRINCIPLE OF METHOD

Aqueous, oxygenated acid solutions of Fe^{2+} may be conveniently used for quantitative detection of e_{aq}^-, H and OH radicals by their production of Fe^{3+}. We find that the photolysis of tryptophan in such solutions exposed to UV light leads to formation of Fe^{3+}. The mechanisms of oxidation of Fe^{2+} are well known from radiation chemistry where such solutions are known as the Fricke dosimeter.[6]

$$e_{aq}^- + H^+ \rightarrow H \tag{1}$$

$$H + O_2 \rightarrow HO_2 \tag{2}$$

$$HO_2 + Fe^{2+} \rightarrow Fe^{3+} + HO_2^- \tag{3}$$

$$HO_2^- + H^+ \rightarrow H_2O_2 \tag{4}$$

$$H_2O_2 + Fe^{2+} \rightarrow Fe^{3+} + OH^- + OH \tag{5}$$

$$OH + Fe^{2+} \rightarrow Fe^{3+} + OH^- \tag{6}$$

It may be seen that each e_{aq}^- and each H will produce three Fe^{3+}, whereas each OH will give one.

To identify the reactive species produced by the photolysis of tryptophan, we added N_2O which reacts rapidly with e_{aq}^-, but is essentially unreactive toward H and OH.[7] The reaction of e_{aq}^- with N_2O leads to formation of one OH,[8] so that one Fe^{3+} is produced instead of three. At intermediate pH, i.e. pH $\sim$ 3, N_2O competes efficiently with H^+ for e_{aq}^- and will consequently cause a reduction of that part of the yield of Fe^{3+} which results from e_{aq}^-. Hence, by measuring the yield of $Fe^{3+}\Phi(Fe^{3+})$ in the presence and absence of N_2O, i.e. in solutions flushed with $\sim 20\%$ O_2 and $\sim 80\%$ N_2O and air ($\sim 20\%$ O_2 and $\sim 80\%$ N_2) respectively, one may calculate $\Phi(e_{aq}^-)$ from

$$\Phi(Fe^{3+}) = \Phi(OH) + 3\Phi(H) + X\Phi(e_{aq}^-) \tag{7}$$

where X is the number of Fe^{3+} resulting from each e_{aq}^-. Thus, in the absence of N_2O $X = 3$, whereas in the presence of N_2O X is some smaller number.

X was determined by measuring the radiolytic yield of Fe^{3+}, i.e. the yields resulting from exposure of the same solutions to ionizing radiation. According to equations (1)–(6), the radiolytic yield is given by

$$G(Fe^{3+}) = 2G(H_2O_2) + G(OH) + 3G(H) + XG(e_{aq}^-) \qquad (8)$$

where the G-values of the various radiolytic species are well known.[6,8] The concentration of Fe^{3+} may be accurately determined by measuring the optical absorption above 300 nm.

3 EXPERIMENTAL

The sample solutions normally contained 1×10^{-4}–5×10^{-4} M L-tryptophan, 1×10^{-3} M $FeSO_4$, 1×10^{-3} M NaCl and H_2SO_4 to make the adequate concentration of H^+. All reagents were Merck *pro analysi* except L-tryptophan which was from Sigma. The water was redistilled from alkaline permanganate. Only freshly made solutions were used.

0·4 ml samples were irradiated and analysed in a 1×0.5 cm quartz cuvette. The sample was bubbled with H_2O saturated, purified air, or with a mixture of $\sim 20\%$ O_2 and $\sim 80\%$ N_2O, for 10 minutes before as well as during the irradiation.

The light source was a D_2-lamp fitted to a 0·25 m Jarrell–Ash grating monochromator with slit widths corresponding to a transmitted bandwidth of 10 nm. The monochromatic light was focused on the sample by a quartz lens.

The intensity of the monochromatic light was measured by a calibrated thermopile. The quantum intensity at intermediate wavelengths was of the order of 1.5×10^{-11} Einstein/s.

The optical absorption spectrum of the samples was measured against an unexposed, identical solution immediately before and after irradiation by means of a Cary 118 scanning spectrophotometer. The irradiation time was usually 15 minutes, and the concentrations of Fe^{3+} resulting from this irradiation ranged downwards from 1.5×10^{-5} M.

The radiolytic yields of Fe^{3+} were measured by exposing the same solutions to 220 k Vp X-rays with a dose producing approximately 1.5×10^{-5} Fe^{3+}.

All experiments were at room temperature.

4 RESULTS

The absorption spectrum of the UV-irradiated samples relative to that of unexposed solutions was identical to that observed for Fe^{3+} in 10^{-5} M solutions of $FeCl_3$ except for a slight decrease in the absorption from tryptophan. This was found to be the case for all wavelengths of the exciting light and for pH = 3 as well as pH = 1. Identical spectra resulted from the X-ray exposures. Exposure with light of 310 nm wavelength, i.e. just above the absorption spectrum of tryptophan, produced no Fe^{3+}. Exposure with lower wavelengths of identical

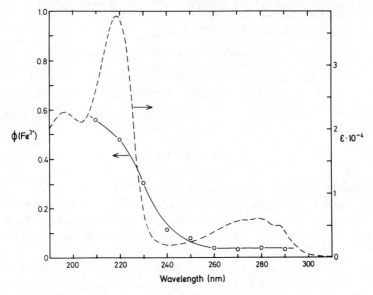

Figure 1 The wavelength dependence of the quantum yield of Fe^{3+} observed by UV exposure of aqueous, aerated solutions containing 5×10^{-4} M tryptophan, 10^{-3} M $FeSO_4$, 10^{-3} M NaCl and $0\cdot1$ M H_2SO_4 (full curve). The bandwidth of the exciting light is ~ 10 nm. The quantum yield of the e_{aq}^- giving rise to the oxidation of Fe^{2+} is given by: $\Phi(e_{aq}^-) = \Phi(Fe^{3+})/3$. The broken curve is the absorption spectrum of tryptophan

solutions, but without tryptophan, showed that direct photo-oxidation of Fe^{2+} as well as any auto-oxidation which might result from impurities were negligible. Hence, the photolytic products giving rise to the production of Fe^{3+} originated exclusively from tryptophan.

The concentration of Fe^{3+} was calculated from the OD at 305 nm which is close to the first absorption maximum of Fe^{3+}. The quantum yield of Fe^{3+} in solutions containing 5×10^{-4} M tryptophan is shown as a function of the wavelength of the exciting light in Figure 1. It can be seen that $\Phi(Fe^{3+})$ was independent of wavelength down to about 250 nm. Below this wavelength $\Phi(Fe^{3+})$ increased rapidly with the excitation energy so that at 210 nm it is more than one order of magnitude larger than the constant value above 250 nm.

Table 1   Relative yields of oxidation of Fe^{2+} (in terms of $OD_{305\,nm}$) in acid solutions of tryptophan and $FeSO_4$ resulting from exposure to 220 nm UV light and 220 k Vp X-rays

Flushing gas	$[H^+]$ M/l	220 nm UV light OD (305 nm)	220 k V X-rays OD (305 nm)
Air	10^{-3}	$0\cdot029 \pm 0\cdot0005$	$0\cdot032 \pm 0\cdot0005$
$80\% N_2O + 20\% O_2$	10^{-3}	$0\cdot021$	$0\cdot027$
Air	$0\cdot1$	$0\cdot028$	
$80\% N_2O + 20\% O_2$	$0\cdot1$	$0\cdot029$	

$\Phi(Fe^{3+})$ was not affected by a fivefold decrease of the tryptophan concentration to 1×10^{-4} M and was independent of acidity between pH = 3 and pH = 1.

Relative values of the photolytic and radiolytic yields of Fe^{3+} in solution flushed with air and N_2O/O_2 mixture, as obtained upon exposure with 220 nm UV light and 220 kVp X-rays respectively, are given in Table 1. It appears that at pH = 3 the presence of N_2O caused a substantial reduction of $\Phi(Fe^{3+})$ demonstrating that a significant fraction of this yield was caused by e_{aq}^-. At pH = 1 N_2O had no effect, indicating that N_2O did not affect the oxidation of Fe^{2+} in other ways than by competing with H^+ for electrons.

Using equations (7) and (8) it is calculated from the relative values of $\Phi(Fe^{3+})$ and $G(Fe^{3+})$ in Table 1 that 100% of the yield of Fe^{3+} at 220 nm is caused by e_{aq}^-. Taking into account uncertainties in the experimental data and in the literature values of the radiolytic radical yields, this result is correct to within 15%. Hence, $\Phi(Fe^{3+})$ appears to be directly proportional to the yield of e_{aq}^- ejected from tryptophan. According to equation (7), $\Phi(e_{aq}^-) = \Phi(Fe^{3+})/3$ in N_2O free solutions.

The moderate intensity of our present light source does not allow a similar identification of the reactive species produced by light of wavelength above 250 nm because of the small yield in this region. However, the experiments of Feitelson,[1] and Hopkins and Lumry[2] clearly indicate that this species is e_{aq}^-.

From Table 1 it may be seen also that in the absence of N_2O $\Phi(Fe^{3+})$, and hence $\Phi(e_{aq}^-)$, was independent of pH. This result may indicate that the e_{aq}^- is not associated with the parent ion, but becomes homogeneously distributed.

Finally, it should be mentioned that tryptophan is an efficient OH scavenger[8] and therefore might be expected to interfere with the oxidation of Fe^{2+} (equation (6)). However, it was found that the radiolytic yield of Fe^{3+} was not affected by the presence of tryptophan. This result implies that the OH adduct of tryptophan oxidizes Fe^{2+} quantitatively. Hence, in effect tryptophan did not affect the oxidation of Fe^{2+}.

5 DISCUSSION

First it should be noted that with the light intensity used in the present experiments only about 2% of the tryptophan molecules absorbed more than one UV photon during the 15 min exposure. Hence, all types of two-photon processes can be disregarded here.

The constant value of $\Phi(e_{aq}^-)$ above 250 nm (Figure 1) demonstrates that in this excitation region the primary step in the formation of e_{aq}^- occurs from the lowest vibrational level of S_1. Two different estimates of the gas phase ionization potential I_g of tryptophan (or indole) give values of 7·86 eV[9] and 8·43 eV[10] respectively. In order to account for the difference between I_g and the excitation energy of S_1 which is about 4 eV, one has to postulate that the formation of e_{aq}^- takes place via a complex between the S_1 molecule and the surrounding water and/or via an intermediate reaction, e.g. a proton transfer as suggested by

Dobson and Grossweiner[11] for the case of phenol. It is not clear whether this S_1–H_2O complex is of exciplex nature, i.e. an interaction primarily with one water molecule, as proposed by Hopkins and Lumry[2] or if it is simply a hydration of the S_1 molecule.

The large increase of $\Phi(e_{aq}^-)$ which is observed below 250 nm (Figure 1) demonstrates that for excitation energies larger than about 5·0 eV the initial step in the formation of e_{aq}^- takes place from levels above the lowest vibrational level of S_1. The onset of the absorption spectrum of the S_2 level of tryptophan appears to be at about 240 nm (see Figure 1). Allowing for the bandwidth of the exciting light in the present experiments it seems reasonable to assume that the formation of e_{aq}^- induced by wavelengths below 250 nm occurs primarily from the S_2 level. However, the steady increase of $\Phi(e_{aq}^-)$ with excitation energy over the entire region above 5·0 eV implies that the probability of formation of e_{aq}^- increases considerably with increasing vibrational energy of the S_2 state. The same conclusion has been reached from studies of the photoionization of naphthols[12,13] and of the ferrocyanide ion.[14]

The mechanism of the formation of e_{aq}^- from S_2 is not understood. The S_2 state probably has a lifetime of the order of 10^{-12} s.[15] The large value of $\Phi(e_{aq}^-)$, i.e. $\Phi(e_{aq}^-) = 0.2$ at 210 nm, means that the primary step in the formation of e_{aq}^- from S_2 must occur on the same time scale in order to compete with internal conversion. Struve and Rentzepis[16] recently reported a solvation time of an excited molecule in ethanol at room temperature of about 4×10^{-11} s. If the hydration time of the excited tryptophan molecule is of the same order of magnitude, the primary step in the formation of e_{aq}^- from S_2 occurs before hydration of the excited molecule can take place. Hence, hydration is probably not involved in this primary step. The lifetime of the S_1 state, on the other hand, is fully sufficient to allow hydration to occur prior to the ionization process.

The onset of the increase of $\Phi(e_{aq}^-)$ at about 250 nm in the present case (Figure 1) corresponds to an excitation energy of 5·0 eV. This energy is still 2·9 to 3·4 eV below I_g, depending on the value chosen for I_g. A direct tunnelling of the electron from the excited orbital into a nearby trap in the surrounding medium seems unlikely since it would require high concentrations of deep, empty traps in water. Studies of trapped electrons in several polar glasses at 4 K[17] indicates that such empty traps do not exist, but are created by the presence of the electron.

From fluorescence quenching studies it is known that electron scavengers may react with the lowest excited state of indoles by an electron transfer reaction.[18] However, such a mechanism cannot explain the present results. Thus, a reaction of tryptophan in the S_2 state with the present concentration of electron scavengers, e.g. 10^{-3} M H^+, would imply a reaction rate constant of the order of 10^{15} M^{-1} s^{-1}.

Hence, some sort of intermediate state or reaction must be involved in the photoionization of tryptophan. Several workers[14,19-21] have proposed that the photoionization occurs via a CTTS state. Very little is known about these CTTS state(s). However, it seems reasonable that the excited electron orbital is delocalized so that there is a considerable dipolar interaction between the electron and the neighbouring water molecules which may result in a gradual solvation of the electron and the positive ion. The quantum yield of photoionization would then be determined by the relative rates of dielectric relaxation of water leading to the hydration of the electron and recombination, i.e. internal conversion from the CTTS state to some lower excited state or to the ground state. This hypothesis is supported by the observation that $\Phi(e_{aq}^-)$ for tryptophan in glycol–water glass is of the order of 0·01 for an excitation energy of 5·9 eV,[22] whereas at room temperature, where the dielectric relaxation is much faster, we find $\Phi(e_{aq}^-) = 0·2$ for the same excitation energy. A similar increase of $\Phi(e_{aq}^-)$ with increasing temperature has been observed by others.[23] It cannot be decided from the present experiments whether the increase of $\Phi(e_{aq}^-)$ with increasing vibrational energy of S_2 reflects the presence of several CTTS states or if the excess vibrational energy is somehow utilized to break down the water structure and thereby speed up the hydration and hence the formation of e_{aq}^-.

A recent study of the recombination luminescence of tryptophan in glycol–water glass indicated in fact the existence of a CTTS-like state with a lifetime of roughly 2×10^{-8} s at 77 K.[24] The probability of recombination appeared to decrease strongly with increasing temperature,[25] i.e. with increasing rate of dielectric relaxation, indicating that the lifetime of the CTTS-like state decreased correspondingly. Rentzepis *et al.*[26] have observed that the time needed for the optical absorption spectrum of e_{aq}^- (produced by photoionization of $Fe(CN)_6^{4-}$)) to develop is about 4×10^{-12} s. As noted by these authors, it is conceivable that what they observe is not the hydration of a 'free electron', but the gradual transfer of a CTTS state into a pair of hydrated ions.

The importance of e_{aq}^- formation from S_1 in the photodecomposition of indole derivatives has been emphasized by several authors.[3–5] It is also believed that this process is an important primary step in the photodenaturation of proteins, partly because of the oxidation of the tryptophan residues and partly because of the effect that the ejected electrons may cause by reacting with other residues, especially cystine.[3] In view of the strong increase of $\Phi(e_{aq}^-)$ below 250 nm reported here, the importance of e_{aq}^- formation is likely to be much larger when higher excited levels of tryptophan are excited. As noted above, however, the actual excitation energy which is necessary to release electrons from higher excited levels may depend on the polarizability of the medium surrounding the tryptophan residues and hence on whether these residues are buried inside the protein or located on its surface. The importance of solvent polarity has recently been verified experimentally.[30]

Very little is known about the role of molecular excitation in the radiolysis of condensed media, although the yield of bound excited levels produced by ionization radiation may be of the same order of magnitude as the yield of ionization. It is commonly assumed that ionizing radiation leads predominantly to excitation of highly excited levels,[27] but that the major fraction of the subsequent decomposition takes place only after internal conversion to the lowest excited levels. Indeed this is often taken for granted in photochemistry as well. The present results indicate, however, that decomposition may well occur from higher excited levels. If the quantum yield of the formation of e_{aq}^- presently observed for tryptophan proves to be typical, decomposition originating from bound, higher excited levels may be of considerable importance in many systems. Thus, photochemical data which are usually obtained by excitation of the lowest excited singlet and triplet level, cannot as a matter of course be applied directly to highly excited molecules formed by radiolysis.

Finally, it should be noted that the present conclusion that the formation of e_{aq}^- from higher excited levels competes efficiently with internal conversion to the fluorescent S_1 level, is in accordance with recent experiments showing that the fluorescence quantum yield of tryptophan (as well as of some other aromatics) drops significantly below 250 nm,[28,29] that is, in the wavelength region where the increase of $\Phi(e_{aq}^-)$ commences.

REFERENCES

1. J. Feitelson, *Photochem. Photobiol.*, **13**, 87 (1971).
2. T. R. Hopkins and R. Lumry, *Photochem. Photobiol.*, **15**, 555 (1972).
3. L. I. Grossweiner and Y. Usui, *Photochem. Photobiol.*, **13**, 195 (1971).
4. R. Santus and L. I. Grossweiner, *Photochem. Photobiol.*, **15**, 101 (1972).
5. V. Subramanyan and G. Tollin, *Photochem. Photobiol.*, **15**, 449 (1972).
6. H. Fricke and E. J. Hart, *Radiation Dosimetry*, Vol. 2, p. 167 (eds. F. H. Attix and W. C. Roesch), Academic Press, London, 1966.
7. M. Anbar and P. Neta, *Int. J. Appl. Radiat. Isotopes*, **18**, 493 (1967).
8. I. G. Draganic and Z. D. Draganic, *The Radiation Chemistry of Water*, Academic Press, London, 1971.
9. I. Fischer-Hjalmars and M. Sundbom, *Acta Chem. Scand.*, **22**, 607 (1968).
10. M. A. Slifkin and A. C. Allison, *Nature*, **215**, 950 (1967).
11. G. Dobson and L. I. Grossweiner, *Trans. Faraday Soc.*, **61**, 708 (1965).
12. C. R. Goldschmidt and G. Stein, *Chem. Phys. Letters*, **6**, 299 (1970).
13. V. K. Kläning, C. R. Goldschmidt, M. Ottolenghi and G. Stein, *J. Chem. Phys.*, **59**, 1753 (1973).
14. M. Shirom and G. Stein, *J. Chem. Phys.*, **55**, 3379 (1971).
15. J. B. Birks, *Photophysics of Aromatic Molecules*, p. 187, Wiley–Interscience, London, 1970.
16. W. S. Struve and P. M. Rentzepis, *J. Chem. Phys.*, **59**, 5014 (1973).
17. T. Higashimura, M. Noda, T. Warashina and H. Yoshida, *J. Chem. Phys.*, **53**, 1153 (1970).
18. R. F. Steiner and E. P. Kirby, *J. Phys. Chem.*, **73**, 4130 (1969).
19. M. Ottolenghi, *Chem. Phys. Letters*, **12**, 339 (1971).
20. G. Stein, *Adv. in Chemistry*, Ser. 50, 230 (American Chemical Soc., Washington D.C.), (1965).
21. L. I. Grossweiner and H.-I. Joschek, *Adv. in Chemistry*, Ser. 50, p. 279 (American Chemical Soc., Washington D.C.), (1965).
22. J. Moan, private communication.
23. M. Shirom and Y. Siderer, *J. Chem. Phys.*, **58**, 1250 (1973).
24. H. B. Steen and O. Strand, *J. Chem. Phys.*, **60**, 5043 (1974).
25. H. B. Steen, *Radiation Res.*, **41**, 268 (1970).

26. P. M. Rentzepis, R. P. Jones and J. Jortner, *J. Chem. Phys.*, **59**, 766 (1973).
27. R. L. Platzman, *Radiation Research*, p. 20 (ed. G. Silini), North-Holland Publ. Co., Amsterdam, 1967.
28. I. Tatischeff and R. Klein, this volume, paper D.1.
29. H. B. Steen, *J. Chem. Phys.*, **61**, 3997 (1974).
30. L. Kevan and H. B. Steen, *Chem. Phys. Letters*, **34**, 184 (1975).

D.15. Luminescent properties of Angiotensin II and its analogues

C. A. Ghiron

Department of Biochemistry, University of Missouri, Columbia, Missouri 65201

F. M. Bumpus

Division of Research, Cleveland Clinic, Cleveland, Ohio 44106

J. W. Longworth

Biology Division, Oak Ridge National Laboratory, Oak Ridge, Tennessee 37830, U.S.A.

SUMMARY

The solution conformation of Angiotensin II (AII) has been studied extensively, but the question whether AII has a single preferred and biologically active conformation in solution has not as yet been resolved. On the basis of structure-activity studies, it has been proposed that the tyrosine$_4$ and phenylalanine$_8$ residues of AII, necessary for pressor activity, are contiguous in the biologically active conformation of this hormone. This proposal implies efficient Phe → Tyr energy transfer, which should be demonstrable by fluorescent measurements. Both residues fluoresce in the octapeptide, and the Phe → Tyr energy transfer is less than 50% efficient. By contrast, at 230 K in an equivolume mixture of glycerol and water, fluorescence anisotropy measurements indicate an almost complete (100%) transfer of electronic energy from Phe to Tyr. The C-terminal penta-, hexa-, and hepta-peptides and the Ala$_6$ analogue of AII were found to give similar results. Equally small transfer was found at 300 K when AII was studied in various other solvent mixtures. Very little Phe → Tyr transfer was detected even at 230 K with the analogue Phe$_3$–Tyr$_8$ AII. The TyrOMe$_8$ and

† Research sponsored by USPHS Grant HL-6835, by a USPHS Special Fellowship, by Oak Ridge Associated Universities and by the U.S. Atomic Energy Commission under contract with the Union Carbide Corporation.

Trp_8 analogue of AII showed efficient transfer of electronic energy between their aromatic residues both in water at 300 K and in glycerol–water at 230 K. Triplet–triplet (Tyr^- → Trp) energy transfer in Trp_8 AII was readily detected and was only partially abolished by the addition of saturating amounts of guanidine hydrochloride. These results support the hypothesis that AII does not possess a preferred conformation in water at 300 K. However, as the size and hydrophobicity of the side chain at position 8 is increased, this molecule is able to attain an ordered conformation characterized by a close association of the side chains at positions 4 and 8. At low temperatures it appears that a stable, ordered conformation is attained by all the peptides studied except $Phe_3–Tyr_8$ AII. Since, under these conditions, the Phe and Tyr residues are contiguous in the ordered structure of AII, we speculate that this conformation may approximate the biologically active one in the receptor.

1 INTRODUCTION

The solution conformation of Angiotensin II (AII) and its asparagine analogue ($Asn_1–Arg_2–Val_3–Tyr_4–Ile_5–His_6–Pro_7–Phe_8$) have been studied by thin-film dialysis, ultracentrifugation, gel filtration, titration, hydrogen–deuterium or hydrogen–tritium exchange, optical rotary dispersion–circular dichroism (ORD–CD), and ultraviolet, infrared, laser Raman and NMR spectroscopy. These methods have yielded structural information that has excluded a number of proposed structures.[1] However, the studies have not resolved the question as to whether AII has or has not a single preferred and biologically active conformation in solution.

This octapeptide hormone has been shown to have a variety of biological activities.[2] Among these, its pressor action has been most extensively investigated. The structure-activity requirements for pressor function can be summarized as follows: (a) there must be at least 6 amino-acids (residues 3–8 of AII) in the peptide chain; (b) the presence of 8 amino-acids residues (AII) yields maximum activity; (c) a tyrosine residue is required in position 4 of AII; (d) the imidazole ring of histidine is required in position 6; (e) proline or an amino-acid with a secondary amino N must be adjacent to position 8; (f) an aromatic amino-acid with free carboxyl must be in position 8.

On the basis of such observations, Smeby et al.[3] proposed that the tyrosine (Tyr_4) and phenylalanine (Phe_8) residues, necessary for pressor activity of AII, are contiguous in the biologically active conformation of the octapeptide. Implicit in this proposal is that there would be efficient transfer of energy from Phe to Tyr. Modification of spectra as a consequence of such a process should be demonstrable through fluorescence measurements. In this study, we have sought to detect evidence for transfer as support for this proposed structural element by determining the luminescent properties of AII and its analogues.

2 MATERIAL AND METHODS

Spectral-quality glycerol and trifluoroethanol were purchased from Matheson–Coleman–Bell. AII, the C-terminal hepta-, hexa- and penta-peptides, the Ala_6 and the position-8 analogues ($TyrOMe_8$ and Trp_8) of AII were synthesized by procedures similar to those described previously.[4] All other chemicals, including some of the AII and the Phe_3–Tyr_8 AII, were obtained from commercial sources and were used without further purification.

The ultraviolet absorption measurements were performed utilizing either a Beckman D.U. monochromator in combination with a Gilford photometer or a Zeiss model PMQII spectrophotometer. The absorption measurements were done with slit widths set so that the bandpass was the same as that employed in the excitation monochromator of the fluorometer. The fluorescence excitation and emission spectral data were obtained utilizing an instrument similar in design to that described by Luk.[5]

The fluorescence polarizations were measured with an instrument built by Knopp and Longworth.[6] The exciting light was monochromatic. The fluorescence was isolated with cut-off glass colour filters that passed light of wavelength greater than 305 nm or 335 nm. Polarization was determined at right angles to the excitation direction with two analyser prisms and two matched photomultipliers (a Cornu polarimeter). The ratio of the light intensity was directly determined electronically through a voltage-to-frequency conversion and a ratio frequency scaling, for both polarizing positions. Therefore, the emission anisotropy (A) is given by

$$A = [(R_V/R_H) - 1]/[(R_V/R_H) + 2]$$

where R equals $I_{\parallel}/I_{\perp}$ for vertical (V) and horizontal (H) excitation. Any residual wavelength-dependent polarization in the exciting light is thus eliminated. The phosphorescence lifetimes and emission measurements were determined with an instrument and by methods described previously.[7,8]

3 RESULTS AND DISCUSSION

The evidence that absorption by Phe contributes to the Tyr fluorescence is summarized in Figure 1. Relative Tyr fluorescence yields were determined at various excitation wavelengths, and the yields for $N - Ac$–$L \cdot Tyr$–NH_2 and for H–$L \cdot Phe$–$L \cdot Tyr$–OH were independent of wavelength while that for AII progressively decreased as the excitation wavelength was decreased from 270 to 255 mm. These results indicate that there is very efficient Phe $\rightarrow$ Tyr singlet–singlet energy transfer in the dipeptide. Transfer is considerably less efficient in AII. Had no energy transfer occurred, the relative fluorescence yield would have declined at wavelengths less than 270 nm and attained values equal to the Tyr fractional absorbance of AII at each particular excitation wavelength.

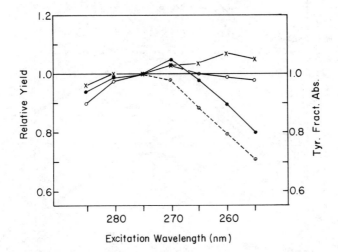

Figure 1 Wavelength dependence of the quantum yield of (○) H–L · Phe–L · Tyr–OH, (×) N-Ac–L · Tyr–NH₂, and (●) Angiotensin II (AII) compared to (⊙) the tyrosinyl fractional absorbance of AII. The fluorescence yield was measured at 304 nm. Solvent, water; temperature, 300 K

The tyrosyl absorbance was calculated by subtracting the molar absorbance of N–Ac–L · Phe–OMe from that of AII. This procedure was deemed appropriate since the absorption maximum of AII is red-shifted 1 nm and its molar absorbance is decreased slightly as compared to that of N–Ac–L · Tyr–NH$_2$, whereas for Phe derivatives these parameters are relatively independent of environment. For example, the fractional absorbances at 255 nm obtained by this procedure were 0·292 and 0·708 for Phe and Tyr, respectively. These values do not differ greatly from those of 0·285 and 0·715 calculated for an equimolar mixture of the two amino-acids. This excellent concurrence indicates that our preparations of AII did not contain significant quantities of non-fluorescent ultraviolet-absorbing impurities, excluding this as a possible explanation for the observed low transfer efficiency.

Fluorescence emission spectra (Figure 2) are consistent with our interpretation. When an equimolar mixture of N–Ac–L · Tyr–NH$_2$ and N–Ac–L · Phe–OMc was excited with 270 nm light, the emission spectrum obtained was that of Tyr, since Phe has a low absorbance at this exciting wavelength. By contrast, when the fluorescence emission spectrum was determined upon excitation with 252 nm light, an appreciable amount of Phe fluorescence was detected with a wavelength maximum at about 280 nm. It should be noted that the two emission spectra were normalized at 305 nm, a wavelength where the Phe fluorescence

490

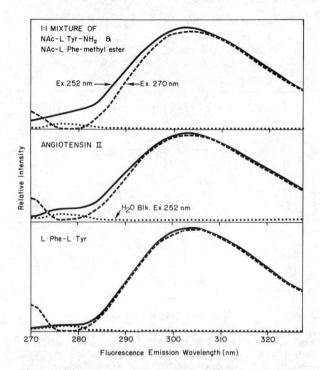

Figure 2 Fluorescence emission of an equimolar mixture of
N-Ac–L · Tyr–NH$_2$ and N-Ac–L · Phe–methyl ester (top
panel), Angiostensin II (middle panel) and L · Phe–L · Tyr
(lower panel). Excitation, 252 and 270 nm; concentration,
$A_{275nm}^{10mm} = 0.10$; temperature, 300 K; solvent, water. The
emission of the water blank excited at 252 nm is also shown
in each panel. Spectra were normalized at 305 nm

makes only a slight contribution to the total emission. When the dipeptide was
studied in an analogous manner, no Phe fluorescence was detected. An intermediate amount of Phe fluorescence was observed when AII was tested. These
results strengthen our interpretation that Phe → Tyr energy transfer occurs
only to a limited extent in AII.

Utilizing several replicates of the experiment depicted in Figure 1, we estimated that the Phe → Tyr transfer efficiency is 30 ± 15 % and concluded that
the transfer is less than 50 % efficient. Using Förster's theory, we calculate the
average distance between the Phe and Tyr side chains in AII to be greater than
1·0 nm, assuming $\kappa^2 = 2/3$, $\phi_D = 0.02$[9] and $R_0 \phi_D^{-1/6} = 1.86$ nm.[10] Small
values of κ^2 are unlikely for the following reasons: (a) Fermandjian et al.[11]
have concluded on the basis of CD measurements that 'the tyrosine side chain
of AII behaved as if it was free of interactions limiting its rotation'; and (b) the
transition dipole moment of Phe has approximately hexagonal symmetry in
the plane of the ring.[12] Thus, any slight bending of the χ_1 and χ_2 bonds of Phe
in combination with a fairly equal distribution of the three staggered conforma-

tions for χ_1 and some rotational freedom for χ_1 should yield a κ^2 for the transfer from Phe side chain which approximates the value as if the Phe were free of any interactions limiting its rotation.[13]

Utilizing the value of 2·6 for the characteristic ratio $\langle r^2 \rangle / xl^2$ (taken from a plot of the characteristic ratios for racemic copolymers of the polyalanine as a function of the stereoregulatory[14]), where the degree of polymerization, $x = 4$; the length of the bond, $l = 0.38$ nm; and the $W_{iso} = 1$ (i.e., stereoregular with all residues L); we calculate a value of 1·23 nm for $\langle r \rangle$, the end-to-end distance of a tetrapeptide with a random conformation. This value is to be compared with 0·73 nm, the distance if the octapeptide were an α-helix.[15] Our results, therefore, do not support the proposal that Phe and Tyr are adjacent in the time-averaged solution conformation of AII at 300 K.

The C-terminal penta-, hexa- and hepta-peptides and the Ala_6 analog of AII were also found to have a low transfer efficiency never exceeding 50 %. Equally small transfer was found at 300 K when AII was studied in the following solvents: trifluoroethanol; ethanol–water (1:1); glycerol–water (1:1); and 0·1 M KCl–water. The fluorescence emission spectra revealed an intermediate amount of Phe fluorescence in all cases except when AII was dissolved in trifluoroethanol, in which only a negligible amount of Phe fluorescence was observed.

No Phe $\rightarrow$ Tyr transfer was detected with the analogue Phe_3–Tyr_8 AII. We estimate the Phe-to-Tyr distance to be 1·36 nm if this analogue were in a random conformation. If the peptide were an α-helix, the distance would be 0·72 nm. Since the Förster critical distance, R_0, for Phe $\rightarrow$ Tyr transfer at 300 K has a value of 0·98 nm, it may be concluded that the Phe and Tyr side chains are not in close proximity in the time-averaged conformation of Phe_3–Tyr_8 AII.

The Tyr fluorescence yield of $TyrOMe_8$ AII, a good agonist, was independent of the exciting wavelength. Since the absorption spectra of Tyr and TyrOMe are almost identical, this result does not allow us to decide whether energy transfer has occurred between these two residues. Had no transfer occurred, however, we should have observed a slight blue shift in the wavelength maximum and an increase in the full-width half-maximum in the emission spectrum of this analogue, since the wavelength maximum of p-methylanisole is blue-shifted approximately 5 nm in comparison to that of p-cresol.[16] The fluorescence emission of $TyrOMe_8$ AII had the identical wavelength maximum and the same full-width half-maximum of 32·8 nm (0.358 μm^{-1}) as AII, indicating that efficient energy transfer occurs from TyrOMe to Tyr. We estimate that R_0, the Förster critical distance, for this donor–acceptor pair is 0·90 nm.† These data suggest that the Tyr and TyrOMe residues are contiguous in the time-averaged conformation of $TyrOMe_8$ AII.

An analogue which also appears to have this structural element is the agonist–antagonist Trp_8 AII. In this case, the tryptophan (Trp) fluorescence yield (wave-

† The following assumptions and calculations were made: $\kappa^2 = 2/3$; $J_{AD} = 0.58$ (10^{-16} M^{-1} cm^6); $n = 1.5$; $R_0 \phi_D^{-1/6} = 1.36$ nm; $\phi_{Tyr} = 0.20$, pH $= 7$, water, 300 K; [17] ϕ_{TyrOMe} in AII $= 0.43 \phi_{Tyr} = 0.09 = \phi_D$; therefore $R_0 = 0.90$ nm.

492

length maximum, 355 nm) was also found to be independent of excitation wavelength (295–265 nm). If photons absorbed by Tyr were not contributing to the Trp fluorescence, the yield would have decreased at exciting wavelengths less than 290 nm. The Trp_8 AII fluorescence emission spectrum upon excitation with 275 nm light (where approximately 20% of the absorbed light is absorbed by Tyr) was superimposable with the one obtained upon 295 nm excitation (a wavelength where only Trp absorbs). Thus, no Tyr fluorescence could be detected, confirming that the Tyr → Trp energy transfer is 100% efficient. We estimate that, on the average, the Tyr-to-Trp side-chain separation must be less than 0·9 nm. †

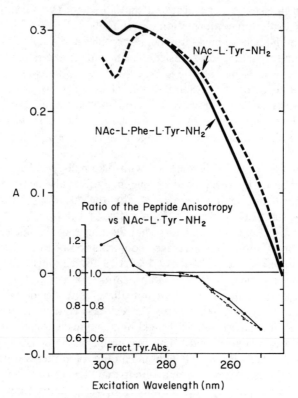

Figure 3 The fluorescence absorption anisotropy (A) of N-Ac–L · Tyr–NH₂ and N-Ac–L · Phe–L · Tyr–NH₂ plotted as a function of excitation wavelength (nm). Solvent, equivolume glycerol–water; temperature, 233 K; cutoff filters, 305 nm. The insert compares the ratio of the A-value of the dipeptide with that of N-Ac–L · Tyr–NH₂, plotted as a function of excitation wavelength to the Tyr fractional absorbance of the dipeptide

† The following assumptions and calculations were made: $\kappa^2 = 2/3$; $J_{AD} = 4\cdot8(10^{-16}\,\mathrm{M}^{-1}\,\mathrm{cm}^6)$; $R_0\phi_D^{-1/6} = 1\cdot92\,\mathrm{nm}$; $\phi_{Tyr} = 0\cdot20$, pH = 7, water, 300 K;[17] ϕ_{Tyr} in AII = 0·35 $\phi_{Tyr} = 0\cdot07 = \phi_D$; $R_0 = 1\cdot23\,\mathrm{nm}$; $r/R_0 = 0\cdot75$ for 0·8 transfer efficiency; $r \approx 0\cdot9\,\mathrm{nm}$.

Fluorescence polarization measurements can also be used to detect transfer of singlet energy. The complicating rotational depolarization effect was eliminated by dissolving the samples in an equivolume glycerol–water mixture and reducing the temperature to ~ 230 K.[6] The Tyr fluorescence emission anisotropy, A, plotted versus the excitation wavelength for N–Ac–L $\cdot$ Tyr–NH$_2$ and for N–Ac–L $\cdot$ Phe–L $\cdot$ Tyr–NH$_2$ is shown in Figure 3. The A-values for these two compounds are equal in the wavelength range 290–270 nm, but at excitation wavelengths less than 270 nm, where Phe contributes to the absorbance, the A value for the dipeptide is significantly less than that obtained for N–Ac–L $\cdot$ Tyr–NH$_2$. As can be seen in the insert of Figure 3, the ratio of the anisotropy of the peptide to that of N–Ac–L $\cdot$ Tyr–NH$_2$ plotted as a function of excitation wavelength parallels the Tyr fractional absorption of the dipeptide. From these observations we conclude that Phe $\rightarrow$ Tyr energy transfer is $\sim 100\%$ efficient and that the portion of the Tyr fluorescence contributed by Phe absorption is completely depolarized.

The Tyr fluorescence emission anisotropy spectrum of AII is very similar to the one obtained for the dipeptide, indicating that Phe $\rightarrow$ Tyr energy transfer is $\sim 100\%$ efficient. Since this exalted efficiency cannot be accounted for solely by the increase in the Phe fluorescence yield to 0·06 on lowering the temperature, we must conclude that AII attains an ordered conformation which permits very efficient energy transfer at 230 K. In addition, this conformation must be quite

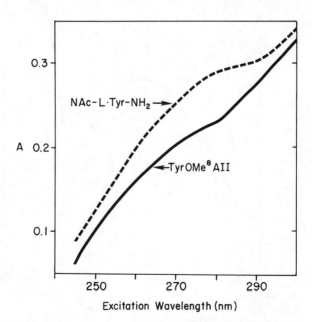

Figure 4 Fluorescence absorption anisotropy (A) of N-Ac–L $\cdot$ Tyr–NH$_2$ and TyrOMe$_8$ AII plotted as a function of excitation wavelength. Solvent, equivolume glycerol–water; temperature, 228 K; cutoff filters, 305 nm

stable, since the addition of guanidine hydrochloride (4·5 M) to the AII solution did not appreciably modify the magnitude of the anisotropy spectrum.

Similar results were obtained for the C-terminal penta-, hexa-, and hepta-peptides and for the Ala_6 analogue of AII. Very little Phe → Tyr energy transfer was detected for the analogue Phe_3–Tyr_8 AII ($R_0 = 1·2$ nm, $\langle r \rangle = 1·36$ nm). Since in the excitation wavelength range of 290 to 270 nm the A-values of AII and its analogues were the same as those measured for N-Ac–L · Tyr–NH_2, it may be concluded that no aggregation occurred in glycerol–water at 230 K. However, the magnitude of the Tyr fluorescence emission anisotropy of $TyrOMe_8$ AII (Figure 4) is substantially less than that of N-Ac–L · Tyr–NH_2 at all excitation wavelengths, confirming that energy transfer occurs between the two aromatic residues of this AII analogue.

A comparison of the Trp fluorescence emission anisotropy spectra of Trp_8 AII and N-Ac–L · Trp–NH_2 (Figure 5) reveals a decrease in the A-value of the peptide at exciting wavelengths less than 285 nm, where the Tyr residue contributes to the absorbance. The fact that the anisotropy ratio of 0·85, upon excitation at 275 nm, is close to 0·80, the Trp fractional absorbance at this wavelength, suggests that Tyr → Trp energy transfer occurs efficiently·

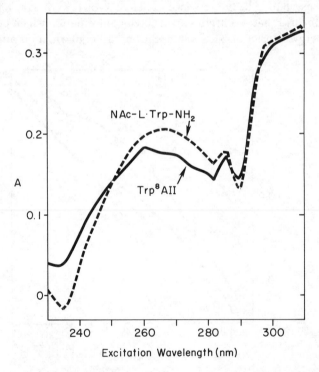

Figure 5 Fluorescence absorption anisotropy (A) of N-Ac–L · Trp–NH_2 and Trp_8 AII plotted as a function of excitation wavelength (nm). Solvent, equivolume glycerol–water; temperature, 220 K; cutoff filters, 335 nm

Support for this contention comes from phosphorescence measurements at 77 K. The phosphorescence emission spectra of the Trp_8 AII excited at 280 and 295 nm were superimposable and nearly identical to that of N–Ac–L · Trp–NH_2. Since no Tyr phosphorescence was detected, the energy transfer is probably $\sim 100\%$ efficient.

Ionized tyrosine-to-tryptophan triplet–triplet energy transfer has recently been demonstrated in trypsin.[8] Since this transfer is mediated by an electron exchange interaction, its demonstration requires that the average distance between the donor and acceptor side chains be less than 1 nm. Figure 6 shows the phosphorescence spectra of Trp_8 AII at 77 K dissolved in an equivolume mixture of glycerol–water at basic pH in the presence and absence of saturating amounts of guanidine hydrochloride. As can be seen, with 309 nm excitation, where the Trp fractional absorbance is quite low, the Trp phosphorescence emission with its vibrational peaks and sharp onset at 398 nm is superimposed upon the ionized Tyr emission with a 360 nm onset. In the presence of GuHCl, the Trp phosphorescence is barely discernible upon the much more prominent ionized Tyr emission.

Confirmatory evidence for triplet–triplet energy transfer was obtained by determining the phosphorescence lifetime of the ionized Tyr residue at 380 nm (excitation 295 nm). If transfer occurs between a donor–acceptor pair, the phosphorescence lifetime of the donor is diminished. In the present case, the lifetime of the phenolate donor was reduced from a normal value of 1·21 to

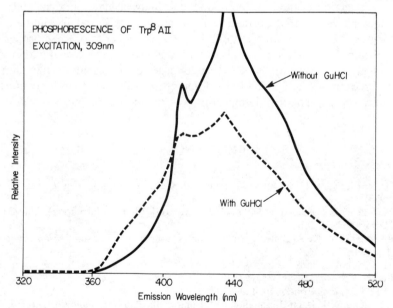

Figure 6 Phosphorescence emission of Trp_8 AII in the presence and absence of saturating amounts of guanidine hydrochloride (GuHCl). Solvent, equivolume glycerol–water at basic pH; excitation, 309 nm; temperature, 77 K; concentration, $A_{280nm}^{10mm} = 0.9$. The sensitivity for the 'without GuHCl' is increased eightfold

0·94 s, and in the presence of GuHCl a value of 1·07 s was obtained. The time dependence of the 380 nm emission in all these cases was exponential. By contrast, when Trp_8 AII was excited with 308 nm light, the 435 nm phosphorescence could be resolved into two exponential components. The long-lived component had a lifetime of 6·6 s both in the presence and absence of GuHCl, confirming that the emission is from the Trp residue. All of these observations permit us to conclude that Trp_8 AII possesses an ordered conformation and indicate that it can, at least, be partially disrupted by saturating amounts of GuHCl.

In summary, the evidence presented in this paper supports the hypothesis that AII does not have a preferred conformation in water at 300 K. Unfortunately, our attempts to increase the Phe → Tyr transfer efficiency by modifying the solvent either failed or gave equivocal results. However, as the size and hydrophobicity of the side chains at position 8 are increased, the molecule attains an ordered conformation characterized by a close association of the side chains at positions 4 and 8.

The Saclay group has presented strong evidence that ordered conformations are predominant for high trifluoroethanol concentrations and at elevated temperatures.[11,18] The present studies show that at low temperatures, in equivolume glycerol–water, a stable ordered conformation is also predominant.

It has been suggested that the biologically active conformation of AII may arise only when it interacts with the receptor.[1,11] We suggest that the agonist–antagonist action of the position-8 analogues may originate in their ability to maintain an ordered conformation in water at 300 K and to simulate the biologically active conformation of AII. As a consequence of these properties, the position-8 analogues may interact with the receptor more readily than AII. Since in glycerol–water at 230 K AII has an ordered conformation characterized by the proximity of its Tyr and Phe side chains, we speculate that this conformation may approximate the biologically active one in the receptor.

REFERENCES

1. G. R. Marshall, H. E. Bosshard and J. D. Glickson, *Nature, New Biol.*, **245**, 125 (1973).
2. M. J. Peach, *Chemistry and Biology of Peptides*, p. 471 (ed. J. Meienhofer), Ann Arbor, Michigan, 1972.
3. R. R. Smeby, K. Arakawa, F. M. Bumpus and M. M. Marsh, *Biochim. Biophys. Acta*, **58**, 550 (1962).
4. M. C. Khosla, R. A. Leese, W. L. Maloy, A. T. Ferreira, R. R. Smeby and F. M. Bumpus, *J. Med. Chem.*, **15**, 792 (1972).
5. C. K. Luk, *Biopolymers*, **10**, 1229 (1971).
6. J. A. Knopp, J. J. ten Bosch and J. W. Longworth, *Biochim. Biophys. Acta*, **188**, 185 (1969).
7. C. Hélène and J. W. Longworth, *J. Chem. Phys.*, **57**, 399 (1972).
8. C. A. Ghiron, J. W. Longworth and N. Ramachandran, *Proc. Nat. Acad. Sci. U.S.A.*, **70**, 3703 (1973).

9. E. Leroy, H. Lami and G. Laustriat, *Photochem. Photobiol.*, **13**, 411 (1971).
10. J. Eisinger, B. Feuer and A. A. Lamola, *Biochemistry*, **8**, 3908 (1969).
11. S. Fermandjian, J. L. Morgat and P. Fromageot, *Eur. J. Biochem.*, **24**, 252 (1971).
12. T. Marzcalek, *Acta Phys. Polon.*, **A37**, 227 (1970).
13. P. L. Wessels, J. Feeney, H. Gregory and J. J. Gormley, *J. Chem. Soc. Perkin II*, 1691 (1973).
14. P. J. Flory, *Statistical Mechanics of Chain Molecules*, p. 285, John Wiley and Sons, New York, 1969.
15. J. J. ten Bosch and J. A. Knopp, *Biochim. Biophys. Acta*, **188**, 173 (1969).
16. I. B. Berlman, *Handbook of Fluorescence Spectra of Aromatic Molecules*, 2nd ed., pp. 158, 160, Academic Press, New York, 1971.
17. R. W. Cowgill, *Biochim. Biophys. Acta*, **200**, 18 (1970).
18. P. Marche, J. L. Morgat and P. Fromageot, *Eur. J. Biochem.*, **40**, 513 (1973).

D.16. Lifetimes of triplet states of aromatic amino-acids and peptides in water

D. V. Bent and E. Hayon

Pioneering Research Laboratory, U.S. Army Natick Laboratories, Natick, Mass., U.S.A.

SUMMARY

A quadrupled neodymium pulsed laser emitting at 265 nm with a pulse duration of ~ 3 ns and 15 ns was used to study the photophysics and photochemistry of aromatic amino-acids and peptides in water at 25 °C. Optical excitation of tyrosine, phenylalanine and tryptophan, and some of the corresponding oligopeptides, produced short-lived transient absorption spectra which have been assigned to triplet–triplet absorptions. The ^{3}Tyr has maxima at ~ 250, ~ 295 and ~ 575 nm with a lifetime of 10 μs at 'zero' tyrosine concentration. The ^{3}Phe has maxima at ~ 245 and ~ 318 nm, and 8 $\tau = 3 \cdot 1$ μs in 3·8 mM solutions. The ^{3}Trp has a maximum at 450 nm and $\tau = 14 \cdot 3$ μs. These triplet excited states are quenched by oxygen and disulphides with $k \sim 4$–5×10^9 $\text{M}^{-1} \text{s}^{-1}$. Electron transfer results from the quenching reaction, with the formation of the superoxide radical $\cdot \text{O}_2^-$ and the RSSR$^-$ radical anion, respectively. Photoionization from Tyr and Phe proceeds mainly from the triplet state via a biphotonic mechanism, while photoionization from Trp proceeds mainly from the singlet excited state. The dependence of the yield of triplet and of e_{aq}^- upon pH was investigated.

1 INTRODUCTION

The study of the electronic excited states of proteins has been actively pursued during the last two decades (see References 1 and 2 for a review). The luminescence of natural proteins has been shown to arise mainly from the ring moieties of the aromatic amino-acids—tryptophan, tyrosine and phenylalanine—present

498

in the protein. Systematic investigations of the fluorescence properties of these aromatic amino-acids and their corresponding oligopeptides have revealed a considerable amount of important basic information and understanding. Based on the energy levels of the lowest singlet excited states, it was concluded that electronic energy transfer Phe → Tyr → Trp was feasible as a result of dipolar resonance coupling.

The structure of proteins leads to energy migration processes. Proteins do, however, phosphoresce and phosphorescence from tyrosine has also been reported. These observations indicate that the chemistry of triplet states of Tyr and Trp may be important. Furthermore, it was suggested recently that long-lived excited states, probably triplet states, were the main precursors in the photoionization and photodissociation of tyrosine[3–5] and phenylalanine[6,7] in water at 20 °C. The quenching of ^{3}Tyr–OH by disulphides (RSSR) was also shown[4] to lead to electron transfer, with the formation of RSSR$^-$

$$^3\text{Tyr–OH} + \text{RSSR} \rightsquigarrow \text{RSSR}^- + \text{Tyr–OH}^+ \qquad (1)$$

The disulphide radical anion usually decays by S–S bond rupture[8]

$$\text{RSSR}^- \rightleftharpoons \text{RS·} + \text{RS}^-(\text{RSH}) \qquad (2)$$

Presented below are the results obtained from a 265 nm laser flash photolysis study of aromatic amino-acids and peptides in water at 25 °C. The triplet–triplet absorption spectra, their lifetimes and quenching reactions by oxygen and disulphides have been studied.

2 EXPERIMENTAL

Optical excitation of the aromatic amino-acids and peptides in water at 25 °C was carried out using a frequency-quadrupled neodymium laser (Holobeam) emitting at 265 nm, with light intensity of ∼ 20 millijoules and single pulses of ∼ 3 and ∼ 15 ns duration. Kinetic absorption spectrophotometry of the transient species produced was carried out at 90° to the laser beam, using the light output from a pulsed 250 watt Xenon lamp. Full details will be published elsewhere.[9] A water-jacketed 10 × 10 mm optical cell was used. The laser beam diameter was ∼ 8 mm.

Solutions were deaerated by bubbling with prepurified Ar gas. Fresh samples were used for each laser pulse. Exposure of the solution to the Xenon lamp was minimized by using a synchronized shutter (open for ∼ 5–10 ms). Borate, phosphate, perchloric acid and potassium hydroxide were used to buffer the solutions. The amino-acids were supplied by Calbiochem, Cyclochemicals, Aldrich and K & K Laboratories.

N_2O was used to convert e_{aq}^- to OH radicals which were scavenged by *tert*-butyl alcohol (see reference 9 for further details).

3 RESULTS AND DISCUSSION

Similar experimental conditions were used for the laser flash photolysis of all the aromatic amino-acids and peptides. The absorbance of the solutions at 265 nm was usually in the range 0·3–0·7.

3.1 Tyrosine

The flash photolysis of aqueous solutions of tyrosine at pH 7·5 showed[3–5] the formation of the characteristic transient optical absorption of the phenoxy radical ($\lambda_{max} \sim 405$ nm) and of the hydrated electron ($\lambda_{max} \sim 720$ nm). The photoionization process was indicated to occur via a long-lived excited state. The lifetime of the singlet ^{1}Tyr* is $\sim 5\cdot 1$ ns,[10] and the dissociation constants of the singlet and triplet excited states of tyrosine have been estimated,[11] pK_{S_1} $\sim 4\cdot 5$ and $pK_{T_1} \sim 8\cdot 5$. In neutral solutions, the absorption of two successive quanta of light was found[5] to be required to photoionize tyrosine and other phenolic compounds.

Similar results were observed in the present work using 265 nm laser photolysis for optical excitation of tyrosine in water. In addition, a new short-lived absorption was found and is shown in Figure 1. In $6\cdot 1 \times 10^{-4}$ M solutions, this transient decays by a first-order process with $k = 1\cdot 8 \pm 0\cdot 1 \times 10^5$ s^{-1} (see Table 1). This spectrum is assigned to a long-lived excited state of tyrosine, probably the triplet state. No triplet emission could be observed under our experimental conditions. The phenoxy radical and e_{aq}^- were not formed from the decay of this triplet state, but were produced within the ~ 15 ns duration of the laser pulse. This is consistent with the biphotonic mechanism[5] for

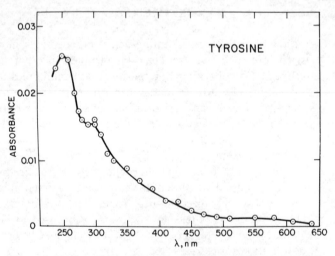

Figure 1 Triplet–triplet absorption spectrum of tyrosine, produced by laser photolysis (~ 15 ns pulse) of aqueous solutions of $6\cdot 1 \times 10^{-4}$ M tyrosine at pH 7·5. Experiment done in presence of N_2O (1 atm) and $1\cdot 0$ M t-butyl alcohol (see text)

Table 1 Lifetimes of triplet states of aromatic amino-acids and peptides in water at 25°C

Aromatic amino-acid	pK$_a$	Concentra-tion, (M)	pH	k (s^{-1})	τ (s)
Tyrosine	2·2, 9·1, 10·1	6·1 × 10^{-4}	6·0	2·5 ± 0·1 × 10^{5a}	4·0 μs
Tyrosyl-glycine		6·1 × 10^{-4}	6·0	2·9 ± 0·1 × 10^5	3·4 μs
Glycyl-tyrosyl-glycine	2·9, 8·5, 10·5	6·1 × 10^{-4}	6·0	2·7 ± 0·2 × 10^5	3·7 μs
Phenylalanine	1·8, 9·1	3·8 × 10^{-3}	7·5	3·2 ± 0·3 × 10^5	3·1 μs
Indole	—	1·5 × 10^{-4}	7·5	8·6 ± 0·4 × 10^4	11·6 μs
Tryptophan	2·4, 9·4	1·5 × 10^{-4}	7·5	7·0 ± 0·7 × 10^4	14·3 μs
N-Methyl-tryptophan		1·5 × 10^{-4}	7·5	7·5 ± 0·8 × 10^4	13·3 μs
N-Acetyl-tryptophan		1·5 × 10^{-4}	7·5	6·1 ± 0·3 × 10^5	16·4 μs

[a] k extrapolated to 'zero' tyrosine concentration is 1·0 ± 0·05 × 10^5 s^{-1}.

photoionization. Using a ~3 ns laser pulse, the concentrations of ϕO· and e_{aq}^- were significantly reduced, as expected, since only a fraction of the ^{1}Tyr* populated the triplet state during this short pulse.

The lifetime of the triplet state was found to be dependent on the concentration of tyrosine, see Figure 2. From this plot, the lifetime of ^{3}Tyr at 'zero' tyrosine concentration is 10 ± 0·1 μs, and $k(^3$Tyr + Tyr) = 2·5 ± 0·1 × 10^8 M^{-1} s^{-1}.

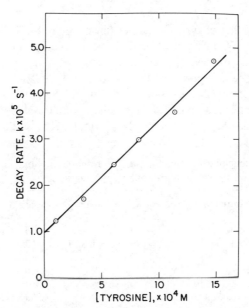

Figure 2 Quenching of the triplet state of tyrosine by ground-state tyrosine. Experiment done at pH 7·5, air-free, and the decay of the triplet state was monitored at 250 nm

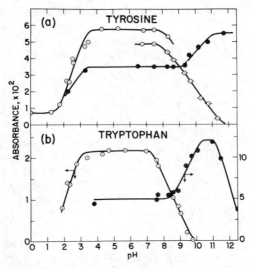

Figure 3 Dependence of the yields of triplet states and hydrated electrons upon pH: (a) 3.4×10^{-4} M tyrosine, triplet monitored at 250 nm (○) and 270 nm (—○—), e_{aq}^- monitored at 650 nm (●); (b) 1.5×10^{-4} M tryptophan, triplet monitored at 460 nm (○) and e_{aq}^- monitored at 650 nm (●)

The quantum yields of tyrosine triplet (monitored at 250 and 270 nm) and e_{aq}^- (monitored at 650 nm) were found to be markedly dependent upon pH (Figure 3a). These changes appear to be similar to the dependence of the fluorescence of tyrosine upon pH, as reported by Feitelson.[12] From the mid-point of the 'titration' curves shown in Figure 3(a), $pK_a \sim 2.5 \pm 0.2$ and $pK_a \sim 9.7 \pm 0.2$ can be derived for ^{3}Tyr. These values are close to the $pK_a^1 = 2.2$ (for —COOH), $pK_a^2 = 9.1$ (for —NH$_3^+$) and $pK_a^3 = 10.1$ (for —OH) for

Table 2 Rate of quenching of triplet states of aromatic amino-acids and peptides by oxygen and disulphides in water at 25 °C

Aromatic amino-acid	Quencher	pH	k, (M^{-1} s^{-1})[a]
Tyrosine (6.1×10^{-4} M)	O$_2$	7.5	$4.8 \pm 0.2 \times 10^9$
Tyrosine (6.1×10^{-4} M)	Lipoate (RSSR)	7.5	$4.0 \pm 0.2 \times 10^9$
Tyrosine	Tyrosine	7.5	$2.5 \pm 0.1 \times 10^8$
Glycyl-tyrosyl-glycine (6.1×10^{-4} M)	O$_2$	6.0	$3.9 \pm 0.3 \times 10^9$
Glycyl-tyrosyl-glycine (6.1×10^{-4} M)	Lipoate (RSSR)	6.0	$3.2 \pm 0.3 \times 10^9$
Phenylalanine (3.8×10^{-4} M)	O$_2$	7.5	$3.3 \pm 0.3 \times 10^9$
Phenylalanine (3.8×10^{-4} M)	Ni^{2+}	7.5	$4.1 \pm 0.2 \times 10^7$
N-Methyl-tryptophan	Lipoate (RSSR)	7.5	$3.1 \pm 0.2 \times 10^9$

[a] Derived from k(s^{-1}) vs quencher concentration plots.

ground state tyrosine. The titration curve of e_{aq}^- gives a pK_a value of 10.1 ± 0.2, in good agreement with the ionization of the phenolic group. The mechanism of photoionization of tyrosinate[3–5,9] is significantly different from that of tyrosine.

Glycyltyrosine and glycyltyrosylglycine form similar T − T absorption spectra, with somewhat reduced yields compared to tyrosine (assuming similar T − T extinction coefficients). Their lifetimes are also very close to that of ^{3}Tyr, (see Table 1).

The quenching of ^{3}Tyr by lipoate ions (a cyclic disulphide) was found to produce RSSR·$^-$, reaction (1), in agreement with earlier work.[4] The quenching rate is $4.0 \pm 0.2 \times 10^9$ M^{-1} s^{-1} (see Table 2). The quenching of ^{3}Tyr by oxygen is also near diffusion-controlled, $k = 4.8 \pm 0.2 \times 10^9$ M^{-1} s^{-1}. It was found (details of experiment will be given elsewhere[9]) that quenching of ^{3}Tyr by oxygen also proceeds by electron transfer with the formation of the superoxide radical $\cdot O_2^-$ (whose optical absorption has a $\lambda_{max} \sim 245$ nm):

$$^3\text{Tyr–OH} + O_2 \rightarrow \text{Tyr–O·} + \cdot O_2^- + H^+ \tag{3}$$

It is important to note the significant role of $\cdot O_2^-$ in enzymic redox reactions.[13]

3.2 Phenylalanine

The flash photolysis of aqueous solutions of phenylalanine demonstrated[6,7] that both photoionization and photodissociation processes were occurring:

$$\text{NH}_3^+\text{CH COO}^- \underset{\underset{\displaystyle \text{CH}_2\text{Ph}}{|}}{\xrightarrow{hv}} \begin{cases} e_{aq}^- + \text{NH}_3^+\dot{\text{C}}\text{HCH}_2\text{Ph} + \text{CO}_2 & (4) \\ \\ \text{PhCH}_2\cdot + \text{NH}_3^+\dot{\text{C}}\text{HCOO}^- & (5) \end{cases}$$

The formation of $^+\text{NH}_3\dot{\text{C}}\text{HCH}_2\text{Ph}$ has not been confirmed.[6,7] In neutral solutions, *both* processes had a long-lived triplet state as a precursor and were biphotonic.

The laser photolysis of Phe gave rise to the transient optical absorptions due to the radicals $\text{PhCH}_2\cdot$, $\text{NH}_3^+\dot{\text{C}}\text{HCOO}^-$ and e_{aq}^-. Superimposed on these absorptions was a much shorter-lived transient whose spectrum is given in Figure 4. This intermediate decays by first-order kinetics with $k = 3.2 \pm 0.3 \times 10^5$ s^{-1}, and is assigned to the T − T spectrum of phenylalanine.

The lifetime of ^{1}Phe* is ~ 6.8 ns.[10,12,14] The formation of the benzyl radical and of e_{aq}^- reached its maximum during the 15 ns laser pulse (no formation of these species from the decay of ^{3}Phe was observed), supporting the biphotonic mechanism previously postulated.[6,7]

The ^{3}Phe is efficiently quenched by oxygen, with $k = 3.3 \pm 0.3 \times 10^9$ M^{-1} s^{-1}. Nickel ions (as nickel perchlorate) also quench ^{3}Phe, with $k = 4.1 \pm 0.2 \times 10^7$ M^{-1} s^{-1}.

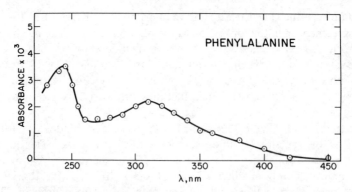

Figure 4 Triplet–triplet absorption spectrum of phenylalanine, produced by laser photolysis (~ 15 ns pulse) of 3.8×10^{-3} M Phe in aqueous solution at pH 7.5. Experiment done in presence of N_2O (1 atm) and 1.0 M t-butyl alcohol (see text)

3.3 Tryptophan

The flash photolysis of aqueous solutions of tryptophan and indole has been examined.[15,16] Photoionization has been observed, as well as the formation of other transient absorptions which have not been categorically characterized.

Santus and Grossweiner[15] have observed a band with $\lambda_{max} = 460$ nm and lifetime $<5\,\mu$s which they assigned to the ^{3}Trp. They could not distinguish between the triplet and the singlet excited states as the precursor for electron ejection. Feitelson[17] has suggested the singlet excited state as the precursor. The lifetime of the singlet excited state of tryptophan is ~ 2–3 ns.[18,19]

The laser photolysis of tryptophan in oxygen-free aqueous solutions at pH 7.5 showed the presence of a short-lived transient absorption with a maximum at 450 nm (see Figure 5). This spectrum is assigned to ^{3}Trp, in agreement with Santus and Grossweiner's[15] conclusion. It decays with $k = 7.0 \pm 0.7 \times 10^4 \text{ s}^{-1}$. A number of other transient absorption bands are observed, and will be presented and discussed elsewhere.[20] These bands do not appear, however, to be produced from the decay of ^{3}Trp.

The change in absorbance of ^{3}Trp with pH, as monitored at 460 nm, is shown in Figure 3(b). From these titration curves, apparent pK_a values of 2.2 ± 0.2 and 8.5 ± 0.2 can be derived, and are to be compared with 2.4 and 9.4 for ground-state tryptophan. It is important to note that the ϕ of tryptophan fluorescence increases sharply, while that of ^{3}Trp decreases. The lifetime of ^{3}Trp in the pH range 2–9 is the same as in neutral solutions. The quantum yield for formation of e_{aq}^- increases over the pH range 8–11 (Figure 3b). At pH > 11 the yield of e_{aq}^- decreases as does the fluorescence yield of tryptophan. These results would seem to indicate that ^{1}Trp* is the precursor in the photoionization of tryptophan in agreement with an earlier[17] suggestion (see also ref. 20).

N-methyl tryptophan, N-acetyl tryptophan and indole give similar triplet states (Table 1), with $\tau \sim 10$–15 μs.

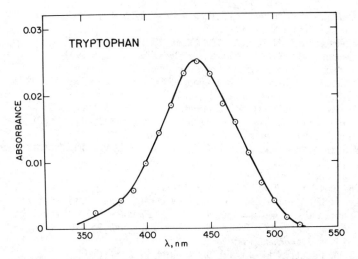

Figure 5 Triplet–triplet absorption spectrum of tryptophan, produced by laser photolysis (~ 15 ns pulse) of 1.5×10^{-4} M Trp in aqueous solution at pH 7.5. Experiment done in presence of N_2O (1 atm) and 1.0 M t-butyl alcohol (see text)

The triplet excited state of N-methyl tryptophan, ^3N-Me–Trp, can be quenched by lipoate ions with a $k = 3.1 \pm 0.2 \times 10^9$ M^{-1} s^{-1}. The quenching mechanism involves electron transfer, as previously observed for ^{3}Tyr, with the formation of the characteristic[8] transient absorption spectrum of the RSSR.$^-$ radical anion

$$^3N\text{-Me–Trp} + \text{RSSR} \rightarrow N\text{-Me–Trp}^+ + \text{RSSR}^{\cdot-} \tag{6}$$

4 CONCLUSIONS

The formation of triplet states of aromatic amino-acids and their oligopeptides may be of considerable importance in the photochemistry of proteins. Their lifetimes are relatively long such that they could interact with various substrates and functional groups. The formation of RSSR.$^-$ from the quenching of triplet states of aromatic amino-acids by disulphides may lead to the rupture of the —S—S— bridges in proteins. The formation of $\cdot O_2^-$ on quenching of the triplet states by oxygen may be significant[13] in enzymic redox reactions. Furthermore, there are indications[21] that singlet oxygen may be produced from the decay of superoxide radicals.

REFERENCES

1. S. V. Konev, *Fluorescence and Phosphorescence of Proteins and Nucleic Acids*, Plenum Press, New York, 1967.

2. *Excited States of Proteins and Nucleic Acids* (eds. R. F. Steiner and I. Weinryb), Plenum Press, New York, 1971.
3. J. Feitelson and E. Hayon, *J. Phys. Chem.*, **77**, 10 (1973).
4. J. Feitelson and E. Hayon, *Photochem. Photobiol.*, **17**, 265 (1973).
5. J. Feitelson, E. Hayon and A. Treinin, *J. Amer. Chem. Soc.*, **95**, 1025 (1973).
6. L. J. Mittal, J. P. Mittal and E. Hayon, *Chem. Phys. Letters*, **18**, 319 (1973).
7. L. J. Mittal, J. P. Mittal and E. Hayon, *J. Amer. Chem. Soc.*, **95**, 6203 (1973).
8. M. Z. Hoffman and E. Hayon, *J. Amer. Chem. Soc.*, **94**, 7950 (1972).
9. D. V. Bent and E. Hayon, *J. Amer. Chem. Soc.*, **97**, 2599, 2606, 2612 (1975).
10. J. Feitelson, *Photochem. Photobiol.*, **9**, 401 (1969).
11. E. Yeargers, *Photochem. Photobiol.*, **13**, 165 (1971); E. Van der Donckt, *Progr. React. Kinet.*, **5**, 273 (1970).
12. J. Feitelson, *J. Phys. Chem.*, **68**, 391 (1964).
13. I. Fridovich, *Acc. Chem. Res.*, **5**, 321 (1972).
14. E. Leroy, H. Lami and G. Laustriat, *Photochem. Photobiol.*, **13**, 411 (1971).
15. R. Santus and L. I. Grossweiner, *Photochem. Photobiol.*, **15**, 101 (1972).
16. V. Subramanyan and G. Tollin, *Photochem. Photobiol.*, **15**, 449 (1972).
17. J. Feitelson, *Photochem. Photobiol.*, **13**, 87 (1971).
18. W. B. De Lauder and Ph. Wahl, *Biochemistry*, **9**, 2750 (1970).
19. J. Feitelson, *Israel J. Chem.*, **8**, 241 (1970).
20. D. V. Bent and E. Hayon, *J. Amer. Chem. Soc.* **97**, 2612 (1975).
21. *Chemiluminescence and Bioluminescence*, p. 131 (eds. M. J. Cormier, D. M. Hercules and J. Lee), Plenum Press, New York, 1973.

Session E
Excited states of visual pigments

E. The excited states of retinals, retinols and visual pigments

Barnett Rosenberg

Department of Biophysics, Michigan State University, East Lansing, Michigan 48824, U.S.A.

1 INTRODUCTION

We see the stars at night and the sun at noon. This commonplace epitomizes the two most astonishing facts of vision; a sensitivity of the receptor cells to photons which is at the limits physically allowed; and the wide dynamic range of the visual apparatus. This is shown by a simple quantitative analysis of the observation. The weakest of the approximately six thousand visible stars has an apparent magnitude of 6 and, on the same scale, the sun is -26.9. From the ratio of these two values I calculate that the sun is approximately 10^{13} times brighter than the weakest visible star. Thus, the eye functions well over a dynamic range of 10^{13}. A rough measure of the vertical component of sunlight, the Solar Constant (~ 0.5 cal/cm^2/min, in the visible range) leads to a value for the vertical component of the dimmest visible star of approximately 10^4 photons/cm^2/s. We know the area of the pupil, the transmission of the eye, the absorbance of the rods in the retina, the saccadic movement of the eye and the imperfect image formation by the lens, so we can guess that single rods are receiving no more than a few photons per second. The rods are single quantum detectors.

A critical, but unresolved, question is whether a single mechanism can explain both the high sensitivity and the large dynamic range (adaptation), or whether two or more mechanisms are necessary. Evidence accrued over the past century has proved that 'bleaching' of the photopigments is a major component in adaptation over the large dynamic range of light levels. There is after all this time still no evidence that 'bleaching', or an intermediate in the process, is the transduction act leading to 'seeing' single photons. This is a discomforting situation which clearly calls for entertaining alternate hypotheses of the transduction mechanism. The chase for the molecular details of how an excited electronic state of rhodopsin produces a visual response is one of the most exciting in the field of sensory physiology today.

510

I interpret the charge of the conference organizers to be the production of a review of the visual system which would allow the reader to appreciate the role that excited states play in vision. I shall start then with a brief description of the current understanding of the structure of the organ of vision, leading from the gross macroscopic down to the molecular level.

2 THE STRUCTURE OF THE VERTEBRATE EYE

I take as example, the human eyeball, shown in a horizontal cross-section in Figure 1. The cornea and lens focus a picture of the world on the thin ($\sim 100\ \mu$m)

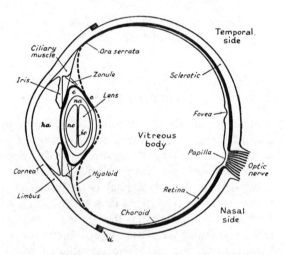

Figure 1 Schematic diagram of a horizontal section of the eye. *ha*—aqueous humour; *c*—capsule of the lens (thickness greatly exaggerated); *e*—shell; *na*—adult nucleus; *ne*—embryonic nucleus; *ic*—central space; *ii*—insertion of the extrinsic muscle. (After Y. LeGrand, *Light, Colour and Vision*, p. 47, Chapman and Hall Ltd., London, 1957)

membrane covering most of the inside surface of the eye, the retina. The optic nerve bundle to the brain passes through the papilla (the 'blind spot') converging from a lateral spread over the forward surface of the retina which enervates the entire membrane with approximately 10^6 fibres. The fovea is a small region ($\sim 10\ \mu$m in diameter) containing a high density ($\sim 6 \times 10^6$) of cone cells. These cells subserve high light intensity (photopic) vision. Each cone contains only one of three possible cone pigments, as has been established by microspectrophotometry,[1] and allows the discernment of colours. The summed absorbances correlate with the photopic sensitivity curve, which has its maximal response at $\lambda_{max} = 550$ nm. No cone pigments have yet been isolated and purified. The convergence of the number of cone output signals onto a single nerve fibre is small, producing a high visual acuity.

The rods form a second class of receptor cells. They are distributed over the retina with a density maximum approximately 20° off the visual axis, falling to a lower density in the fovea and in the regions beyond 80° off-axis. The rods number approximately 125×10^6, so that the output of hundreds of rods converge on a single optic nerve fibre. They contain only one pigment, rhodopsin, which has an absorption band that correlates well with the low light intensity (scotopic) visual sensitivity curve, which peaks at about $\lambda_{max} = 506$ nm. Scotopic vision is colourless.

The retina is a complex system of detecting cells (rods and cones), information processing cells (the secondary neurons, bipolar, amacrine and horizontal cells) and transmitting cells (optic nerve ganglia). The layered structure, in cross-section, is shown in Figure 2. The left side is a stained section, while the right side is a schematic representation of the various elements and their synaptic connections. The light absorbing pigments are located in the outer segments of the receptor cells (the receptor layer). In the vertebrate eye, the direction of the

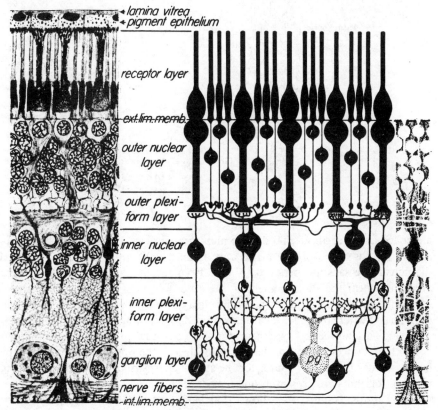

Figure 2 Cross-section of the human retina. (After G. L. Walls, *The Vertebrate Eye and its Adaptive Radiation*, p. 43, Hafner Publishing Co., New York, 1963)

incident light is from below, thus some filtering of the light occurs due to various pigments in the cells and circulating blood supply. In some species of birds, coloured oil droplets containing carotenoid pigments intervene in the pathway, producing marked spectral filtering of the light.

3 THE RECEPTOR CELLS

A schematic diagram of the rod cell is shown in Figure 3. The outer segment (o) is cylindrically shaped, and covered with the plasma membrane. It contains

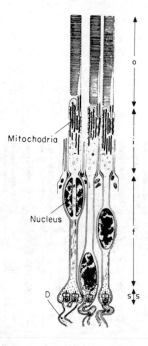

Figure 3 Schematized vertebrate retinal rod cells. (After J. J. Wolken, *Vision*, p. 56), Charles C. Thomas, Springfield, Illinois, 1966

a stacked array of about 500 double membraned discs (or sacs) which contain rhodopsin, the visual pigment. It is approximately 20 μm long and a few μm in diameter. The sacs are believed to arise from the invagination of the plasma membrane near the stalk connecting the outer and inner segments. These sacs are continually created, and the entire array moves toward the pigment epi- thelium end where the older sacs are discarded. In mammals, the average lifespan of a sac is a few weeks. The reason for this continual production is unknown.

The high concentration of the mitochondria in the inner segment below the neck provides the energy source for the stored energy released by the triggering photon, for the syntheses and aggregation of new sacs, and possibly also for sodium pumps. Hagins[2] has shown that Na^+ ions in the dark permeate from the extracellular fluid through the plasma membrane, down through the inner segment to the synaptic junction with the secondary neurons and back through the extracellular fluid. This circulation of Na^+ ions must be driven by an energy source ('sodium pump' apparatus) which may be located in these mitochondria rich regions.

Early electron microscopy studies of Sjöstrand established that the outer segments of the rod were filled with a stack of triple layered discs, as shown in Figure 4, the two opaque layers of which are fused along their rims. These are the sacs. The thickness of the sacs, which may vary as an artifact of specimen preparation, particularly dehydration, ranges from 100 to 240 Å depending upon the species and type of receptor. These sacs are the 'unit structure' of the outer segment. The thickness of each membrane is 30–40 Å, and the intrasaccular space is 70–80 Å. In recent years, low angle X-ray diffraction studies of the outer segments, fixed or intact (even wet), have yielded additional evidence on the membrane structure. However, unresolved disagreements on the interpretation of the data have diluted the usual authoritative value of X-ray results. Tentatively, we may specify some characteristics here. These are primarily the work of Worthington, Blasie, Wilkins and Blaurock.[3] The repeat distance between sacs in frog outer segments is about 300 Å. The space between two neighbouring sacs is about 150 Å and is a water environment containing polysaccharides. The sacs have a plane of symmetry between the upper and lower membranes. The sac is about 150 Å thick. The intrasaccular space is about 73–85 Å thick. The sac membranes consist of about 50% protein and 50% lipid. Rhodopsin constitutes 80–90% of the total protein. Phospholipids make up 75% of the lipids—a uniquely high value, and contain about 4% cholesterol—a very low value. The phospholipids are mainly phosphatidyl-choline, phosphatidyl ethanolamine and phosphatidyl serine in a ratio of approximately 45:37:11. These results suggest a fairly fluid membrane (see below).

The rhodopsin protein appears to be a roughly spherical molecule of about 40 Å diameter, molecular weight 40,000 g/mole, and a molar absorbance of about 43,000. Blasie and Worthington[4] have identified a particle of 40–50 Å diameter in the plane of the membrane which appears to be rhodopsin. In the membrane, the rhodopsin is in a planar liquid array, with a nearest-neighbour distance of about 60 Å, and an equivalent unit cell size of 70 Å. Apparent crystallinity in electron micrographs may be artifactual.

Analysis of the data of Blaurock and Wilkins show that they are consistent with the assumption that the lipids in the membrane are in the form of a bimolecular leaflet (bilayer), with a 40 Å separation between the two lines of high electron density (lipid head groups and protein). The rhodopsin appears to be floating in the bilayer with about 20 Å protruding into the hydrophilic region outside the bilayer. Apparently the molecule sinks a further 7 Å into the membrane after extensive radiation. It is not known for certain, whether rhodopsin occurs on both sides of the membrane, or even whether there is the same amount of rhodopsin on both sides of the sac. One possible model, due to Vanderkooi and Sundaralingam,[5] is shown in Figure 5. Here the two membranes of the sac are identical, and each membrane is symmetrical.

514

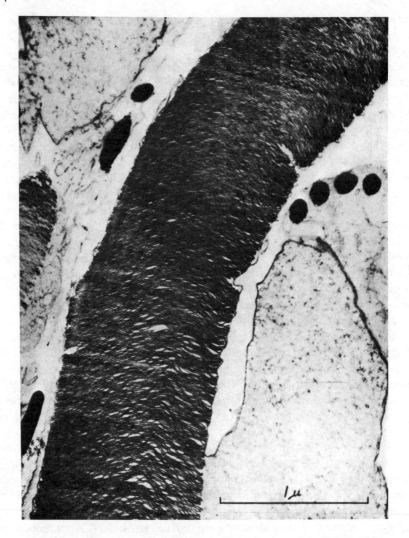

Figure 4 Electron micrograph of retinal rod outer segment (× 59,000). (After F. S. Sjöstrand, in *Biophysical Science—A Study Program*, p. 305 (eds. J. L. Oncley *et al.*), John Wiley and Sons Inc., New York, 1959)

Hagins and Jennings[6] proposed that free Brownian rotation of the chromophoric groups in rhodopsin in the plane of the sac membrane could account for the lack of induced photodichroism when the rods were bleached with polarized light. Recently, both Brown[7] and Cone[8] have proved this hypothesis; the first by fixing rhodopsin using gluteraldehyde, and showing that photodichroism then results from a polarized light bleach, the second by measuring the transient photodichroism using a fast photometer. The relaxation time of the transient is about 20 μs at room temperature, and has a Q_{10} of 3. From these results

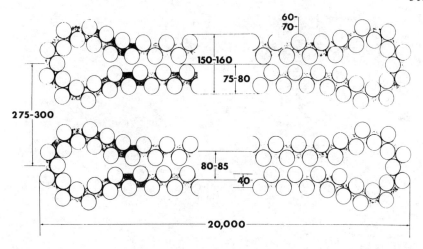

Figure 5 Schematic cross-section of rod sacs. All measurements are given in angstroms. The circles represent the protein molecules, while the stippled region represents the lipid bilayer. (After G. Vanderkooi and M. Sundaralingham, *Proc. Nat. Acad. Sci.,* **67**, 233 (1970))

Cone has calculated a 'viscosity' of the lipid membrane of about 2 poises, i.e. about the same as olive oil. He concludes that the lipid membrane might best be described as highly fluid. In addition to the ability of the rhodopsin to rotate freely (about an axis perpendicular to the membrane surface) Cone[9] has also shown, by a microspectrophotometric bleaching technique, that a segment of bleached rhodopsin molecules on one side of the rod cylinder can diffuse to the other side of the cylinder. The lateral diffusion has a half-time of 35 s at room temperature and a Q_{10} of 3·7 (in frog rods). These results are also consistent with a 'viscosity' of about 1 poise.

The molecular weights of rhodopsins range about 35,000–41,000 g/mole. The gross morphology of the molecule is assumed to be roughly spherical of diameter ~ 40 Å (although an ellipsoidal shape has been suggested by Wu and Stryer,[10] with a major axis of ~ 75 Å). The exterior, or a large part of the exterior, is assumed to be hydrophobic, and may have a more or less loose association with lipids and polysaccharides. The sequence of amino-acids in the protein is still unknown, and no crystals have yet been grown. Therefore definitive X-ray structural analysis of the protein molecule may be safely assumed to be at least a decade away.

The prosthetic chromophoric group in all visual pigments is a C_{20} carotenoid (a haplocarotene) now called retinal. The structures of retinal are shown in Figure 6. The aldehyde end group of retinal replaces the alcohol end group of vitamin A (retinol). Because of the numerous formal single and double bonds in the structure, it is capable of assuming a large variety of *cis–trans* isomers by rotations about the single and double bonds. The isomer that reacts in the dark with the protein, opsin, to form all known rod pigments is the 11-*cis* isomer, the

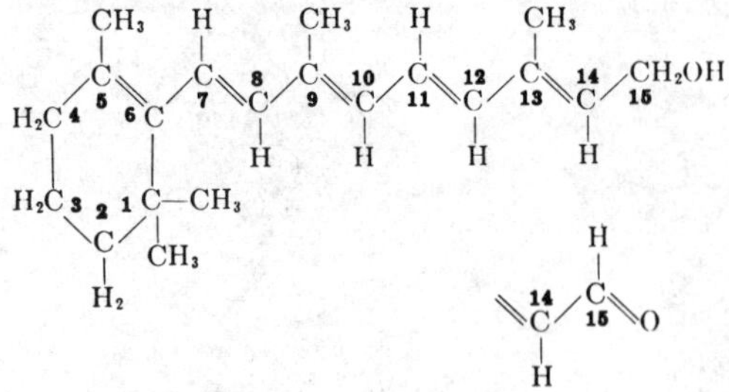

all-*trans* Vitamin A

Retinene

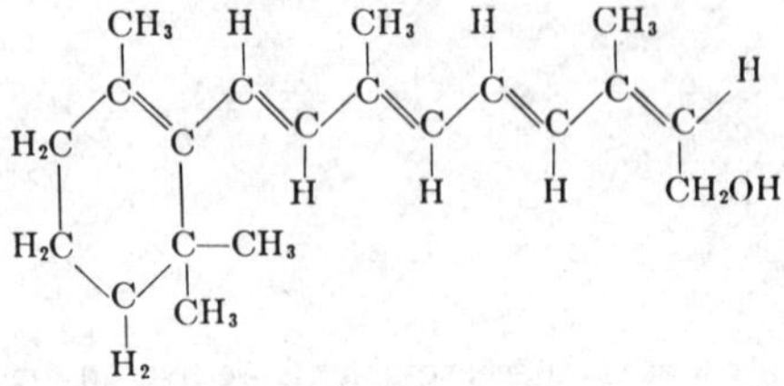

13-*cis* (*neo*-a)

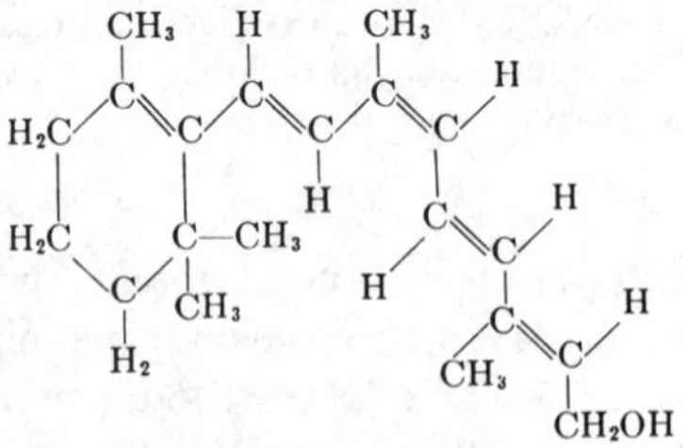

9-*cis* (*iso*-a)

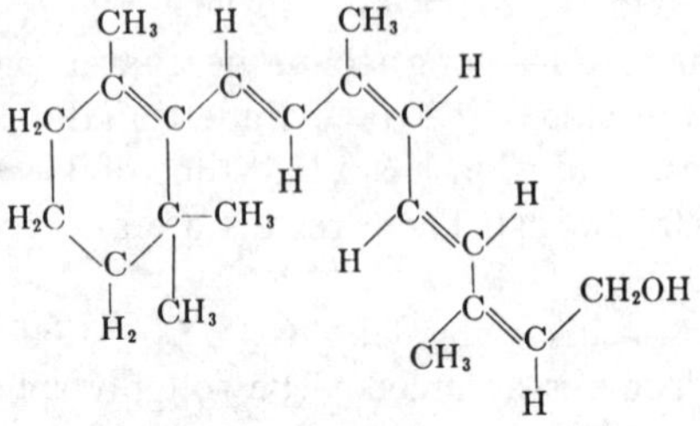

9,13-dicis (*iso*-b)

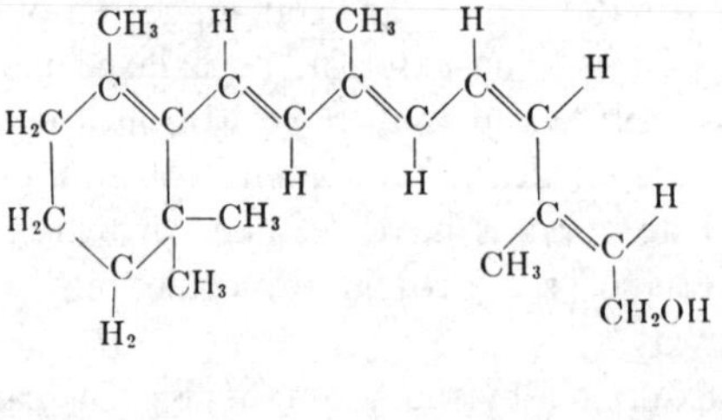

11-*cis* (*neo*-b)

Figure 6 The structures of retinal (retinene) and retinol (vitamin A). Geometric isomers of vitamin A and retinene. A similar series of structures, differing only in possessing an extra double bond in the 3,4 position, represents vitamin A_2 and retinene$_2$. The upper four structures represent the unhindered, and hence most probable, configurations. The lowermost structure is the hindered *cis* isomer from which all known visual pigments are derived. (After G. Wald, in *Light and Life*, p. 728, The Johns Hopkins Press, Baltimore, 1961) structure of which has been determined from X-ray crystallography by Sperling *et al.*[11] and is shown in Figure 7. We will return to the details of this later.

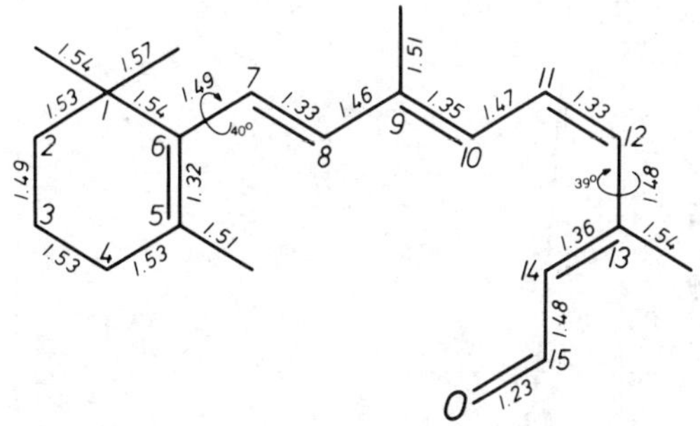

Figure 7 The molecular structure of 6s-*cis*, 11-*cis*, 12s-*cis* retinal determined by X-ray crystallography. (After W. Sperling, in *Biochemistry and Physiology of Visual Pigments*, p. 21 (ed. H. Langer), Springer-Verlag, New York, 1973)

The retinols have broad, almost featureless, absorption bands with the longest wavelength at $\lambda_{max} = 325$ nm. The retinals have broad, featureless bands with the longest wavelength band at $\lambda_{max} = 380$ nm. Rhodopsin absorption bands are shown in Figure 8, and in humans the peak at longest wavelength is at $\lambda = 506$ nm. As shown in the figure, the longest wavelength band coincides in position and breadth with the photobleaching sensitivity curve, and has also been shown to agree with the visual sensitivity curve for scotopic vision.

After irradiation of rhodopsin, the 506 nm band disappears, and a new band appears at $\lambda_{max} = 380$ nm. This change is the 'bleaching' of rhodopsin. Years of intense effort have shown that the 'bleaching' is a complicated process which occurs in a number of stages, each of which has a stable structure below a given temperature. The full course is illustrated in Figure 9 after Yoshizawa.[12] The intermediates are detected by the spectral changes. At 4 K only the initial transients occur on irradiation. This is illustrated in the spectra of Figure 10. Illumination by yellow light blue-shifts the spectrum from rhodopsin (curve 1) to hypsorhodopsin: blue light shifts the spectrum to the red (curve 4) to the bathorhodopsin (earlier called prelumirhodopsin). It has been suggested that absorption of a photon of light causes a photoisomerization of the retinal chromophore from the 11-*cis* configuration. There is no accompanying significant change in the protein conformation. In the dark, at higher temperatures, additional changes occur leading through the stages shown in Figure 9. It must be noted that all the stages prior to the metarhodopsin II conversion to pararhodopsin are photoreversible, and that the metarhodopsin II is only partially photoreversible, primarily as a slow dark reaction after photon absorption. This suggests a major protein conformational change in the *meta* I → *meta* II

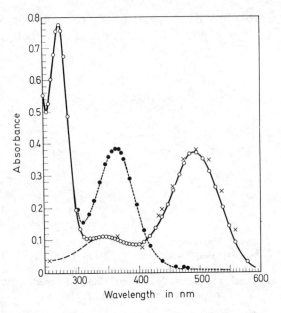

Figure 8 The spectral bands of rhodopsin. The spectral absorbance of a nearly pure extract of cattle rhodopsin at pH 9·2 (curve through plain circles); and of the bleached extract (curve through filled circles). Crosses give the photosensitivities of frog rhodopsin (according to Goodeve, Lythgoe and Schneider) adjusted to correspond to the absorbance curve at 500 nm. The dashed interpolation gives the probable contribution to absorbance below 320 nm by the rhodopsin chromophore. (After F. D. Collins, R. M. Love and R. A. Morton, *Biochem J.*, **51**, 292 (1952)

reaction, but no major changes prior to this. This raises the question of the cause of the spectral shifts of the various intermediates, and indeed of the rhodopsin itself.

Retinal has a peak absorption at $\lambda = 380$ nm; rhodopsin at $\lambda = 506$ nm. It is known that the reaction of the retinal with the protein is to form a Schiff base between the aldehyde end group on retinal with the ε-amino group of lysine of the protein. Model Schiff base reactions, however, produce a blue shift to about $\lambda_{max} = 340$ nm. Since protonation of these model compounds yields products with $\lambda_{max} = 440$ nm, it is thought that the final coupling is a protonated Schiff base. There is no conclusive evidence yet available on this point. Numerous investigators have suggested non-specific protein charge group interactions with the π-electrons of the conjugated retinyl system, or with the nitrogen of the Schiff base as being necessary to produce the additional red shift to 506 nm. Thus, it is at least reasonable that changes in the geometry of the retinyl moiety

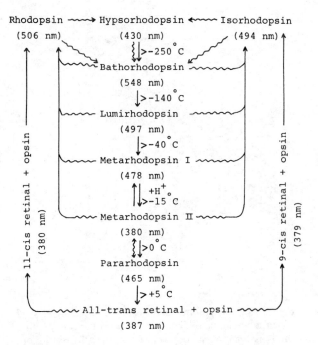

Figure 9 The bleaching cycle of cattle rhodopsin. The
wavy lines represent photoreactions, the straight lines are
thermal reactions. (After T. Yoshizawa and S. Horiuchi, in
Biochemistry and Physiology of Visual Pigments, p. 73 (ed.
H. Langer), Springer-Verlag, New York, 1973)

will modulate the interactions with the protein, causing perturbations of the
electronic energy levels and the absorption spectra.

4 THE EXCITED STATES OF RETINYL PIGMENTS

In considering the excited states of visual pigments we have a variety of model
systems to analyse, running the gamut from the *cis* and *trans* isomers of simple
polyenes, retinol, retinal, Schiff bases of retinal, protonated Schiff bases of
retinal, rhodopsin solubilized in detergent micelles, and finally rhodopsin in
sac membranes. These compounds are now becoming of great interest to
spectroscopists, and some progress in understanding the fate of their excited
state energy is possible. Thus far, we have implicated only the photoisomerization
of rhodopsin as an early photophysical step initiating the bleaching process.
Photoisomerization may occur directly through the excited singlet state or
through a triplet intermediate. A schematic diagram of such pathways, based
upon rotation of the ethylenic double bond, is shown in Figure 11. It is apparent
that for the triplet pathway, a minimum of the potential energy curve occurs at

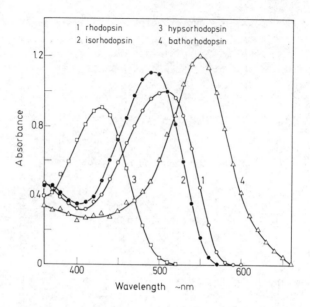

Figure 10 Early spectral changes in the bleaching process of cattle rhodopsin. The spectral absorbances of cattle rhodopsin (curve 1), isorhodopsin (curve 2), hypsorhodopsin (curve 3) and bathorhodopsin (curve 4) at liquid-helium temperature. The maximum absorbance of rhodopsin ($\lambda_{max} = 506$ nm) is plotted arbitrarily at 1·0, and the other curves stand in quantitative relation. Isorhodopsin ($\lambda_{max} = 494$ nm), hypsorhodopsin ($\lambda_{max} = ca$ 430 nm) and bathorhodopsin ($\lambda_{max} = ca$ 548 nm) of equivalent concentration thus possess maximum absorbances about 1·12, 0·91 and 1·21 times that of rhodopsin. (After T. Yoshizawa and S. Horiuchi, in *Biochemistry and Physiology of Visual Pigments*, p. 71 (ed. H. Langer), Springer-Verlag, New York, 1973)

a 90° rotation, with a low barrier to reversion or completion. Thus, if the 11-*cis* isomer is the configuration for rhodopsin, and all-*trans* is the configuration for lumirhodopsin, then the half-rotational form may be the low-temperature stable form for the hypso- and batho-rhodopsin pigments. This was suggested by Abrahamson[13] as shown in Figure 12, which is a result of semi-empirical quantum-mechanical calculation. Since the spatial correlation of the coupled electrons is different in the singlet and triplet states, it may be that at half-rotation there is a mixed configuration of singlets, doublets and triplets, and excitation from this ground state to higher levels may not be easy to unravel, but intrinsically there is nothing to prevent it accounting for the batho- and hypso-chromic shifts as well as the changes in oscillator strength. At this time it is purely a speculative mechanism.

Another speculation, due to Honig and Karplus,[13a] is that the retinyl moiety

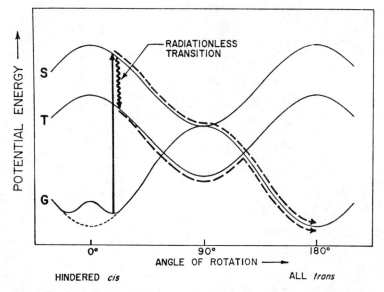

Figure 11 Schematic potential energy curves for rotations of an ethylenic bond. Schematic diagram of the potential energy curves of the ground-singlet, excited-singlet and triplet-π electron states as a function of the angle of rotation of a conjugated bond. The two minima near 0° correspond to the departure from planarity in the hindered-*cis* isomer. The two dashed lines show the two possible pathways for photoisomerism to the all-*trans* isomer. (After B. Rosenberg, in *Advances in Radiation Biology*, Vol. 2, p. 217 (eds. L. G. Augenstein, R. Mason and M. R. Zelle), Academic Press, New York, 1966)

of rhodopsin is 11-*cis*, 12 S-*cis* (see Figure 7), which upon photoexcitation undergoes the progressive changes, first to 11-*trans*, 12 S-*cis*, then to 11-*trans*, 12 S-*trans*. They associate bathorhodopsin with the first and lumirhodopsin with the second. The difficulty with all hypotheses associating bond rotations with the early intermediates in bleaching is that Busch *et al.*[14] have shown a rise time, at room temperature, for species absorbing in the bathorhodopsin region, of 6 picoseconds. It is questionable that the required rotations of a part of a molecule with a large moment of inertia can occur in such short times.

Using flash photolysis technique, Abrahamson *et al.*[15] have found a quantum efficiency of ~ 0.1 for intersystem crossing, in all-*trans* retinal, and no detectable effects in its protonated Schiff base. They assume the necessity for an n–π^* intermediate for efficient intersystem crossing which is available in the aldehyde, but not in the protonated Schiff base. Since this is a lower efficiency than was commonly assumed for photoisomerization of rhodopsin (~ 0.6), it tended to cast doubt on the possible involvement of the triplet state as an intermediate in photoisomerization. The difficulties with this interpretation are obvious: (1) the retinal chromophore of rhodopsin is 11-*cis* and not all-*trans*; (2) the prosthetic attachment is via a Schiff base (protonated or non-protonated) and

522

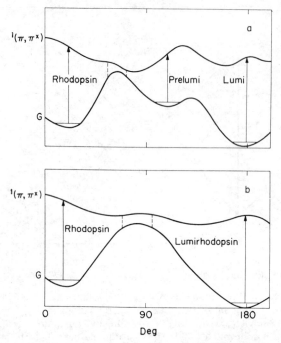

Figure 12 Calculated potential energy curves for rhodopsin. Potential curves for the transition rhodopsin → prelumirhodopsin (*prelumi*) → lumirhodopsin (*lumi*); G = ground state,[1] $(\pi, \pi*)$ = excited singlet state: (a) at $-195\,°C$; (b) at room temperature. (After E. W. Abrahamson and S. E. Ostroy, *Progress in Biophysics and Molecular Biology*, **17**, 179 (1967))

does not appear to have lone pair electrons available (although some of the canonical structures of the protonated Schiff base allows less perturbed lone pair electrons and a lower $n-\pi*$ transition): (3) the photoisomerization quantum efficiencies for the various isomers of retinal and model systems were not well known: (4) the quantum efficiencies for triplet state photoproduction of the various isomers of retinal and model systems were not well known; (5) the techniques for measuring quantum efficiencies in general are notoriously imprecise and non-reproduceable; and (6) the results depend upon solvent (polar *vs* non-polar, hydrogen bonding capacity, dielectric constant, etc.), temperature, environment (solution, detergent micelles, protein envelope), pH, concentration (dimer formation, excimer formation, etc.) and purity. Thus, it is an ambitious, but necessary, undertaking to sort out all of these factors.

A number of laboratories have begun the attack. Bensasson *et al.*[16] re-evaluated the quantum efficiency for intersystem crossing, Φ_{ISC}, to the triplet state of all-*trans* retinal recently, with the result that $\Phi_{ISC} = 0.60 \pm 0.10$, a much higher value than that of Abrahamson *et al.* More recent results, reported at this

Table 1 The photoisomerization of retinal. Quantum efficiencies (γ) for the photoisomerization at 365 nm of all-*trans* retinal and the *monocis* isomers in hexane at 25 and $-65\,^\circ$C (columns II and III), compared with the averaged quantum efficiencies for the photoisomerization at 450, 500 and 550 nm of the retinylidene chromophores of rhodopsin (11-*cis*), isorhodopsin (9-*cis*) and metarhodopsin I (all-*trans*) in aqueous digitonin at 20 °C

I	II	III	IV
Reaction	Retinal	Retinal	Chromophore
	γ_{25}	γ_{-65}	γ_{20}
13-*cis* to *trans*	0·4	0·1	
11-*cis* to *trans*	0·2	0·6	0·6
9-*cis* to *trans*	0·5	0·25	0·2
All-*trans* to average *cis*: maximum value	0·2	0·005	
minimum value	0·06	0·002	
All-*trans* to 11-*cis*			0·3
All-*trans* to 9-*cis*			0·06

conference,[17] produced a somewhat lower value, but for the case of 11-*cis* retinal they conclude that the triplet state is a major route for photoisomerism to the all-*trans* isomer. Rosenfeld *et al.*[18] also have studied the generation of triplet states by sensitized energy transfer as well as by direct pulse radiolysis and laser photolysis. They find a markedly enhanced Φ_{ISC} for the unprotonated Schiff bases compared to the pure retinals, and also derived evidence for the efficient photocreation of a retinylic cation. Meanwhile, Kropf and Hubbard[19] have measured the quantum efficiencies, (γ), for photoisomerization of the isomers of retinals and found the values to be significantly lower than expected ($\sim 0\cdot2$ for 11-*cis* to all-*trans* at 25 °C). They report an intriguing temperature dependence, in that upon cooling the solutions to -65 °C in hexane, the γs markedly decrease, except for the 11-*cis* to all-*trans* conversion, which increases by a factor of 3. The results are shown in Table 1. Thus, the situation is the reverse of that of 10 years ago, and is likely to oscillate a few more times before damping out to a steady state set of comparable values of Φ_{ISC} and γ for the retinals and model systems which could then allow an unambiguous statement of the role of the triplet states in the photoisomerization and photochemistry of these polyene pigments.

5 FLUORESCENCE OF RETINOLS AND RETINALS

Considering the high degree of flexibility of the retinyl polyenes, with its concomitant high probability of radiationless deactivation of the excited states, it is surprising that fluorescence emission of some of these molecules are readily detectable. The first extended study of the factors determining the fluorescence efficiency is that of Thomson.[20] The molecules investigated contained the retinyl carbon framework with five different end groups in the C_{15} position, and four stereoisomers of retinol. In no case was he able to detect a long-lived (> 10 ms) emission. The absorption and fluorescence emission bands

Table 2   Quantum efficiencies of fluorescence at room temperature

	Solvent	
	3-methylpentane– methylcyclohexane	Methanol
trans-Retinol$_1$[a]	0.02(20 ×)[b]	0.006(55 ×)
trans-Retinyl acetate[a]	0.02(20 ×)	0.004(50 ×)
Retinyl oxime	0.0006	0.0001
Retinyl acid	Not detectable at room temperature	
Retinylidene methylamine	Not detectable at room temperature	
All-*trans*-Retinol	0.02	0.006
13-*cis*-Retinol	0.01	0.003
11-*cis*-Retinol	0.006	0.002
9-*cis*-Retinol	0.007	0.002

[a] Values determined by comparison with quinine bisulphate. The remainder were determined by comparison with *trans*-retinol$_1$ under comparable conditions of geometry and absorbance.
[b] Values in parentheses are approximate factors by which quantum yield increases on cooling to 77 K.

were unusually broad and devoid of structure. The spectra were corrected and analysed according to the techniques of Birks and Dyson[21] for investigating mirror-image symmetry in weakly emitting substances. This technique yields direct information on the nuclear configurations of the ground and emitting states and the position of the $0 - 0$ band. The mirror-image symmetry for the corrected absorption and emission bands is largely maintained for these molecules, with a very large Stokes' shift ($\sim 5000 \text{ cm}^{-1}$). Increasing the electronegativity of the end groups induces a progressively increasing failure of the mirror-image symmetry. Isomerization from the all-*trans* to the *cis* isomers decreases the symmetry as well. Thomson found that he could adequately account qualitatively for the various phenomena with the assumption that the optical transitions involved are Franck–Condon forbidden. The forbiddenness arises from bond alternation in the ground state leading to gross changes in geometry on excitation, and also from steric hindrances in the retinyl carbon framework.

The absolute values of the quantum yields of fluorescence determined by Thomson are quite low. These are shown in Table 2. The low values, the temperature dependences and the solvent effects are all consistent with a Franck–Condon forbidden transition. There also appears to be a correlation between the progressive failure of the mirror-image relation and the decrease in quantum efficiency, and he concludes, 'The failure of the mirror image relation is most significant in determining the quantum efficiency of fluorescent transitions between Franck–Condon states.' Birks had earlier shown that the radiative lifetimes of excited states in diphenylpolyenes increased as the mirror-image relation progressively failed. I will return to this point shortly.

A weak fluorescence has been detected in retinal isomers by Thomson[22] and by Balke and Becker[23] (in the latter paper the quantum efficiencies are rather high, $\sim 0.17-0.71$). Thomson found fluorescent emission maxima at $\lambda = 490$ nm for the all-*trans*, 9-*cis*, 11-*cis* and 13-*cis* isomers in hydrocarbon solvents at 77 K. No room-temperature emission could be detected. However, because the excitation spectra did not meet his major criterion of agreeing closely with the absorption spectra, he did not publish his complete results. He found that the action spectrum for emission peaked on the long wavelength side of the absorption band. This wavelength dependence of excitation was independently reported by Balke and Becker, and more recently by Moore and Song,[24] and Christensen and Kohler.[25] Thomson speculated that the emitting state in retinal was an n-π^* state, energetically just below the main π-π^* band. Balke and Becker assume it to be a π-π^* transition: Moore and Song suggest dimers as the emitting species (a π-π^* transition again): Christensen and Kohler postulate a low-lying forbidden ^1A_g state as the emitting state. Moore and Song have measured the polarization of the exciting (not absorbing) and emitting transitions and concluded they are parallel. Thus, the transitions of absorption and emission are likely to occur between just two states. This suggests one further simplistic possibility, that the absorbing and emitting transitions are between the ground and n-π^* states ($^1A_g \rightleftarrows n$-π^*). This would be inconsistent with the high quantum efficiencies reported by Balke and Becker, since absorption directly to the n-π^* state should have a lower transition probability than the $^1A_g \rightarrow {}^1B_u$ main band. However, Moore and Song report a fluorescence quantum efficiency (~ 0.02), much lower than that of Balke and Becker (0.71), more consistent with the results of Thomson and, therefore, in consonance with this suggestion.

The wavelength dependence of the fluorescence of retinals are attributed by each author to a unique, but different cause. Balke and Becker utilize a competition in the higher vibrational levels of the ^1B_u main band between a fast photochemistry and thermal equilibration with the lower vibrational levels. Moore and Song assume exciton splitting of the main excited state of the retinal dimers in solution, with only excitation to the lower band leading to emission. Christensen and Kohler assume a fast radiationless process (possibly intersystem crossing) originating in the ^1n-π^* which competes with the internal conversion to the emitting, forbidden ^1A_g state. However, it is not clear how this could cause the wavelength dependence of the excitation spectrum, since they assume that the absorbing transition is $^1A_g \rightarrow {}^1B_u$ and not $^1A_g \rightarrow {}^1n$-π^*. In my added suggestion (which I do not take too seriously) the wavelength dependence arises from the restriction that only direct excitation to the ^1n-π^* state leads to emission, and that the n-π^* is likely to be buried under the long-wavelength tail of the main ^1B_u band. Direct excitation of the ^1B_u state may, by internal conversion, contribute some to the population of the ^1n-π^* states, but competing processes must dominate. It should by now be obvious that the wavelength dependence of the retinal fluoresence is by no means clearly understood.

6 RADIATIVE LIFETIMES OF POLYENE EXCITED STATES

Quantum statistics relate the probabilities of absorption and emission between two states through the Einstein coefficients. Strickler and Berg[26] have extended this to molecular spectroscopy by integrating the relations over the broad bands of the electronic transitions. They required three assumptions to do this: (1) absorption and emission result from the same electronic transition; (2) the nuclear configuration of the absorbing and emitting states are very similar; and (3) the dipole transition moments are identical for all involved nuclear configurations. Under these assumptions, the radiative lifetime, τ_0, is given by:

$$\frac{1}{\tau_0} = 2\cdot 88 \times 10^{-9} n^2 \langle v_f^{-3} \rangle_{Av}^{-1} \int \frac{\varepsilon(v)}{v} dv \tag{1}$$

where

$$\langle v_f^{-3} \rangle_{av}^{-1} = \frac{\int f(v) dv}{\int f(v) v^{-3} dv} \tag{2}$$

and $f(v)$ is the fluorescence spectrum. Thus, from the integrals of the fluorescence spectrum and the extinction coefficients one can calculate the natural radiative lifetime of an excited state. If non-radiative processes compete with radiative emissions, the measured lifetime τ, of the state, will be shorter than τ_0, and the quantum efficiency, Φ_f, will be less than one. The simple relation is:

$$\Phi_f = \frac{\tau}{\tau_0} \tag{3}$$

Thomson had measured Φ_f for a variety of retinols and retinyl polyenes. Dalle and Rosenberg[27] then measured τ for these molecules under identical conditions, and also evaluated τ_0 from the Strickler and Berg relation. This permitted a comparison of the calculated and measured radiative lifetimes. The experi-

Table 3 Spontaneous emission rate constants evaluated from absorption and from emission, expressed in 10^6 s^{-1}

Polyene molecule	Absorption		Emission	
	non-polar	polar	non-polar	polar
(1) All-*trans* retinol	370	360	5	1·6
(2) All-*trans* retinyl acetate	360	340	4	0·9
(3) 9-*cis* Retinol	290	300	1·2	0·8
(4) 13-*cis* Retinol	290	280	1·8	0·6
(5) All-*trans* diphenylhexatriene	530	530	60	60

mental results are shown in Table 3 for the calculated ('absorption') spontaneous emission rate constants (reciprocal of the lifetime) and the measured ('emission') rate constants. The conclusion is clear that the predicted values of the radiative lifetimes from (1) are about a factor of 100 shorter than the measured values. This implies a serious breakdown of the utility of the Strickler and Berg relation for the retinyl polyenes in predicting excited state lifetimes. Since Birks and Dyson[21] and Nikitina[28] had also reported smaller, but definite, discrepancies for simpler polyenes, it would appear that in general the Strickler and Berg relation breaks down when applied to polyenes. The magnitude of the discrepancies between theory and experiment has been shown to be related to the Stokes' shifts for these molecules, as shown in Figure 13. If further work substantiates this linear relation between the log of the discrepancy ratio and the Stokes' shift, I propose that it be called 'Dalle's rule', after the discoverer. Dalle and Rosenberg excluded the possibility of a non-correspondence of the absorption and emission transitions by polarization studies. This has more recently

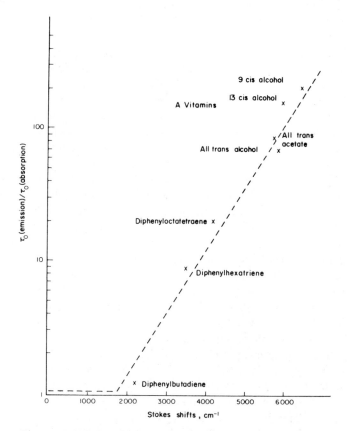

Figure 13 Relation between Stokes' shift and the discrepancy ratio of the radiative lifetimes of polyenes. (After J. P. Dalle and B. Rosenberg, *Photochemistry and Photobiology*, **12**, 158 (1970))

been verified by Moore and Song.[29] These two results, the proportionality between the discrepancy and the Stokes shift, and the parallelism of the absorption and emission transition dipoles provide a clue as to which of the three assumptions involved in the derivation are wrong, namely the identity of the nuclear configurations in the ground and excited states. Dalle and Rosenberg, therefore, suggested that after a vertical transition by absorption or emission of a photon, the nuclear configuration of the molecule relaxes. They also showed that the radiative lifetime is strongly temperature dependent. This is contrary to the last assumption of Strickler and Berg that the transition moment is constant for all occupied vibrational sublevels of the ground and excited states. A model was proposed consistent with these results, which suggested that after thermal equilibrium was established in the vibro-rotational levels of the excited states, emission occurs from the lower levels and non-radiative processes occur from levels (which could be related to twisting modes of the molecule), about 0·5 kcal/mole above the lowest level.

In synopsis then, the experiments described above suggest that the first assumption used in deriving the Strickler and Berg relation is valid for these polyenes, but the latter two assumptions are not. It is somewhat surprising, therefore, that Hudson and Kohler[30] proposed an explanation for the lifetime discrepancies based upon the existence of a lower lying (below ^1B_u) forbidden (oscillator strength ~ 0.06) singlet state as the emitting state. Their experiments primarily show the appearance of a weak absorption shoulder to the red of the main absorption band, at low temperature in α, ω, diphenyloctatetraene dissolved in a host crystal of bibenzyl. A theoretical justification for this new state was provided by Schulten and Karplus,[31] who identified the state as the forbidden $^1A_g^-$. The calculation was a Pariser–Parr–Pople semi-empirical type, but included all double excited configurations. The inclusion of these configurations was essential, as calculations of this type involving only single-excited configurations put the forbidden $^1A_g^-$ energy level above, and not below, the ^1B_u level. Hudson and Kohler[32] have generalized these results to be applicable to all polyenes. There is a certain degree of attractiveness in their proposal, in that such a simple model is consistent with a large amount of data, such as the large Stokes' shifts, the anomalously long lifetimes, the progressive failure of mirror-image symmetry, and the differences in solvent shift behaviour of the absorption and emission spectra. It also places another state in the energy region below the ^1B_u, with a long radiative lifetime and, therefore, a possible source for competing photoisomerism or photochemistry. The final energy level diagram emerging from their work is shown in Figure 14. It is, therefore, essential to clarify the status of the $^1A_g^-$ state. Evidence against it being the emitting state in polyenes comes from the polarization studies referred to above, which show that the absorbing and emitting polarization vectors are parallel. This cannot be true for a $^1A_g \rightarrow {}^1B_u$ absorption and a $^1A_g^- \rightarrow {}^1A_g$ emission. Gavin, Risemberg and Rice[33] were unable to detect any weak absorption bands on the red side of the main $^1A_g \rightarrow {}^1B_u$ transition in cis- and trans-1, 3, 5-hexatriene in a high-resolution study in the gas phase. Moore and Song[29]

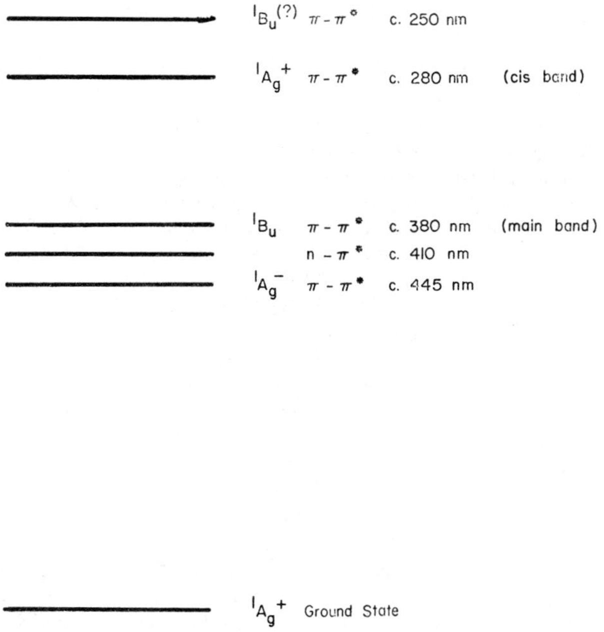

Figure 14 Suggested assignments and positions of the singlet excited states of all-*trans*-retinal. (After B. Honig and T. G. Ebrey, *Annual Review of Biophysics and Bioengineering*, **3**, 163 (1974))

did detect a slight inflection on the red edge of the major absorption band of 1,8-diphenyl-1,3,5,7-octatetraene, but no depolarization trend in this region (the figures and legends apparently are interchanged in their paper). Song (personal communication) believes that the weak, low-lying absorption band apparent in polyenes in condensed phases can result from aggregation (dimers, etc.) even at low concentrations ($\sim 10^{-6}$ M). This suggests that vapour phase absorption studies on diphenyloctatetraene should be tried, to confirm the bands found by Hudson and Kohler in the condensed phase. More recently Gavin and Rice[34] studied the transitions to higher excited states of the 1,3,5-hexatriene isomers and conclude that the ordering of states predicted from the inclusion of double-excited configurations in the Pariser–Parr–Pople calculations is less plausible than using the single-excited states only. I conclude that the status of the $^1A_g^-$ state as the lowest singlet excited state in polyenes is, at present, very ambiguous.

I thank the organizers for their kind invitation to participate in this conference. It encouraged me to return to the field after a four-year absence, and to read with growing pleasure of the many new, exciting developments and controversies that are reenergizing the excited states of polyenes.

REFERENCES

1. P. A. Liebman, *Handbook of Sensory Physiology*, Vol. VII/1, p. 481 (ed. H. J. A. Dartnell), Springer, Heidelberg, 1972.
2. S. Yoshikami and W. A. Hagins, *Biochemistry and Physiology of Visual pigments*, p. 245 (ed. H. Langer), Springer, Heidelberg, 1973.
3. C. R. Worthington, *Annual Review of Biophysics and Bioengineering*, **3**, 53 (1974).
4. J. K. Blasie and C. R. Worthington, *J. Molec. Biol.*, **39**, 417 (1969).
5. G. Vanderkooi and M. Sundaralingam, *Proc. Nat. Acad. Sci.*, **67**, 233 (1970).
6. W. A. Hagins and W. H. Jennings, *Disc. Faraday Soc.*, **27**, 180 (1960).
7. P. K. Brown, *Nature New Biology*, **236**, 35 (1972).
8. R. A. Cone, *Nature New Biology*, **236**, 39 (1972).
9. M. Poo and R. A. Cone, *Nature*, **247**, 438 (1974).
10. C. W. Wu and L. Stryer, *Proc. Nat. Acad. Sci.*, **69**, 1104 (1972).
11. R. Gilardi, I. L. Karle, J. Karle and W. Sperling, *Nature*, **232**, 187 (1971).
12. T. Yoshizawa and S. Horiuchi, *Biochemistry and Physiology of Visual Pigments*, p. 69 (ed. H. Langer), Springer, Heidelberg, 1973.
13. E. W. Abrahamson and S. E. Ostroy, *Progr. Biophys. Mol. Biol.*, **17**, 179 (1967).
13a. B. Honig and M. Karplus, *Nature*, **229**, 558 (1971).
14. G. E. Busch, M. L. Applebury, A. A. Lamola and P. M. Rentzepis, *Proc. Nat. Acad. Sci.*, **69**, 2808 (1972).
15. E. W. Abrahamson, R. G. Adams and V. J. Wulff, *J. Phys. Chem.*, **63**, 441 (1959).
16. R. Bensasson, E. J. Land and T. G. Truscott, *Photochem. Photobiol.*, **17**, 53 (1973).
17. R. Azerad, R. Bensasson, M. B. Cooper, E. A. Dawe and E. J. Land, paper E1, this Volume.
18. T. Rosenfeld, A. Alchalel and M. Ottolenghi, paper E2, this Volume.
19. A. Kropf and R. Hubbard, *Photochem. Photobiol.*, **12**, 4 (1970).
20. A. J. Thomson, *J. Chem. Phys.*, **51**, 4106 (1969).
21. J. B. Birks and D. J. Dyson, *Proc. Roy. Soc.*, **A275**, 135 (1963).
22. A. J. Thomson, *Abstr. Biophys. Soc.*, **7**, 124 (1967).
23. D. E. Balke and R. S. Becker, *J. Amer. Chem. Soc.*. **89**, 5061 (1967).
24. T. A. Moore and P. Song, *Nature New Biology*, **243**, 30 (1973).
25. R. L. Christensen and B. E. Kohler, *Photochem. Photobiol.*, **19**, 401 (1974).
26. S. J. Strickler and R. A. Berg, *J. Chem. Phys.*, **37**, 814 (1962).
27. J. P. Dalle and B. Rosenberg, *Photochem. Photobiol.*, **12**, 151 (1970).
28. A. N. Nikitina, G. S. Ter Sarkisian, B. M. Mikhailov and L. E. Minchenkova, *Opt. Spectry*, **14**, 347 (1963).
29. T. A. Moore and P. Song, *Chem. Phys. Letters*, **19**, 128 (1973).
30. B. S. Hudson and B. E. Kohler, *Chem. Phys. Letters*, **14**, 299 (1972).
31. K. Schulten and M. Karplus, *Chem. Phys. Letters*, **14**, 305 (1972).
32. B. S. Hudson and B. E. Kohler, *J. Chem. Phys.*, **59**, 4984 (1973).
33. R. M. Gavin, S. Risemberg and S. A. Rice, *J. Chem. Phys.*, **58**, 3160 (1973).
34. R. M. Gavin and S. A. Rice, *J. Chem. Phys.*, **60**, 3231 (1974).

E.1. Singlet–triplet quantum efficiencies of retinal isomers

R. Azerad

Université de Paris XI, Institut de Biochimie, 91405-Orsay, France

R. Bensasson and M. B. Cooper

Université de Paris VI, E.R. 98 Laboratoire de Chimie Physique, 91405-Orsay, France

E. A. Dawe

Department of Structural Chemistry, The University, Bradford BD7 1DP, U.K.

E. J. Land

Paterson Laboratories, Christie Hospital and Holt Radium Institute, Manchester M20 9BX, U.K.

SUMMARY

Nanosecond pulse radiolysis and laser flash photolysis are being used to study the triplet excited states of several isomers of retinal, the chromophore of the visual pigment rhodopsin. Absorptions and lifetimes of the triplet states of 9-, 11-, 13-*cis* and all-*trans* isomers of retinal have been studied. Their triplet–triplet extinction coefficients were measured against biphenyl triplet as standard, by a development of the energy transfer method which allows for the short lifetime of these polyene triplets. These extinctions were employed in the determination of the corresponding singlet–triplet crossover efficiencies (ϕ_T) for 265 nm excitation, using naphthalene as standard. ϕ_T was found to vary considerably from one isomer to another.

1 INTRODUCTION

The primary photochemical reaction causing photobleaching of the visual pigment rhodopsin is believed to be an 11-*cis* to all-*trans* isomerization of the retinal chromophore.[1] This isomerization causes a subsequent modification

531

in the conformation of the protein opsin, attached to retinal via a Schiff base linkage, which triggers a complex series of processes leading ultimately to the transmission of an impulse through the optic nerve.[2] In the primary photochemical event, absorption of light by the chromophore would be expected to lead to the formation of singlet, and subsequently triplet, excited states. Isomerization might then occur via one or both of these types of state.

The first and simplest model of rhodopsin to be studied from the point of view of triplet states was all-*trans* retinal.[3] Flash photolysis of retinal resulted in the observation of an intense absorption, with peak around 450 nm, which was identified as the triplet state. Measurements of its extinction coefficient ($7 \cdot 58 \times 10^4 \, \text{M}^{-1} \, \text{cm}^{-1}$ in methyl cyclohexane), via the singlet depletion method, were used to lead to an estimate of $\phi_T = 0 \cdot 11$, based on ferrioxalate actinometry. This low value for ϕ_T, may be compared with the value for the quantum yield of photobleaching of rhodopsin determined to be $0 \cdot 67$.[4] The discrepancy between these two yields has been an important factor in arguments against the involvement of the triplet state of retinal in the early stages of vision.[5]

The application of lasers to flash photolysis has made possible the development of a new method of determining triplet quantum efficiencies.[6,7] This new method is more accurate than the technique previously employed[3] which used conventional flash photolysis, since actinometry is inherently much simpler with the single emission lines of a laser than with the continua of xenon flash lamps. Both methods also require the relevant extinction coefficients of triplet–triplet transitions (ε_T) and recently more reliable methods for determining these coefficients have become available.[8] We now give a progress report of our measurements of ε_T and ϕ_T for four isomers of retinal, namely the all-*trans*, 9-*cis*, 11-*cis* and 13-*cis* isomers. Our redetermination of ϕ_T for the all-*trans* isomer has been described briefly in a previous communication.[9]

2 EXPERIMENTAL

The laser flash photolysis equipment was as previously described.[7] Solutions, deaerated by bubbling with argon, were irradiated with 30 ns flashes of 265 nm radiation in a cell of optical path 1 cm. The resulting changes in light absorption were monitored as a function of time at each wavelength using photoelectric detection. Monochromator bandwidths were normally 3·0 nm or less. Photomultiplier signals were displayed on an oscilloscope and photographed on polaroid film. Changes in the laser light intensity were monitored by deflecting $\sim 10\%$ of the laser radiation on to a separate diode whose output was also displayed on the oscilloscope. The pulse radiolysis equipment was also as previously described.[10] Argon-flushed solutions were irradiated in cells of optical path 2·5 cm with 100 or 200 ns pulses of 9–12 MeV electrons. The resulting absorption changes were monitored photoelectrically. Variations in radiation dose from pulse to pulse were allowed for using a calibrated secondary emission chamber.

11-*cis* retinyl acetate was kindly donated by Dr Montavon of Hoffman La Roche and Co., Basel. 11-*cis* retinal was prepared from the acetate as follows: the acetate was hydrolysed to 11-*cis* retinol by KOH treatment in methanol. The retinol was then purified by thin-layer chromatography and oxidized to 11-*cis* retinal by MnO_2. 9-, 13- and all-*trans* retinal were used as supplied by Sigma. The ratio $\varepsilon_{365\,nm}/\varepsilon_{254\,nm}$ was used as a criterion of isomeric purity. Other chemicals were the purest available commercially. Hexane was BDH spectroscopic grade.

3 RESULTS

3.1 Laser flash photolysis and pulse radiolysis of retinal solutions

Laser flash photolysis or pulse radiolysis of hexane solutions of the 9-*cis*, 11-*cis*, 13-*cis* and all-*trans* isomers of retinal led to the observation of transient absorptions with peaks around 440–450 nm. The changes in optical density immediately after laser flash photolysis are shown in Figure 1. The absorptions decayed by first-order kinetics, the decay constants being listed in Table 1. In the case of the all-*trans* isomer, the absorption has been assigned to the triplet excited state, particularly since this transient species sensitizes the formation of triplet β-carotene.[9] In view of the similarity between all four transient spectra, it seems likely that the absorptions derived from the *cis* isomers are likewise due to the corresponding triplet excited states.

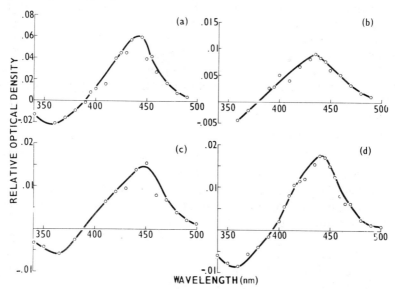

Figure 1 Transient absorption spectra immediately after laser flash photolysis of retinal isomers in hexane ($\sim 10^{-5}$ M) (a) 9-*cis*, (b) 11-*cis*, (c) 13-*cis*, (d) all-*trans*. The number of exciting photons was constant within each spectrum but varied from one spectrum to another

Table 1   Data for retinal isomers[a]

Isomer	λ_{max} (nm)	decay constant $(s^{-1} \times 10^{-5})$	Triplet maximum extinction $(M^{-1} cm^{-1} \times 10^{-4})$ pulse radiolysis	flash photolysis	Singlet–triplet crossover efficiency ϕ_T (265 nm excitation)
9-cis	445 ± 5	1·2 ± 0·1	5·6 ± 0·6	6·3 ± 1·5	0·71 ± 0·15
11-cis	440 ± 5	1·2 ± 0·1	6·2 ± 0·6	5·7 ± 1·4	0·13 ± 0·03
13-cis	445 ± 5	1·3 ± 0·1	5·6 ± 0·6	4·4 ± 1·1	0·18 ± 0·04
All-trans	445 ± 5	1·1 ± 0·1	7·8 ± 0·8	6·9 ± 1·7	0·40 ± 0·06

[a] the values given are the means of three or more independent experiments.
The error limits cover the range of values obtained for individual determinations.

In the pulse radiolysis of all-trans and 13-cis retinal additional transient peaks around 600 nm were observed. These decayed by second-order kinetics and are probably due to radical ions.[11] We have not yet studied the 9-cis and 11-cis isomers in this region of the spectrum. Since in this paper we are only concerned with triplet states, these 600 nm absorptions will not be referred to again.

A small final change in absorption detected at 365 nm, the wavelength maximum of the singlet ground state, tens of microseconds after laser excitation when the triplet had decayed, was consistent with photoisomerization of 11-cis retinal to all-trans retinal. Similar but even smaller final changes in absorption were also observed for 9-cis, 13-cis and all-trans retinal. Although these final changes are undoubtedly real, quantitative measurements which would lead to isomerization yields are difficult with our present apparatus, being designed primarily for submicrosecond studies, the base-line light stability over tens of microseconds being insufficient. Such measurements are also hindered by the reduction in analysing light intensity caused by the retinal ground state absorption at 365 nm.

3.2 Extinction coefficients

These were obtained by the energy transfer 'comparative' technique,[12] which basically consists of comparing an unknown triplet with a known one, in this case biphenyl ($\varepsilon_{361}^{T} = 42{,}800\ M^{-1}\ cm^{-1}$). The procedure involves taking a solution of biphenyl (donor, D) + one of the retinal isomers (acceptor, A) and exciting as far as possible the biphenyl alone. Comparison of the biphenyl and derived retinal triplet optical densities then leads to a comparison of biphenyl and retinal triplet extinctions.

In such experiments three different reactions have to be taken into account:

$$^3D \xrightarrow{k_1} {}^1D$$

$$^3D + {}^1A \xrightarrow{k_2} {}^1D + {}^3A$$

$$^3A \xrightarrow{k_3} {}^1A$$

The overall decay of ^3D, k_2, is equal to $k_1 + k_Q[A]$. If k_1 and k_3 were negligible as compared with $k_Q[A]$, then the total retinal triplet concentration (OD^A) would have equalled the total biphenyl concentration (OD^D), and

$$\frac{OD^A}{OD^D} = \frac{\varepsilon^A}{\varepsilon^D} \tag{1}$$

However for most of the retinal–biphenyl mixtures studied, k_1 and k_3 were not negligible as compared with $k_Q[A]$. Furthermore, in some mixtures a little retinal triplet was formed instantaneously by direct excitation of retinal. Three corrections were therefore necessary which modified relationship (1).

(a) Biphenyl triplet decay by means other than energy transfer

This causes the total concentration of acceptor retinal triplet observed (OD^A observed) to be less than the donor biphenyl triplet concentration by the proportion $k_Q[A]/(k_1 + k_Q[A]) = (k_2 - k_1)/k_2$. If transfer had been complete, the observed acceptor optical density would have been greater by the amount $k_2/(k_2 - k_1)$, i.e.

$$OD^A(\text{corrected}) = OD^A(\text{observed}) \times \frac{k_2}{k_2 - k_1} \tag{2}$$

(b) Retinal triplet decay during its formation

This causes the maximum observed acceptor retinal triplet to be further reduced. It can be shown[13] that the maximum acceptor optical density observed, $OD^{A(\text{max})}$, is related to the total acceptor optical density formed, OD^A, by the expression:

$$\frac{OD^{A(\text{max})}}{OD^A} = \exp\left\{ -\frac{\ln k_2/k_3}{k_2/k_3 - 1} \right\} \tag{3}$$

For each isomer of retinal, the total acceptor optical density formed was thus estimated from data obtained at several concentrations of acceptor, the relevant k_2 being deduced from the k_2 previously found in (a) for low concentrations of retinal. k_3, which does not vary with retinal concentration, is the triplet decay rate obtained on laser flash photolysis or pulse radiolysis of retinal alone in hexane (Table 1).

Since $OD^{A(\text{max})}$ is attained at a time when the donor triplet has not decayed completely, overlap of donor and acceptor triplet absorption must also be considered. In this case, unlike the biphenyl–quinone system previously described,[14] the overlap is insignificant and so no correction was necessary.

(c) Retinal triplet formation by direct excitation

Using pulse radiolysis excitation no correction is necessary as the concentration of biphenyl was always at least 10^3 times higher than the concentration of retinal. Using laser excitation, much more dilute solutions are necessary for homogeneity of excitation. However, sufficient acceptor concentrations must be retained for efficient transfer to occur. Under such conditions a significant amount of acceptor triplet may be formed by direct excitation. The directly formed acceptor triplet optical density, OD^A(direct), for a given biphenyl–retinal mixture, was obtained by exciting a solution of retinal alone at the same concentration as chosen for the mixture. The relation (1) becomes:

$$\frac{OD^D}{OD^A(\text{observed}) - OD^A(\text{direct})} = \frac{\varepsilon^D}{\varepsilon^A}$$

The triplet extinctions obtained using flash photolysis and pulse radiolysis after application of all the corrections described for 9-, 11-, 13-*cis* and all-*trans* retinal are given in Table 1. The values obtained using pulse radiolysis tend to be the more accurate since fewer corrections are necessary. Such values were therefore used in the determinations of ϕ_T (see below).

3.3 Singlet → triplet quantum efficiencies

The principle of the method is to compare the concentration of triplets formed on laser excitation of a solution of retinal with the concentration of triplets formed by the same number of quanta from a standard solution with a known triplet extinction and crossover efficiency. It is arranged that both solutions have the same singlet ground state optical density at the laser excitation wavelength (265 nm) and so equal numbers of singlets are excited. The ratio of the numbers of triplets formed is then equal to the ratio of the crossover efficiencies:

$$[\text{Singlet}]^R = [\text{Singlet}]^S \quad R = \text{retinal}$$

$$\frac{[\text{Triplet}]^R}{\phi_T^R} = \frac{[\text{Triplet}]^S}{\phi_T^S} \quad S = \text{standard}$$

$$\frac{OD_T^R}{\phi_T^R \varepsilon_T^R} = \frac{OD_T^S}{\phi_T^S \varepsilon_T^S}$$

Therefore

$$\phi_T^R = \frac{OD_T^R \phi_T^S \varepsilon_T^S}{OD_T^S \varepsilon_T^R}$$

where OD_T^R and OD_T^S are the maximum optical densities for retinal and standard triplet absorptions, ε_T^R and ε_T^S are the maximum triplet–triplet extinction coefficients for retinal and standard, and ϕ_T^R and ϕ_T^S are the singlet→triplet quantum efficiencies for retinal and standard.

The values of ϕ_T for 265 nm excitation thus obtained for the four isomers of retinal studied are given in Table 2. The standard used was naphthalene ($\phi_T = 0.71$,[15] $\varepsilon_T = 24,500$ M^{-1} cm^{-1} at 415 nm[12]). It is important in this type of laser experiment to ensure that molecules in their excited states do not absorb

Table 2 Measurements of ϕ_T for all-*trans*-retinal using various laser excitation intensities and solute concentrations

Percentage of full laser intensity (265 nm excitation)	Retinal concentration (M × 10⁴)	ϕ_T
10	1·25	0·37
	2·5	0·43
30	1·25	0·46
	2·5	0·39
100	0·5	0·34
	1·25	0·43
	2·5	0·38

significant amounts of exciting light, otherwise this would lead to intensity-dependent quantum yields.[16] For this reason low numbers of exciting photons and dilute solutions were used, resulting in small singlet→triplet conversions (< 10%) and homogeneous concentrations of triplets, also avoiding self-absorption of any fluorescent light. In our experiments optical densities at the excitation wavelength of 265 nm were usually < 0·02 in the path of the exciting beam under observation (0·2 mm). This condition is also normally adopted in relative fluorescence quantum yield measurements.[17] The quantum efficiencies reported here were independent of the number of incident photons and of the solute concentration (see Table 2 for results for all-*trans* retinal).

4 DISCUSSION

4.1 Triplet absorption spectra, extinction coefficients and lifetimes

The triplet absorption profiles of all four retinal isomers studied are similar (Figure 1). However their extinction coefficients (Table 1) at the maximum around 450 nm (Figure 1) differ, as found with the corresponding ground state singlet absorptions. The present value for the extinction of the all-*trans* isomer agrees well with the conventional flash photolysis determination of Dawson and Abrahamson,[3] $\varepsilon_{445} = 75{,}800 \text{ M}^{-1} \text{ cm}^{-1}$ in methyl cyclohexane using the singlet depletion method, and the pulse radiolysis measurements of Rosenfeld, Alchalel and Ottolenghi,[18] $\varepsilon = 75{,}000 \text{ M}^{-1} \text{ cm}^{-1}$ in hexane using the energy transfer technique. Our estimate for the 11-*cis* isomer ($62{,}000 \text{ M}^{-1} \text{ cm}^{-1}$) appears to be significantly lower than the previous estimate of Rosenfeld *et al.*[18] ($75{,}000 \text{ M}^{-1} \text{ cm}^{-1}$). Like many other polyenes, the triplet lifetimes of the retinals are short, being in the microsecond region (Table 1), there being little difference between the four isomers studied. The decay constants found here agree with previous determinations: for the all-*trans* isomer ($1·1 \times 10^5 \text{ s}^{-1}$,[3,19] $1·2 \times 10^5 \text{ s}^{-1}$[18]), for the 11-*cis* isomer ($1·2 \times 10^5 \text{ s}^{-1}$[18]). Data for the 9-*cis* and 13-*cis* triplets have not previously been reported.

4.2 Singlet–triplet crossover efficiencies

The quantum efficiencies found ranged between 0·13 and 0·71 (see Table 1). All these retinals thus exhibit crossover efficiencies much higher than most polyenes studied previously,[19] many of which do not have $n\pi^*$ states. Variations in relative positions of lowest $n\pi^*$ and $\pi\pi^*$ states[5,20–22] may contribute to the variation in ϕ_T from one retinal isomer to another. The value obtained here for all-*trans* retinal of 0·40 ± 0·06 is above the estimate (0·11) by Dawson and Abrahamson.[3] The present value is on the other hand lower than the estimate of 0·7 ± 0·1 obtained recently by Rosenfeld et al.[18] using 337·1 nm laser excitation, and our previous estimate of 0·60 ± 0·1. We believe that our earlier value may be somewhat too large due to the laser intensities being too high. The only previous estimate for 11-*cis* retinal was by Rosenfeld et al.[18] who found 0·6 ± 0·1 using 337·1 nm excitation, based on an extinction of triplet 11-*cis* retinal of 75,000 M^{-1} cm^{-1}. Use of our present extinction estimate of $\varepsilon = 62,000$ for 11-*cis* retinal would increase the ϕ_T estimate of Rosenfeld et al. to 0·73, which is clearly different from our estimate of $\phi_T = 0·13$. This could be due to a wavelength dependency, which has already been found for the fluorescence quantum yields[20,22] and photoisomerization yields[23] of retinals.

4.3 Relevance to the visual pigment rhodopsin

It is clear that the magnitude of the singlet–triplet intersystem crossing efficiency of retinal very much depends upon the isomeric state of the molecule. It may also be a function of wavelength, temperature, viscosity and nature of the medium in which it is contained. In rhodopsin the 11-*cis* retinal chromophore is believed to be encapsulated by the protein opsin. The protein therefore probably applies considerable constraints upon the chromophore which may affect its singlet–triplet crossover efficiency. In addition, 11-*cis* retinal is connected chemically to opsin via a lysine Schiff base linkage which also modifies some of the physical characteristics of the chromophore. Thus, although the singlet–triplet crossover efficiency found here for the isolated 11-*cis* retinal chromophore in hexane at room temperature (0·13) is much lower than the photobleaching yield of 0·67 found for rhodopsin itself, the triplet state of the chromophore could still be involved in the early stages of vision. Experiments aimed at improving the precision of the estimates of ε_T and ϕ_T already gained, at measuring ϕ_T as a function of wavelength, at measuring photoisomerization yields and similar studies of the *n*-butylamine and lysine Schiff bases of retinals, closer models of rhodopsin, are in progress.

ACKNOWLEDGEMENTS

We are grateful to Professor M. Bessis, Director of the Institut de Pathologie Cellulaire (INSERM), Kremlin Bicêtre, Paris, for the use of the laser and other facilities, and to M. R. Lajeunie for technical assistance. We also thank Dr J. P.

Keene and Mr B. W. Hodgson for their constant supervision of the accelerator and pulse radiolysis apparatus in Manchester. We are also grateful to Hoffman La Roche and Co. Basel, for generously donating a sample of 11-*cis* retinyl acetate, and to N.A.T.O. for a travel grant (No. 699). This work was supported by grants from the Cancer Research Campaign and the Medical Research Council.

REFERENCES

1. R. Hubbard and A. Kropf, *Proc. US Nat. Acad. Sci.*, **44**, 130 (1958).
2. H. J. A. Dartnall (ed.), *Handbook of Sensory Physiology*, Vol. VII/I, *Photochemistry of Vision*, Springer-Verlag, Berlin, 1972.
3. W. Dawson and E. W. Abrahamson, *J. Phys. Chem.*, **66**, 2542 (1962).
4. H. J. A. Dartnall, *Vision Research*, **8**, 339 (1968).
5. E. W. Abrahamson and S. E. Ostroy, *Progr. Biophys. Mol. Biol.*, **17**, 181 (1967).
6. J. T. Richards and J. K. Thomas, *Trans. Faraday Soc.*, **66**, 621 (1970).
7. R. Bensasson, C. Chachaty, E. J. Land and C. Salet, *Photochem. Photobiol.*, **16**, 27 (1972).
8. H. Labhart and W. Heinzelman, *Organic Molecular Photophysics*, Vol. 1, p. 297 (ed. J. B. Birks), J. Wiley, London and New York, 1973.
9. R. Bensasson, E. J. Land and T. G. Truscott, *Photochem. Photobiol.*, **17**, 53 (1973).
10. J. P. Keene, *J. Sci. Instrum.*, **41**, 493 (1964).
11. C. M. Lang, J. Harbour and A. V. Guzzo, *J. Phys. Chem.*, **75**, 2861 (1971).
12. R. Bensasson and E. J. Land, *Trans. Faraday Soc.*, **67**, 1904 (1971).
13. C. Capellos and B. H. J. Bielski, *Kinetic Systems*, Wiley–Interscience, New York, 1972.
14. E. Amouyal, R. Bensasson and E. J. Land, *Photochem. Photobiol.*, **20**, 415 (1974).
15. C. A. Parker and T. A. Joyce, *Trans. Faraday Soc.*, **62**, 2785 (1966).
16. S. Speiser, R. van der Werf and J. Kommandeur, *Chem. Phys.*, **1**, 297 (1973).
17. J. H. Chapman, T. Forster, G. Kortum, C. A. Parker, E. Lippert, W. H. Melhuish and G. Nebbia, *Appl. Spectroscopy*, **17**, 171 (1963).
18. T. Rosenfeld, A. Alchalel and M. Ottolenghi, *J. Phys. Chem.*, **78**, 336 (1974).
19. T. G. Truscott, E. J. Land and A. Sykes, *Photochem. Photobiol.*, **17**, 43 (1973).
20. R. S. Becker, K. Inuzuka, J. King and D. E. Balke, *J. Amer. Chem. Soc.*, **93**, 43 (1971).
21. R. L. Christensen and B. E. Kohler, *Photochem. Photobiol.*, **18**, 293 (1973).
22. R. L. Christensen and B. E. Kohler, *Photochem. Photobiol.*, **19**, 401 (1974).
23. A. Kropf and R. Hubbard, *Photochem. Photobiol.*, **12**, 249 (1970).

E.2. Intersystem crossing, ionic dissociation and cis–trans isomerization mechanisms in the photolysis of retinol and related molecules

T. Rosenfeld, A. Alchalel and M. Ottolenghi

*Department of Physical Chemistry, The Hebrew University,
Jerusalem, Israel*

SUMMARY

Photochemical experiments, using pulsed N_2 laser excitation and continuous irradiation methods, are carried out in polar and non-polar solutions of all-*trans* retinol (ROH), retinyl-acetate (RAc) and retinyl-*n*-butyl-amine (RBA). The effects of dissolved oxygen and solvent polarity on the yields of (a) intersystem crossing, (b) all-*trans*→11-*cis* photoisomerization and (c) ionic photodissociation (e.g.

$$RAC \xrightarrow{h\nu} R^+ + Ac^-)$$

are measured and analysed. In *n*-hexane photoisomerization is enhanced by dissolved oxygen. The effect is associated with oxygen-induced intersystem crossing. The drastic enhancement of isomerization upon increasing the solvent polarity leads to a mechanism involving the retinylic cation R^+. The presently available yields for the primary photoprocesses of molecules related to the visual chromophore are discussed.

1 INTRODUCTION

The understanding of the primary steps in the photochemistry of rhodopsin is closely associated with the photochemical behaviour of simple molecules related to the visual chromophore such as retinal, the corresponding alcohol (retinol or vitamin A), as well as the protonated and unprotonated Schiff's bases. Since it is generally accepted that the primary step in vision is associated with the 11-*cis*→all-*trans* isomerization of the rhodopsin chromophore,[1] we have previously paid special attention to both direct, and triplet-sensitized, photo-isomerizations of retinal[2] and its n-butyl-amine Schiff base.[3]

The application of pulsed laser photolysis methods to retinol (ROH) systems has recently led[4] to the detection of a low-yield triplet state ($\lambda_{max} = 405$ nm). A second (conducting) transient, absorbing around 590 nm, was identified as the retinylic cation R^+, thus establishing ionic photodissociation as a third primary photochemical step, in addition to ISC[4] and $cis \rightarrow trans$ isomerization.[5] In the present work we have studied the effects of dissolved oxygen and solvent polarity on the photochemistry of retinol and of the closely related molecules, retinyl acetate (RAc) and retinyl-n-butyl-amine (RBA). The data presented lead to the detection of new photoisomerization paths in retinol and related molecules, involving oxygen-quenching of the fluorescent state (non-polar solutions) and the retinylic cation R^+ (polar solutions).

2 EXPERIMENTAL

All-*trans* retinal, retinol and retinyl-acetate (Sigma Chemicals Co.), as well as biacetyl (BDH) were used without further purification. Retinyl-n-butylamine was prepared by reducing the corresponding Schiff base of retinal with sodium borohydride.[6] N,N,N',N', tetramethyl-p-phenylenediamine (TMPD) and biphenyl, both Eastman Kodak, were zone refined. n-Hexane, n-butanol, *iso*propanol, ethanol and methanol were all Fluka, spectrograde reagents.

Absorption spectra were recorded using a Cary-14 spectrophotometer. Emission and excitation spectra were measured on a Turner Mod. 210 fluori-meter. Continuous irradiations around 313 nm were carried out using a medium-pressure Hg arc, or a Cd lamp, both filtered by $NiSO_4$–$CoSO_4$ and potassium biphthalate solutions, and a 7–54 Corning glass filter. Actinometry was carried out using uranyl-oxalate solutions.

Fluorescence lifetimes were measured using the apparatus described by Ofran and Feitelson.[7]

The pulsed photolysis technique, using the 337·1 nm (10 ns, 0·5 mJ) pulse of an Avco–Everett laser, and the pulse-radiolysis apparatus were both previously described.[8,9] Solutions were deaerated or oxygen-saturated by bubbling correspondingly nitrogen (or argon) and oxygen. All experiments were carried out under deep red light using freshly prepared solutions.

3 RESULTS

(i) The fluorescent state

When submitted to pulsed laser excitation, solutions of ROH, RAc and RBA exhibit a similar short-lived transient, absorbing around 435 nm, and decaying within a few nanoseconds ($\tau_{1/2} < 8$ ns). Both spectra and decay patterns in the cases of RAc and RBA (Figure 1) are essentially identical to those previously

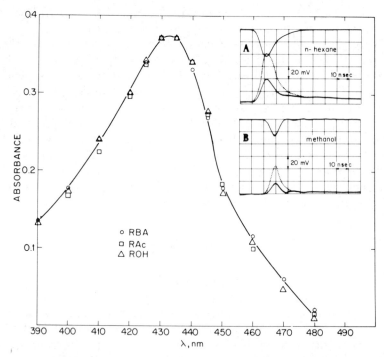

Figure 1 Characteristic oscillograms and spectra due to the excited singlet fluorescent state. The spectra for **RBA, RAc** and **ROH**, recorded 10 ns after pulsing 10^{-4} M *n*-hexane solutions, have been normalized to the same maximum absorbance intensity. In the oscillograms, showing the effect of the solvent polarity on the fluorescence and on the excited singlet–singlet absorption, the upper trace is recorded without the monitoring light beam (fluorescence only). The dashed line is the net excited singlet absorbance, obtained by subtracting the corresponding fluorescence contribution from the lower traces recorded with the monitoring beam (fluorescence + absorption)

reported by us for retinol,[4] leading to a common identification of the 435 nm species as the fluorescent singlet state, $^1M^*$. As expected, the 495 nm fluorescence intensity and the excited singlet–singlet 435 nm absorbance exhibit the same sensitivity to dissolved oxygen (Table 1 and Figure 2) and to the solvent polarity (Figure 1). Similarly (see Table 1) the three molecules are found to exhibit the same fluorescence yield, for which the absolute value of 0·02 was measured by Thomson[10] (for ROH and RAc) in aerated non-polar solutions.

We have also measured the corresponding fluorescence lifetimes (τ), obtaining for ROH and RAc in deaerated *n*-hexane the values 4·5 and 4·6 ns, which fairly agree with those (4·6 and 4·8 ns) previously reported by Dalle and Rosenberg.[11] For retinol in air-saturated and oxygen-saturated *n*-hexane we obtained $\tau = 3·3$ and $\tau = 2·5$ ns, correspondingly. Increasing the solvent polarity leads to a marked decrease in the fluorescence yields and lifetimes. The corresponding data are presented in Table 2.

Table 1 Relative yields of fluorescence, intersystem crossing and all-*trans* → 11-*cis* photoisomerization in the photochemistry of ROH, RAc and RBA in *n*-hexane

Solution	Fluorescence (ϕ_F^r)[a]			Triplet (ϕ_T^r)			Isomerization (ϕ_{ISO}^r)		
	ROH	RAc	RBA	ROH	RAc	RBA	ROH	RAc	RBA
Deaerated	1	1	1	1[c]	1[c]	1[c]	1[d]	1[d]	1[d]
Air-saturated	0·75 ± 0·05[b]	0·75 ± 0·05	0·75 ± 0·05	10 ± 1	11 ± 1	10 ± 1	3·0 ± 0·3	2·4 ± 0·3	3·5 ± 0·3
O$_2$-saturated	0·55 ± 0·05	0·55 ± 0·05	0·55 ± 0·05	25 ± 2	27 ± 2	24 ± 2	5·3 ± 0·3	4·5 ± 0·3	6·0 ± 0·3

[a] The values of ϕ_F^r were determined from the relative maxima of fluorescence signals measured at 495 nm on the laser apparatus, from relative fluorescence intensities measured at 470 nm on the steady-state fluorimeter, or from areas under the excited singlet–singlet absorbance profile.

[b] The absolute value of the fluorescence yield in aerated *n*-hexane reported by Thomson[10] is $\phi_F = 0.020$.

[c] The absolute value determined by us for all three molecules in deaerated *n*-hexane (see text) is $\phi_T = 0.03 \pm 0.01$.

[d] The absolute value measured by us for all three molecules in aerated *n*-hexane (see text) is $\phi_{ISO} = 0.032 \pm 0.002$, assuming for all three molecules an all-*trans* → 11-*cis* transition with the same difference in extinction coefficients[10]: $\Delta\varepsilon = 1.71 \times 10^4$ cm^{-1} M^{-1}.

Table 2 Solvent dependence of primary processes in the photochemistry of retinyl-acetate and retinol[a]

Solvent	Retinyl acetate								Retinol	
	ϕ_T	D_{R^+} [b]	ϕ_{ISO}^r [c]	ϕ_F^r [d]	$\tau^d_{(ns)}$	$k_q \times 10^{-8}$ s^{-1}	$\dfrac{D_{R^+}^{meth}}{D_{R^+}}$	$\dfrac{k_q^{meth}\tau^{meth}}{k_q^i \tau^i}$	ϕ_F^d	τ^d ns
Hexane	0·3	<0·001	0·112	1	4·50[e]	—	—	—	0·026[g]	4·5
n-Butanol	0·07	0·072	0·45	0·62	2·85[f]	1·3	2·39	2·22	—	—
Ethanol	0·035	0·150	0·68	0·30	1·37[f]	5·1	1·15	1·18	—	—
Methanol	0·028	0·172	1·0	0·17	0·80[e,f]	10·3	1	1	0·0072[h]	1·3

[a] The fluorescence parameters ϕ_F, ϕ_F^d and τ^d are in deaerated solutions. ϕ_T, D_{R^+} and ϕ_{R0} were measured in aerated systems.

[b] D_{R^+} is the initial absorbance change at 590 nm observed in the pulse laser photolysis of 10^{-4} M solutions of R·Ac.

[c] Relative all-trans → 11-cis photoisomerization yields measured (see text) in 46×10^{-5} M solutions of RAc.

[d] Relative fluorescence yields measured on the laser apparatus (10^{-4} M solutions).

[e] Values measured directly.

[f] Values determined from the corresponding ϕ_F values assuming a constant ϕ_F/τ (see text).

[g] Derived from the data of Figure 1 and the absolute value of $\phi_F = 0·02$ reported by Thomson[10] in aerated solutions.

[h] Derived from the value 0·006 reported by Thomson[10] in aerated methanol and using the factor $\phi_F^d/\phi_F^{air} = 1·2$ measured by us on the steady-state fluorimeter or estimated from the areas under the corresponding singlet–singlet absorbance profiles.

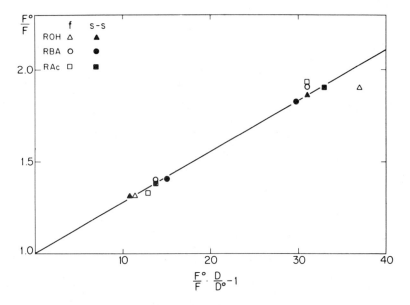

Figure 2 Plot of oxygen quenching data for ROH, RBA and RAc in 10^{-4} M n-hexane solutions according to equation (I). D/D° was estimated from the corresponding initial triplet absorbance changes at 405 nm. The relative fluorescence parameter was estimated from the areas under the excited singlet–singlet absorption profile (s–s) or from the maxima of the fluorescence signals (f), measured at 495 nm on the laser apparatus or on the steady-state fluorimeter

(ii) The triplet state: Extinction coefficients and quantum yields

In addition to the 435 nm excited-singlet absorption, laser-pulsed retinol solutions exhibit a long-lived absorbance change around 405 nm, which has been previously attributed to the triplet state, ^{3}M*.[4] An identical transient is also observed in the case of RAc and RBA (Figure 3). Triplet extinction coefficients (ε_T) have been determined according to previously described photosensitization procedures,[2,3,12] populating ^{3}M* via energy transfer from the triplet state of biphenyl (^{3}B*):

$$^3B^* + M \xrightarrow{k_{ET}} {}^3M^* + B \qquad (k_{ET} = 3 \times 10^9 \, \text{M}^{-1} \, \text{s}^{-1}) \qquad (1)$$

$$(M \equiv ROH, RAc, RBA)$$

^{3}B* was produced by pulse radiolysis in a 0.1 M biphenyl solution in deaerated n-hexane, or by laser photolysis via the TMPD–B exciplex[2–4] in a 5×10^{-3} M TMPD–0.1 M biphenyl system. Both procedures led to the values: $\varepsilon_T(ROH) = 5.75(\pm 0.05) \times 10^4 \, \text{cm}^{-1} \, \text{M}^{-1}$ and $\varepsilon_T(RAc) = 6.4(\pm 0.1) \times 10^4 \, \text{cm}^{-1} \, \text{M}^{-1}$.

Triplet quantum yields were determined as described by Bensasson et al.,[13] calibrating the laser pulse intensity by measuring the absorbance changes due to the triplets of pyrene or anthracene. The value $\phi_T = 0.3 \pm 0.05$ was obtained

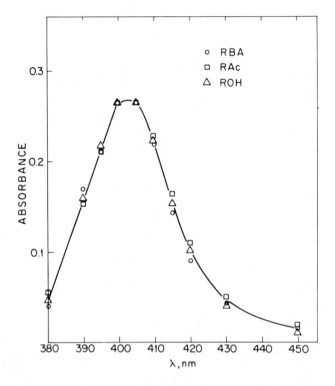

Figure 3 Triplet spectra recorded (~ 20 ns after pulsing) for air-saturated, 10^{-4} M, n-hexane solutions of RBA, RAc and ROH

for all three molecules in aerated n-hexane. As in the case of retinol,[4] deaeration decreases the triplet quantum yields of RAc and RBA by a factor of ~ 10. Assuming[4,14] that O_2 acts uniquely via:

$$^1M^* + O_2 \xrightarrow{k_{O_2}} {}^3M^* + O_2 \qquad (2)$$

we may apply the expression of Medinger and Wilkinson[15]

$$\frac{F^0}{F} = 1 + \phi_T^d \left(\frac{D_T}{D_T^0} \frac{F^0}{F} - 1 \right) \qquad (I)$$

where F^0 and F are, respectively, the fluorescence intensities in the absence and in the presence of O_2. D_T^0 and D_T are the corresponding absorbance changes due to the triplet state and ϕ_T^d is the triplet yield in the absence of a quencher. A plot of F^0/F versus $(F^0/F)(D_T/D_T^0) - 1$ (Figure 2) leads, for all three molecules, to $\phi_T^d = 0.03$, in agreement with the results of the previous procedure. The previously calculated value of ϕ_T in aerated solutions is also consistent with that (0.26 ± 0.05) obtainable from the expression:

$$\phi_T = k_{O_2}[O_2]/([k_f + k_{O_2}[O_2]) \qquad (II)$$

(which neglects the small contribution of ISC from non-quenched molecules) where $k_f = 1/\tau$ (in deaerated solutions), $[O_2] = 2 \times 10^{-3}$ M in air-saturated n-hexane, and using the value $k_{O_2} = 3 \pm 10^{10}$ M^{-1} s^{-1} derived from the measured lifetimes in aerated systems (see above).

The triplet quantum yields are markedly reduced by increasing the solvent polarity. The data for RAc in aerated solutions are shown in Table 2.

(iii) The retinylic cation

The third, 590 nm, transient observed[4] in laser-photolysed polar solutions of ROH, RAc and RBA has been identified as the retinylic cation R^+ formed according to

$$\text{ROH (or RAc, RBA)} \xrightarrow{h\nu} R^+ + OH^- \text{ (or } CH_3COO^-, RNH^-) \tag{3}$$

The solvent dependence of the relative absorbance changes (D_{R^+}), associated with the ionic photodissociation process, is given, for RAc, in Table 2. An exact determination of the absolute photodissociation quantum yields (ϕ_{R^+}) could not be carried out in view of the uncertainty in the value of the extinction coefficient of R^+ which depends on the solvent and on the acid concentration.[16] However, a rough estimate of $\phi_{R^+} \simeq 0.5$, for RAc in methanol, was obtained using the value of $\varepsilon_{R^+} = 4.4 \times 10^4$ M^{-1} cm^{-1} (measured by Blatz and Pippert,[16] in a 51% sulphuric acid–methanol mixture), by applying the technique employed for the determination of ϕ_T, using the triplets of anthracene or pyrene as laser intensity monitors. The relative value of D_{R^+} is also sensitive to the particular molecule involved, decreasing when passing from RAc to ROH and RBA, the corresponding ratios being $1:0.24:0.14$.

(iv) The photoisomerization process

Exposure of (all-*trans*) ROH, RAc and RBA solutions in n-hexane to the ($\lambda > 300$ nm) radiation of the medium-pressure Hg (or Cd) sources filtered by a 0–52 Corning glass filter, results in a decrease of the main 325 nm absorbance, accompanied by a growing-in of the ('*cis*') band around 250 nm. Such a process, characteristic of all-*trans*→11-*cis* photoisomerizations of retinyl-polyenes[17] was found to be markedly affected by increasing the oxygen content in non-polar solutions as well as by increasing the solvent polarity. The relevant data for the three molecules are presented in Figure 4 and Table 1. Moreover, in the same polar medium (methanol), the photoisomerization yield (ϕ_{ISO}) is greater for the molecule exhibiting a higher D_{R^+} value, the relative values of ϕ_{ISO} for RAc, ROH and RBA being $1:0.5:0.26$, (Figure 4). We should finally record that, due to difficulties associated with the choice of suitable triplet energy donors, we have been unsuccessful in providing in the present system clear evidence for a triplet-sensitized photoisomerization, similar to that recently investigated for retinal[2] and its Schiff bases.[3]

548

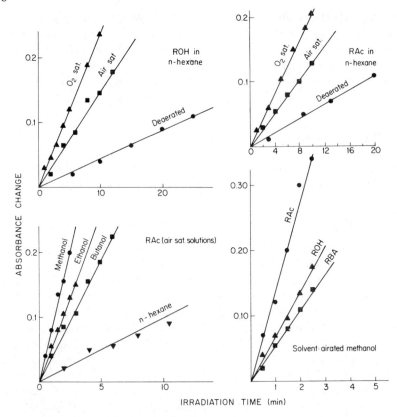

Figure 4 Solvent and oxygen effects on the initial (negative) absorbance changes at 320 nm, characteristic of the all-*trans* → *cis* isomerization of ROH, RAc and RSB. Concentrations: 5×10^{-5} M

4 DISCUSSION

(i) Photoisomerization in aerated and deaerated non-polar solutions

An examination of Table 1 shows that the O_2-quenching process enhances all-*trans* → 11-*cis* isomerization to an extent which is common to all three molecules. The quantitative agreement of our quenching data with the expressions (I) and (II) establishes reaction (2) as a unique route of the oxygen fluorescence quenching process. The data are therefore not inconsistent with an isomerization mechanism proceeding via the (O_2-induced) triplet state, as suggested by Gordon Walker and Radda.[18a] However, an alternative explanation to the O_2 effect may involve a path in which isomerization takes place within a (charge-transfer) complex $^1(M^+.O_2^-)^*$, probably acting as a short-lived intermediate in reaction (2),[14,19,20] rather than from $^3M^*$ itself.

Although not being discriminative about the details of the O_2-induced isomerization process, the O_2 quenching data of Table 1 rule out a mechanism in which isomerization (in deaerated or aerated solutions) proceeds solely via the thermalized triplet state. Such a mechanism implies a constant ϕ_T^r/ϕ_{ISO}^r ratio independent of the amount of O_2 in the solution which is at variance with the data of Table 1. Our experimental observations are explainable in terms of the following expressions:

$$\phi_{ISO}^d = k_{ISO}\tau^d \tag{III}$$

$$\phi_{ISO}^{O_2} = k_{ISO}\tau^{O_2} + A\phi_T \tag{IV}$$

where ϕ_{ISO}^d and $\phi_{ISO}^{O_2}$ are the all-*trans*→11-*cis* isomerization yields observed correspondingly in the absence and in the presence of oxygen, τ^d and τ^{O_2} are the corresponding observed fluorescence lifetimes, and ϕ_T is the triplet yield in the presence of O_2. In writing the above expressions we have assumed that isomerization in deaerated solution takes place, with a rate k_{ISO}, from the thermalized singlet state in competition with fluorescence and that the O_2-quenching process (producing the triplet state) is associated with an isomerization yield A. We have also assumed that ϕ_T measures the yield of the O_2-quenching process (i.e. that ϕ_T is given by equation (II)), thus neglecting the small amount of ISC due to the O_2-independent mechanism. An alternative possibility is that—resembling the case of the retinal triplet state[2]—isomerization from $^1M^*$ occurs before thermalization, i.e. preferentially from higher vibronic states. In such a case expression (IV) will be replaced by:

$$\phi_{ISO}^{O_2} = \phi_{ISO}^d + A\phi_T \tag{V}$$

The validity of the above expressions can be tested by showing that A is the same for the two oxygen concentrations (2×10^{-3} M in air-saturated and 10^{-2} in O_2-saturated, *n*-hexane). Using the measured values of $\phi_{ISO}^{O_2}$, ϕ_T, τ^{O_2}, ϕ_{ISO}^d and τ^d (see previous text and Table 1), we obtain for ROH, from equations (III) and (IV), $k_{ISO} = 2\cdot5 \times 10^6$ s^{-1} and $A = 0\cdot082 \pm 0\cdot008$ for both $[O_2]$ values. Using equation (V), the A-values obtained are in the range $0\cdot07 \pm 0\cdot007$ for ROH and $0\cdot064 \pm 0\cdot006$ for RAc. The inaccuracy in A, due to the corresponding inaccuracies in $\phi_{ISO}^{O_2}$ and ϕ_T, do not permit a discrimination between the two (pre-thermalized and thermalized) mechanisms. We may conclude however that our experimental data are in quantitative agreement with the superposition of an O_2-independent process and an O_2-enhanced isomerization path of yield A. The details of the isomerization mechanism in deaerated systems are still unclear. Isomerization may occur directly from $^1M^*$ (thermalized or non-thermalized). Alternatively, it may proceed indirectly, via $^3M^*$. In such a case, since ϕ_{ISO} is not simply proportional to ϕ_T, our data would imply that triplet isomerization is a complicated process, depending on the (vibronic) mode of triplet preparation.

(ii) Photoisomerization and ionic-photodissociation in polar solvents

We have previously reported (Tables 1 and 2) the values ϕ_F^d(hexane) = $0\cdot026$, ϕ_F^d(methanol) = $0\cdot0072$, τ^d(hexane) = $4\cdot5$ ns and τ^d(methanol) = $1\cdot3$ ns obtained, correspondingly, for the fluorescence yields and lifetimes of retinol in the two deaerated solvents. Within the limits of our experimental accuracy the ratio $\phi_F^d/\tau^d = (5.65 \pm 0.10) \times 10^6 \text{ s}^{-1}$ is independent of the solvent polarity, indicating that the drop in the value of ϕ_F^d, when passing from n-hexane to methanol, is quantitatively accounted for by the shortening of the fluorescence lifetime due to a polarity-enhanced deactivation process (rate constant k_q) competing with the emission of fluorescence. Table 2 shows identical behaviour and conclusions in the case of RAc.

An examination of Table 2 shows that the shortening of the fluorescence lifetime with increased solvent polarity is accompanied by a parallel increase of the relative absorbance of R^+, suggesting the identification of k_q with the rate constant of an ionic dissociation process, such as:

$$^1RAc^* \xrightarrow{k_q} R^+ + A\bar{c}^- \tag{4}$$

Since the generation of R^+ occurs during the lifetime of the laser pulse, ionization cannot proceed via the (long-lived) triplet state, being thus consistent with reaction (4). The assumption that R^+ originates from the thermalized $^1M^*$ state in competition with fluorescence emission, rather than from non-relaxed (pre-thermalized) states of $^1M^*$, is in keeping with the constancy of the parameter ϕ_F^d/τ^d (see above). A more quantitative test, showing that ionic photodissociation is actually the process responsible for the shortening of the fluorescence lifetime in polar solvents, may be derived from the expressions:

$$1/\tau = k_d + k_q = 1/\tau(\text{hexane}) + k_q \tag{VI}$$

and

$$\phi_{R^+} = k_q\tau \tag{VII}$$

where it is assumed that the deactivation rate constant k_d (representing fluorescence emission, intersystem crossing, etc.) is solvent-independent. Equation (VII) leads to:

$$\phi_{R^+}^i/\phi_{R^+}^j = k_q^i\tau^i/k_q^j\tau^j \tag{VIII}$$

where the superscripts i and j denote two polar solvents such as butanol and methanol. An examination of Table 2, where k_q is calculated using equation (VI), shows that, assuming $\phi_{R^+}^i/\phi_{R^+}^j \simeq D_{R^+}^i/D_{R^+}^j$, equations (VI)–(VIII) are qualitatively in keeping with our experimental data. A better agreement is obtained when it is taken into account that the extinction coefficient of R^+ increases when passing from methanol to ethanol and butanol.[16] Unfortunately the extinction coefficients reported by Blatz and Pippert[16] in alcohol–sulphuric acid mixtures are both solvent and acid-dependent and cannot be applied in the present acid-free photochemical systems.

Attention should be finally paid to the marked solvent dependence of the RAc photoisomerization yields (ϕ_{ISO}) (Table 2) indicating an efficient polarity-enhanced isomerization path. Since D_{R^+} and ϕ_{ISO} in polar solvents are roughly

Table 3 Triplet (ϕ_T), fluorescence (ϕ_F), ionic photodissociation (ϕ_{ion}) and photoisomerization (ϕ_{ISO}^D, ϕ_{ISO}^T) yields of molecules related to the visual chromophore

Molecule R≡	Ref.	Isomer	Solvent	ϕ_T deaerated	ϕ_T aerated	ϕ_F (deaerated)	ϕ_{ion}	ϕ_{ISO}^D [g] deaerated	ϕ_{ISO}^D [g] aerated	ϕ_{ISO}^T [g] (biphenyl sensitized)
Retinal (RCHO)	2	11-cis	n-Hexane	0·6 + 0·1		c	c, d	0·13 ± 0·01[e]	0·13 ± 0·01[e]	0·13 ± 0·02
		All-trans	n-Hexane	0·7 ± 0·1		c	c, d	0·10 ± 0·01[e]	0·10 ± 0·01[e]	<0·02
n-butyl-amine Schiff base (RSB) RCH=NR'	3	11-cis	n-Hexane	c		c	c, d	(4 ± 1) × 10⁻³		0·45
		All-trans	n-Hexane	c		c	c, d	(2 ± 1) × 10⁻³		0·02–0·05
		9-cis	n-Hexane	c		c	c, d	—		0·06
		13-cis	n-Hexane	c		c	c, d	—		0·08
Retinol R—CH₂—OH	a	All-trans	n-Hexane	0·03	0·30	0·026	c	0·011	0·034	—
		All-trans	Methanol	0·03	0·035	0·0072	~0·1	0·15	0·14	—
Retinyl acetate R—CH₂—OCOCH₃	a	All-trans	n-Hexane	0·03	0·3	0·026	c	0·012	0·032	—
		All-trans	Methanol	0·03	0·03	0·004	~0·5	0·28	0·28	—
Retinyl-n-butyl-amine (RBA) R—CH₂—NHR'	a	All-trans	n-Hexane	0·03	0·3	0·026	c	0·01	0·034	—
		All-trans	Methanol	—	0·03	0·005	~0·08	0·092	0·0925	—
Retinal oxime R—CH=N—OH	b	11-cis	n-Hexane	(8 ± 2) × 10⁻³	0·07	—	c	—	0·017	0·2 ± 0·1
		11-cis	Methanol	—	0·02	—	f	—	0·05	—
		All-trans	n-Hexane	(9 ± 2) × 10⁻³	0·075	6 × 10⁻⁴	c	0·002	0·002	0·012 ± 0·006
		All-trans	Methanol	—	0·075	1 × 10⁻⁴	f	—	0·01	—

[a] This work.
[b] Unpublished data in this laboratory.
[c] Undetectable yields.
[d] Undetectable by absorbance or conductivity methods in methanol.
[e] Independent of solvent and polarity[17] and of oxygen content.
[f] A species absorbing around 610 nm (OD ≃ 0·01) is only observed in polar solutions of the oxime and is tentatively attributed to the RCH=N⁺ ion.
[g] ϕ_{ISO}^D and ϕ_{ISO}^T are correspondingly the direct-excitation and the triplet-sensitized photoisomerization yields. All values reported are for continuous excitation at 313 nm.

proportional to each other it may be concluded that ionic dissociation is accompanied by isomerization according to:

$$\text{RAc(all-}trans) \xrightarrow{h\nu} R^+ + Ac^- \rightarrow \text{RAc(11-}cis)$$

This is also in keeping with the fact that D_{R^+} and ϕ_{ISO} increase when passing (in the same solvent) from RBA to ROH and RAc (see above and Table 3). The ionic isomerization mechanism is plausible in view of the decrease in the multiplicity of essential double bonds which is most probably associated with the formation of R^+, with a structure represented by:

The high efficiency of the ionic photoisomerization in retinol and related molecules obviously calls for a careful consideration of possible charge-transfer interactions in rhodopsin,[21] which may play an important role in the isomerization of the pigment chromophore taking place in the early stages of the visual process.

(iii) Isomerization mechanisms in molecules related to the visual chromophore

Table 3 summarizes some presently available parameters associated with primary photoprocesses in molecules related to the visual chromophore. It is evident that *cis–trans* photoisomerization may take place according to the following mechanisms:

(a) A path involving the triplet state. The participation of the triplet in isomerization processes is unambiguously demonstrated by energy-transfer experiments in the cases of retinal, RSB and the oxime. Evidence consistent with triplet-sensitized isomerization was also presented for retinol using riboflavin as energy donor.[18a] However, a definite confirmation of this conclusion awaits the analysis of the complicating effects associated with the irreversible photochemical decomposition of riboflavin,[18b] probably via a radical mechanism. Special attention should be paid to the exceptionally high value of ϕ_{ISO}^T observed for the process 11-*cis*→all-*trans*, and to the relative inefficiency of the reverse *trans*→*cis* photoreaction.

(b) A mechanism involving ionic photodissociation. This process, characteristic of the 'fluorescent' molecules ROH, RAc, RBA and the oxime, is associated with the generation of the retinylic cation in polar media. Its high efficiency is demonstrated by the marked dependence of ϕ_{ISO}^D on the solvent polarity.

(c) An O_2-induced mechanism, due to oxygen fluorescence quenching. This process, which is not observed in the cases of the non-fluorescent retinal and RSB molecules, may involve the above mechanism (a), i.e. triplet states produced via O_2-quenching (rather than the previously mentioned triplet–triplet energy transfer). Alternatively, isomerization can occur within short-lived charge-transfer states (e.g. $^1(ROH^+.O_2^-)^*$) acting as intermediates in the oxygen-induced intersystem crossing.

Excited-singlet mechanisms, not involving ionic intermediates or triplet states are also plausible. Such processes may contribute to the direct-excitation isomerization yields observed in non-polar solutions of the fluorescent systems (ROH, RAc and RBA), as well as in the cases of retinal[2] and RSB.[3] However, their exact role in the direct-excitation isomerization of each molecule is still unclear.

It is interesting to note (see Table 3) that substantial photoisomerization (i.e. $\phi_{ISO} > 0.1$) is observed only in systems exhibiting either relatively high ISC yields, or substantial ionic photodissociation yields. Moreover, as we have previously pointed out, the (11-*cis*) → (all-*trans*) conversion (which is that associated with the primary step in vision) is exceptionally efficient in the triplet-sensitized paths (ϕ_{ISO}^T). As long as the present systems are considered as models reflecting the early stages in the photochemistry of rhodopsin ($\phi_{ISO} = 0.67$), such observations would call for the consideration of triplet states or of ionic species (or excited charge-transfer states) as photoactive intermediates in the stages preceding the formation of prelumi-rhodopsin (PLR).

REFERENCES

1. G. Wald, *Science*, **162**, 230 (1968).
2. T. Rosenfeld, A. Alchalel and M. Ottolenghi, *J. Phys. Chem.*, **78**, 336 (1974).
3. T. Rosenfeld, A. Alchalel and M. Ottolenghi, *Photochem. Photobiol.*, **20**, 121 (1974).
4. T. Rosenfeld, A. Alchalel and M. Ottolenghi, *Chem. Phys. Letters*, **20**, 291 (1973).
5. R. Hubbard, *J. Amer. Chem. Soc.*, **78**, 4662 (1956).
6. (a) D. Bounds, *Nature*, **216**, 1178 (1967); (b) G. Wald, *Nature*, **205**, 254 (1965).
7. M. Ofran and J. Feitelson, *Chem. Phys. Letters*, **19**, 425 (1973).
8. C. R. Goldschmidt, M. Ottolenghi and G. Stein, *Israel J. Chem.*, **8**, 29 (1970).
9. D. Zehavi and J. Rabani, *J. Phys. Chem.*, **75**, 1738 (1971).
10. A. J. Thomson, *J. Chem. Phys.*, **51**, 4106 (1969).
11. J. P. Dalle and B. Rosenberg, *Photochem. Photobiol.*, **12**, 151 (1970).
12. R. Bensasson and E. J. Land, *Trans. Faraday Soc.*, **67**, 1904 (1971).
13. R. Bensasson, E. J. Land and T. G. Truscott, *Photochem. Photobiol.*, **17**, 53 (1973).
14. R. Potashnik, C. R. Goldschmidt and M. Ottolenghi, *Chem. Phys. Letters*, **9**, 24 (1971).
15. T. Medinger and F. Wilkinson, *Trans. Faraday Soc.*, **61**, 620 (1965).
16. P. E. Blatz and D. L. Pippert, *J. Amer. Chem. Soc.*, **90**, 1296 (1968).
17. A. Kropf and R. Hubbard, *Photochem. Photobiol.*, **12**, 249 (1970).
18. (a) A. Gordon Walker and G. K. Radda, *Nature*, **215**, 1483 (1967); (b) N. Lasser, private communication.

19. (a) H. Tsubomura and R. S. Mulliken, *J. Amer. Chem. Soc.*, **82**, 5966 (1960); (b) H. Linschitz and L. Pekkarinen, *J. Amer. Chem. Soc.*, **82**, 2411 (1960).
20. T. Brewer, *J. Amer. Chem. Soc.*, **93**, 775 (1971).
21. (a) P. E. Blatz, *J. Gen. Physiol.*, **48**, 753 (1965); (b) F. J. Grady and D. C. Brog, *Biochem.*, **7**, 675 (1968); (c) C. M. Land, J. Harbour and A. V. Guzzo, *J. Phys. Chem.*, **75**, 2861 (1971); (d) R. Mendelson, *Nature*, **243**, 22 (1973).
22. A. M. Schaffer, T. Yamaoka and R. S. Becker, *Photochem. Photobiol.*, **21**, 297 (1975).
23. L. Salem, presented at 25th Intern. Congress Pure and Appl. Chem., Jerusalem, July 1975.

Note added in proof. The formation of the retinylic cation following the treatment of (ground state) retinol with strong acids proceeds via the addition of a proton to the hydroxyl oxygen, followed by a loss of water.[16] The present data, showing ionic photodissociation in alcoholic solvents which are very weak acids, are therefore indicative of a dramatic increase in the basicity of the hydroxyl group in the thermalized fluorescent state of the molecule. This observation is analogous to the very recently reported change, upon excitation, of the nitrogen basicity in the Schiff base of retinal.[22] A drastic excited state polarization as a general feature of retinyl polyenes (associated with pK changes when appropriate terminal functional groups are present) has been theoretically predicted by Salem and Bruckmann.[23]

E.3. A theoretical treatment of olefin cis–trans *photoisomerization*

S. J. Formosinho

The Chemical Laboratory, University of Coimbra, Portugal

SUMMARY

The tunnel-effect theory of radiationless transitions in large molecules is applied to the evaluation of the rates of *cis–trans* photoisomerization in stilbene and retinal. Good agreement is found with experimental data. The results are discussed with regard to the primary process of photobleaching in vision.

1 INTRODUCTION

The important reaction of *cis–trans* photoisomerization, one of the simplest types of rearrangement, provides a useful model for understanding the chemistry of excited molecules. Olson[1] suggested that such rearrangements might be understood in terms of the potential energy (p.e.) curves of the electronic excited states, an idea which was followed up by other workers.[2,3] Such a simple approach does not give good agreement with experiment, and although more sophisticated models have appeared,[4] these are conceptual rather than numerical. A tunnel-effect theory has been applied sucessfully to the qualitative and quantitative understanding of radiationless transitions in aromatic molecules,[5] and is now extended to treat *cis–trans* photoisomerizations. The theory is initially developed for the much studied isomerization of stilbenes[4,6] and is then applied to the photoisomerization of retinals and similar compounds relevant to the primary photoprocess in vision.[7]

2 THEORETICAL FORMULATION

For an isoenergetic transition, the rate of conversion between two electronic states is

$$k = v \exp \left\{ -(2\pi/h)[2\mu(D - Ev)]^{1/2} \, \Delta x/2\eta \right\} \tag{1}$$

where the symbols are as in Reference 5. With aromatic molecules, the radiation-less transitions are dominated by CH stretch vibrations and η is the relative number of hydrogen atoms.[5] However, when there are large changes in other bonds, and consequently strong relative displacements of the potential curves in different electronic states, other vibrations become important in the evaluation of rates. In *cis–trans* isomerizations, the p.e. of the molecule varies with the angle of rotation along a CC bond. If there is an appreciable energy barrier to rotation, the rate of isomerization (determined by the slowest radiationless process) is given by the rate of crossing from the p.e. curve of one isomer to the curve of the excited or ground state of the other isomer, and the effective vibration is the torsional mode.[8] If there is no energy barrier to rotation in an excited state, the molecule will wander about on the rotational p.e. curve, and can cross to any of the isomeric forms of a lower electronic state. Here the rate-determining step is radiationless conversion between the two states, which is controlled by CH stretching modes. Since the CH frequency is much higher than the frequency of rotation, the probability of the transitions can be determined for the CH modes at each angle of rotation (Figure 1).

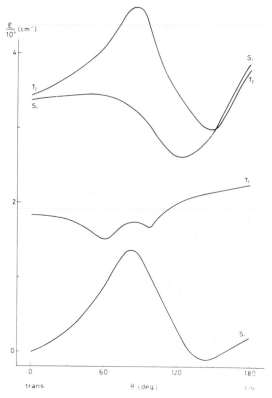

Figure 1 Potential energies of the electronic states of stilbene shown as a function of the angle of twist, θ

Since excited state *cis–trans* isomerization can only compete with other radiationless processes if there is no appreciable energy barrier to isomerization, which is not the case where the torsional mode is the effective vibration (i.e. v torsional is small, μ rotor is high), then rates will be controlled as previously[5] by the CH stretching modes. No displacement is assumed for these modes because no appreciable changes on CH bond lengths occur on electronic excitation.[9]

3 PHOTOISOMERIZATION OF STILBENE

trans-Stilbene fluoresces strongly ($\phi_F = 0.75$) at low temperatures (123 K), but the yield decreases on increasing the temperature with an apparent activation energy of 750–1000 cm^{-1}.[10,11] *cis*-Stilbene does not fluoresce appreciably even at low temperatures.[11] Both stilbenes isomerize on direct excitation; the yields of isomerization of the *trans* form are strongly temperature dependent ($\phi_{t \to c} = 0.002$ at 93 K, $\phi_{t \to c} = 0.5$ above 243 K), but the *cis* photoisomerization is much less so ($\phi_{c \to t} \simeq 0.35$ above 173 K, decreasing to 0.1 at 93 K).[2,11,12] The isomerization temperature dependence is independent of solvent polarity and follows closely the temperature variation of the fluorescence of *trans*-stilbene. The photoisomerization is thought to proceed via the lowest triplet state[13] (lifetime of the order of a few μs) in agreement with the work of Hammond *et al.*[6] on photosensitized isomerization of stilbenes. Thermal isomerization is also observed with an activation energy of 14,000 cm^{-1} and a pre-exponential factor of 10^{10} to 10^{12} s^{-1}.[14]

The p.e. curves of stilbene as a function of angle of phenyl group rotation around the CC double bond are presented for several electronic states (Figure 1). The curves are based on the calculation of Borrell and Greenwood[15] modified according to experiment. Radiationless rates between the various electronic states are determined from equation (1) by measuring Δx on the CH potential energy curves[5] at the electronic energy gaps of the *trans*, *cis* and perpendicular

Table 1 Non-radiative rate constants in stilbene

Transition	$k_{nr}(s^{-1})$			
	trans	perp		*cis*
$S_1 \to S_0$	2×10^6	5×10^7		7×10^5
$S_1 \to T_1$	2.5×10^5	7×10^7		1.5×10^5
$S_1 \to T_2$	4×10^7			5×10^8
$T_2 \to T_1$	1×10^9			2×10^9
$T_1 \to S_0$	3×10^2	3.5×10^5	6×10^5	10^2

forms (Figure 1). Δx is further corrected by the relative number of hydrogen atoms η. Values of $\nu_{CH}3000 \, \text{cm}^{-1}$ and $D35,000 \, \text{cm}^{-1}$ were used. Transitions occur mainly on the attractive side of the potential energy curves except for $S_1 \to S_0$ internal conversion (i.c.), where a $D \simeq 90,000 \, \text{cm}^{-1}$ was used. The spin forbidden factor for $S_1 \to T_2$ is 10^5 and for $S_1 \to T_1$ 10^{-6}.[5] For calculated rates, see Table 1.

From the estimated rates the isomerization route for direct excitation is a fast intersystem crossing (i.s.c.) $S_1 \to T_2$ followed by i.c. to T_1, with isomerization taking place in this state. Isomerization cannot occur in S_1 or T_2 because of the high energy barriers to rotation. The possibility of isomerization occurring as an integral part of the $S_1 \to S_0$ conversion is ruled out because it would be effected by the torsional mode, and as the most favourable CH modes give $S_1 \to S_0$ rates much smaller than other non-radiative rates from S_1, a fortiori, torsional modes provide even slower rates.

The temperature dependence of fluorescence and isomerization in trans-stilbene could be caused by a temperature dependence of the $S_1 \to T_1$ transition,[5] if no other triplet was located below S_1. However, such temperature dependences give activation energies ca 50–$100 \, \text{cm}^{-1}$, smaller than experimental values. A possibility is that T_2 is below S_1 in the trans form but rises above S_1 at some twisted configuration near the trans.[15] Although this could provide higher activation energies, it would give i.s.c. rates ca 10^8 even at low temperatures and therefore no strong increase in fluorescence should be observed. Therefore T_2 of trans-stilbene must be above S_1 as shown in Figure 1. The small activation energy in the isomerization of cis-stilbene indicates that T_2 in this configuration is almost isoenergetic with S_1, or even lower.

We have disregarded the crossing of the p.e. curve of S_0 in perpendicular configuration with the p.e. curve of T_1,[15] firstly because the activation energy for thermal isomerization is at most $14,000 \, \text{cm}^{-1}$ whilst the lowest energy of photosensitization of stilbene is $15,000 \, \text{cm}^{-1}$, secondly because crossing of the two curves should allow thermal isomerization via the triplet T_1 and therefore the pre-exponential factor should be low[16], which is not the case. The presence of a double minimum in the T_1 p.e. curve corresponds to Hammond's suggestion[17] of the existence of two stereoisomeric triplets of some polyene chain compounds and was introduced to explain the dependence of the photostationary ratio mixtures of isomerization on the energy of the triplet photosensitizers. Once the molecule is in T_1, the stereoisomeric forms should be populated almost equally and the rate constants from the two forms will be near the photostationary mixtures at high energies of sensitization or via direct excitation. The position of the minimum was adjusted to the experimental values $[cis]/[trans] = 1.7$.

The yield of conversion $\phi_{t \to c} = 0.5$ at room temperature allows estimation of the i.s.c. yield in trans-stilbene, ϕ_T^{trans}, because

$$\phi_{t \to c} = \phi_T^{trans} \frac{k_{cis}}{k_{cis} + k_{trans}}$$

where k_{cis} and k_{trans} are the rates of conversion of T_1 to the ground state cis- and trans-stilbene. A value $\phi_T^{trans} = 0.79$ gives a radiative rate of 8×10^6 s^{-1}, which, although low, agrees with the long radiative lifetimes (ca 10^{-7} s) of the polyene molecules.[18-20] With this rate the i.s.c. yield at low temperatures, where T_2 is not populated, is 0.024 and the isomerization yield drops to 0.015. The fluorescence yield decreases from 0.78 at low temperatures to 0.15 at room temperature. In contrast the fluorescence yield from cis-stilbene is 0.016 and the isomerization yield is 0.37. These calculated results agree with experiment. No account was taken of change of radiative rate with temperature, as, in contrast to retinol,[20] there is no evidence for this. Excited singlet lifetimes at room temperature are 2 ns for cis-stilbene and 25 ns for the trans isomer. The trans singlet lifetime of 1.5-3.3 ns of Shorygin and Ivanova[10] from fluorescence intensity measurements seems too short, but this derives from assumed values for the radiative rate constant. The lifetime of the twisted triplet T_1 is ca 2-3 μs.

In agreement with experiment,[21] deuteration is not expected to change isomerization yields much, as the $S_1 \rightarrow T_2$ barrier widths are quite small and any change in the decay of T_1 (ca 2-3 times longer) will not affect the ratio of decay into cis and trans ground states.

Triplet–triplet energy transfer is known to sensitize isomerization of olefins,[6] with a cis–trans stationary ratio varying with sensitizer energy. A simple scheme

$$^3X + trans \xrightarrow{k_1} \, ^1X + T_1$$

$$^3X + cis \xrightarrow{k_2} \, ^1X + T_1$$

$$T_1 \xrightarrow{k_{trans}} S_0^{trans}$$

$$T_1 \xrightarrow{k_{cis}} S_0^{cis}$$

gives a stationary state ratio of

$$\frac{[cis]}{[trans]} = \frac{k_1}{k_2} \cdot \frac{k_{cis}}{k_{trans}} \tag{2}$$

^3X and ^1X are the triplet and ground state of the sensitizer. The energy transfer can go to the trans- or cis-T_1, where it reverts quickly to the twisted forms, or the twisted triplet forms can be directly sensitized through a slower transfer process involving rotation in the T_1 state.[6] With sensitizer energies below cis-T_1, vertical energy transfer is mainly to the trans form and favours cis-stilbene in the stationary state. For energies above the cisoid triplet both forms are assumed (to be) equally populated, but between this energy and 15,000 cm^{-1} only the transoid triplet can be sensitized (see Figure 1) and all the cis-stilbene is converted into trans. The stationary state ratio was calculated from equation (2) assuming diffusion-controlled vertical energy transfer for triplet energies 1000 cm^{-1} above T_1 and a Boltzmann distribution dependence for lower energies. The maximum non-vertical energy transfer rates were taken as ten times slower than the diffusion-controlled rate and other values follow a Boltzmann distribution. The estimated ratio (Figure 2) follows the experimental points.[6] Some quinone sensitizers are off the curve, but this could be because sensitization occurs via a chemical reaction.

The non-stationary ratio for each triplet sensitizer should depend slightly on temperature because the transoid triplet is of lower energy than the cisoid, and it is not surprising that photostationary mixtures become slightly more *cis*-rich with increasing temperature.[6]

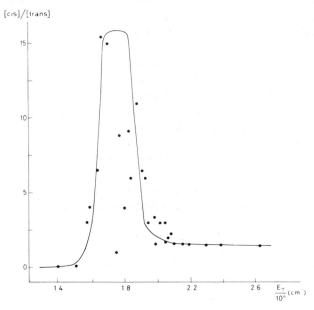

Figure 2 Stationary states in the photosensitized isomerization of the stilbenes in function of the triplet energy, E_T

4 PHOTOISOMERIZATION OF RETINAL AND RETINOL

Rhodopsin is a polyene aldehyde, 11-*cis*-retinal, attached to a protein opsin. Retinal has five carbon–carbon double bonds and a carbon–oxygen double bond in the same conjugated system. When retinal is bound in the native state, the carbon–oxygen bond is replaced by the carbon–nitrogen bond of a Schiff's base linkage.[22] The events leading to vision are initiated by the photoisomerization of 11-*cis*-retinal to the all-*trans* polyene. Since retinal, or the reduced form retinol are the chromophores of visual pigments, there is considerable interest in their photochemistry.[23,24] To improve our understanding of the photochemistry of retinal and related compounds, tunnel-effect theory is applied to the calculation of their isomerization rates.

Retinals start to absorb around 450 nm with a maximum close to 360 nm (ε *ca* $3 \times 10^4 \, M^{-1} \, cm^{-1}$). The spectra are sensitive to temperature and concentration.[25,26] Retinols show similar spectral features but are slightly shifted to shorter wavelengths. All-*trans*-retinal fluoresces at 77 K ($\phi_F = 0.7$) on excitation at 440 nm,[27] but the yield decreases on decreasing the excitation wavelength or on changing the concentration, e.g. at 77 K and low concentrations (10^{-6} mol. 1^{-1}), Moore and Song[26] found $\phi_F = 0.02$. The yield decreases on increasing the temperature. The 11-*cis*-retinal has a very small temperature independent fluorescence ($\phi_F = 0.001$).[27] For all-*trans*-retinol in non-polar solvents ϕ_F is 0.02 at room temperature and *ca* 0.4 at 77 K; the fluorescence yield of 11-*cis*-retinol (0.006) is temperature independent.[19] The yields of both compounds decrease by a factor of 3 in methanol.

The singlet lifetime of *trans*-retinol at room temperature is 4.4 ns[28] and the radiative lifetimes are about 300 ns. This long rate, typical of polyenic structures, may result from an electronic forbidden transition,[29,30] although this does not explain the change of radiative lifetime of all-*trans*-retinol from 50 ns at 100 K to 250–300 ns at 300 K.[20] Similar variations have been found in 11-*cis*-retinal.[31] The common interpretation is that excited state geometry changes caused by thermal population of torsional vibrations will lead to less favourable Franck–Condon factors for the radiative transitions. [19,20,31] The radiationless processes of these states are also temperature dependent.[20]

Dawson and Abrahamson[32] determined an i.s.c. yield for *trans*-retinal in a non-polar solvent of 0.11. More recently Bensasson *et al.*[33] determined $\phi_{i.s.c.}$ 0.6 ± 0.1 of the same order as the yield (0.5–0.67) of photobleaching in rhodopsin.[23,34] The triplets of retinals have been observed on direct flash excitation with a first-order decay rate $9.6 \times 10^4 \text{ s}^{-1}$.[32,35] In contrast, retinol triplet state can only be observed on flash photolysis in the presence of a sensitizer; the decay rate is $6 \times 10^4 \text{ s}^{-1}$.[36] Triplet–triplet absorption spectra in retinals and retinols have been attributed to the twisted form of the lowest triplet state. Abrahamson[23] attributed the difference between retinal and retinol to the lack of formation of the triplet state retinol directly from S_1. This has been confirmed by Ottolenghi and coworkers[28] who found for *trans*-retinol $\phi_{i.s.c.} = 6 \pm 3 \times 10^{-2}$. To explain the formation of T_1 in retinal Abrahamson[23] suggested the presence of a $^1(n, \pi^*)$ state below $^1(\pi, \pi^*)$ providing a pathway for i.s.c.; in retinol the (n, π^*) states are much higher in energy and T_1 is not formed directly from S_1. Although $^1(n, \pi^*)$ is calculated to be slightly lower than the $^1(\pi, \pi^*)$, Becker *et al.*[27] favour $^1(\pi, \pi^*)$ as being the lowest excited singlet state in *trans*-retinal. In retinal and retinol the lowest triplet state is unquestionably a (π, π^*) state.[23]

Kropf and Hubbard[37] have studied the quantum efficiencies of isomerization in retinals. In hexane the yield for all-*trans*-retinal is 0.2 at 298 K and 0.005 at 208 K; the apparent activation energy is 1700 cm^{-1}. In ethanol the yield is 0.1 and has a much smaller activation energy (150 cm^{-1}). The yields depend slightly on excitation wavelength. The yield of isomerization from the 11-*cis*-retinal to all-*trans* is 0.2 at 298 K and 0.6 at 208 K in hexane. In ethanol $\phi_{c \rightarrow t}$ is considerably lower, with yields also depending on excitation wavelength. Similarly retinols undergo direct and sensitized isomerization,[38] but no yields have been reported.

Pullman and Pullman[39] calculated localization energies for isomerization in retinals, although later workers[27,40,41] have preferred calculating torsional potential energy curves for retinol ground and excited configurations. In the ground state there is a strong barrier to rotation with a maximum displaced towards the *cis*-retinol to reproduce the $S_1 \rightarrow cis/S_1 \rightarrow trans$ decay ratio. The S_0 potential curve crosses the energy curve of T_1, in agreement with the results of Hubbard[38] for the thermal isomerization of retinal which gave $\Delta H^\dagger$ of 8000 cm^{-1} and $\Delta S^\dagger$ of -21 to -10 e.u. The large negative value suggests a thermal isomerization via a triplet state.[16] The thermal isomerization is catalysed by iodine with $\Delta S^\dagger + 0.7$ e.u. suggested to be caused by an enhanced i.s.c.[23] The p.e. curve has no barrier to rotation in agreement with the isomerization by triplet sensitizers. The twisted configuration corresponds probably to one or two minima of energy because retinol triplet is formed with diffusion controlled rate using sensitizers of energy above the lowest triplet of anthracene, but with much slower rates with other sensitizers (e.g. pentacene).[36] This has been attributed to a non-vertical energy transfer process.[6]

The potential energy curve of the lowest $^1(\pi, \pi^*)$ state has been calculated to have a rotational energy barrier $ca\ 7000 \text{ cm}^{-1}$.[27] However, the rotational barrier in retinols should not be very different from the ones in carotenoids, where isomerization is known to occur via direct excitation or via photosensitization.[42] Since the triplet states of these molecules are not formed directly from S_1 it appears that isomerization can occur from the lowest excited singlet. In our model this implies that there are no appreciable energy barriers to rotation in the lowest $^1(\pi, \pi^*)$ state. The potential curves for several electronic states of retinol and retinal (including the (n, π^*) states) presented in Figure 3 are based on theoretical calculations[27,41] modified according to the experimental evidence previously discussed. A triplet (n, π^*) state with a singlet–triplet splitting of 2000 cm^{-1} is also included. Higher triplet (π, π^*) states calculated to be almost isoenergetic with $^1(\pi, \pi^*)$ will be shown to be higher than the lowest excited singlets. Nevertheless all the p.e. curves should be considered as qualitative rather than quantitative.

For retinol there are no low lying (n, π^*) states, thus the only relevant states are the (π, π^*) levels of Figure 3. Rates of radiationless conversion in retinol at several angles of rotation were estimated as for stilbene with $\eta = 0.5$ (Figure 4). At low temperatures the molecule has insufficient energy to overcome the small potential energy barrier of rotation in $^1(\pi, \pi^*)$ and stays in the *trans* form. With a radiative rate constant of $2 \times 10^7 \text{ s}^{-1}$ [20] we can estimate a fluorescence yield of 0.5, an i.c. yield of 0.5 and a yield of triplet formation of 0.004. At this low temperature no direct isomerization should occur in retinol because the molecule is trapped in one of the isomeric forms and the yield of triplet formation is too low to provide any alternative route. Similar considerations apply to the *cis* form. At room temperature the molecule contains enough energy to overcome the small energy barrier to rotation giving all possible twisted configurations between *trans* and *cis*. From the change of the non-radiative and radiative rates[20] with the angle of rotation we can calculate quantum yields of the various physical processes. Assuming all configurations equally probable, integration along the angle of rotation for the quantum yields gives $\phi_T = 0.06$, $\phi_F = 0.05$

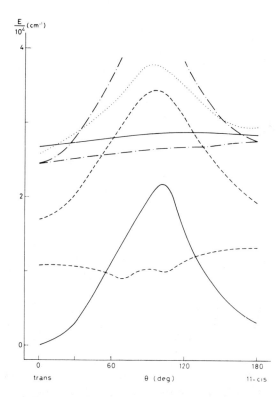

Figure 3 Potential energies of the electronic states of retinal shown as a function of the angle of twist, θ, around the 11–12 double bond $(-\,^1(\pi, \pi^*);$ $---\,^3(\pi, \pi^*); \ldots\,^1(n, \pi^*); -\cdot-\cdot^3(n, \pi^*))$

and $\phi_{S_1 \to S_0} = 0.9$; the calculated singlet lifetime is 7 ns. All these results are in reasonable agreement with experimental data in nonpolar solvents. However the presence of any $^3(\pi, \pi^*)$ close to $^1(\pi, \pi^*)$ would favour i.s.c., contrary to experimental observations.

At room temperature the isomerization yield should be high, depending on the ratio of decay of $^1(\pi, \pi^*)$ to the *cis* and *trans* isomers. The angle of rotation for maximum energy of S_0 was adjusted to give a ratio of $k_{cis}/k_{trans} = 2/3$. With this ratio $\phi_{t \to c} = 0.4$ and $\phi_{c \to t} = 0.6$.

Retinal is more complex because in addition to the (π, π^*) states there are singlet and triplet (n, π^*) levels lying near the $^1(\pi, \pi^*)$ (Figure 3). To explain the solvent effects on isomerization yields we favour the proposition[23] that $^1(n, \pi^*)$ is the lowest singlet in all-*trans*-retinal; in 11-*cis*-retinal the lowest excited singlet is (π, π^*). The radiationless rates between the various electronic states were estimated as before with a spin forbidden factor of 10^{-4} between the $^1(\pi, \pi^*)$ and $^3(n, \pi^*)$ and 10^{-6} between $^1(n, \pi^*)$ and $^3(n, \pi^*)$.[43] Figure 4 shows the variation of non-radiative rates from (π, π^*) singlet as a function of the angle of twist. With a

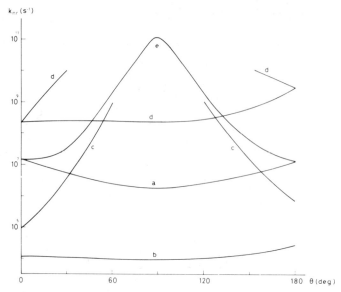

Figure 4 Radiative and non-radiative rate constants of the lowest
$^1(\pi, \pi^*)$ state in retinal as a function of the angle of twist, θ (a) radiative
transition; (b) $^3(\pi, \pi^*)$ (T$_1$); (c) $^3(\pi, \pi^*)$ (T$_2$); (d) $^3(n, \pi^*)$; (e) internal
conversion to S$_0$

radiative rate constant 10^6 s^{-1} for $^1(n, \pi^*)$ we can estimate at low temperatures
$\phi_F = 0.03$, $\phi_T = 0.12$ and $\phi_{t \to c} = 0.04$ for *trans*-retinal.

In *trans*-retinal at high temperatures, since the $^1(n, \pi^*)$ state has a high barrier
to rotation the $^1(\pi, \pi^*)$ state will be appreciably populated. An important process
from this state is i.s.c. to $^3(n, \pi^*)$ for which there are no estimations of the p.e.
curve as a function of the angle of rotation. If the potential curve has no appreciable barrier to rotation as in $^1(\pi, \pi^*)$ state, the calculated yields are $\phi_T = 0.58$,
$\phi_F = 0.01$ and the i.c. yield is 0.4. However, if the $^3(n, \pi^*)$ state has a high energy
barrier of rotation, the triplet yield drops to 0.35 and $\phi_{i.c.}$ increases to 0.65; the
fluorescence yield is 0.003. Since isomerization occurs on the singlet $^1(\pi, \pi^*)$ and
the lowest triplet $^3(\pi, \pi^*)$ manifolds after internal conversion from $^3(n, \pi^*)$, the
isomerization yield is the same ($\phi_{t \to c} = 0.4$) in both cases. The first hypothesis
agrees with the triplet yield (0.6) of Bensasson *et al.*[33] but gives too high a value
for ϕ_F. The 0.11 yield of Dawson and Abrahamson seems to be too low.

In 11-*cis*-retinal the lowest singlet state is probably $^1(\pi, \pi^*)$. At low temperatures $\phi_T \simeq 1.0$, $\phi_F = 0.004$ and all the isomerization goes via the lowest triplet
state with $\phi_{c \to t} = 0.6$. On raising the temperature the molecule will travel
along the p.e. curve of $^1(\pi, \pi^*)$ with no appreciable change in $\phi_F(0.003)$ if the
potential curve of the $^3(n, \pi^*)$ presents an energy barrier to rotation. However,
any thermal population of the close $^1(n, \pi^*)$ state which provides a route for
internal conversion ($\phi = 0.88$) to *cis* ground state, could decrease the yield of
isomerization. Experimentally the yield of isomerization decreases to 0.2 at
25 °C.[37]

In polar solvents the (n, π^*) states rise and the (π, π^*) decreases in energy with a relative displacement of 1000–1500 cm^{-1}. An increase in polarity for *trans*-retinal will make the $^1(n, \pi^*)$ and $^1(\pi, \pi^*)$ states almost isoenergetic and no increase on the isomerization yield ($\phi_{t \to c} = 0.4$) is observed on raising temperature. With 11-*cis*-retinal, $^3(n, \pi^*)$ could rise above $^1(\pi, \pi^*)$ causing a decrease in isomerization yield at low temperatures as seems to be observed.[37] The concentration effects on the fluorescence of *trans*-retinal can be understood also in terms of the relative positions of the electronic excited states. Moore and Song[26] proposed that *trans*-retinal dimerizes at concentration $ca\ 10^{-3}$ mol l^{-1}. Dimerization will decrease the energy of the (π, π^*) states and will increase the energy of the (n, π^*) levels because association of the molecules involves the non-bonding electrons of the oxygen in the carbonyl group. Furthermore, dimerization hinders rotation even in the $^1(\pi, \pi^*)$ state. Therefore if the (n, π^*) states are above $^1(\pi, \pi^*)$ and this presents some barrier to rotation, the fluorescence yield, even at room temperature, should rise to 0.5. An increase of the wavelength of excitation might decrease ϕ_F if the energy of the $^3(n, \pi^*)$ state is attained.

5 DISCUSSION

Taking the photoisomerization of 11-*cis*-retinal to all-*trans*-retinal or similar compounds as the chemical model of the primary photoprocess in vision the present theoretical model leads to the question of the nature of the Schiff's base-protein in rhodopsin. If rhodopsin *in vivo* is a protonated Schiff's base its photochemical behaviour should be similar to retinol, and a very efficient isomerization is expected to occur at body temperature via S_1 with a yield $\phi_{c \to t} = 0.6$ matching the photobleaching yield;[23,24] the time response associated with the isomerization process will be a few ns. If rhodopsin is in an unprotonated form photoisomerization may occur in both singlet and triplet states.[23] In this event two time responses will be expected associated with the primary photoisomerization; one of the order of ns and the other in the region of few μs. The thermal population of the lowest $^1(n, \pi^*)$ causes a drop in isomerization yield to 0.2, a value smaller than the yield of photobleaching of rhodopsin. Although this might favour rhodopsin as a protonated Schiff's base, it must be pointed out that if in the native rhodopsin the $^1(n, \pi^*)$ state is slightly shifted (~ 200 cm^{-1}) to higher energies no appreciable population of the (n, π^*) singlet can be attained at room temperature. Recent Raman studies of the photoreceptor-like pigment of halobacteria halobium[44] reveal this to be an unprotonated Schiff's base, but no extrapolation can be made for rhodopsin. The nature of the Schiff's base in rhodopsin and the yield of triplet formation seem not to be determining factors of the photobleaching in vision. However, the photoconductivity theory for vision,[45] which requires a high population of a triplet state, can only be compatible with rhodopsin in an unprotonated form.

ACKNOWLEDGEMENTS

I am grateful to Dr H. D. Burrows for many helpful discussions and suggestions.

The financial support of SMN is gratefully acknowledged. This work is included in the research project CQ-2 of the CEQNR (Chemical Laboratory, University of Coimbra).

REFERENCES

1. A. R. Olson, *Trans. Faraday Soc.*, **27**, 69 (1931).
2. G. N. Lewis, T. T. Magel and D. Lipkin, *J. Amer. Chem. Soc.*, **62**, 2973 (1940).
3. J. L. Magee, W. Shand Jr. and H. Eyring, *J. Amer. Chem. Soc.*, **63**, 677 (1941).
4. W. M. Gelbart and S. A. Rice, *J. Chem. Phys.*, **50**, 4775 (1969).
5. S. J. Formosinho, *J.C.S. Faraday II*, **70**, 605 (1974).
6. G. S. Hammond, J. Saltiel, A. A. Lamola, N. J. Turro, J. S. Bradshow, D. O. Cowan, R. C. Counsell, V. Vogt and C. Dalton, *J. Amer. Chem. Soc.*, **86**, 3197 (1964).
7. R. Hubbard and A. Kropf, *Proc. Natl. Acad. Sci. U.S.*, **44**, 130 (1958).
8. S. J. Formosinho and J. D. da Silva, *Mol. Photochem.*, **6**, 409 (1974).
9. E. F. McCoy and I. G. Ross, *Austral. J. Chem.*, **15**, 573 (1962); G. Herzberg, *Molecular Spectra and Molecular Structure*, Vol. III, pp. 612, 634, D. Van Nostrand, New York, 1966.
10. P. P. Shorygin and T. M. Ivanova, *Sov. Phys. Dokl.*, **3**, 764 (1958).
11. R. H. Dyck and D. S. McClure, *J. Chem. Phys.*, **36**, 2326 (1962).
12. S. Malkin and E. Fischer, *J. Phys. Chem.*, **66**, 2482 (1962); S. Malkin and E. Fischer, *J. Phys. Chem.*, **68**, 1153 (1964); E. Fischer, *Fortschr. Chem. Forsch.*, **7**, 605 (1967); D. Gegiou, K. A. Muskat and E. Fischer, *J. Amer. Chem. Soc.*, **90**, 12 (1968).
13. T. Förster, *Z. Electrochem.*, **56**, 716 (1952).
14. R. B. Cundall, *Prog. Reaction Kinetics*, **2**, 165 (1964).
15. P. Borrell and H. H. Greenwood, *Proc. Roy. Soc.*, A**298**, 453 (1967).
16. R. S. Mulliken, *Phys. Rev.*, **41**, 751 (1932).
17. W. G. Herkstroeter, J. Saltiel and G. S. Hammond, *J. Amer. Chem. Soc.*, **85**, 482 (1963).
18. J. B. Birks and D. J. Dyson, *Proc. Roy. Soc.*, A**275**, 135 (1963).
19. A. J. Thomson, *J. Chem. Phys.*, **51**, 4106 (1969).
20. J. P. Dalle and B. Rosenberg, *Photochem. Photobiol.*, **12**, 151 (1970).
21. K. A. Muskat, D. Gegiou and E. Fischer, *J. Amer. Chem. Soc.*, **89**, 4814 (1967).
22. D. Bownds, *Nature*, **216**, 1178 (1967); M. Akhtar, *Chem. Comm.*, 631 (1967); R. P. Poincelot, P. G. Millar, R. L. Kimbel Jr. and E. W. Abrahamson, *Nature*, **221**, 256 (1969).
23. E. W. Abrahamson and S. E. Ostroy, *Prog. Biophys. Mol. Biol.*, **17**, 179 (1967).
24. E. W. Abrahamson and S. M. Japar, *Handbook of Sensory Physiology*, Vol. II (ed. H. J. A. Dartnall), Springer, Berlin, 1972.
25. L. Jurkowitz, *Nature*, **184**, 614 (1959); J. N. Loeb, P. K. Brown and G. Wald, *Nature*, **184**, 617 (1959).
26. T. A. Moore and P. S. Song, *Nature*, **243**, 30 (1973).
27. R. S. Becker, K. Inuzuka, J. King and D. E. Balke, *J. Amer. Chem. Soc.*, **93**, 43 (1971).
28. T. Rosenfeld, A. Alchalar and M. Ottolenghi, *Chem. Phys. Letters*, **20**, 291 (1973).
29. B. S. Hudson and B. E. Kohler, *Chem. Phys. Letters*, **14**, 299 (1972); *J. Chem. Phys.*, **59**, 4984 (1973).
30. See however T. A. Moore and P. S. Song, *Chem. Phys. Letters*, **19**, 128 (1972).
31. W. Sperling and C. N. Rafferty, *Nature*, **224**, 591 (1969).

32. W. Dawson and E. W. Abrahamson, *J. Phys. Chem.*, **66**, 2542 (1962).
33. R. Bensasson, E. J. Land and T. G. Truscott, *Photochem. Photobiol.*, **17**, 53 (1973); a more recent value *ca* 0·4 was reported by Land and coworkers in paper E1 at this Conference.
34. H. J. A. Dartnall, *Vision Research*, **8**, 339 (1963).
35. K. H. Grellmann, R. Menning and R. Livingston, *J. Amer. Chem. Soc.*, **84**, 546 (1962).
36. A. Sykes and T. G. Truscott, *Trans. Faraday Soc.*, **67**, 679 (1971).
37. A. Kropf and R. Hubbard, *Photochem. Photobiol.*, **12**, 249 (1970).
38. R. Hubbard, *J. Amer. Chem. Soc.*, **78**, 4662 (1956).
39. A. Pullman and B. Pullman, *Proc. Natl. Acad. Sci. U.S.*, **47**, 7 (1961).
40. K. Inuzuka and R. S. Becker, *Nature*, **219**, 383 (1968).
41. J. R. Wiesenfeld and E. W. Abrahamson, *Photochem. Photobiol.*, **8**, 487 (1968).
42. L. Zechmeister, *Cis-Trans Isomeric Carotenoids, Vitamins A and Arylpolyenes*, Springer, Vienna, 1962.
43. M. El-Sayed, *J. Chem. Phys.*, **38**, 2834 (1963).
44. E. Mendelsohn, *Nature*, **243**, 22 (1973).
45. B. Rosenberg, *Adv. Radiation Biol.*, **2**, 193 (1966).

E.4. Pararhodopsin: an 'excited state' of the visual pigment molecule?

Aubrey Knowles

MRC Vision Unit, University of Sussex, Falmer, Brighton, BN1 9QG, Sussex, U.K.

SUMMARY

The visual pigment molecule of most species consists of a conjugated polyene, 11-*cis*-retinal, linked to a protein that is at least 100 times its molecular weight. There is considerable electronic interaction between the two parts of the molecule for the absorption maximum of the retinal moiety is red-shifted by about 100 nm on combining with the protein. Electronic excitation of the pigment molecule causes *cis–trans* isomerization of the retinal and the resulting deformation of the polyene chain upsets the stability of the whole molecule, the electronic excitation being transformed into a steric strain which eventually leads to the disruption and loss of the visible absorption band of the molecule. The dark reactions following the isomerization process are remarkably slow and changes can be detected several hours after the initial excitation. Thus the presence of the protein adds a new phase in the photochemistry of visual pigments that is not normally encountered with simpler molecules.

Several slowly-decaying transients can be observed during the dark reaction, and this paper will describe studies on one of the longest-lived, '*para*rhodopsin'. About 30 per cent of the pigment molecules that are excited eventually pass through this state, which is remarkable not only for its longevity—the half-life ranging from 4·7 min at 37 °C to 3 days at 11 °C—but also for the fact that its absorption maximum is near that of the parent visual pigment. This suggests that despite the isomerization of the chromophoric group and the disruption of the molecule, some of its original structure is maintained. The significance of *para*rhodopsin in the visual excitation process will be discussed.

1 DISCUSSION

The other contributions to this Conference might leave the impression that retinal is the only part of the rhodopsin molecule that has any bearing on its photochemistry, but it should be stressed that the visual pigment has many interesting and unique properties not shown by isolated retinal, nor with any of the model compounds so far prepared. In the pigment molecule, retinal is closely associated with 'opsin', which is basically a protein of molecular weight about 40,000 and a similar mass of phospholipid. Although the mass of the protein is 140 times that of the retinal, the protein is spheroidal with a diameter of about 40 Å while the retinal is roughly linear with a chain length of about 20 Å; if the retinal is pictured fitting into a specially-shaped cleft in the protein, a very large proportion of the protein amino-acid groups could lie within range of interaction with the retinal. This multiple interaction has a profound effect upon the electronic structure of the retinal and leads to a major red-shift of the long-wavelength absorption band of the retinal, displacing it by 60 to 190 nm in the rhodopsins from different species.

Similarly, the photochemistry of the rhodopsin molecule is quite unlike that of free retinal. Electronic excitation causes isomerization of the chromophoric group; the resulting change in shape makes it break out of its niche in the opsin, the electronic interaction is lost, and so the visible absorption band is also lost. This process of detachment is slow, and even at 37 °C has a half-life of the order of minutes. During this period the molecule passes through a series of discrete states, each with a characteristic absorption spectrum. While passing through one of these intermediates, displacement of charge takes place in the receptor cell as the first step in vision.

Thus the electronic excitation energy put into the molecule is converted into a potential energy due to the strain of the isomerized chromophoric group, and this is progressively released as the molecule relaxes into its final dissociated colourless form. These intermediates have been characterized both in the retina and in extracted rhodopsin and I have been studying one of the longest-lived of them, called 'pararhodopsin'.

Lythgoe[1] in 1937 noted that illumination of rhodopsin cooled on ice gave a transient orange-coloured intermediate, but it was not until 1963 that a species of $\lambda_{max} = 465$ nm was recognized as a component of Lythgoe's orange solution.[2] This was thought to be a minor artifact, but Ostroy, Erhardt and Abrahamson[3] later concluded that it should be placed in the main decay sequence of photo-excited rhodopsin. The species has been recognized in the retinas of rabbit, frog, rat and man, and in rhodopsin extracts from frog and ox. It is difficult to obtain an accurate absorption spectrum for pararhodopsin because it is usually found with other species that have overlapping spectra, namely, unbleached rhodopsin, isorhodopsin, metarhodopsin II—the preceding intermediate—and the products of pararhodopsin decay. It has generally been assumed that the extinction coefficient of the pararhodopsin peak is the same as that of rhodopsin and there is some uncertainty about the actual λ_{max}, and so I will describe an attempt to determine these two quantities for extracted bovine rhodopsin.

Figure 1 shows the absorption spectrum of bovine rhodopsin in digitonin at neutral pH. This was illuminated with a photographic flash unit ($t_{0.5} = 2$ ms) through an orange filter (transmitting less than 1 per cent below 510 nm) and

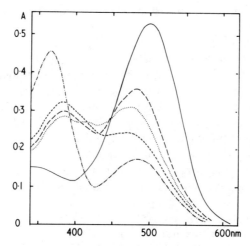

Figure 1 Absorption spectrum of bovine rhodopsin in digitonin and phosphate buffer at pH 7, 25 °C: (i) —— before flash, (ii) – – – 30–90 s after flash, (iii) .... 1 hr after flash, (iv) – – – – 6 hr after flash and (v) – · – after addition of hydroxylamine. Flash of wavelengths > 510 nm, half-life 2ms

the spectrum recorded between 30 and 90 s after the flash. The intensity of the flash is great enough for all of the molecules to absorb at least one quantum, and some many more. Some of the molecules absorbing second quanta will undergo photoreversal to give rhodopsin (λ_{max} = 500 nm) or *iso*rhodopsin (λ_{max} = 486 nm). Thus spectrum (ii) represents a mixture of several species, principally rhodopsin, *iso*rhodopsin, *meta*rhodopsin II (λ_{max} = 380 nm) and *para*rhodopsin. During the following six hours, the absorbance at 470 nm falls steadily as *para*rhodopsin decays, while that at 380 nm at first falls as *meta*rhodopsin II decays, and then rises as the final product accumulates. At the end of this dark period, hydroxylamine was added to the solution. This reacts with any remaining transients and with the final product to give retinylidene oxime (λ_{max} = 366 nm) but does not affect the rhodopsin and *iso*rhodopsin remaining in the solution.

The processes can be followed by monitoring the extract at three wavelengths after it is flashed (Figure 2). The deviations from the simple first-order growth and decay reactions suggest that there are other minor reactions taking place at the same time during the first hour, but after this, it can be assumed that the only change occurring is the decay of *para*rhodopsin to the final products. Figure 3 shows the difference spectra for the extract over the period 1–6 hr, the positive part being due to the loss of *para*rhodopsin and the negative part due to the accumulation of the products. An absolute spectrum for *para*rhodopsin can be obtained from this by the addition of the spectrum of the products, since the latter can be calculated from the difference spectrum for the addition of

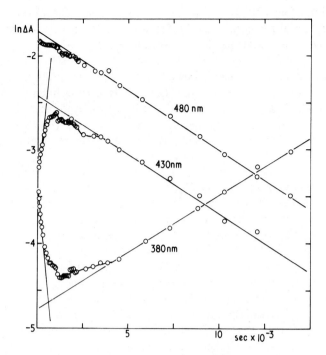

Figure 2 First-order plot of absorbance changes after flashing rhodopsin, measured at 480, 430 and 380 nm. Experimental conditions were the same as Figure 1

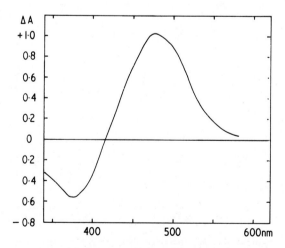

Figure 3 Difference spectrum between spectra at 1 hr and 6 hr from Figure 1. Loss of absorbance shown as positive difference

hydroxylamine. The final products have considerable absorption at 470 nm and, despite careful pH control, the spectrum of the products has been found to vary between experiments. Many workers describe the final product of bleaching extracted rhodopsin as 'free retinal', but the end product in these reactions at neutral pH appears to be a mixture of compounds of retinal with opsin which varies slightly in composition between experiments. The actual conditions of illumination affect the end product, for if a tungsten light source is used instead of the flash unit, the same reaction sequence is seen, and the *para*rhodopsin behaves in the same way, but the final product spectrum is slightly different.

Kinetic measurements of the decay of *para*rhodopsin in the retina are, in fact, slightly easier, for enzyme systems are present that reduce the decay product to retinol, and this does not absorb above 400 nm. This extra reaction does not affect the decay process of *para*rhodopsin and my data for cattle rhodopsin solutions agree with those of Baumann[4] on frog retina, $k_1 = 1.4 \times 10^{-3} \, s^{-1}$ at 21 °C; Baumann and Bender[5] on human retina, $k_1 = 4 \times 10^{-3} \, s^{-1}$; and Ernst and Kemp[6] on rat retina, $k_1 = 0.22 \times 10^{-3} \, s^{-1}$ at 24 °C.

The absorption spectrum of the products in this experiment is shown in Figure 4, together with the computed spectrum of *para*rhodopsin. The absorption spectrum of a hypothetical rhodopsin of $\lambda_{max} = 470$ nm is shown for comparison. This reference spectrum was calculated from a nomogram prepared by Dartnall,[7] that has proved very accurate in predicting the shape of the absorption band of a visual pigment of given λ_{max}. This demonstrates that *para*rhodopsin has the rather broad absorption band typical of native visual pigments and so its electronic—and hence structural—make up is quite similar. Although the chromophoric group of *para*rhodopsin is thought to be in the all-*trans*-con-

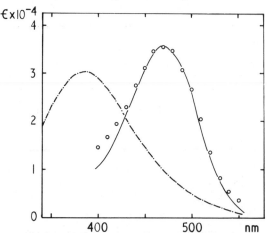

Figure 4 — · — Spectrum of final products calculated from curve (v), Figure 1. ∞ Derived absorption spectrum of *para*rhodopsin and —— spectrum of a rhodopsin of $\lambda_{max} = 470$ nm calculated from the Dartnall nomogram

figuration, it thus seems to be capable of entering the close interaction with opsin seen with 11-*cis*-retinal in the original pigment. This could mean that the opsin has also undergone some conformational change in order to accommodate the all-*trans* chromophoric group, which is borne out by the Arrhenius plot of Figure 5. The activation energy for *para*rhodopsin decay is quite large, and is comparable with the value of 50 kcal mole^{-1} found by Baker and Williams[8] for the thermal decay of *rhodopsin* in the dark. Both processes involve the release of the chromophoric group from opsin and so this could mean that it is bound in the same way in both rhodopsin and *para*rhodopsin.

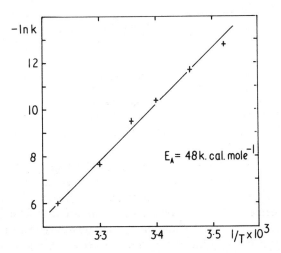

Figure 5 Arrhenius plot of first-order decay constants
for *para*rhodopsin

The extinction coefficient at 470 nm for *para*rhodopsin shown in Figure 4 is based on that for rhodopsin at 500 nm, being 40,600 M^{-1} cm^{-1}.[9] It has been assumed by other workers that ε_{470} for *para*rhodopsin is about the same as that of rhodopsin at 500 nm, but in fact the value from these experiments is considerably lower. This means that concentrations quoted in earlier experiments were probably larger than realized. The actual amounts of *para*rhodopsin generated at various temperatures are given in Figure 6. This shows the proportions of the original rhodopsin molecules found at the end of the experiment as rhodopsin and *iso*rhodopsin, and those which are found as bleached products after either passing through the *para*rhodopsin (PR) stage, or by some other route:

$$\text{rhodopsin} \xrightarrow{h\nu} R*\underset{h\nu}{\overset{\nearrow}{\rightsquigarrow}} \frac{meta\text{rhodopsin II} \rightarrow para\text{rhodopsin}}{\text{rhodopsin} + iso\text{rhodopsin}} \searrow \text{final products}$$

The amounts of PR formed at the lower temperatures are small, but by the time that physiological temperatures are reached, more than half of the bleached

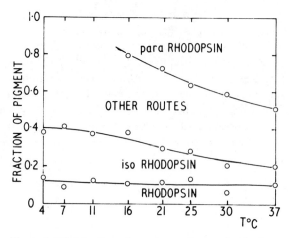

Figure 6 Distribution of reaction pathways for rhodopsin extracts flashed at various temperatures. The decay
of *para*rhodopsin at temperatures below 16 °C was too
slow to permit accurate assessment of the amount
present

molecules have passed through the PR stage. This shows that PR has some
importance in the solution photochemistry of the retina, and since it has been
seen in the living retinas of man, rat and rabbit, it seems probable that it has
an equal importance in the function of mammalian retinas. The rate of accumulation of PR is too slow for it to have any function in the generation of the visual
neural signal, but it might have one of two other functions: (i) to regulate the
response of the retina under bright lighting conditions, or (ii) to act as an intermediate in the regeneration of rhodopsin. The adaptation of the retina to bright
lights is thought to occur primarily at the receptor level; under such conditions,
PR will tend to accumulate in the receptor, and if it limits the response of the
receptor to further quanta, an adaptation related to the light intensity would
result. Ernst and Kemp[6] have found such a relationship between the PR content
of an isolated rat retina and the magnitude of the electrical response of the retina
to small test flashes. When the intensity of illumination is reduced, the PR
concentration will fall and the sensitivity of the eye return. Ripps and Weale[10]
have measured the kinetics of dark adaptation of the human eye and find that
these can be explained on the basis of normal pigment regeneration plus the
decay of an intermediate of λ_{max} about 470 nm and with a decay rate of
8×10^{-3} s at 37 °C, which is similar to my value for the decay of bovine PR.

The absorption spectrum of PR has considerable overlap with that of rhodopsin and so the regulatory mechanism could be a simple screening of rhodopsin
molecules by PR. However, despite the high yield of PR on bleaching rhodopsin,
the accumulated concentration would not be high enough to produce the
observed drop in sensitivity of the receptor. A more probable mechanism is
based on the current picture of the primary generation of the electrical response

of the retina. Excitation of one rhodopsin molecule in a receptor cell will open a channel in an internal membrane to allow the passage of a large number of ions and so polarize the cell. This takes place during the time taken for the excited pigment molecule to reach the PR stage, but, having reached it, this single PR molecule would prevent the whole cell from generating another neural signal until it has decayed. Further quanta would be absorbed by pigment molecules in the cell while it was inactive, thus maintaining the level of PR and the inactivation but not causing any further visual excitation.

2 CONCLUSIONS

*Parar*hodopsin is a relatively stable intermediate in photobleaching of rhodopsin. Although it appears late in the sequence of bleaching reactions, its properties are remarkably like those of the original rhodopsin. It is not excited in an electronic sense, but decays spontaneously because of steric strain in the molecule resulting from the initial absorption of light. It is present in the retina under normal lighting conditions and the data presented in this paper show that its concentration has been underestimated in the past. Its function is probably one of light-adaptation of the retina, rather than being involved in the generation of the neural signal.

ACKNOWLEDGEMENT

I would like to thank Mrs Anne Priestley for her excellent technical assistance.

REFERENCES

1. R. J. Lythgoe, *J. Physiol.* (*Lond.*), **89**, 331 (1937).
2. R. G. Matthews, R. Hubbard, P. K. Brown and G. Wald, *J. Gen. Physiol.*, **47**, 215 (1963).
3. S. E. Ostroy, F. Erhardt and E. W. Abrahamson, *Biochim. Biophys. Acta*, **112**, 265 (1966).
4. Ch. Baumann, *J. Physiol.* (*Lond.*), **222**, 643 (1972).
5. Ch. Baumann and S. Bender, *J. Physiol.* (*Lond.*), **235**, 761 (1973).
6. W. Ernst and C. M. Kemp, *Vision Res.*, **12**, 1937 (1972).
7. H. J. A. Dartnall, *Brit. Med. Bull.*, **9**, 24 (1953).
8. B. N. Baker and T. P. Williams, *Vision Res.*, **8**, 1467 (1968).
9. G. Wald and P. K. Brown, *J. Gen. Physiol.*, **37**, 189 (1953).
10. H. Ripps and R. A. Weale, *Nature*, **222**, 775 (1969).

Session G
Energy transfer in biological molecules

G. What has energy transfer done for biochemistry lately?

J. Eisinger and R. E. Dale†

Bell Laboratories, Murray Hill, New Jersey 07974, U.S.A.

SUMMARY

The physical basis and history of the non-radiative transfer of excitation energy is reviewed. Emphasis is given to the long-range (Förster) energy transfer whose rate has an inverse sixth power dependence on the donor–acceptor separation and therefore is potentially a most useful tool for determining intramolecular distances in biological macromolecules. The transfer rate also depends on the relative donor and acceptor orientations and their freedom of motion, so that estimates of the orientation factor (κ^2) are required before this potential can be realized. Experimental and theoretical methods for obtaining upper and lower limits of κ^2 by the use of polarization spectroscopic techniques are outlined.

1 DISCUSSION

While *absorption spectroscopy* has been routinely accepted by biochemists for a long time and the popularity of *emission spectroscopy* has recently been on the increase, *excitation energy transfer* still tends to be regarded as a somewhat exotic phenomenon which most biochemists consider as existing only in the deep recesses of the biophysicist's or even the physicist's mind. In my lecture today I will try to seek out the monster in its lair without offering rigorous derivations but by relying on intuitive arguments backed up by the relevant references to review papers on the subject. The only *new* results which I will have to offer will be concerned with the orientation factor for long-range energy

---

† Present address: Merganthaler Laboratory for Biology, The Johns Hopkins University, Baltimore, Maryland 21218, U.S.A.

transfer and I will devote the last third of my talk to this work. In my concluding remarks I will try to answer the question posed in the title and finally I will attempt to predict the future developments in the field.

In using *absorption* or *emission spectroscopy* to study a chromophore belonging to a macromolecule we ignore all electrons except the *least* strongly held ones for an eminently practical reason. These electrons have the lowest excitation energy and their transitions can therefore be observed with little or no interference from the absorption of light by the thousands of less volatile electrons in the molecule, thereby providing us well localized probes on a macromolecule.† If the chromophore electrons and an incident electromagnetic wave (or photons) are *directionally* and *energetically compatible*, i.e. have similar polarization vectors and 'resonate' by virtue of having the same energy (i.e. frequency), then the two may interact in a way which leaves the chromophore, at least temporarily, in an excited state.

Whatever our point of view the absorption process is essentially a local phenomenon. Whether we see it as a classical or quantum-mechanical event, it is useful to be aware that the role of the light wave can be played by a nearby but distinct excited chromophore (D*) whose excited state is again *directionally* and *energetically* suitable for interacting with the first chromophore A. If there are no other chromophores in the vicinity the processes

$$A + D^* \rightarrow A^* + D \text{ (energy transfer (ET))}$$

and

$$A + h\nu \rightarrow A^* \qquad \text{(absorption)}$$

are thus formally equivalent. The second is intuitively more acceptable to most scientists, simply because the wave propagation of light through space is a more familiar phenomenon than the relatively short-range interaction between two chromophores which is normally hidden from our view. This interaction is nonetheless quite real and where it is a Coulomb interaction (generally approximated by a dipole–dipole interaction), its theoretical foundation is well understood. Only when it takes the form of an *exchange* interaction do we have to enter the purely quantum-mechanical world, incidentally one in which our theoretical understanding is still very limited. We will ignore the trivial (i.e. radiative) ET which refers to the emission of a photon followed by its reabsorption by another chromophore.

Even after a theoretical understanding of ET has been achieved, the experimental observation and interpretation of ET is somewhat more complicated than that of the emission and absorption processes in that it must generally be inferred from changes in the emission properties of *two* chromophores and can

---

† We ignore for the time being non-localized perturbations of absorption and emission spectra, e.g. hypochromism or exciplex formation.

therefore be considered as second-order spectroscopy. Since energy transfer is effective only over distances of up to tens of angstroms and such distances are of the order of magnitude of biological macromolecular dimensions, it is not surprising that this phenomenon has intrigued increasing numbers of molecular biologists who are willing to tackle the theoretical and experimental rigours of energy transfer to help them in their quest for an understanding of molecular structure and interactions.

The earliest examples of transfer of excitation energy from one atom to another without mediation of a photon that I am aware of are just over 50 years old and are exemplified by the sensitized fluorescence of gases reported by J. Franck[1] and by G. Cario.[2] This phenomenon was later referred to as 'collisions of the second kind'[3] but it is now clear that only a close approach and no actual collision is required. J. Perrin[4] was the first to refer to the transfer of energy (transfert d'activation) in 1924 when he developed his theory of concentration depolarization which illustrates the intimate connection between ET and polarization which will be discussed later. A general theory of energy transfer which lends itself to all forms of the interaction energy between donor and acceptor was developed by Förster[5] some 25 years ago. Since then numerous theoretical treatments[6-10] and reviews[11-13] of ET have appeared and solutions for the most interesting limiting cases exist. These limiting cases are often classified according to the relative magnitudes of the following energies:

(1) ΔE, the difference in the excitation energies of the donor and acceptor chromophores (D and A);

(2) U, the interaction energy between D and A which can formally be written $U = \langle \psi_{D*} \psi_A | H_c | \psi_D \psi_{A*} \rangle$, H_c being the coupling Hamiltonian and the ψs being the D and A wavefunctions with the asterisk denoting the excited state; and

(3) Δ, the bandwidths of the electronic transitions of D and A, also known as the Franck–Condon energies.

The vibrational structures of the D and A spectra also play an important part, but since the molecular biologist is almost always concerned with solvated chromophores whose spectra show at best poorly defined vibrational structure, we will sacrifice completeness for the sake of simplicity and confine ourselves to spectra whose vibrational structure is not resolved.

If the interaction energy exceeds the widths of the electronic transitions and their separation, i.e. $U \gg \Delta E, \Delta$, one is dealing with the so-called *strong coupling limit*. This situation manifests itself by profound changes in absorption and emission spectra of the chromophore pair compared to the superposition of the spectra of the isolated chromophores. These changes are often referred to as exciton splitting (of the order of $2|U|$). The excitation energy oscillates between the two chromophores at a rate of

$$k = 2\pi |U|/\mathbf{h} \tag{1}$$

but since it becomes impossible to define a donor and acceptor, we prefer not to consider exciton theory as belonging to the domain of energy transfer. Exciton theory covers polymer and crystal structures as well as pairs of chromophores and has been reviewed in several papers, e.g. Reference 11 and references therein.

If the interaction energy is small compared to both the bandwidth and energy gap of the donor–acceptor spectra ($U \ll \Delta E, \Delta$), the so-called *very weak coupling limit* applies. The rate of energy transfer is then given by

$$k_T \sim \frac{32|U|^2}{h} J \tag{2}$$

where J is the overlap integral between the donor emission and acceptor absorption spectrum which ensures that energy remains conserved. The form of J depends on the relative rate of energy transfer and vibronic relaxation. If transfer occurs before vibronic relaxation the overlap may increase or decrease depending on the relative energies of donor emission and acceptor absorption bands. Fluorescence lifetimes are generally of the order of 10^{-9} s and the lifetimes of a vibrational state of a molecule in solution has only recently been measured for the first time.[14] It turns out to be of the order of 10^{-11} s. Therefore, if pretransfer relaxation increases the overlap integral by two orders of magnitude 'before vibronic relaxation' energy transfer becomes comparable to the classical 'after relaxation transfer'. Explicit expressions for J corresponding to these two limits are given elsewhere.[15]

While equation (2) shows that the ET rate, k_T, is proportional to the square of the interaction energy, and the overlap integral J ensures that energy conservation is taken into account, the dominant interaction energy must be established before the dependence of k_T on the donor-acceptor separation and orientation is known.

Two cases are of considerable practical importance: The first applies to D–A pairs which have overlapping electronic wavefunctions, which generally implies that the two chromophores are nearest neighbours with no intervening solvent molecules. Molecular orbital wavefunctions are not sufficiently well known to permit precise calculations of the distance dependence of the *exchange interaction* energy U_{ex} which is proportional to the overlap integral between the D and A wavefunctions, but it has been estimated that for parallel conjugated π-electron systems, k_{ex} is of the order of 10^{10} s^{-1} for a D–A separation of 3·4 Å, the usual stacking spacing.[16,17]

The *coulombic interaction* is the other important interaction and at distances greater than a few angstroms it far exceeds U_{ex}.[17] Since the coulombic interaction is generally dominated by the electric dipole term

$$U_{dip} \sim \frac{\mu_D \cdot \mu_A}{r^3} \tag{3}$$

(μ_A, μ_D being the transition dipoles of the virtual transitions which occur in A and D during transfer and r being the D–A separation) it is clear from equation

(2) that the energy transfer rate is proportional to the inverse sixth power of r. When the relative orientation of μ_A and μ_D are properly taken into account by the orientation factor κ^2, when the magnitudes of μ_A and μ_D are expressed in terms of the donor lifetime τ_D, the donor quantum yield ϕ_D and the acceptor molar extinction coefficient $\varepsilon_A(v)$ and the effect of the dielectric constant of the medium intervening between A and D is expressed in terms of the index of refraction, n, the transfer rate for electric dipole, or *long-range*, or *Förster, energy transfer* may be written as

$$k_T = \frac{8 \cdot 8 \times 10^{-25} \kappa^2 \phi_D}{n^4 \tau_D r^6} J' \qquad (4)$$

where

$$J' = \int_0^\infty F_D(v) \varepsilon_A(v) v^{-4} \, dv \qquad (5)$$

where $F_D(v)$ is the normalized donor fluorescence spectrum. (Here it is assumed that energy transfer *follows* vibronic relaxation of the donor.) The so-called Förster distance R_0 is that value of r at which Förster transfer is equal to the de-excitation rate of the isolated donor, i.e. $k_T \tau_D = 1$. It follows from equation (4) that

$$R_0^6(\text{cm}) = 8 \cdot 8 \times 10^{-25} \kappa^2 n^{-4} \phi_D J' \qquad (6)$$

Experimentally the energy transfer efficiency T is more accessible than k_T and it is readily seen to be given by

$$T = \frac{k_T}{\tau_D^{-1} + k_T} = \frac{R_0^6}{r^6 + R_0^6} \qquad (7)$$

Förster distances of the order of 10–20 Å are not uncommon for energy transfer between pairs of naturally occurring chromophores with reasonable spectral overlap, for instance the aromatic amino-acids, nucleotides, etc. Typical values which often depend on the environment of the particular chromophore which can introduce appreciable spectral shifts have been calculated and tabulated.[17,18] R_0 for singlet energy transfer from tyrosine to tryptophan is typically 15 Å and the blue fluorescence of proteins is emitted primarily from tryptophan. For dyes used as extrinsic fluorescent labels R_0 values approaching 100 Å are possible.

It is clear from equations (4), (6) and (7) that even a rough determination of k_T or T leads to a fairly precise value of the interchromophore distance. This has intrigued many experimenters who have made intramolecular distance measurements by the use of energy transfer.[19–26] A particular example will be discussed later, but it is worth noting that energy transfer occurs not only in the biochemical laboratory but is *widespread* in nature. For instance, it has been shown that only a small fraction of the chlorophyll molecules in a chloroplast are photochemically active. These form the so-called reaction centres where the primary photoprocess occurs. The vast majority of chlorophyll molecules, exceeding the reaction centres by a factor of several hundred, act as an antenna to absorb light and transfer the excitation energy among themselves until it is trapped by a reaction centre where a photochemical reaction occurs.[27]

In considering the applications of ET to biochemical problems† it is useful to separate the ones which deal with single donor–acceptor pairs and those involving a multiplicity of donor and/or acceptor. The latter class of experiments include demonstrations of ET from the aromatic amino-acids of a protein to a bound luminophore as a criterion of binding[37,38] or in order to locate the binding site.[39] Here we wish to emphasize the other class of problems which are exemplified by attempts to elucidate the solution structure of a macromolecule containing a donor and an acceptor situated at unique sites.[19–26] ET used in this way has been referred to as a 'spectroscopic ruler'.[40]

Such applications rely heavily on the now well-established experimental verification of the validity of both the inverse sixth-power[19,40] and overlap integral[41] dependence of the rate of ET in such systems.

Since the transfer rate is also directly proportional to the orientation factor κ^2, it has been necessary in analysing these experiments to assign a more or less arbitrary value to this parameter since the relative orientations of donor and acceptor are usually unknown. The most common choice has been that corresponding to complete randomization of both D and A orientations over all space during the transfer time, i.e. $\kappa^2 = (2/3)$.[42] While this is an excellent assumption for *inter*molecular transfer in sufficiently fluid solutions, it has never been shown to be applicable in a case of *intra*molecular transfer.

The orientation factor κ^2, which may vary from 0 to 4, is defined in terms of the angles between the donor and acceptor transition moments (the transfer angle θ_T) and between these and the vector joining them (θ_D, θ_A):

$$\kappa^2 = (\cos \theta_T - 3 \cos \theta_D \cos \theta_A)^2 \tag{8}$$

A luminophore attached to a macromolecule is of course expected to have some restricted freedom of motion with respect to the macromolecular framework. Under normal solution conditions this motion will be comparable with the transfer rate (dynamic averaging regime) and an *appropriate* average value of κ^2 will determine the transfer rate and thereby the transfer efficiency.‡ In order to estimate the effect of luminophore motion on $\langle \kappa^2 \rangle$ we may use the following model:[44,48] Each transition dipole moment vector is restricted to lie within a cone of half-angle ψ. The cone *axes* which correspond to the mean directions of the donor and acceptor transition dipoles may of course have arbitrary directions and a general expression for $\langle \kappa^2 \rangle$ may be formulated. For the sake of simplicity and for heuristic reasons we shall restrict ourselves to those cases for which the two cone axes are either parallel or perpendicular to each other and to the separation vector. Each cone half-angle ψ may take any value between

---

† Many reviews of ET in various biological macromolecular systems such as oligo- and polypeptides and proteins[28–33] and nucleic acids[29,34–36] have appeared.

‡ The opposite limit, that of static averaging, obtains when the distribution of luminophore orientations does not change during the transfer time. An average value of κ^2 is not generally appropriate in this limit as discussed elsewhere.[43]

zero and π, the latter (as well as $\pi/2$) corresponding to complete orientational isotropy and to $\langle \kappa^2 \rangle = \frac{2}{3}$. When $\psi = 0$, the transition dipole is fixed with respect to the macromolecule and the extreme values of κ^2 correspond to such cases: $\kappa^2 = 0$ for transition moments perpendicular to each other and the separation vector and $\kappa^2 = 4$ when they are parallel-in-line. As ψ increases and the cones are allowed to open up the corresponding $\langle \kappa^2 \rangle$ values continue to correspond to extrema with $\langle \kappa^2 \rangle_{max}$ decreasing from 4 and $\langle \kappa^2 \rangle_{min}$ increasing from 0 (see Figure 1). To clarify the use of this model we will re-examine Beardsley and Cantor's determination of the distance between the anticodon loop and the CCA terminus of $\mathrm{tRNA}^{Phe}_{yeast}$ in solution.[21] These authors measured the efficiency of ET from the fluorescent Y-base in the anticodon loop to an acridine dye label attached to the 3′ terminal ribose and estimated the interchromophore separation to be 46 Å when $\langle \kappa^2 \rangle$ was assumed to be $\frac{2}{3}$. In order to apply the model described above, it is necessary to estimate the reorientational freedom of both luminophores. This can be done by measuring the depolarization of luminophore

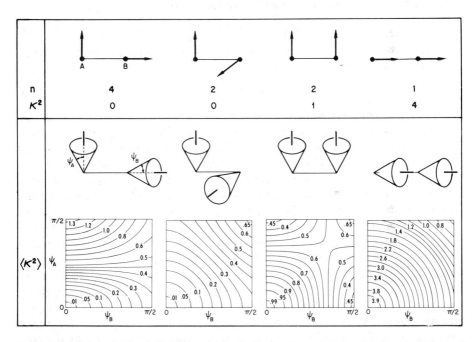

Figure 1 The upper part of the diagram illustrates the nine possible relative orientations of two transition dipoles each of which is fixed and can lie along either the x, y or z axis of a Cartesian triad. The corresponding κ^2 values are shown along with their statistical weights (n) and they are seen to lead to an average for κ^2 of 2/3, the same as for isotropically random orientations of the transition dipole moments. The lower part of the figure illustrates how these $\langle \kappa^2 \rangle$ values change as the transition dipole directions are permitted orientational freedom within cones of half-angles ψ_A and ψ_B. Note that $\langle \kappa^2 \rangle$ departs quite slowly from its fixed minimum and maximum values (0 and 4) as the two cones open up and that when each cone half-angle is $\pi/2$, corresponding to an isotropic distribution of the transition dipole directions, $\langle \kappa^2 \rangle$ is equal to 2/3 for each of the cases considered

fluorescence when each is excited directly.[44,49] The rotational motion of the tRNA molecule as a whole contributes to the overall depolarization, but this can be allowed for experimentally. The donor and acceptor half-cone angles turned out to be about 30°, and the maximum and minimum values of $\langle \kappa^2 \rangle$ were about 3 and 0·12 respectively (see Figure 1). The corresponding estimate for the interluminophore separation range is 34–60 Å. It is interesting to note that recent X-ray diffraction studies of the same tRNA indicated the anticodon–CCA terminus separation to be about 80 Å.[45] Taking into account the uncertainty in the separation of the dye and Y-base introduced by linkage of the dye to the free CCA terminal triplet, the upper limit of 60 Å thus determined is consistent with this crystal structure.

While the analysis presented above can eliminate some of the uncertainty in $\langle \kappa^2 \rangle$ resulting from luminophore motion, the uncertainty resulting from the unknown relative orientation of the *cone axes* remains. It can be reduced somewhat by measurement of the depolarization associated with the transfer itself. Defining the transfer depolarization factor as

$$d_T = \tfrac{3}{2} \cos^2 \theta_T - \tfrac{1}{2} \tag{9}$$

its dynamically averaged value can be shown as a consequence of Perrin's law of isotropic depolarization,[46] to be

$$\langle d_T \rangle = (\tfrac{3}{2} \cos^2 \bar{\theta}_T - \tfrac{1}{2}) \langle d_D \rangle \langle d_A \rangle \tag{10}$$

where $\bar{\theta}_T$ is the transfer angle subtended by the D and A cone *axes* and $\langle d_D \rangle$, $\langle d_A \rangle$ are the donor and acceptor depolarization factors for direct excitation of these luminophores.[44,48] Measurement of $\langle d_T \rangle$ requires the use of *polarized* emission spectroscopy, but this is in any case highly desirable since both steady-state and time-dependent emission measurements are generally polarization sensitive in conventional fluorometers in which the emission is viewed in a direction perpendicular to that of excitation (see e.g. Reference 47). Since transfer depolarization depends only on the angle θ_T between the donor and acceptor transition dipole moments, while as pointed out earlier κ^2 depends on *three* angular variables there unfortunately exists no one-to-one correspondence between these two parameters. Nevertheless, a measurement of the transfer depolarization is useful in that it permits the testing of particular models for donor–acceptor geometries. To illustrate this point, consider the geometries corresponding to the maximum and minimum values of $\langle \kappa^2 \rangle$ which have already been discussed in connection with the Beardsley–Cantor experiment. Once ψ_A and ψ_D are known, a particular relative orientation of the cone axes corresponds to a particular transfer depolarization, $\langle d_T \rangle$. Thus, as ψ_A and ψ_D range independently from 0 to $\pi/2$, $\langle d_T \rangle$ takes on values between $-0·5$ and 0 if the two cone axes are mutually perpendicular while it lies correspondingly between 1 and 0 if they are parallel-in-line (see Figure 2).

What sobering or encouraging thoughts do we wish to impart on the occasion of the 50th anniversary of science becoming aware of energy transfer? On the one hand it is gratifying that a theory exists which allows us to describe with

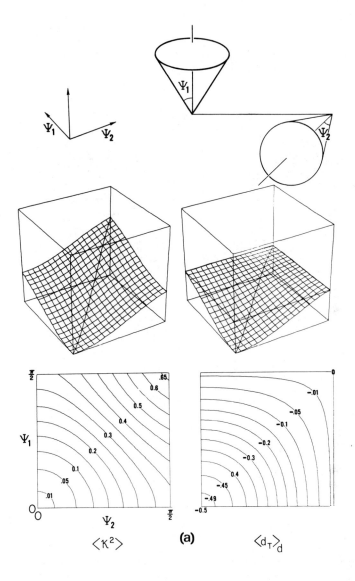

$$\langle \kappa^2 \rangle \qquad \text{(a)} \qquad \langle d_T \rangle_d$$

confidence events whose origin and termination are separated by angstroms in space and nanoseconds in time. Practical applications of this new understanding have on the other hand been on a much smaller scale than for many other spectroscopic tools of the biochemist. The *empirical* use of ET phenomena such as fluorescence and polarization titrations in binding studies has been widespread and their popularity is bound to increase as more of the needed apparatus becomes commercially available. The use of energy transfer for elucidating macromolecular structure, however, which began with such high hopes about ten years ago, and whose practicality has been firmly established

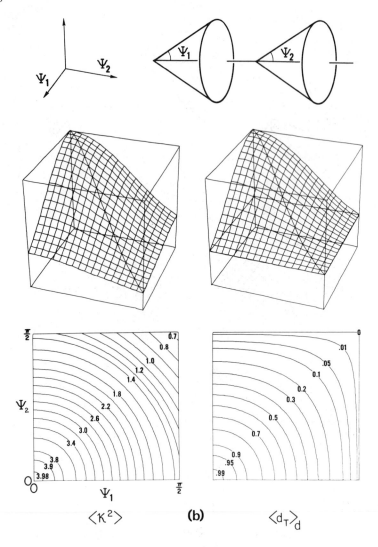

Figure 2 (a) and (b) correspond to the second and fourth models shown in Figure 1, but include contour and tridimensional plots for $\langle d_T \rangle$ as well as $\langle \kappa^2 \rangle$. Similar graphical solutions for many additional models with the transition moments of the donor and acceptor lying on the surface or within cones with various half-angles and for various relative orientations of the cone axes are presented in References 44 and 48. Among the orthogonal models given in Figure 1, 2(a) leads to minimum and 2(b) to maximum values of $\langle \kappa^2 \rangle$, for any pair of values of the cone half-angles ψ_1 and ψ_2

since then,[19-26] never caught on, probably because our joy in being able to measure absolute intramolecular distances spectroscopically was diminished by the gnawing suspicion that the uncertainty in $\langle \kappa^2 \rangle$ was a sword of Damocles hanging over us. It is possible that our new understanding of the intimate connection between $\langle \kappa^2 \rangle$ and the donor, acceptor and transfer depolarization factors which we have discussed, along with the increased use of steady state and time-dependent *polarized* emission spectroscopy, will lead to renewed interest in these, the most elegant applications of ET. If they indeed prove useful, energy transfer will offer a sorely needed experimental technique capable of complementing X-ray diffraction studies by extending crystalline state results to macromolecules in solutions.

REFERENCES

1. J. Franck, *Z. Physik*, **9**, 859 (1922).
2. G. Cario, *Z. Physik*, **10**, 185 (1922).
3. H. Kallmann and F. London, *Z. Phys. Chem.*, **B2**, 207 (1928).
4. J. Perrin, 2me *Conseil de Chim. Solvay*, p. 322, Gauthier-Villars, Paris, 1925.
5. Th. Förster, *Ann. Phys.* (*Leipzig*), **2**, 55 (1948).
6. D. L. Dexter, *J. Chem. Phys.*, **21**, 836 (1959).
7. Th. Förster, *Disc. Faraday Soc.*, **27**, 7 (1959).
8. Th. Förster, 'Excitation transfer', in *Comparative Effects of Radiation*, p. 300 (eds. M. Burton, I. S. Kirby-Smith and I. L. Magee), Wiley, New York, 1960.
9. Th. Förster, *Radiation Res. Suppl.*, **2**, 326 (1960).
10. G. W. Robinson and R. P. Frosch, *J. Chem. Phys.*, **38**, 1187 (1963).
11. Th. Förster, 'Delocalized excitation' and 'Excitation transfer', in *Modern Quantum Chemistry*, Pt. III, p. 93 (ed. O. Sinanoglu), Academic Press, New York, 1965.
12. R. G. Bennett and R. E. Kellogg, in *Progress in Reaction Kinetics*, Vol. IV, p. 215 (ed. G. Porter), Pergamon Press, Oxford, 1965.
13. A. A. Lamola, in *Organic Photochemistry*, p. 17, Vol. XII of *Techniques of Organic Chemistry* (ed. A. Weissberger), Wiley, New York, 1969.
14. A. Laubereau, D. von der Linde and W. Kaiser, *Phys. Rev. Letters*, **28**, 1162 (1972).
15. M. Gueron, J. Eisinger and R. G. Shulman, *J. Chem. Phys.*, **47**, 4077 (1967).
16. R. S. Sommer and J. Jortner, *J. Chem. Phys.*, **49**, 3919 (1968).
17. J. Eisinger, B. Feuer and A. A. Lamola, *Biochemistry* (*Easton*), **8**, 3908 (1969).
18. G. Karreman and R. H. Steele, *Biochim. Biophys. Acta*, **25**, 280 (1957).
19. S. A. Latt, H. T. Cheung and E. R. Blount, *J. Amer. Chem. Soc.*, **87**, 995 (1965).
20. J. Eisinger, *Biochemistry* (*Easton*), **8**, 3902 (1969).
21. K. Beardsley and C. R. Cantor, *Proc. Nat. Acad. Sci.* (*U.S.*), **65**, 39 (1970).
22. S. A. Latt, D. S. Auld and B. L. Vallee, *Proc. Nat. Acad. Sci.* (*U.S.*), **67**, 1383 (1970).
23. T. C. Werner, J. R. Bunting and R. E. Cathou, *Proc. Nat. Acad. Sci.* (*U.S.*), **69**, 795 (1972).
24. P. W. Schiller, *Proc. Nat. Acad. Sci.* (*U.S.*), **69**, 975 (1972).
25. C-W. Wu and L. Stryer, *Proc. Nat. Acad. Sci.* (*U.S.*), **69**, 1104 (1972).
26. J. R. Bunting and R. E. Cathou, *J. Mol. Biol.*, **77**, 223 (1973).
27. R. K. Clayton, *Light and Living Matter*, McGraw-Hill, New York, 1972.
28. L. Stryer, *Radiation Res. Suppl.*, **2**, 432 (1960).
29. S. V. Konev, *Fluorescence and Phosphorescence of Proteins and Nucleic Acids* (trans. S. Udenfriend), Plenum Press, New York, 1967.
30. I. Z. Steinberg, *Ann. Rev. Biochem.*, **40**, 83 (1971).
31. I. Weinryb and R. F. Steiner, in *Excited States of Proteins and Nucleic Avids*, p. 277 (eds. R. F. Steiner and I. Weinryb), Plenum Press, New York, 1971.

32. W. D. Dandliker and A. J. Portman, in *Excited States of Proteins and Nucleic Acids*, p. 199 (eds. R. F. Steiner and I. Weinryb), Plenum Press, New York, 1971.
33. J. W. Longworth, in *Excited States of Proteins and Nucleic Acids*, p. 319 (eds. R. F. Steiner and I. Weinryb), Plenum Press, New York, 1971.
34. M. Gueron and R. G. Shulman, *Ann. Rev. Biochem.*, **37**, 571 (1968).
35. J. Eisinger and A. A. Lamola, in *Excited States of Proteins and Nucleic Acids*, p. 107 (eds. R. F. Steiner and I. Weinryb), Plenum Press, New York, 1971.
36. J. Eisinger and A. A. Lamola, in *Methods of Enzymology*, p. 24 (eds. L. Grossman and K. Moldave), Academic Press, New York, 1971.
37. J. E. Churchich, *Biochemistry (Easton)*, **4**, 1405 (1965).
38. R. F. Chen and J. C. Kernohan, *J. Biol. Chem.*, **242**, 5813 (1967).
39. R. A. Badley and F. W. J. Teale, *J. Mol. Biol.*, **58**, 567 (1971).
40. L. Stryer and R. P. Haugland, *Proc. Nat. Acad. Sci. (U.S.)*, **58**, 719 (1967).
41. R. P. Haugland, J. Yguerabide and L. Stryer, *Proc. Nat. Acad. Sci. (U.S.)*, **63**, 23 (1969).
42. Th. Förster, *Fluoresenz Organischer Verbindungen*, Vanderhoeck and Rupprecht, Göttingen, 1951.
43. J. Eisinger and R. E. Dale, *J. Mol. Biol.*, **84**, 643 (1974).
44. R. E. Dale and J. Eisinger, 'Polarized excitation energy transfer', Ch. 4 in *Concepts in Biochemical Fluorescence* Vol. 1 (eds. R. F. Chen and H. Edelhoch), Marcel Dekker Inc., New York (1975).
45. S. H. Kim, G. J. Quigley, F. L. Suddath, A. McPherson, D. Sneden, J. J. Kim, J. Weinzierl and A. Rich, *Science*, **179**, 285 (1973).
46. F. Perrin, *Ann. Phys. (Paris)*, **12**, 169 (1929).
47. M. Shinitzky, *J. Chem. Phys.*, **56**, 5979 (1972).
48. R. E. Dale and J. Eisinger, *Biopolymers*, **13**, 1573 (1974).
49. W. E. Blumberg, R. E. Dale, J. Eisinger and D. M. Zuckerman, *Biopolymers* **13**, 1607 (1974).

G.1. A unified approach to excited state homotransfer in biological systems

Robert M. Pearlstein †

Biology Division, Oak Ridge National Laboratory,‡ Oak Ridge, Tennessee 37830

Katja Lindenberg§

Department of Chemistry, University of California at San Diego, La Jolla, California 92093

Richard P. Hemenger †

Biology Division, Oak Ridge National Laboratory,‡ Oak Ridge, Tennessee 37830

SUMMARY

We apply the exciton–phonon theory of exciton motion in organic crystals to bio-organic systems of identical molecules. The theory is then used to describe the general phenomena of fluorescence depolarization and impurity quenching that result from singlet homotransfer. We show that exciton–phonon theory is an improvement over Förster hopping theory in the description of fluorescence depolarization and sensitized dye fluorescence for nucleic acids. We suggest new experiments to test the theory.

1 INTRODUCTION

Since the discovery of the photosynthetic unit some forty years ago, electronic excited state energy transfer has been a topic of considerable interest in biology. While certain aspects of such energy transfer, particularly those involving

---

† Present address: Chemistry Division, Oak Ridge National Laboratory.
‡ Operated by the Union Carbide Corporation for the U.S. Atomic Energy Commission.
§ Formerly Katja Lakatos-Lindenberg. Supported in part by the Oak Ridge Associated Universities and the Oak Ridge National Laboratory, and by a grant from the Academic Senate of the University of California.

591

bimolecular heterotransfer, are farily well understood, others are not. Foremost among the latter is multiple homotransfer, that is energy transfer through a system of identical molecules, i.e. chromophores. There are a number of systems in which multiple homotransfer is feasible, if not actually demonstrated, most notably photosynthetic systems and nucleic acids. In both of these there are apparent contradictions and unexplained phenomena.

Theoretical estimates of the resonance interaction energy J between identical chromophores are based on individual chromophore absorption properties.[1,2] These theories usually assume dipole–dipole interactions, for which $J \propto R^{-3}$, where R is the distance between the interacting dipoles. However, the resonance interaction energy that determines the extent of hypochromism and related optical effects is J times a sum of Franck–Condon factors.[3] For biological systems of interest such as stacked nucleic acids even this reduced resonance interaction energy is large, $\sim 100 \, \text{cm}^{-1}$, corresponding to a circular frequency $\sim 2 \times 10^{13} \, \text{s}^{-1}$. It is generally accepted that optical phenomena occurring prior to vibrational relaxation are coherent, i.e. phase relations established upon absorption among the chromophores sharing the excitation energy are maintained. Thus, extensive, coherent excitation transfer is possible in these biological systems prior to vibrational relaxation, which occurs typically in times of order $10^{-12} \, \text{s}$. The existence of pre-relaxation transfer has not been demonstrated experimentally in these systems. We therefore wish to discuss here the properties of post-relaxation coherent transfer, the possibility of which is not generally recognized. We stress that a comprehensive theory of excitation transfer that holds for all times, pre- as well as post-relaxation, does not yet exist.

In this paper we re-examine multiple homotransfer phenomena for the nucleic acid singlet states in the light of recent theoretical developments concerning post-relaxation coherent transfer that are based upon a fundamental description of the interaction between the electronic excitation and molecular vibrations.[4] This so-called 'exciton–phonon interaction' crucially affects multiple homotransfer, or exciton motion, in organic molecular systems.[5] We extend the exciton–phonon theory developed for organic crystals to construct a unified picture of exciton motion in biological systems.

We begin by describing the main features of exciton–phonon theory. We then apply the theory to the general phenomena of fluorescence depolarization and impurity quenching of fluorescence. Finally, we reinterpret these phenomena for singlet excitons in nucleic acids.

2 EXCITON–PHONON THEORY

Isolated organic molecules can vibrate in a number of distinct ways referred to as intramolecular vibrational modes. In aggregates of such molecules these modes are broadened and additional modes appear. However, we believe that in biological systems the broadening and additional modes have little effect on exciton motion, except possibly at very low temperatures. For clarity we neglect all intermolecular vibrational effects. Our final results are not affected by this simplification.

The exciton–phonon theory of Grover and Silbey leads to a density matrix equation[5] that describes the motion of the dressed exciton:

$$d\rho_{nm}/dt = i\tilde{J}(\rho_{n+1,m} + \rho_{n-1,m} - \rho_{n,m+1} - \rho_{n,m-1})$$
$$- F[2\rho_{nm} - \delta_{nm}(\rho_{n+1,m+1} + \rho_{n-1,m-1})] \qquad (1)$$
$$+ F(\delta_{n,m+1} + \delta_{n,m-1})\rho_{mn}$$

The diagonal part of the density matrix, $\rho_{nn}(t)$, is the probability that molecule n is excited at time t. The off-diagonal elements $\rho_{nm}(t)$, $n \neq m$, contain information about the relative phases of probability amplitudes at molecules n and m. Equation (1) as written is for a one-dimensional translationally invariant system with only nearest-neighbour interactions.[6] All these restrictions can be easily removed. $\tilde{J}$ is the resonance interaction energy between adjacent chromophores in the vibrationally relaxed state. The term proportional to $\tilde{J}$ in equation (1) describes the free-particle motion of the exciton. The terms proportional to F describe effects of thermal fluctuations which tend to destroy all phase relations, i.e. they cause the off-diagonal elements of the density matrix to decay with time at a rate proportional to F. However, the total probability of excitation is conserved in equation (1), i.e. $\sum_n \rho_{nn}(t) = 1$. The parameters $\tilde{J}$ and F are discussed in more detail later.

Equation (1) implies two distinct types of exciton motion depending on the relative sizes of $\tilde{J}$ and F.[5,7a] If $\tilde{J} \ll F$, the motion is strictly of the hopping type, i.e. diffusion-like, and is described by a master equation. If $\tilde{J} \gg F$, the motion is coherent or wavelike within M molecules of the starting site, where $M \sim \tilde{J}/F$, and is diffusion-like over larger distances. In the bio-organic literature, exciton motion is usually described strictly as a hopping process.[8,9] As we have just noted, such a description is only valid when $\tilde{J} \ll F$. However, we believe that the opposite condition, $\tilde{J} \gg F$, is the one that occurs in many systems of interest, especially biological ones. Indeed, our main purpose is to explore this assertion and work out its consequences. When $\tilde{J} \gg F$ there is a simple heuristic picture[7a] of the exciton motion: an initially coherent exciton state decays at a rate of about $3F$. At the same time, to compensate for this decay, each site receives excitation at the same rate but with a phase that is random with respect to that of the excitation received by any other site. This incoherent process redistributes excitation over all coherent states but by itself produces very little net motion of the exciton. Exciton motion at each instant is governed almost entirely by the coherent part of equation (1), i.e. the term proportional to $\tilde{J}$.

Because any excitation transfer prior to vibrational relaxation is necessarily coherent, the rate of that relaxation is an upper limit on the size of the parameter F in equation (1) that causes loss of coherence. Thus, $F \lesssim 10^{12} \, s^{-1}$. Therefore, if $\tilde{J} \gtrsim 10 \, cm^{-1}$, $\tilde{J} > F$ and the exciton motion is, contrary to the prevalent hopping assumption, quite coherent and wavelike. We believe this result to be rather general for bio-organic systems of identical chromophores. Although it has been conjectured that coherence is a consequence of reversibility,[10] such conjectures have been controversial.[11a]

3 KINETIC PHENOMENA

In biological systems, kinetic phenomena involving exciton motion such as fluorescence depolarization and impurity quenching of the excited state have received considerable study.[9,12] The interpretation of experimental results depends on the nature of the exciton motion. In this section we investigate the consequences of exciton–phonon theory for the interpretation of fluorescence depolarization and impurity quenching.

We begin by considering fluorescence depolarization. We first discuss a coherent system, i.e. one for which $F = 0$. We then consider the effect of small F on these results, and finally consider the consequences of a hopping model.

In a hopping model individual molecules are treated as dipoles which absorb and emit polarized light independently. At the other extreme, in the Simpson–Peterson strong coupling case,[11b] i.e. with the resonance interaction energy greater than the Franck–Condon bandwidth, coherent exciton states of the entire system absorb and emit polarized light independently. Each delocalized state can then be treated as a dipole whose oscillator strength and vector moment are the coherent sums over the whole system of the oscillator strengths and vector moments of the individual molecules. This fully coherent picture implies fluorescence spectra and lifetimes that are considerably different from those of the isolated molecules. Stacked cyanine dyes are examples of this altered fluorescence behaviour.[13] However, for the cases we consider, system fluorescence spectra and (unquenched) lifetimes are similar to those of the corresponding isolated molecules. Even though the spectra in these cases may be continuous, we believe the situation corresponds to Simpson–Peterson weak coupling,[11b] with thermal corrections, rather than to Förster very weak coupling.[11a] Thus, for these cases, even though the dressed exciton motion may be quite coherent ($\tilde{J}/F$ large), fluorescence is more nearly a process in which the molecular dipoles radiate independently. In exciton–phonon theory, therefore, the distribution of excitation within the system is determined by the interplay of $\tilde{J}$ and F, and the fluorescence polarization in these cases is in general different from what would be predicted by a hopping model.

By the usual optical selection rules for well-ordered systems, very few coherent exciton states are initially excited. In the limit $F \to 0$ no excitation is subsequently transferred to any of the other exciton states. The fluorescence polarization p is then large for most system geometries, with a theoretical maximum value of $\frac{1}{2}$ when only a single exciton state is initially excited. (We neglect the small differences in direction between absorption and emission dipoles within the same electronic transition.) In a system with non-zero $F \ll \tilde{J}$, on the other hand, even if only one state is initially excited, F 'mixes in' all other exciton states with random phases during the course of the excited-state lifetime

τ.[7a] A fraction, $\sim 1/(1 + 3F\tau)$, of the total fluorescence yield comes from the initially excited exciton states and the remainder from an equal mixture of all the exciton states. In general, the polarization of the light emitted by the equal mixture is lower than that emitted by the initial states. Hence a non-zero F is expected to diminish the polarization, which approaches a lower limit for $F \gg 1/\tau$.

As an idealized example, we mention a collection of infinite rigid polymers with polymer axes randomly distributed in space. If on each polymer the dipoles are all mutually parallel, $p = \frac{1}{2}$ irrespective of the value of F. If each polymer has helical symmetry with molecular dipoles perpendicular to the polymer axis and the angle between adjacent molecular dipoles is $\lesssim 45°$, calculations show[7b] that $p = \frac{1}{2}$ in the limit $F \to 0$ and decreases to $p = \frac{1}{7}$ for $F \gg 1/\tau$.

The hopping model of exciton transfer is usually described by a master equation,[14] i.e. an equation formally identical to the diagonal part of equation (1), but with $\tilde{J} = 0$ and F replaced by a quantity $\mathscr{F}$ that is evaluated from the overlap of the molecular absorption and emission spectra according to the Förster formula.[11a] In general $\mathscr{F} \gg F$. One possible exception is the nucleic acids (see §4). If both F and $\mathscr{F} \gg 1/\tau$, both models predict the same polarization. If, however, $F \lesssim 1/\tau$ (by coincidence or by design), then usually $\mathscr{F} \gg 1/\tau$ and the hopping model in general predicts a substantially lower polarization than does exciton–phonon theory. We wish to stress that if $F \lesssim 1/\tau$, it is possible to obtain a relatively high polarization even in the presence of extensive coherent excitation transfer (large $\tilde{J}$) and in geometries for which a hopping model predicts low polarization.

We turn now to impurity quenching of the excited state, which includes the phenomenon of sensitized fluorescence. The excitation energy in a molecular system is said to be quenched when the energy is irreversibly removed from that system. Quenching by an impurity is the removal of excitation from the system through those few molecues in the immediate vicinity of the impurity. For simplicity we specialize to the case that each impurity causes the removal of excitation from a single molecule with rate constant Γ. This simplification has no substantial effect on our conclusions (see §4). For each molecule j that interacts directly with an impurity, a term, $-(\Gamma/2)(\delta_{nj} + \delta_{mj})\rho_{nm}$, is added to the right-hand side of equation (1).[7a] From this modified form of equation (1) we find that the excitation in the system as a function of time, $P(t)$, exhibits several types of kinetic behaviour that result from the interplay of coherent, incoherent and quenching terms.[7a] $P(t)$, of course, decays in time, and does so faster with impurity quenching than without. The consequently shortened fluorescence lifetime τ, as is well known, can therefore be varied experimentally as a function of impurity quenching. This fact may prove to be useful for fluorescence depolarization experiments, since, as we have already noted, exciton–phonon theory and hopping theory usually predict substantially different results when $F \lesssim 1/\tau$.

For a one-dimensional system with only nearest-neighbour interactions the kinetic behaviour of $P(t)$ is understood in some detail. For present purposes we need consider only the situation of weak quenching,[15] i.e. $\Gamma \ll \tilde{J}$. We assume optical excitation and standard selection rules. In this weak quenching case, the results are very simple:

$$P(t) = \exp\left[-(\Gamma c + \hat{\tau}^{-1})t\right], \tag{2}$$

and thus

$$\tau^{-1} = \Gamma c + \hat{\tau}^{-1}, \tag{3}$$

where $\hat{\tau}$ is the actual excited-state lifetime in the absence of quenching and c is the fractional concentration of quenchers. The results for $P(t)$ and τ are the same in hopping theory if $\Gamma \ll \mathscr{F}$. The more complicated, strong quenching case, which for present purposes we need not discuss, is treated elsewhere.[7a,15b]

4 NUCLEIC ACID SINGLET EXCITONS

A number of experiments have been done with single-stranded helical poly-riboadenylic acid (poly rA), as well as with other nucleic acids, to detect singlet exciton energy transfer.[12,16] These include measurements of the depolarization of fluorescence and fluorescence quenching experiments. Interpretations of the results of these experiments based on a hopping model have led to the conclusion that transfer takes place over at most one or two bases during the lifetime of the excited state. Here we reinterpret these results in terms of exciton–phonon theory and show that they are consistent with extensive, coherent energy transfer.

The value of J at the peak of the 260 nm band of poly rA has been estimated as several hundred wavenumbers.[12] If the peak value of $\tilde{J} \sim 100\ \mathrm{cm}^{-1}$, then since the Franck–Condon factor for the $0-0$ transition is about 10 times smaller than the Franck–Condon factor for the peak,[17] one would estimate $\tilde{J}$ for the $0-0$ transition to be $\sim 10\ \mathrm{cm}^{-1}$, corresponding to a circular frequency of $\sim 2 \times 10^{12}\ \mathrm{s}^{-1}$.

The results of experiments on fluorescence depolarization at 210 K show that the value of the polarization for poly rA is about the same as that for the di-nucleotide ApA, and about half that for the mononucleotide AMP.[18] These results, which indicate no depolarization of the polymer relative to the dimer, have been taken as the strongest evidence against efficient transfer. One is forced to this interpretation in a hopping picture because the ApA depolarization places a lower limit on $\mathscr{F}$ and the poly rA result places an upper limit on $\mathscr{F}$. Since the two limits are nearly equal, $\mathscr{F} \sim 1/\tau \sim 10^{11}\ \mathrm{s}^{-1}$. In hopping theory $\mathscr{F}$ is the only homotransfer parameter so that the conclusion that transfer is inefficient follows immediately from the fact that $\mathscr{F}$ is not much bigger than $1/\tau$ at this temperature. However, in exciton–phonon theory the situation is quite different. There are two homotransfer parameters, $\tilde{J}$ and F, and these depolarization experiments place no restriction on $\tilde{J}$ and only an upper limit on F. In this case the dinucleotide results place no lower limit on F. This is because, even if $F = 0$, both dinucleotide exciton states are excited and the polarization can be substantially less than that of the mononucleotide. The polymer result certainly places an upper limit on F, i.e. $F \lesssim 1/\tau \sim 10^{11} \, s^{-1}$, which is consistent with the upper limit given in §2. Since essentially one exciton state is excited in an infinite rigid polymer then for $F = 0$ the monomer and polymer polarizations should coincide. Because the actual polymers bend, more than one exciton state may be excited and there may be some depolarization even if $F = 0$. The polarization experiments that have been done therefore place no lower limit on F. Our analysis implies that if the depolarization of the polymer is due primarily to bending, then an oligomer of length 10–20 would have a higher polarization than either the dimer or the polymer. It also implies that at lower temperatures, for which $\tau \gg 10^{-11} \, s$, exciton–phonon theory predicts a higher polymer polarization than does hopping theory if $F \ll \mathcal{F} \sim 10^{11} \, s^{-1}$.

Fluorescence quenching experiments have been done both at room temperature and near liquid-nitrogen temperature.[12,16] The room-temperature experiments[19] have all used exogenous dyes, both intercalating and non-intercalating, as quenchers of the excited singlet state in various nucleic acids. At low temperature the quenching of excited singlets in oligo rA and poly rA by exciplex formation[20] and by addition of mercuric ions[21] has been studied.

We discuss the dye-quenching experiments first. Experiments done with various dyes and nucleic acids show that UV-excitation of the nucleic acid leads to sensitized fluorescence from the dye. In all cases the sensitized fluorescence has been interpreted as due to a one-step Förster transfer from excited nucleic acid bases directly to the dye. In each case results were interpreted to imply some effective number of bases that transfer excitation to each dye molecule within the nucleic acid singlet state lifetime. The number of bases depends on the nucleic acid and dye chosen, and there is some disagreement between observers,[19] but it is generally in the range 4–20. This interpretation in terms of one-step, direct transfer to the dye implies that homotransfer cannot extend over more than a few bases during the excitation lifetime.

The interpretation of these experiments in terms of exciton–phonon theory is rather different. The dye undergoes resonance interactions with a few of its neighbouring bases. Like the base–base interactions, these dye–base interactions are coherent and reversible, and by themselves produce no quenching. Quenching occurs as a result of relaxation processes in the dye that degrade the latter's excitation from the S_2 to the S_1 level, thereby destroying dye–base resonance interactions. Even though the rapid relaxation condition for Förster transfer is clearly not satisfied for these base–dye transfers, the effective quenching rate constant Γ for each dye molecule must depend on the nucleic acid fluorescence–dye absorption overlap integral. Using equation (3), one deduces that the reciprocal of the concentration (base/dye ratio for single-stranded polymers; base pair/dye ratio for double-stranded polymers) at which half the nucleic acid singlet excitation is quenched by the dye is $\Gamma \hat{\tau}$. This quantity is what in previous interpretations has been called 'the effective number of bases that transfer excitation to each dye molecule within the nucleic acid singlet state lifetime'. The dependences of this quantity on the various parameters are consistent with experimental results.[19] We note that if similar sensitized dye fluorescence experiments can be done at lower temperature (with $\tilde{J}$ and Γ not much changed), and hence longer unquenched poly rA singlet lifetimes $\hat{\tau}$, homotransfer ought to substantially increase the effective number of bases quenched by each dye molecule.

The fluorescence from random copolymers of A and C comes partially from the adenine singlet and partially from the AC exciplex. The fluorescence yield of adenine increases and that of the exciplex decreases with increasing n in poly (A_nC). For $n = 1.05$ the fluorescence comes almost entirely from the exciplex, and for $n = 6.5$ most of the fluorescence comes from the adenine.[20] This quenching of the adenine singlet in poly (A_nC) by exciplex formation is more difficult to analyse than the dye experiments. It appears that neither exciton–phonon theory nor a hopping model consistent with the fluorescence polarization results can account for quenching by exciplex formation if the latter is assumed to be a straightforward quenching process of rate constant Γ independent of n. A possible explanation would be that Γ decreases as n increases, a dependence that may result from the stacking properties of the bases in poly (A_nC). The sensitivity of excited-state complex formation to the details of stacking is well illustrated by the fact that AA excimers form in poly dA but not in poly rA.[16]

Mercuric ions also quench the fluorescence of nucleic acids. The data do not appear to be inconsistent with exciton–phonon theory, but no detailed analysis has yet been made.

5 CONCLUSION

We have applied the exciton–phonon theory of exciton motion in an organic crystal to a bio-organic system of identical molecules whose aggregate absorption spectrum has no well-resolved vibrational structure and is close in shape to that of its molecular components. We then used the theory to describe the general phenomena of fluorescence depolarization and impurity quenching that result from singlet homotransfer. This new description provides a theoretically consistent and unified framework for the understanding of these phenomena. Predictions of exciton–phonon theory are generally different from those of the widely used hopping theory, which describes only one limit of exciton transport.

Specifically, we have re-examined the problem of singlet homotransfer in nucleic acids. We have indicated several experimental situations in each of which the prediction of exciton–phonon theory differs from that of hopping theory. These include the deliberate shortening of fluorescence lifetime in order to increase polarization, the possibility of a maximum value of the fluorescence polarization as a function of chain length in oligo rA, and the possibility of extending, through homotransfer, the range of bases that sensitizes the fluorescence of each dye molecule bound to poly rA.

There are two phenomena in nucleic acids that we have not discussed, which may have some bearing on homotransfer: an apparent, weak, hyperchromic electronic transition on the red edge of the main 260 nm band, and the anomalously long lifetime of adenine fluorescence.[12] These two phenomena engender contradictory explanations at present. In any case, we believe that the exciton–phonon picture presented here will remain essentially unchanged.

It is likely that exciton–phonon theory can be applied fruitfully to other biological situations in which homotransfer is thought to occur, most notably to triplet states of nucleic acids and chlorophyll singlets in photosynthetic systems.

ACKNOWLEDGEMENT

We are grateful to Professor R. Silbey for valuable discussions.

REFERENCES AND NOTES

1. M. Weissbluth, *Quart. Rev. Biophys.*, **4**, 1 (1971).
2. W. Rhodes, *Spectroscopic Approaches to Biomolecular Conformation*, p. 123 (ed. D. W. Urry), American Medical Assoc., Chicago, 1970.
3. G. W. Robinson and R. P. Frosch, *J. Chem. Phys.*, **38**, 1187 (1963).
4. S. Fischer and S. A. Rice, *J. Chem. Phys.*, **52**, 2089 (1970); M. K. Grover and R. Silbey, *J. Chem. Phys.*, **52**, 2099 (1970).
5. M. Grover and R. Silbey, *J. Chem. Phys.*, **54**, 4843 (1971).
6. The effect of 'local' scattering, which may be important in some situations, has been neglected for simplicity in equation (1). It is a trivial matter to include this effect as a parameter in the formalism (see Reference 7). However, one of the most important sources of local scattering, fluctuations of the solvent shift, is likely to be much smaller in biological systems than it is in solutions.
7. (a) R. P. Hemenger, K. Lakatos-Lindenberg and R. M. Pearlstein, *J. Chem. Phys.*, **60**, 3271 (1974); (b) R. M. Pearlstein, K. Lindenberg and R. P. Hemenger, unpublished results.
8. J. B. Birks, *Photophysics of Aromatic Molecules*, Wiley–Interscience, London and New York, 1970.
9. R. S. Knox, *Bioenergetics of Photosynthesis* (ed. Govindjee), Academic, New York, 1975.
10. G. W. Robinson, *Brookhaven Symposium in Biology*, No. 19, 16 (1967).
11. (a) Th. Förster, *Modern Quantum Chemistry, Part III: Action of Light and Organic Crystals*, p. 93 (ed. O. Sinanoğlu), Academic, New York, 1965; (b) W. T. Simpson and D. L. Peterson, *J. Chem. Phys.*, **26**, 588 (1957).
12. C. Hélène, *Physico-Chemical Properties of Nucleic Acids*, Vol. 1, p. 119 (ed.

13. A. S. Davydov, *Theory of Molecular Excitons*, McGraw-Hill, New York and London, 1962.
14. R. P. Hemenger, R. M. Pearlstein and K. Lakatos-Lindenberg, *J. Math. Phys.*, **13**, 1056 (1972).
15. (a) R. M. Pearlstein, *J. Chem. Phys.*, **56**, 2431 (1972); (b) R. P. Hemenger and R. M. Pearlstein, *Chem. Phys.*, **2**, 424 (1973).
16. J. Eisinger and A. A. Lamola, *Excited States of Proteins and Nucleic Acids*, p. 107 (ed. R. F. Steiner and I. Weinryb), Plenum, New York and London, 1971.
17. M. Guéron, J. Eisinger and A. A. Lamola, *Basic Principles in Nucleic Acid Chemistry*, Vol. 1, p. 311 (ed. P. O. P. Ts'o), Academic, New York, 1974.

18. J. W. Longworth, unpublished results. At 210 K the fluorescence lifetime $\tau \sim 10^{-11}$ s.
19. B. M. Sutherland and J. C. Sutherland, *Biophys. J.*, **9**, 1045 (1969); M. Kaufman and G. Weill, *Biopolymers*, **10**, 1983 (1971).
20. T. Montenay-Garestier, C. Hélène and A. M. Michelson, *Biochim. Biophys. Acta*, **182**, 342 (1969).
21. R. O. Rahn, M. D. C. Battista and L. C. Landry, *Proc. Nat. Acad. Sci. U.S.A.*, **67**, 1390 (1970).

G.2. Electronic energy transfer in some class B proteins: trypsin, lysozyme, α-chymotrypsin and chymotrypsinogen A

J. Lerner and H. Lami

*Laboratoire de Biophysique, U.E.R. des Sciences pharmaceutiques,
Université Louis Pasteur, 67083 Strasbourg, France*

SUMMARY

The tyrosine–tryptophan electronic energy transfer has been investigated in trypsin, lysozyme, α-chymotrypsin and chymotrypsinogen A, using a method based on the demonstration of the sensitization of the tryptophan emission by tyrosine. The microenvironment (polar or non-polar) of the residues implied in the process was taken into account. Only in the case of trypsin was the transfer found to be important (transfer efficiency: 0·27). On the contrary, intertryptophan energy migration was evidenced in all four proteins by analysing fluorescence absorption polarization spectra.

1 INTRODUCTION

Electronic energy transfer in proteins has been extensively investigated, most often in order to obtain some information on the protein structure. Considering the particular case of class B proteins (containing tryptophan) it is possible, from the observed sequence of singlet energies, to predict the following transfers: Phe → Tyr → Trp, as a result of dipolar resonance coupling. However, owing to its small fractional absorption, the interference of phenylalanine can be safely discarded. Hence only two processes are important, namely Tyr → Trp transfer and intertryptophan energy migration which is possible as a result of the important overlap of absorption and emission spectra of this residue.

The Tyr–Trp transfer has been evidenced in numerous proteins.[1,2] On the other hand, intertryptophan energy migration is still a controversial process.[2,3]

The aim of the present work was to reconsider the electronic energy transfer in four proteins of known structure: trypsin, lysozyme, α-chymotrypsin and chymotrypsinogen A. These proteins have already been studied by Teale,[5]

Weber,[4] Konev[3] and Longworth[2] demonstrating tyrosine–tryptophan transfer in trypsin[2] and inter-tryptophan transfer in trypsin and α-chymotrypsinogen A.[3]

In the present study two methods were used successively. The first one is based on the demonstration of the sensitization of the tryptophan emission by tyrosine by means of a technique already used by one of us in the case of alkaline phosphatase.[6] The second consists of measuring the polarization of emission in order to show a depolarization anisotropy which indicates the presence of the above mentioned transfer processes.

2 MATERIALS AND METHODS

2.1 Materials

N-acetyl-L-tyrosine amide, N-acetyl-L-tryptophan amide (Cyclo Chemical), bovine pancreas trypsin, α-chymotrypsin and chymotrypsinogen A, and hen egg white lysozyme (lots TRL 100, CDS and LYSF, respectively, from Worthington), were used without further purification.

Trypsin and lysozyme were dissolved in a phosphate buffer ($M/15$ Na_2HPO_4 and $M/10$ KH_2PO_4, Merck) at pH $= 5\cdot5$, α-chymotrypsin and chymotrypsinogen A in an aqueous buffer ($0\cdot1$ M, Tris, HCl) at pH $= 7\cdot0$. Both buffers were used for the amino-acid models. As a non-polar solvent, dioxane (spectroscopic grade, Merck) was chosen.

2.2 Measurements

Absorption spectra have been recorded with a Cary 15 spectrometer. Fluorescence spectra were obtained, using a Zweiss MM 12 monochromator associated with a 500 watt Hg vapour lamp for excitation, and a 500 mm grating Jarrell–Ash monochromator and a 6256S EMI photomultiplier for detection. The intensity of the lamp was monitored by a second photomultiplier receiving a constant fraction of the excitation beam. Fluorescence emission from the front surface of the $1\cdot0$ cm quartz cell was detected. All the spectra were corrected for the spectral sensitivity of the monochromator–photomultiplier system and the fluorescence intensities transformed into quantum units per unit wavelength interval. Quantum yields were measured by reference to a tryptophan neutral aqueous solution, assuming its yield, at 300 K and for an excitation wavelength of 280 nm, to be equal to $0\cdot14$.[7] The effect of the refractive index of the solvent was also taken into account.

The fluorescence absorption polarization spectra were recorded with an apparatus described by Weber,[8] on solutions in glycerol (redistilled, Merck) at $-70\,°C$. The bandwidths employed were 7 nm for excitation and observation.

3 LOCATION OF TRYPTOPHAN AND TYROSINE RESIDUES

The demonstration of electronic energy transfer with the method described below requires knowledge of the location within the protein of the different residues implied in the process. In this work we shall distinguish between external (m) and internal (m') located tryptophan residues and external (n) and internal located (n') tyrosine residues. A survey of the literature gave us numerous data, listed in Table 1. Methods as different as solvent perturbation difference spectroscopy,[9,10] thermal perturbation difference spectroscopy,[11] quenching by iodide ions,[12] iodination,[13,14] reaction with N-acetylimidazole and cyanuric fluoride[15] have been considered. Some recent results of Burstein[16] concerning tryptophan are also included.

Table 1   Location of tryptophan and tyrosine residues

Protein	Tryptophan		Tyrosine		Ref
	m	m'	n	n'	
Trypsin	2·6	1·4	7·3	2·7	9
			6	4	15
			6·7	3·3	
	3	1	2	8	2
	0·4	3·6			16
Lysozyme	4·26	1·74			10
	3·3	2·7			12
	3·5–3·9	2·1–2·5	1	2	11
			2	1	13
	3	3	2	1	2
	4·2	1·8			16
α-Chymotrypsin	3·04	4·96			10
			3	1	15
			3	1	14
	3	5	2	2	2
	3·2	4·8			16
	4	4			
Chymotrypsinogen A	3·44	4·56			10
	2·4	5·6			11
			2·5	1·5	15
			3	1	14
	2	6	2	2	2
	1·6	6·4			16
	0	8			

Critical examination of these data and comparison with the results of X-ray analysis led us finally to adopt the following values of the number of tryptophan and tyrosine residues in external and internal locations:

Trypsin	$m = 2·6$	$m' = 1·4$	$n = 7·3$	$n' = 2·7$
Lysozyme	$m = 4·26$	$m' = 1·74$	$n = 2$	$n' = 1$
α-Chymotrypsin	$m = 3·04$	$m' = 4·96$	$n = 3$	$n' = 1$
Chymotrypsinogen A	$m = 3·44$	$m' = 4·56$	$n = 2·75$	$n' = 1·25$

The fractional numbers express the degree of accessibility of the different residues.

4 TYROSINE FLUORESCENCE

Before measuring transfer efficiencies it is important to show the absence of direct tyrosine fluorescence emission. This has been done applying a matrix analytic procedure developed by Weber.[17] For all four proteins, the fluorescence intensity was measured at three wavelengths: 310 nm, 380 nm and λ_{max} of the emission spectrum, for three different excitation wavelengths: 270 nm, 280 nm and 295 nm. The results were displayed in a 3×3 matrix, whose 2×2 determinants all approached zero within the experimental error. Hence, according to Weber's analysis, no direct emission from tyrosine is present in the case of these proteins.

5 TYROSINE–TRYPTOPHAN ENERGY TRANSFER

5.1 Principle of the method

The method used to determine the tyrosine–tryptophan energy transfer is essentially based on the demonstration of the sensitization of the acceptor (tryptophan) by the donor (tyrosine). Neglecting any initial transfer from phenylalanine and any direct emission from tyrosine for the reasons mentioned above, the fluorescence quantum yield $\phi_p(\lambda)$ of the proteins considered is given by:

$$\phi_p(\lambda) = \phi_{Tp}(\lambda)(f_{Trp}(\lambda) + \eta f_{Tyr}(\lambda)) \tag{1}$$

where η is the wavelength-independent efficiency with which excitation energy is transferred to tryptophan from tyrosine, $\phi_{Tp}(\lambda)$ is the fluorescence quantum yield of tryptophan when incorporated in the protein, and $f_{Trp}(\lambda)$ and $f_{Tyr}(\lambda)$ are the fractions of light absorbed by tryptophan and tyrosine, respectively, at wavelength λ.

Since the absorption of phenylalanine is neglected, we may write:

$$f_{Tyr} = 1 - f_{Trp} \tag{2}$$

the fractional absorption of tryptophan being defined in the following manner:

$$f_{Trp}(\lambda) = \frac{m\varepsilon_{Trp}(\lambda) + m'\varepsilon'_{Trp}(\lambda)}{m\varepsilon_{Trp}(\lambda) + m'\varepsilon'_{Trp}(\lambda) + n\varepsilon_{Tyr}(\lambda) + n'\varepsilon'_{Tyr}(\lambda)} \tag{3}$$

Here a distinction is made between residues in a polar microenvironment (m tryptophan residues and n tyrosine residues with molar extinction coefficients $\varepsilon_{Trp}(\lambda)$ and $\varepsilon_{Tyr}(\lambda)$, respectively) and in a non-polar microenvironment (m' tryptophan residues with ε'_{Trp} and n' tyrosine residues with ε'_{Tyr}, respectively).

The fluorescence quantum yield $\phi_{Tp}(\lambda)$ is not directly measureable, but we can assume that its value is proportional to $\phi_t(\lambda)$ defined by the relation:

$$\phi_t(\lambda) = \frac{m\phi(\lambda) + m'\phi'(\lambda)}{m + m'} \tag{4}$$

where $\phi(\lambda)$ and $\phi'(\lambda)$ represent the fluorescence quantum yields of tryptophan in a polar and a non-polar medium, respectively. Hence:

$$\phi_{Tp}(\lambda) = c\phi_t(\lambda)$$

c being a constant, and equation (1) can be written as

$$\phi_p(\lambda) = c\phi_t(\lambda)(f_{Trp}(\lambda) + \eta f_{Tyr}(\lambda))$$

Finally, substituting equation (2) in this expression, we obtain:

$$\phi_p(\lambda)/\phi_t(\lambda) = cf_{Trp}(\lambda)(1 - \eta) + c\eta$$

Thus, plotting the ratio $\phi_p(\lambda)/\phi_t(\lambda)$ versus $f_{Trp}(\lambda)$ should yield a straight line whose slope s and intercept i can be used to get the value of η:

$$\eta = \frac{i}{i + s}$$

5.2 Results and discussion

In Figure 1 are shown plots of $\phi_p(\lambda)$, $\phi_t(\lambda)$ and $f_{Trp}(\lambda)$ versus wavelength for trypsin, lysozyme, α-chymotrypsin and chymotrypsinogen A, respectively. The values of $\phi_t(\lambda)$ were obtained from equation (4). As models of the tryptophan residue in polar and non-polar media, solutions of N-acetyl-L-tryptophan amide in water and dioxane were chosen and used to obtain $\phi(\lambda)$ and $\phi'(\lambda)$ (Figure 2). For arguments in favour of these models, see Reference 18. The fraction of exciting light absorbed by tryptophan was calculated using the values of m, m', n and n' given below, and the molar extinction coefficients observed on solutions of N-acetyl-L-tyrosine amide and N-acetyl-L-tryptophan amide in water and dioxane.

Plots of the ratio $\phi_p(\lambda)/\phi_t(\lambda)$ versus $f_{Trp}(\lambda)$ are shown in Figure 3. As a result of the preponderance of tryptophan residues in lysozyme, α-chymotrypsin and chymotrypsinogen A, the experimental points pertaining to these proteins are narrowly clustered and do not allow a very precise determination of the transfer efficiency. As can be seen in Table 2, which lists the values of s, i and η obtained by use of the method of least squares, negative values for the transfer efficiencies of these proteins are obtained. We therefore conclude that there is no tyrosine–tryptophan transfer in these macromolecules, a result which is qualitatively in agreement with the values published by Longworth (see Table 2). As regards trypsin, the method yields a much lower value than the ones reported by Kronman and Holmes, and Longworth (see Table 2).

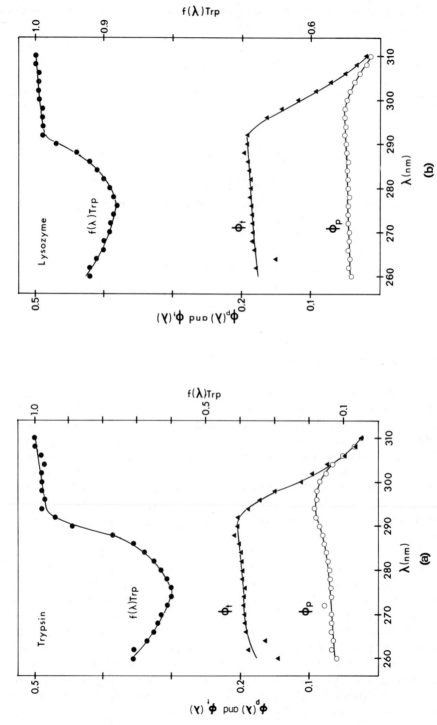

Figure 1 Protein fluorescence quantum yield $\phi_p(\lambda)$ (—○—), fractional absorption of tryptophan $f_{\text{Trp}}(\lambda)$ (—●—) and calculated tryptophan quantum yield $\phi_t(\lambda)$ (—▲—) versus excitation wavelength λ. (a) Trypsin. (b) Lysozyme. (c) α-Chymotrypsin. (d) Chymotrysinogen A

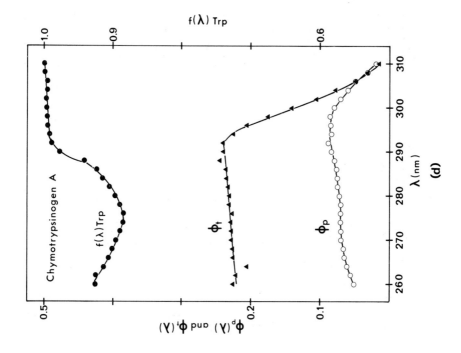

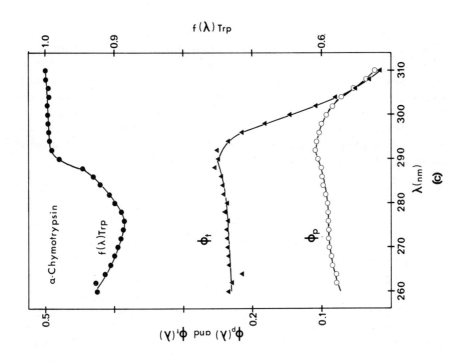

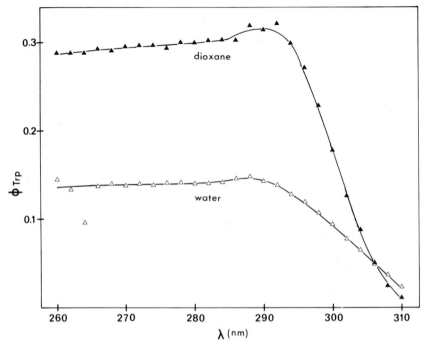

Figure 2 *N*-Acetyl-tryptophan amide. Fluorescence quantum yield versus excitation wavelength in water (–△–) and dioxane (–▲—)

Finally we must note some weaknesses of the present method. First the transfer efficiency η was assumed to be wavelength independent. In view of Weber's 'red edge effect',[19] this may not be the case for long-wavelength excitation. A more serious defect originates from equation (4). This relation contains the implicit assumption that the microenvironment of the tryptophan residues which emit is identical with that of the whole set of tryptophan residues present in the protein. Since obviously not all tryptophan residues emit, this is probably not the case. Some improvement of the method is certainly possible by a better estimation of the parameters m and m' of equation (4).

Table 2 Tyrosine–tryptophan energy transfer efficiencies

Protein	s	i	η	η in ref.
Trypsin	0·354	0·134	0·274	$\begin{cases} 0\cdot78^2 \\ 1\cdot07^1 \end{cases}$
Lysozyme	0·351	−0·07	−0·249	0^2
α-Chymotrypsin	0·460	−0·036	−0·084	0^2
Chymotrypsinogen A	0·578	−0·213	−0·584	0^2

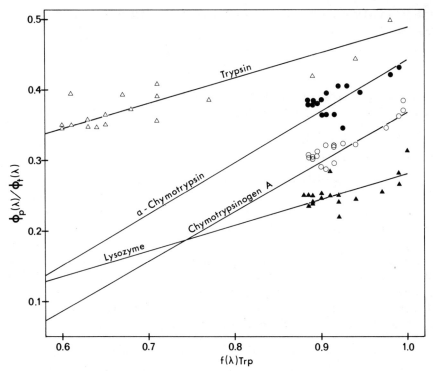

Figure 3 $\phi_p(\lambda)/\phi_t(\lambda)$ versus tryptophan fractional absorption $f_{Trp}(\lambda)$, for trypsin $(-\triangle-)$, lysozyme $(-\blacktriangle-)$ α-chymotrypsin $(-\bullet-)$ and chymotrypsinogen A $(-\bigcirc-)$.

6 INTER-TRYPTOPHAN ENERGY TRANSFER

6.1 Principle of the method

Inter-tryptophan energy transfer can be demonstrated by measuring, in a rigid medium, the fluorescence polarization spectrum of the protein and comparing it to that obtained from tryptophan under the same conditions. With linear polarized excitation (at wavelength λ), the fluorescence is partly depolarized, the residual polarization $p(\lambda)$ being given by:

$$p(\lambda) = \frac{I_{\parallel} - I_{\perp}}{I_{\parallel} + I_{\perp}}$$

where $I_{\parallel}$ and $I_{\perp}$ are the components of the emission respectively parallel and perpendicular to the electric vector of the exciting light. If we denote by $p_0(\lambda)$ the value of $p(\lambda)$ obtained with the emitting residues in the absence of any energy transfer, and by $p_t(\lambda)$ the value measured for the protein, the efficiency of the

610

energy transfer is given, according to Galanin (see Reference 3, p. 131) by the relation:

$$\eta(\lambda) = \frac{p_0(\lambda) - p(\lambda)}{p_0(\lambda)} \tag{5}$$

If two different transfer processes are present, as is possible in trypsin for which we have observed a tyrosine–tryptophan transfer, $\eta(\lambda)$ represents a mean value.

6.2 Results and discussion

The fluorescence absorption polarization spectra of lysozyme, α-chymotrypsin and chymotrypsinogen A, measured on solutions in 50% propylene glycol–water at $-70\,^\circ$C, have been published by Weber.[4] Taking for $p_0(\lambda)$ the values obtained by the same author[20] with tryptophan in propylene glycol at $-70\,^\circ$C and using equation (5) yields the results shown in Figure 4. The curve relating to trypsin was obtained by measuring $p(\lambda)$ and $p_0(\lambda)$ on solutions (10^{-4} M) of the enzyme and N-acetyl tryptophan amide respectively, in glycerol at $-70\,^\circ$C. In all cases an important intertryptophan energy migration is seen to be present. However the most important result is related to the trypsin curve which displays a distinct increase of the value of η between 265 and 285 nm. This domain

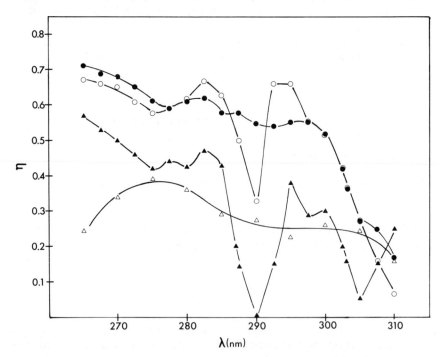

Figure 4 Intertryptophan energy transfer efficiency η versus excitation wavelength for trypsin (-△-), lysozyme (-▲—), α-chymotrypsin (-●-) and chymotrypsinogen A (-○-)

corresponds to the first absorption band of tyrosine and hence the effect of tyrosine–tryptophan energy transfer is unambiguously demonstrated.

ACKNOWLEDGEMENTS

The authors would like to thank Professor G. Laustriat and D. Gerard for helpful discussions and E. Leroy for her assistance with the computation.

This research was supported by the CNAMTS (Caisse Natonale de l'Assurance Maladie des Travailleurs Salariés), Paris.

REFERENCES

1. M. J. Kronman and L. G. Holmes, *Photochem. Photobiol.*, **14**, 113 (1971).
2. J. W. Longworth, *Excited States of Proteins and Nucleic Acids* (eds. R. F. Steiner and I. Weinryb), Macmillan, London, 1971.
3. S. V. Konev, *Fluorescence and Phosphorescence of Proteins and Nucleic Acids*, Plenum Press, New York, 1967.
4. G. Weber, *Biochem. J.*, **75**, 345 (1960).
5. F. W. J. Teale, *Biochem. J.*, **76**, 381 (1960).
6. D. Gerard, H. Lami and G. Laustriat, *Biochem. Biophys. Acta*, **263**, 496 (1972).
7. J. Eisinger and F. Navon, *J. Chem. Phys.*, **50**, 2069 (1969).
8. G. Weber, *J. Opt. Soc. Amer.*, **46**, 962 (1956).
9. G. B. Villanueva and T. T. Herskovits, *Biochem.*, **10**, 3358 (1971).
10. E. J. Williams, T. T. Herskovits and M. Laskowski Jr., *J. Biol. Chem.*, **240**, 3574 (1965).
11. J. Bello, *J. Biochem.*, **9**, 3562 (1970).
12. S. S. Lehrer, *Biochem.*, **10**, 3250 (1971).
13. J. Wolff and I. Covelli, *Biochem.*, **5**, 867 (1966).
14. A. N. Glazer and F. Sanger, *Biochem. J.*, **90**, 92 (1964).
15. M. J. Gorbunoff, *Biochem.*, **8**, 2591 (1969).
16. E. A. Burstein, N. S. Vedenkina and M. N. Ivkova, *Photochem. Photobiol.*, **18**, 263 (1973).
17. G. Weber, *Nature*, **27**, 4770 (1961).
18. D. Gerard, G. Laustriat and H. Lami, *Biochem. Biophys. Acta*, **263**, 482 (1972).
19. G. Weber and M. Shinitzky, *Proc. Natl. Acad. Sci. U.S.*, **65**, 823 (1970).
20. G. Weber, *Biochem. J.*, **75**, 335 (1960).

G.3. Energy transfer in multi-acceptor proteins and model systems

C. E. Owens and F. W. J. Teale

Department of Biochemistry, The University, Birmingham B15 2TT, U.K.

SUMMARY

Competitive energy transfer from protein to coenzyme in some multiple site enzymes leads to a non-linear dependence of donor fluorescence with respect to the fractional occupation of binding sites. There is often a close approximation to constant fractional quenching of donors by the successively bound acceptors. Computations based on the Förster mechanism show that such behaviour is also closely approximated in systems containing evenly distributed spectroscopically identical donors transferring independently to acceptors arranged in simple symmetry. The known structures and spectroscopic parameters of dehydrogenases are examined in terms of these requirements. The non-linear quenching observed in other protein systems which do not meet these criteria is also discussed.

1 INTRODUCTION

Occupation of the binding sites of proteins by chromophoric ligands is often accompanied by a decrease in luminescence of the intrinsic aromatic amino-acid residues of the macromolecule. When other possible quenching processes, such as contact quenching and induced conformational changes, can be excluded independently this effect of ligand binding can be attributed to long-range resonance energy transfer. The best criterion for the Förster process is protein-sensitized ligand fluorescence, which measures both the efficiency of transfer from the total excited donor population and the emission heterogeneity of the donors. Correspondence between the efficiencies of protein quenching and ligand sensitization indicates a relatively uniform donor quantum yield. In

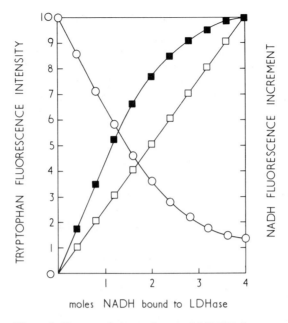

Figure 1 Change of tryptophan and NADH fluorescence intensities with coenzyme site saturation in pigheart lactate dehydrogenase. Tryptophan fluorescence ○, NADH direct fluorescence change □ and tryptophansensitized NADH fluorescence ■.

Continuous curve through the points ○ is the tryptophan intensity calculated for 'geometric quenching'

multisite protein systems the transfer quenching of protein fluorescence produced by the successive binding of ligands will depend, among other things, on the overall geometry of the array of donors and ligands in each species. When the ligand sites act independently in transfer quenching of the donors a linear decrease in protein fluorescence with site occupancy will be observed. On the other hand, when the ligands are so placed that they compete for transfer from the same donors then a non-linear quenching is seen, in which the rate of quenching decreases with ligand saturation as shown in Figure 1. This behaviour has been reported in several protein systems, especially some dehydrogenases binding NADH.[1] In these latter enzymes, the effect of successive filling of NADH sites corresponds closely to an apparently simple mechanism whereby each NADH as it binds reduces the existing fluorescence yield by the same fraction, a process which has been termed 'geometric quenching'.[2-4] Undoubtedly there are many structures and mechanisms of different degrees of complexity which could give rise to this quenching behaviour, but in the following communication we have examined in model systems the simplest geometrical and spectroscopic parameters which will suffice.

2 MODEL SYSTEMS

Constant values of R_0, the donor–acceptor separation at which Förster transfer and emission are equally probable, were used throughout in our calculations. This implies that (i) the donors are spectroscopically identical and have equal fluorescence yields, (ii) the orientation factor is averaged, (iii) there is no interdonor transfer and (iv) the ligands are spectroscopically identical. In addition, the ligand binding sites were considered independent, and identical in environment and affinity, so that a binomial distribution of ligands among the total sites could be assumed. The ligand sites were placed in some simple symmetry on, or near, the surface of a sphere, or group of spheres within which the donors were uniformly distributed. Among the models investigated were (a) two ligand sites at various points on a sphere, (b) two sites, each on the surface of contiguous spherical subunits, (c) four sites in square planar or tetrahedral array on a sphere or four spherical subunits and (d) six sites arranged octa-hedrally on a spherical surface or on six spherical subunits. In addition, models where the sites were grouped together were also considered.

Calculation of the average donor fluorescence yield as a function of ligand site saturation was carried out in two stages. First the fractional fluorescence of the species with M ligands was calculated as

$$\left[\frac{F}{F_0}\right]_M = \frac{1}{1 + \sum_{j=1}^{M} \sum_{i=1}^{D} \left(\frac{R_0}{R_i}\right)^6}$$

where R_i is the distance of the ith donor from the jth ligand and D is the total donor population. Secondly the total fractional fluorescence intensity was obtained by summing the product of the concentration of the M liganded species and the appropriate $[F/F_0]_M$ values for that species. At a site, saturation values of α, the fractional concentrations of the various species were given by the expansion of $[\alpha + (1 - \alpha)]^n$ where n is the number of equivalent sites to be occupied. By variation of the R_0 parameter the total quenching produced in a particular model could be set at any desired level.

The values of R_i were calculated by two different approaches.

2.1 Discontinuous donor distribution

A cubic lattice, with donors at the lattice points, is bounded by a spherical surface containing the ligand sites. R_i values were expressed as hkl lattice point indices, and the competitive effect of more than one ligand calculated by trans-formation of axes. This method was useful when considering cases in which only a few donors were contained in a spherical volume.

2.2 Continuous distribution of donors

Using polar coordinates to express R_i, the $[\bar{F}/F_0]_M$ is given by numerical integration of the expression.

$$\cfrac{1}{1 + \sum_{j=1}^{M} \sum_{\theta_i=0}^{\pi} \sum_{\phi_i=0}^{2\pi} \sum_{Ri=0}^{L} \cfrac{R_0^6}{\{L^2 + R^2 - 2RL[\sin\theta_j\sin\theta_i\cos(\phi_j - \phi_i) + \cos\theta_j\cos\theta_i]\}^3}}$$

where θ_j, ϕ_j, L are the coordinates of the jth ligand site, and θ_i, ϕ_i, R_i are the coordinates of the ith donor.

The value of $[\bar{F}/F_0]$ calculated for a particular model was higher using the discontinuous as against the continuous model, but the values converged when a sufficiently large number of lattice points was considered. In one particular case (two polar sites on a sphere) a straightforward integration of the Förster rate equation could be made, as described elsewhere.[5,6]

3 RESULTS

In Figure 2 the quenching curves computed for the four-ligand case are shown, together with that actually obtained with ox-heart lactate dehydrogenase. Exactly comparable families of curves were also computed for the two-site and six-site systems. The upper curve shows the very close correspondence between

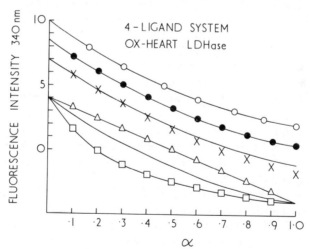

Figure 2 Change in tryptophan fluorescence intensity with NADH site saturation α in ox-heart lactate dehydrogenase compared with various model systems.
Enzyme fluorescence $\bigcirc$, square model $\bullet$, tetrahedral model $\times$, independent ligand sites $\triangle$ and four adjacent sites $\square$. The four similar continuous curves are calculated from
$$F/F_0 = [(1 - x)\alpha - 1]^4 \text{ putting } x = 0.67$$

the observed quenching points and the continuous curve calculated from $F/F_0 = [(1 - x)\alpha - 1]^4$ where α is the fractional site saturation, and x is the constant fractional quenching produced by each ligand. This latter curve is compared in turn with

(1) that calculated for a square planar arrangement of sites on a spherical surface,
(2) a tetrahedral arrangement and
(3) grouped sites and the case of linear quenching.

In contrast with the tetrahedral system, the square arrangement has two different diliganded species, adjacent and diagonal, which were assumed present in equal concentrations for the purposes of our calculations. The enzyme solutions were excited at 298 nm and the emission at 340 nm recorded, so that the protein fluorescence could be almost entirely attributed to the tryptophan content of the subunits. The coenzyme fluorescence increment increased linearly with site saturation, while the tryptophan-sensitized increment of coenzyme fluorescence showed a non-linearity with site saturation which was generally complementary to the tryptophan quenching curve.

4 DISCUSSION

The results shown for the four-ligand model and those calculated for similar two and six-ligand models show that these systems give quenching curves very similar to those observed in the enzyme–coenzyme interactions by an appropriate choice of site geometry and the R_0 parameters.

The square arrangement gave more curvature than the tetrahedral arrangement, as would be expected since in the latter geometry there was less competition between the ligands for the donor transfer. For similar reasons compact groups of subunits gave relatively more quenching than the spherical model of equivalent volume, since a larger fraction of the total donors lay within the effective quenching radius of each binding site. The rapid initial quenching and subsequent slow approach to the minimum seen with bunched ligand sites is also to be expected, as all the ligands compete together for transfer from neighbouring donors, while more remote donors are beyond the effective range of all. A linear fluorescence decrease corresponds to independence of the ligand quenching, so that there is no effective overlapping of the quenching range of different ligands. This linear behaviour can also, of course, be observed in single-site systems, and also in multisite systems where the binding is strongly positively cooperative.[3]

In the enzymes investigated, the linear rise in direct coenzyme fluorescence with site saturation was taken to support the virtual independence of the binding sites. In addition the complementary non-linear curves of tryptophan quenching and tryptophan-sensitized coenzyme fluorescence can be taken as proof that the main quenching mechanism was energy transfer, rather than perhaps a

Table 1

	Residues per ligand site		τ(ns)	Q
Tryptophan	—	—	3·0	0·2
Pig-heart LDHase	2W	4Y	4·5	0·35
Ox-heart LDHase	5W	8Y	4·8	0·38

ligand-induced conformational change which reduced the quantum yield of tryptophan, and also implied a relatively uniform tryptophan quantum yield. This was supported by the fluorescence lifetime and quantum yield data shown in Table 1. Since the average lifetime is intensity weighted, the similar ratio τ/Q for these enzymes and tryptophan itself indicates a relatively homogeneous population of tryptophan. Addition of ligand produced a more rapid decrease in average quantum yield than average lifetime, as would be expected from a system having a wide range of donor–acceptor separations.[5] The quenching profile computed by discontinuous integration changed only slowly as the number of lattice points, (or total number of donors) was reduced, and persisted down to low values (~ 10) comparable with the tryptophan numbers shown in Table 1, provided the donor locations were widely distributed within the volume of integration. In summary, it is suggested that in these enzymes the observed non-linear quenching of tryptophan fluorescence by bound coenzyme may simply reflect the essential symmetry imposed by the grouping of a few similar subunits in the protein complex. This factor, combined with a favourable spatial distribution of relatively unquenched tryptophan leads, through the resonance energy transfer mechanism, to the observed behaviour of these systems. It is not suggested however that non-linear quenching cannot also arise from induced conformational change[7] or other processes unrelated to resonance energy transfer.

REFERENCES

1. H. Theorell and K. Tatemoto, *Arch. Biochem. Biophys.*, **142**, 69 (1971).
2. J. J. Holbrook and H. Gutfreund, *FEBS Lett.*, **31**, 157 (1973).
3. J. J. Holbrook, *Biochem. J.*, **128**, 921 (1972).
4. J. J. Holbrook, *et al.*, *Biochem. J.*, **128**, 933 (1972).
5. R. A. Badley and F. W. J. Teale, *J. Mol. Biol.*, **58**, 567 (1971).
6. R. A. Badley and F. W. J. Teale, *J. Mol. Biol.*, **44**, 71 (1969).
7. I. Iweibo and H. Weiner, *Biochemistry*, **11**, 1003 (1972).

G.4. Fluorescence anisotropy spectra of proteins and energy transfer†

J. W. Longworth

Biology Division, Oak Ridge National Laboratory, Oak Ridge, Tennessee 37830, U.S.A.

C. A. Ghiron

Department of Biochemistry, University of Missouri, Columbia, Missouri 65201, U.S.A.

ABSTRACT

Transfer of singlet electronic energy between aromatic amino-acids within a protein may lead to a depolarization of fluorescence, since it is unlikely that any two neighbouring residues will have identical molecular coordinates. The anisotropy of the fluorescence of tyrosine depends upon the exciting wavelength, decreasing in value at shorter wavelengths, but lacks any fine structure. The corresponding spectrum for trytophan is distinctly fine structured, but nevertheless attains a plateau value at its long-wavelength beginning. Tryptophan can be studied separately from tyrosine in a combination, as tryptophan absorbs to longer wavelengths.

Homotransfer causes an equal depolarization at all exciting wavelengths. This is observed in two proteins that lack any tryptophan residues, bovine pancreatic ribonuclease A and trypsin inhibitor. The anisotropy spectra are uniformly depolarized between 260 and 295 nm, compared to the spectrum for tyrosine. Therefore, extensive tyrosine-to-tyrosine transfer occurs in these proteins.

In our experience inter-tryptophan transfers are uncommon and anisotropy of fluorescence between 300 and 310 nm of a variety of proteins is comparable to that found in tryptophan. Extensive depolarization can be found in the antibiotic gramicidine, which possesses a series of continuous tryptophan

† Research sponsored by USPHS Fellowship, Oak Ridge Associated Universities, and the U.S. Atomic Energy Commission under contract with the Union Carbide Corporation.

residues. At acidic pH values, α-trypsin possesses a depolarized tryptophan spectrum, but this transfer is absent at alkaline pH values in the presence of inhibitor. Hen lysozyme, like human lysozyme, is not depolarized at 305 nm, yet its fluorescence is believed to originate from W 62 and W 108 (the human only from W 108). In hen lysozyme, residue W 63 is 0·5 nm distant and at an angle of 45°; we suggest that the absence of depolarization indicates extensive quenching of W 63 luminescence.

Between 260 and 285 nm there is little variation in the anisotropy of tryptophan. A typical spectrum for many proteins shows depolarization in this wavelength region, which we attribute to tyrosine-to-tryptophan transfers. In subtilisin BPN′, the anisotropy depolarization depends upon wavelength similarly to the fractional absorption of tyrosine.

The anisotropy at 305 nm is due only to tryptophan absorption and typically attains values equal to those for the monomer. We have studied the variation of anisotropy of human lysozyme as a function of temperature-dependent variations in viscosity. Fluorescence lifetimes were determined at 220 and 300 K, and fluorescence intensity was continuously monitored. The Perrin plot, corrected for the lifetime variation, disclosed two correlation times. The shorter time (approximately 2 ns) was attributable to a side-chain relaxation.

G.6. Kinetics of singlet energy migration between dyes bound to nucleic acids

Ph. Wahl, D. Genest, and J. L. Tichadou

Centre de Biophysique Moleculaire, 45045 Orleans Cedex, France

ABSTRACT

The kinetics of energy migration between ethidium bromide (EB) molecules bound to nucleic acid molecules in double helix has been followed by measurements of the decays of the polarized fluorescence. We used a Monte Carlo method, in order to simulate the energy migration.

In the models of ethidium bromide–nucleic acid complexes, used for this calculation, we took into account the normal structure of the nucleic acid duplex and the perturbation induced in that structure by intercalation of the dye.

Comparison between computed and experimental curves leads to the conclusion that EB molecules unwind the nucleic acid helix. The unwinding angle relative to the EB–thymus–DNA and EB–polyd(AT) is 16°, while with polyrA-rU it is equal to 36°.

This new method could be extended to other systems and give accurate information on the spatial distribution of an array of chromophores.

Author Index

(Contributors and contributions are shown in *italics*)

Abkowicz, M., 352
Abrahamson, E. W., 520, 521, 522, 530, 537, 538, 539, 561, 564, 566, 567, 575
Adamczyk, A., 39, 43
Adams, R. G., 530
Agosto, G. M., 466
Ake, R. L., 332
Akhtar, M., 566
Albrecht, A. C., 44, 127, 446, 450, 454, 455
Alchalel, A., 540, 530, 537, 539, 553, 566
Alfano, R. R., 6, 13
Algar, B. E., 91
Allen, F., 91
Allen, M. B., 260
Allen, R. C., 80, 83
Allison, A. C., 484
Aloj, S. M., 456, 467
Alpert, B., 425, 433
Amann, H., 260
Amesz, J., 260
Amouyal, E., 341, 539
Amsel, G., 232
Amster, R. L., 313
Anbar, M., 484
Andersen, J. M., 134
Anderson, S. R., 374
Angus, J. G., 60, 61
Antheunis, D. A., 313
Antonini, E., 433
Aplin, R. T., 164
Appel, W., 260
Applebury, M. L., 271, 530

Arai, S., 433
Arakawa, K., 496
Arecchi, T. T., 232
Argauer, R. J., 379, 387
Arian, Sh., 475
Arneson, R. M., 79, 80, 83
Arnon, D. I., 260
Arrio, B., 399
Arvis, M., 34,
Ascoli, F., 205
Aubailly, M., 445, 165, 423, 455
Auchet, J. C., 410
Augenstein, L., 216, 376, 387
Auld, D. S., 589
Ausländer, W., 261
Avigliano, L., 374
Avron, M., 261
Azerad, R., 531, 530
Azumi, T., 43, 61, 206

Badley, R. A., 410, 590, 617
Bahl, O. P., 466, 467
Bailey, J. L., 260
Baird, S. L., Jr., 206
Baker, B. N., 573, 575
Bakshiev, G. N., 365, 366, 374
Balazs, E. A., 115
Balke, D. E., 141, 142, 525, 530, 539, 566
Ball, J. K., 206
Balny, C., 197
Banerjee, R., 433
Bar, V., 91
Barat, F., 34

621

624

626

632

634

Subject Index